RECENT DEVELOPMENTS IN INFINITE-DIMENSIONAL ANALYSIS AND QUANTUM PROBABILITY

Papers in Honour of Takeyuki Hida's 70th Birthday

Edited by

LUIGI ACCARDI, HUI-HSIUNG KUO, NOBUAKI OBATA,
KIMIAKI SAITO, SI SI, and LUDWIG STREIT

Reprinted from Acta Applicandae Mathematicae, Volume 63, 2000

Springer Science+Business Media, B.V.

Library of Congress Cataloging-in-Publication Data

Recent developments in infinite-dimensional analysis and quantum probability: papers in honour of Takeyuki Hida's 70th birthday / edited by Luigi Accardi ... [et al.].

p. cm.

"Reprinted from Acta applicandae mathematicae, volume 63, 2000."

Includes bibliographical references and index.

ISBN 978-94-010-3842-3 ISBN 978-94-010-0842-6 (eBook)

DOI 10.1007/978-94-010-0842-6

1. Dimensional analysis. 2. Quantum theory. 3. Probabilities. I. Accardi, L. (Luigi), 1947-II, Hida, Takeyuki, 1927-III. Acta applicandae mathematicae.

QC20.7.D55 R43 2001

530.8- -dc21

2001029815

ISBN 978-94-010-3842-3

Printed on acid-free paper

Table of Contents

Takeyuki Hida.

Preface

This volume of *Acta Applicandae Mathematicae* is more than a collection of articles, in fact, in it the reader will find some consistent editorial work devoted to attempting to obtain a unitary picture from the different contributions and to give a comprehensive account of important recent developments in contemporary white-noise analysis and some of its applications. For this reason, not only the latest results, but also motivations, explanations, and connections with previous work have been included.

The wealth of applications, from number theory to signal processing, from optimal filtering to information theory, from the statistics of stationary flows to quantum cable equations, show the power of white-noise analysis as a tool. Beyond these, we have tried to emphasize its connections with practically all branches of contemporary probability, including stochastic geometry, the structure theory of stationary Gaussian processes, Neumann boundary value problems, large deviations, etc.

Among the highlights of the volume we mention the extension of the usual space of Hida distributions as a new and powerful tool to prove regularity theorems in infinite-dimensional analysis; the efforts to extend white-noise theory to random fields (the importance of such an extension has been repeatedly emphasized by Professor Hida); the recent developments in the theory of the Levy Laplacian, a notion that found its natural formulation in terms of white-noise analysis and whose connections with the Yang–Mills equations make it a basic object of study for contemporary mathematics; the contemporary trend towards the merging of two lines of research, white-noise analysis and quantum probability, that is taking place the so-called white noise approach to classical and quantum stochastic calculus, a new line of research that has emerged from the stochastic limit of quantum theory. The systematic exposition of this new approach contained here is probably the first attempt to deal with this topic.

LUIGI ACCARDI

HUI-HSIUNG KUO

NOBUAKI OBATA

KIMIAKI SAITO

SI SI

LUDWIG STREIT

Acta Applicandae Mathematicae **63:** 3–25, 2000.

A White-Noise Approach to Stochastic Calculus

LUIGI ACCARDI[1], YUN GANG LU[1,2] and IGOR V. VOLOVICH[1,3]
[1] *Centro Matematico Vito Volterra, Università di Roma, Rome, 00133, Italy.*
e-mail: accardi@volterra.mat.uniroma2.it
[2] *Dipartimento di Matematica, Università di Bari, Bari, 70125, Italy*
[3] *Steklov Mathematical Institute, Russian Academy of Sciences, Vavilov St. 42, 117966, Moscow, Russia*

(Received: 8 August 2000)

Abstract. During the past 15 years a new technique, called *the stochastic limit of quantum theory*, has been applied to deduce new, unexpected results in a variety of traditional problems of quantum physics, such as quantum electrodynamics, bosonization in higher dimensions, the emergence of the noncrossing diagrams in the Anderson model, and in the large-N-limit in QCD, interacting commutation relations, new photon statistics in strong magnetic fields, etc. These achievements required the development of a new approach to classical and quantum stochastic calculus based on white noise which has suggested a natural nonlinear extension of this calculus. The natural theoretical framework of this new approach is the white-noise calculus initiated by T. Hida as a theory of infinite-dimensional generalized functions. In this paper, we describe the main ideas of the white-noise approach to stochastic calculus and we show that, even if we limit ourselves to the first-order case (i.e. neglecting the recent developments concerning higher powers of white noise and renormalization), some nontrivial extensions of known results in classical and quantum stochastic calculus can be obtained.

Mathematics Subject Classification (2000): 60H40.

Key words: quantum probability, white noise, stochastic limit.

1. The Main Idea of the Stochastic Limit of Quantum Theory

Quantum stochastic differential equations are now widely used to construct phenomenological models of physical systems, for example in quantum optics, in quantum measurement theory, etc. However, the fundamental equation of quantum theory is not a stochastic equation but a usual Schrödinger equation. Therefore, the problem of understanding the physical meaning of these phenomenological models naturally arose.

The stochastic limit of quantum theory was developed to solve this problem and its main result can be concisely formulated as follows: *stochastic equations are limits, in an appropriate sense, of the usual Hamiltonian equations of quantum physics.*

Thus, the stochastic limit provides a derivation of the phenomenological stochastic equations from the fundamental quantum laws. In particular, this gives a microscopic interpretation of the coefficients of these equations and proves that the most important examples of quantum Markov flows arise in physics from the stochastic limit of Hamiltonian models.

From the mathematical point of view, the stochastic limit suggested a new interpretation of the usual stochastic equations, both classical and quantum, as *normally ordered Hamiltonian white noise equations*. In this section, we give a short illustration of the basic ideas of the stochastic limit and show how this naturally leads to the identification of normally ordered first-order white-noise Hamiltonian equations with stochastic differential equations in the sense of Hudson and Parthasarathy.

The starting point of the stochastic limit is not a stochastic equation but a usual Schrödinger equation in interaction representation, depending on a parameter λ

$$\partial U_t^{(\lambda)} = -i\lambda(DA_t^+(S_t g) - D^+A_t(S_t g))U_t^{(\lambda)} \tag{1.1}$$

describing a system S with state space $\mathcal{H}_S$ interacting with a field with creation and annihilation operators $A_t^+(g)$, $A_t(g)$, and D, D^+ are operators on a Hilbert space $\mathcal{H}_S$. One rescales the time parameter according to the law $t \to t/\lambda^2$. This rescaling is motivated both by mathematics (central limit theorem) and by physics (Friedrichs–van Hove rescaling). After the rescaling, one arrives to an equation of the form

$$\partial U_{t/\lambda^2}^{(\lambda)} = (Da_t^{(\lambda)+} - D^+a_t^{(\lambda)})U_{t/\lambda^2}^{(\lambda)}, \tag{1.2}$$

where

$$a_t^{(\lambda)} = \lambda \int_0^{t/\lambda^2} \mathrm{d}s A(S_s g).$$

It was proved in [1] that, as $\lambda \to 0$, the iterated series solution of this equation converges, in a sense which is the natural generalization of the notion of quantum convergence in law, to the solution of the QSDE

$$\mathrm{d}U_t = \left(D\,\mathrm{d}B_t^+ - D^+\,\mathrm{d}B_t + \left(-\frac{\gamma}{2}D^+D + i\alpha D^+D\right)\mathrm{d}t\right)U_t, \tag{1.3}$$

where B_t^+, B_t is the Fock Brownian motion with variance γ acting on the Boson Fock space $L^2(\mathbf{R})\otimes\mathcal{K}$, $H = \kappa D^+D$ and $\kappa, \gamma > 0$, are real numbers, $\mathcal{K}$ is a Hilbert space, whose explicit structure is described in terms of the original Hamiltonian model.

In [3, 4] it was proved that the iterated series solution of this equation converges term by term, in the same limit, and in the same sense as above, to the iterated series solution of the distribution equation

$$\partial_t U_t = (Db_t^+ - D^+b_t)U_t, \tag{1.4}$$

where b_t^+, b_t are the annihilation and creation operators of the Boson Fock white noise with variance γ (> 0) which is characterized, up to unitary equivalence, by the algebraic relations

$$[b_t, b_s^+] = \gamma\delta(s-t), \quad t, s \in \mathbf{R}, \tag{1.5}$$

$$b_t\Phi = 0, \tag{1.6}$$

where Φ is the Fock vacuum. It is therefore natural to conjecture that Equations(1.3) and (1.4) are just two different ways of writing the same equation. To prove this conjecture we have to develop a purely analytical white noise approach to the standard, classical, and quantum Itô calculus and, in particular, a white-noise formulation of the Itô table, based on the general white-noise theory initiated by Hida [10] and developed in [12, 13, 16].

2. Notations on Fock Spaces

We begin by describing a concrete representation of the Fock space which, being well suited for explicit calculations, is most often used in the physical literature. Such a representation can be used whenever the 1-particle space is concretely realized as an L^2-space over some measure space (finite or σ-finite) (S, μ). In this case, the n-particle space can be realized as the space $L^2_{\text{sym}}(S^n, \otimes^n\mu)$ of all the symmetric, square integrable functions on the product space

$$S^n := S \times S \times \cdots \times S \quad (n\text{-times})$$

with the measure $\otimes^n\mu$, which is the product of n copies of the measure μ. In the following we shall fix the choice

$$S = \mathbf{R}^d; \qquad \mu = \text{Lebesgue measure.}$$

Let $\mathcal{F}_1 = L^2(\mathbf{R}^d)$ be the Hilbert space of functions on $\mathbf{R}^d$ with the inner product

$$(f, g) = \int_{\mathbf{R}^d} \overline{f}(s)g(s)\,\mathrm{d}s, \quad f, g \in \mathcal{F}_1 \tag{2.1}$$

and $\mathcal{F}_n = L^2_{\text{sym}}(\mathbf{R}^{nd})$, $n = 1, 2, \ldots$ be the Hilbert space of square integrable functions of n-variables in $\mathbf{R}^d$, symmetric under the permutation of their arguments. The elements of $\mathcal{F}_n$ are called *n-particle vectors*. For an element $\psi_n \in \mathcal{F}_n$ we write $\psi_n = \psi_n(s_1, \ldots, s_n)$, $s_i \in \mathbf{R}^d$ and one has $\psi_n(s_1, \ldots, s_n) = \psi_n(s_{\pi(1)}, \ldots, s_{\pi(n)})$ for any permutation π.

DEFINITION. The *symmetric representation* of the scalar Boson Fock space $\mathcal{F}$ is the direct sum of the Hilbert spaces $\mathcal{F}_n$

$$\mathcal{F} = \bigoplus_{n=0}^{\infty} L^2_{\text{sym}}(\mathbf{R}^{dn}) = \mathcal{F} = \bigoplus_{n=0}^{\infty} \mathcal{F}_n. \tag{2.2}$$

Here we set $\mathcal{F}_0 = \mathbf{C}$. So an element of the Boson Fock space $\mathcal{F}$ is a sequence of functions

$$\Psi = \{\psi_0, \psi_1, \psi_2, \ldots\},$$

where $\psi_0 \in \mathbf{C}$, $\psi_n \in \mathcal{F}_n$, $n = 1, 2, \ldots$ and

$$\|\psi\|^2 = \sum_{n=0}^{\infty} \|\psi^{(n)}\|^2_{L^2(\mathbf{R}^{dn})} < \infty. \tag{2.3}$$

More explicitly

$$\|\psi\|^2 = |\psi^{(0)}|^2 + \sum_{n=1}^{\infty} \int_{\mathbf{R}^{dn}} |\psi^{(n)}(s_1, \ldots, s_n)|^2 \, \mathrm{d}s_1 \ldots \mathrm{d}s_n. \tag{2.4}$$

The inner product of elements $\Psi = \{\psi_n\}_{n=0}^{\infty}$ and $\Phi = \{\phi_n\}_{n=0}^{\infty}$ from $\mathcal{F}$ is given by

$$\begin{aligned}(\Psi, \Phi) &= \sum_{n=0}^{\infty} (\psi_n, \phi_n) \\ &= \overline{\psi}_0 \phi_0 + \sum_{n=1}^{\infty} \int_{\mathbf{R}^{nd}} \overline{\psi_n(s_1, \ldots, s_n)} \phi_n(s_1, \ldots, s_n) \, \mathrm{d}s_1 \ldots \mathrm{d}s_n. \end{aligned} \tag{2.5}$$

The vector $\Psi_0 = (1, 0, 0, \ldots)$ is called the *vacuum vector*. It describes the state of a system in which no particle is present.

3. Annihilator and Creator Densities

Define

$$\mathcal{D}_{\mathcal{S}} := \{\psi \in \mathcal{F} \mid \psi^{(n)} \in \mathcal{S}(\mathbf{R}^{dn})\}. \tag{3.1}$$

In the remaining of this section, unless otherwise specified, all the n-particle vectors shall belong to $\mathcal{D}_{\mathcal{S}}$. Define, moreover,

$$\mathcal{D}^o_{\mathcal{S}} := \{\psi \in \mathcal{D}_{\mathcal{S}} \mid \psi^{(n)} = 0 \text{ for almost all } n \in \mathbf{N}\}, \tag{3.2}$$

$$\mathcal{D}(a) := \left\{\psi \in \mathcal{D}_{\mathcal{S}} \ : \ \sum_{n=1}^{\infty} n \|\psi^{(n)}\|^2 < \infty\right\} \tag{3.3}$$

and notice that $\mathcal{D}(a)$ is a vector space containing both the number and the exponential vectors with test functions in $\mathcal{S}$. Define the *annihilation density*[(1)] a_s

$$(a_s\psi)^{(n)}(s_1, \ldots, s_n) = \sqrt{n+1}\,\psi^{(n+1)}(s, s_1, \ldots, s_n); \quad s \in \mathbf{R}^d, \ n \in \mathbf{N}. \tag{3.4}$$

The right-hand side of (3.4) is well defined whenever it makes sense to speak of the values $\psi^{(n)}$ on any point, for example when $\psi^{(n)}$ is in the L^2-equivalence class of a continuous function for each n, and the sequence of functions $\{(a_s\psi)^{(n)}\}$ defines an element of $\mathcal{F}$. This is surely the case if ψ is in $\mathcal{D}(a)$. Thus, for any $t \in \mathbf{R}^d$, the annihilator a_t is a densely defined operator which maps $\mathcal{D}(a)$ into $\mathcal{F}$.

From (8) it follows that the map a_s is weakly measurable and therefore, for any square integrable function g, the integral

$$A(g) = \int_{\mathbf{R}^d} \mathrm{d}s\, \overline{g}(s) a_s \tag{3.5}$$

called the *annihilation operator* is well defined as a Bochner integral on $\mathcal{D}(a)$. Proposition 1 below shows that it is a preclosed operator. The explicit action of $A(g)$ on vectors in $\mathcal{D}(a)$ is deduced from (8) to be, for $n \in \mathbf{N}$,

$$\begin{aligned}(A(g)\psi)^{(n)} &= \int_{\mathbf{R}^d} \mathrm{d}s\, \overline{g}(s)(a_s\psi)^{(n)}(s_1, \ldots, s_n) \\ &= \sqrt{n+1} \int_{\mathbf{R}^d} \mathrm{d}s\, \overline{g}(s)\psi^{(n+1)}(s, s_1, \ldots, s_n).\end{aligned} \tag{3.6}$$

For example, the explicit action of $A(g)$ on the exponential vectors is also deduced from (3.4):

$$A(g)\psi_f = \int_{\mathbf{R}^d} \mathrm{d}s\, \overline{g}(s) a_s \psi_f = \int_{\mathbf{R}^d} \mathrm{d}s\, \overline{g}(s) f(s) \psi_f = \langle g, f\rangle \psi_f. \tag{3.7}$$

The *creation density* a_s^+ is defined for $\psi \in \mathcal{D}_{\mathcal{E}}$ by

$$(a_s^+\psi)^{(n)}(s_1 \ldots s_n) = \frac{1}{\sqrt{n}} \sum_{i=1}^{n} \delta(s - s_i)\psi^{(n-1)}(s_1 \ldots \hat{s}_i \ldots s_n). \tag{3.8}$$

The δ-function on the right-hand side of (3.8) shows that the creation density $a^+(t)$ is not an operator but a sesquilinear form on the number vectors.

PROPOSITION 1. *For any square integrable function g there exists a preclosed operator $A^+(g)$, defined on the n-particle vectors, represented by continuous functions, by the relation*

$$(A^+(g)\psi)^{(n)}(s_1, \ldots, s_n) := \frac{1}{\sqrt{n}} \sum_{i=1}^{n} g(s_i)\psi^{(n-1)}(s_1, \ldots, \hat{s}_i, \ldots, s_n). \tag{3.9}$$

Moreover, on the n-particle space, $A^+(g)$ is bounded with norm less or equal to $n^{1/2}\|g\|$ (L^2-norm of g) and, on $\mathcal{D}(a)$, $A^+(g)$ satisfies the relation

$$\langle A^+(g)\psi, \psi'\rangle = \langle \psi, A(g)\psi'\rangle. \tag{3.10}$$

Proof. Let ψ be as in the statement. Then, using the definition (3.9) of $A^+(g)$:

$$\|(A^+(g)\psi)^{(n)}\|^2 = \frac{1}{n} \sum_{i,j=1}^{n} \langle g_i\psi_i^{(n-1)}, g_j\psi_j^{(n-1)}\rangle,$$

where

$$g_i(s_1, \ldots, s_i, \ldots, s_n) := g(s_i);$$
$$\psi_i^{(n-1)}(s_1, \ldots, s_i, \ldots, s_n) := \psi^{(n-1)}(s_1, \ldots, \hat{s}_i, \ldots, s_n).$$

Since $\psi^{(n-1)}$ is a symmetric function, it follows that

$$|\langle g_i\psi_i^{(n-1)}, g_j\psi_j^{(n-1)}\rangle| \leqslant \|g_i\psi_i^{(n-1)}\|\|g_j\psi_j^{(n-1)}\| = \|g\|^2\|\psi^{(n-1)}\|^2$$

and therefore

$$\|(A^+(g)\psi)^{(n)}\|^2 \leqslant n\|g\|^2\|\psi^{(n-1)}\|^2.$$

This shows that $A^+(g)$ is a well defined operator on the domain $\mathcal{D}(a)$, bounded on each n-particle space. To prove (3.10) we compute

$$\begin{aligned}
&\langle\psi, A(g)\psi'\rangle \\
&= \sum_n \langle\psi^{(n)}, (A_g\psi')^n\rangle = \sum_n \sqrt{n+1}\int ds\overline{g}(s)\langle\psi^{(n)}, \psi'^{(n+1)}(s,\cdot)\rangle \\
&= \sum_n \sqrt{n+1}\int ds\overline{g}'_s \int \overline{\psi}^{(n)}(s_1, \ldots, s_n)\psi'^{(n+1)}(s, s_1, \ldots, s_n) \\
&= \sum_n \sqrt{n+1}\int ds\overline{g}_s \int \overline{\psi}^{(n)}(s, \ldots, s_n)\psi'^{(n+1)}(s, s_1, \ldots, s_n) \\
&= \sum_n \int ds \int ds_1 \ldots \int ds_n\left[\frac{1}{\sqrt{n+1}}\sum_{i=1}^{n+1} g_{s_i}\psi^{(n)}(s_1, \ldots, \hat{s}_i, \ldots, s_{n+1})\right]\times \\
&\quad \times \psi'^{(n+1)}(s_1, s_2, \ldots, s_{n+1}) \\
&= \langle A^+(g)\psi, \psi'\rangle.
\end{aligned}$$

□

LEMMA 2. *The following formulae hold on* $\mathcal{D}_a$*:*

$$\begin{aligned}
&(a(t_1)a^+(t_2)\psi)^{(n)}(s_1, \ldots, s_n) &(3.11)\\
&= \sum_{i=1}^{n} \delta(t_2 - s_i)\psi^{(n)}(s_1, \ldots, \hat{s}_i, \ldots, s_n, t_1) + \delta(t_2 - t_1)\psi^{(n)}(s_1, \ldots, s_n)\times \\
&\quad \times(a^+(t_1)a(t_2)\psi)^{(n)}(s_1, \ldots, s_n) \\
&= \sum_{i=1}^{n} \delta(t_2 - s_i)\psi^{(n)}(s_1, \ldots, \hat{s}_i, \ldots, s_n, t_1). &(3.12)
\end{aligned}$$

Proof. Define

$$\begin{aligned}
&\phi_{t_2}^{(n+1)}(s_1, \ldots, s_{n+1}) \\
&:= (a^+(t_2)\psi)^{(n+1)}(s_1, \ldots, s_{n+1}) \\
&= \frac{1}{\sqrt{n+1}}\sum_{i=1}^{n+1} \delta(t_2 - s_i)\psi^{(n)}(s_1, \ldots, \hat{s}_i, \ldots, s_{n+1}). &(3.13)
\end{aligned}$$

$\phi^{(n+1)})_{t_2}$ is a distribution with values in $\mathcal{F}_{n+1}$, i.e., for any $g \in \mathcal{S}$,

$$\left[\int \mathrm{d}t_2 g(t_2)\phi_{t_2}\right]^{(n+1)} = \frac{1}{\sqrt{n+1}} \sum_i g(s_i)\psi^{(n)}(s_1, \ldots, \hat{s}_i, \ldots, s_{n+1}).$$

Now, applying a_{t_1}, as defined by (3.8), we find

$$\begin{aligned}
&\left[a_{t_1} \int \mathrm{d}t_2 g(t_2)\phi_{t_2}\right]^{(n)} (s_1, \ldots, s_n) \\
&\quad = g(t_1)\psi^{(n)}(s_1, \ldots, s_n) + \sum_{i=1}^{n} g(s_i)\psi^{(n)}(t_1, \ldots, \hat{s}_i, \ldots, s_n) \\
&\quad = \int \mathrm{d}t_2 g(t_2)\delta(t_2 - t_1)\psi^{(n)}(s_1, \ldots, s_n) + \\
&\qquad + \sum_{i=1}^{n} \delta(t_2 - s_i) g(t_2)\psi^{(n)}(t_1, \ldots, \hat{s}_i, \ldots, s_n) \\
&\quad = \int \mathrm{d}t_2 g(t_2)\sqrt{n+1}\phi_{t_2}^{(n+1)}(t_1, s_1, \ldots, s_n).
\end{aligned}$$

Therefore, in the sense of distributions,

$$(a(t_1)\phi_{t_2})^{(n)}(s_1, \ldots, s_n) = \sqrt{n+1}\phi_{t_2}^{(n+1)}(t_1, s_1, \ldots, s_n) \tag{3.14}$$

and from (3.7) we get that (3.12), (3.12) are proved in a similar way. □

Remark. Comparing (3.12) and (3.12), one deduces the *Boson commutation relations* for a scalar Boson white noise

$$(t_1)a^+(t_2) - a^+(t_2)a(t_1) = \delta(t_2 - t_1).$$

4. Stochastic Integrals with Respect to the Boson Fock White Noises

In this section we shall discuss white noises and stochastic integrals in $\mathbf{R}^d$ rather than in $\mathbf{R}$ because exactly the same formulae are valid in the 1- and in the d-dimensional case.

We have defined the operators

$$A(F) = \langle F, A\rangle = \int_{\mathbf{R}^d} \mathrm{d}s F_s a_s; \qquad A^+(F) = \langle F, A^+\rangle = \int_{\mathbf{R}^d} \mathrm{d}s F_s a_s^+, \tag{4.1}$$

when F is a complex-valued function on $\mathbf{R}$. The generalization of these integrals to the case when F is an operator valued function are called *right stochastic integrals* with respect to a_s (resp. a_s^+). One has also to define the *left stochastic integrals*

$$\langle A, F\rangle = \int_{\mathbf{R}^d} \mathrm{d}s a_s F_s; \qquad \langle A^+, F\rangle = \int_{\mathbf{R}^d} \mathrm{d}s a_s^+ F_s \tag{4.2}$$

and the *two-sided stochastic integrals*

$$\int_{\mathbf{R}^d} \mathrm{d}s\, F_s a_s^{\pm} G_s; \qquad \int_{\mathbf{R}^d} \mathrm{d}s a_s^+ F_s a_s. \tag{4.3}$$

Let $\mathcal{D}_S^0$ be as in (5.6) and let $\mathcal{L}$ be a space of maps from $\mathbf{R}^d$ to linear operators from a dense subspace of $\mathcal{D}_{\mathcal{S}}^o$ to $\mathcal{F}$ with the property that the maps

$$s \mapsto \langle \psi, F_s \varphi \rangle; \qquad s \mapsto \|F_s \psi\|^2; \quad \varphi, \psi \in \mathcal{D}_{\mathcal{S}}^0,$$

are locally integrable. Clearly $s \mapsto a_s$, then $a \in \mathcal{L}$, while $s \mapsto a_s^+$ is not in $\mathcal{L}$.

If P_n denotes the projection onto the n-particle space of the Fock space, then for any t, we can write

$$F_t = \sum_{n,k} P_n F_t P_k =: \sum_{n,k} F_t^{(n,k)}.$$

Remark. By inspection from formulae (4.2) and (4.3), one can guess that even if the *integrand* F_s is bounded, in general the stochastic integrals will not be bounded operators. So a precise definition of the notion of stochastic integral should always specify the domain of vectors where this inegral is defined. The general scheme we shall adopt to define stochastic integrals is the following. If G_s denotes any of the integrands in formulae (4.1) or (4.2) or (4.3), I denotes the corresponding stochastic integral and ψ an arbitrary vector, then I will be characterized by the following two properties:

(i) The n-particle component of $I\psi$ is the Bochner integral of the n-particle component of $G_s\psi$:

$$\left(\int_{\mathbf{R}^d} \mathrm{d}s G_s \psi\right)^n := \int_{\mathbf{R}^d} \mathrm{d}s (G_s \psi)^n.$$

(ii) The n-particle component of $G_s\psi$ is explicitly computed using the rules of the previous section.

In the following sections we shall show how these general principles work in concrete applications.

5. Right Annihilator Integrals

Let $\mathcal{D}_S^0 =: \mathcal{D}$ and $\mathcal{L} := \mathcal{L}(\mathcal{D})$ be as in Equation (3.6).

DEFINITION 3. The right annihilator stochastic integral of $F \in \mathcal{L}$ is the operator

$$\psi = \int F_s a_s \psi \,\mathrm{d}s = \langle F^*, A \rangle \psi := \int F_s A_s \psi \,\mathrm{d}s, \tag{5.1}$$

where the integral is meant as a Bochner integral in the Fock space. It is defined for each $\psi \in \mathcal{D}_{\mathcal{S}}^o$ such that $a_s\psi$ is in the domain of F_s for each s and the vector-valued function $s \in \mathbf{R}^d \mapsto F_s a_s \psi$ is Bochner integrable.

The explicit form of the right annihilator stochastic integral on the n-particle vectors can be easily obtained by using the same technique as in Section 3. In fact, because of definition (3.8), one has that

$$(a_s\psi)^{(n)} = \sqrt{n+1}\psi^{(n+1)}(s,\cdot), \tag{5.2}$$

where $\psi^{(n+1)}(s,\cdot)$ is the function

$$(s_1,\dots,s_n) \in \mathbf{R}^{dn} \mapsto \psi^{(n+1)}(s,s_1,\dots,s_n). \tag{5.3}$$

Therefore, (5.1) is equivalent to

$$\int F_s a_s \psi \, \mathrm{d}s = \sum_{n\geq 0} \sqrt{n+1} \int \mathrm{d}s\, F_s \psi^{(n+1)}(s,\cdot). \tag{5.4}$$

In particular, on the exponential vectors this explicit form is

$$\int F_t a_t \, \mathrm{d}t \psi_f = \int \mathrm{d}t\, F_t f(t) \psi_f, \tag{5.5}$$

where the right-hand side of (5.1) is a usual Bochner integral. The right-hand side of (5.5) is defined on the set of the exponential vectors ψ_f with test function in $\mathcal{H}_1$ such that the vector valued function $s \mapsto f(s)F(s)\psi_f$ is Bochner integrable. From definition (5.1) we have that

$$\langle F^*, A\rangle := \int F a_s \, \mathrm{d}s. \tag{5.6}$$

In the case where $S = \mathbf{R}$ and $F = \chi_I F$ with $I = [0,t]$ we shall simply write

$$\langle F, A_t\rangle := \int_0^t F_s a_s \, \mathrm{d}s. \tag{5.7}$$

Thus the right annihilator integral maps functions $F\colon \mathbf{R}^d \to \mathcal{L}(\mathcal{D})$ into elements of $\mathcal{L}(\mathcal{D})$.

From (5.4) and (5.5) we deduce the estimate

$$\begin{aligned} \left\| \int F_s a_s \psi \, \mathrm{d}s \right\| &\leqslant \sum_{n\geqslant 0} \sqrt{n+1} \int \mathrm{d}s \, \| F_s \psi^{(n+1)}(s,\cdot)\| \\ &= \int \mathrm{d}s \, \| F_s (N+1)^{1/2} \psi^{(n+1)}_{(s,\cdot)} \| (s,\cdot) \| \end{aligned} \tag{5.8}$$

and $A(\chi_{I_j})$.

The definition of exponential vector implies that

$$(N+1)^{1/2} \psi_f^{(n+1)}(s,\cdot) = f(s) \psi_f^{(n)}, \tag{5.9}$$

therefore (5.5) implies that for any exponential vector ψ_f one has

$$\left\| \int F_s a_s \, \mathrm{d}s \psi_f \right\| \leqslant \int |f(s)| \cdot \| F_s \psi_f \| \, \mathrm{d}s. \tag{5.10}$$

A sufficient condition for the finiteness of the right-hand side of (5.10) is that the vector-valued function $s \mapsto f(s)F(s)\psi(s)$ is Bochner integrable.

6. The Left Creator Stochastic Integral

DEFINITION 1. The definition of left creator stochastic integrals is the natural extension of formula (3.13) for the scalar case

$$\left(\int a_t^+ F_t \psi \, \mathrm{d}t\right)^{(n)}(s_1, \ldots, s_n)$$
$$= \frac{1}{\sqrt{n}} \sum_{i=1}^{n} (F(s_i)\psi)^{(n-1)}(s_1, \ldots, \hat{s}_i, \ldots, s_n). \tag{6.1}$$

Definition 1 has a meaning for any measurable function F_s and, given such an F, the natural domain of its left creator stochastic integral is

$$\mathcal{D}\left(\int a_t^+ F_t \, \mathrm{d}t\right) = \left\{\psi \,\middle|\, \sum_{n=1}^{\infty} \left\| \left(\int a_t^+ F_t \psi \, \mathrm{d}t\right)^{(n)} \right\|^2 < \infty\right\} \tag{6.2}$$

or, more explicitly, a vector ψ is in $\mathcal{D}(\int a_t^+ F_t \, \mathrm{d}t)$ if and only if

$$\sum_{n=1}^{\infty} \frac{1}{n} \int_{\mathbf{R}^{dn}} \left| \sum_{i=1}^{n} (F(s_i)\psi)^{n-1}(s_1, \ldots, \hat{s}_i, \ldots, s_n) \right|^2 \mathrm{d}s_1 \ldots \mathrm{d}s_n < \infty. \tag{6.3}$$

We want now to obtain an estimate on the norm of $\int a_s^+ F_s \, \mathrm{d}s \psi$ which guarantees that the stochastic integral exists. This is given by the following lemma:

LEMMA 1. *Let $\psi^{(n-1)}$ belong to $\mathcal{D}(F_s)$ for all $s \in \mathbf{R}^d$. Then one has, for each $n \in \mathbf{N}$,*

$$\left\| P_n\left(\int \mathrm{d}s a_s^+ F_s \psi\right) \right\|^2 \leqslant n \int \mathrm{d}s \| P_{n-1}(F_s\psi) \|^2$$
$$= \int \mathrm{d}s \left\| \sqrt{(N+1)} (F_s\psi)^{(n-1)} \right\|^2. \tag{6.4}$$

In particular

$$\left\| \int \mathrm{d}s a_s^+ F_s \psi \right\|^2 \leqslant \int \mathrm{d}s \| (\sqrt{N+1}) F_s \psi \|^2. \tag{6.5}$$

Proof. The norm square of (6.1) is

$$\int \ldots \int \mathrm{d}s_1 \ldots \mathrm{d}s_n \frac{1}{n} \sum_{i,j} \langle (F_{s_i}\psi)^{(n-1)}(s_1, \ldots, \hat{s}_i, \ldots, s_n),$$
$$(F_{s_j}\psi)^{(n-1)}(s_1, \ldots, \hat{s}_j, \ldots, s_n) \rangle$$

$$\leqslant \frac{1}{n}\sum_{ij=1}^{n}\int ds_1 \ldots ds_n \|(F_{s_i}\psi)^{(n-1)}(s_1,\ldots,\hat{s}_i,\ldots,s_n)\| \times$$
$$\times \|(F_{s_j}\psi)^{(n-1)}(s_1,\ldots,\hat{s}_j,\ldots,s_n)\|$$
$$= \frac{n^2}{n}\int_{\mathbf{R}^d} ds \int_{\mathbf{R}^{d(n-1)}} ds_2 \ldots ds_n \|(F_s\psi)^{(n-1)}(s_2,\ldots,s_{n-1})\|^2$$
$$= n\int_{\mathbf{R}^d} ds \|(F_s\psi)^{(n-1)}\|^2. \square$$

COROLLARY 2. *Let L_s be a function with values in $\mathcal{B}(\mathcal{F})$ such that*

(i) *for any* $0 < T < +\infty \ \sup_{s\in[0,T]} \|L_s\|_\infty < \sqrt{C_T}$,
(ii) L_s *and* L_s^+ *commute with every* a_t, a_t^+, P_k $t \in \mathbf{R}$, $k \in \mathbf{N}$.

Then

$$\left\| P_n \int ds a_s^+ L_s F_s \psi \right\|^2 \leqslant C_T n \int ds \| P_{n-1} F_s \psi \|^2. \tag{6.6}$$

Proof. From Lemma 1 the left-hand side of (6.6) is less than or equal to

$$C_T n \int ds \| P_{n-1} L_s F_s \psi \|^2$$

and, using (ii) and (i), the thesis follows. □

7. The Normally Ordered Two-Sided Integral

DEFINITION 1. The two-sided (normally ordered) integral $\int ds b_s^+ F_s b_s$ is defined, weakly on the exponential or number vectors by

$$\langle \xi, \int ds b_s^+ F_s b_s \eta \rangle = \int ds \langle b_s \xi, F_s b_s \eta \rangle. \tag{7.1}$$

In particular, on exponential vectors one has

$$\langle \psi_f, \int ds b_s^+ F_s b_s \psi_g \rangle = \int ds \overline{f}(s) g(s) \langle \psi_f, F_s \psi_g \rangle. \tag{7.2}$$

LEMMA 2. *For any $n \in \mathbf{N}$ and for any exponential vector ψ_f, one has the estimate*

$$\left\| \left(\int ds b_s^+ F_s b_s \psi_f \right)^{(n)} \right\|^2 \leqslant n \int ds |f(s)|^2 \| (F_s \psi_f) \|^2. \tag{7.3}$$

In particular,

$$\left\| \int ds b_s^+ F_s b_s \psi_f \right\|^2 \leqslant \int ds |f(s)|^2 \| (N+1)^{1/2} \psi_f \|^2. \tag{7.4}$$

Proof. Using $b_s\psi_f = f(s)\psi_f$, the left-hand side of (7.3) becomes

$$\left\| \left(\int \mathrm{d}s b_s^+ f(s) F_s \psi_f \right)^{(n)} \right\|$$

and, because of (8.4), this is

$$\leqslant n \int \mathrm{d}s \|(F_s\psi_f)^{(n)}\|^2 |f(s)|^2,$$

i.e. (1). (2) is obtained from (1) by summing over all n. □

8. Differential Calculus

Usually in stochastic calculus one considers the differentials only as symbolic expressions for the corresponding integrals. We want to develop a differential calculus directly in analogy with classical analysis.

Let us first consider the differentiability properties, with respect to t, of the Brownian motion operators B_t, B_t^+.

THEOREM 1. *Let $\psi \in \mathcal{D}_S^o$ be such that, for each n, $\psi^{(n)}$ is continuous with compact support. Then, with ΔB_t defined by* (6.5) *below, one has the following:*

(i) $$\lim_{\Delta t \to 0} \left\| \left(\frac{\Delta B_t}{\Delta t} - b(t) \right) \psi \right\| = 0, \tag{8.1}$$

where the operator $b(t)$ is defined in (6.5).

(ii) *The strong limit, as $\Delta t \to 0$, of $\Delta B_t^+/\Delta t - b^+(t)$ does not exist on the number vectors. However, the weak limit of this expression on $\mathcal{D}_S^o$ does exist, i.e. $\forall \psi_1, \psi_2 \in \mathcal{D}_S^o$,*

$$\begin{aligned} \lim_{\Delta t \to 0} \left(\psi_1, \frac{\Delta B_t^+}{\Delta t} \psi_2 \right) &= (\psi_1, b^+(t)\psi_2) \\ &\equiv \frac{1}{\sqrt{n}} \sum_{i=1}^{n} \int_{\mathbf{R}^{n-1}} \mathrm{d}s_1 \ldots \mathrm{d}\hat{s}_i \ldots \mathrm{d}s_n \times \\ &\quad \times \psi_1^{(n)}(s_1, \ldots, t, s_n)\psi_2^{(n-1)}(s_1, \ldots, \hat{s}_i, \ldots, s_n). \end{aligned} \tag{8.2}$$

(iii) $$\lim_{\Delta t \to 0} \left\| \left(\frac{\Delta B_t \cdot \Delta B_t^+}{\Delta t} - 1 \right) \psi \right\| = 0. \tag{8.3}$$

(iv) $$\lim_{\Delta t \to 0} \left\| \frac{\Delta B_t^+}{\Delta t} \Delta B_t \psi \right\| = 0. \tag{8.4}$$

Proof. One has

$$(\Delta B_t \psi)^{(n)}(s_1, \ldots, s_n) = \sqrt{n+1} \int_t^{t+\Delta t} \psi^{(n+1)}(s, s_1, \ldots, s_n)\, \mathrm{d}s, \tag{8.5}$$

$$(\Delta B_t^+ \psi)^{(n)}(s_1, \ldots, s_n) = \frac{1}{\sqrt{n}} \sum_{i=1}^{n} \chi_{[t,t+\Delta t]}(s_i) \psi^{(n-1)}(s_1, \ldots, \hat{s}_i, \ldots, s_n) \tag{8.6}$$

and

$$(\Delta B_t \Delta B_t^+ \psi)^{(n)}(s_1, \ldots, s_n) = \psi^{(n)}(s_1, \ldots, s_n) \cdot \Delta t + \sum_{i=1}^{n} \chi_{[t,t+\Delta t]}(s_i) \int_t^{t+\Delta t} \psi^{(n)}(s_1, \ldots, \hat{s}_i, \ldots, s_n, t_1)\, \mathrm{d}t_1. \tag{8.7}$$

From (8.2), we have

$$\left\| \left(\frac{\Delta B_t}{\Delta t} - b(t) \right) \psi^{(n)} \right\|^2 = (n+1) \int_{\mathbf{R}^{dn}} \mathrm{d}s_1 \ldots \mathrm{d}s_n \times \left| \frac{1}{\Delta t} \int_t^{t+\Delta t} \psi^{(n+1)}(s, s_1, \ldots, s_n)\, \mathrm{d}s - \psi^{(n+1)}(t, s_1, \ldots, s_n) \right|^2. \tag{8.8}$$

Because $\psi^{(n+1)}$ are continuous functions with compact support, one can go to the limit $\Delta t \to 0$ under the integral over $\mathrm{d}s_1, \ldots, \mathrm{d}s_n$ and we get (1).

To see that the limit in (ii) does not exist let us take $\psi^{(0)} = \Phi$. Then one has

$$(\Delta B_t^+ \psi)^{(1)}(s_1) = \chi_{[t,t+\Delta t]}(s_1) \tag{8.9}$$

and

$$\left\| \frac{(\Delta B_t^+ \psi)^{(1)}}{\Delta t} \right\|^2 = \frac{1}{(\Delta t)^2} \int \chi_{[t,t+\Delta t]}(s_1)\, \mathrm{d}s_1 = \frac{1}{\Delta t}. \tag{8.10}$$

From (8.10) it is clear that the limit when $\Delta t \to 0$ does not exist. However, there exist the limit of the bilinear form (8.2). Now, from (8.7) one has

$$\left\| \left(\frac{\Delta B_t \cdot \Delta B_t^+}{\Delta t} - 1 \right) \psi^{(n)} \right\|^2 = \int_{\mathbf{R}^{dn}} \mathrm{d}s_1 \ldots \mathrm{d}s_n \left| \frac{1}{\Delta t} \sum_{i=1}^{n} \chi_{[t,t+\Delta t]}(s_i) \int_t^{t+\Delta t} \psi^{(n)}(s_1, \ldots, \hat{s}_i, \ldots, s_n, t_1)\, \mathrm{d}t \right|^2$$
$$= \sum_{i,j=1}^{n} \frac{1}{(\Delta t)^2} \int_t^{t+\Delta t} \mathrm{d}s_i \int_t^{t+\Delta t} \mathrm{d}t_1 \int_t^{t+\Delta t} \mathrm{d}t_1' \int_{\mathbf{R}^{n-1}} \mathrm{d}s_1 \ldots \mathrm{d}\hat{s}_i \ldots \mathrm{d}s_n \times \overline{\psi^{(n)}}(s_1, \ldots, \hat{s}_i, \ldots, s_n, t_1) \psi^{(n)}(s_1, \ldots, \hat{s}_j, \ldots, s_n, t_1') \tag{8.11}$$

which ten ds to zero when $\Delta t \to 0$. This ends the proof of the theorem. □

Remark 1. One can symbolically write the relation (8.1) in the form

$$\mathrm{d}B_t = b(t)\,\mathrm{d}t \tag{8.12}$$

and the relation for bilinear form (8.2) as

$$\mathrm{d}B_t^+ = b^+(t)\,\mathrm{d}t. \tag{8.13}$$

Formulas (8.3), (8.4) look like the Itô rules

$$\mathrm{d}B_t\,\mathrm{d}B_t^+ = \mathrm{d}t, \tag{8.14}$$

$$\mathrm{d}B_t^+\mathrm{d}B_t = 0. \tag{8.15}$$

However, we emphasize that we get them as differential relations (8.3), (8.4) and not as integral relations like in the Itô calculus.

9. Mutual Quadratic Variation

In this section we prove that the limit relation (8.3) is true in a topology much stronger than the one given by strong convergence in a dense subspace of $\mathcal{F}$.

For a real number $\Delta t > 0$, we shall denote

$$\Delta B_t^\pm := B^\pm_{(0,t+\Delta t)} - B^\pm_{(0,t)}. \tag{9.1}$$

From (8.4), (8.5) we deduce that

$$(\Delta B_t \psi)^{(n)}(s_1,\ldots,s_n) = \sqrt{n+1}\int_t^{t+\Delta t} \psi^{(n+1)}(s,s_1,\ldots,s_n)\,\mathrm{d}s, \tag{9.2}$$

$$(\Delta B_t^+ \psi)^{(n)}(s_1,\ldots,s_n) = \frac{1}{\sqrt{n}}\sum_{i=1}^{n} \chi_{(t,t+\Delta t]}(s_i)\cdot \psi^{(n-1)}(s_1,\ldots,\hat{s}_i,\ldots,s_n). \tag{9.3}$$

From (8.6), one deduces the identity, for any real number Δt,

$$\begin{aligned}(\Delta B_t \Delta B_t^+ \psi)^{(n)}(s_1,\ldots,s_n) &= \sum_{i=1}^{n} \chi_{(t,t+\Delta t]}(s_i)\int_t^{t+\Delta t} \psi^{(n)}(s_1,\ldots,\hat{s}_i,\ldots,s_n,t_1)\,\mathrm{d}t_1 + \\ &\quad + \Delta t\cdot\psi^{(n)}(s_1,\ldots,s_n).\end{aligned} \tag{9.4}$$

The Itô multiplication table is obtained from (9.4) when $\Delta t \to 0$ and has the form

$$(\Delta B_t\cdot\Delta B_t^+\psi)^{(n)}(s_1,\ldots,s_n) = \psi^{(n)}(s_1,\ldots,s_n)\Delta t + \mathrm{o}(\Delta t), \tag{9.5}$$

where $o(\Delta t)$ means something that, when summed over all the intervals $(t, t+\Delta t)$ of a partition of a fixed interval (S, T), tends to zero as $\Delta t \to 0$. In order to make this statement precise, one has to choose a topology and this can be done in a multitude of ways. In the following, we prove some estimates which show that some topologies arise quite naturally in our context. For example, if the function $\psi^{(n)}$ is measurable and bounded, then one has the estimate

$$|(\Delta B_t \cdot \Delta B_t^+ \psi)^{(n)}(s_1, \dots, s_n) - \psi^{(n)}(s_1, \dots, s_n)\Delta t| \leqslant \|\psi^{(n)}\|_\infty \Delta t \sum_{i=1}^{n} \chi_{(t,t+\Delta t]}(s_i). \tag{9.6}$$

LEMMA 1. *Assume that $\psi^{(n)}$ is bounded, fix a bounded interval (S, T) and consider the partition of (S, T) into intervals of equal width Δt. Then, if $\sum_t$ denotes summation over the intervals of the partition, one has*

$$\left|\sum_t (\Delta B_t \Delta B_t^+ \psi)^{(n)}(s_1, \dots, s_n) - (T-S) \cdot \psi^{(n)}(s_1, \dots, s_n)\right| \leqslant n\Delta t \cdot \|\psi^{(n)}\|_\infty. \tag{9.7}$$

In particular, the limit

$$\lim_{\Delta t \to 0} \sum_t (\Delta B_t \Delta B_t^+ \psi)^{(n)}(s_1, \dots, s_n) = (T-S) \cdot \psi^{(n)}(s_1, \dots, s_n) \tag{9.8}$$

holds uniformly in $s_1, \dots, s_n$.

Proof. Summing the identity (9.4) over all the intervals of the partition, we obtain

$$\sum_t (\Delta B_t \Delta B_t^+ \psi)^{(n)}(s_1, \dots, s_n) = \sum_{i=1}^{n} \sum_t \chi_{(t,t+\Delta t]}(s_i) \int_t^{t+\Delta t} \psi^{(n)}(s_1, \dots, \hat{s}_i, \dots, s_n, t_1)\, \mathrm{d}t_1 + (T-S) \cdot \psi^{(n)}(s_1, \dots, s_n).$$

But

$$\left|\sum_{i=1}^{n} \sum_t \chi_{(t,t+\Delta t]}(s_i) \int_t^{t+\Delta t} \psi^{(n)}(s_1, \dots, \hat{s}_i, \dots, s_n, t_1)\, \mathrm{d}t_1\right| \leqslant \Delta t \cdot \|\psi^{(n)}\|_\infty \sum_{i=1}^{n} \sum_t \chi_{(t,t+\Delta t]}(s_i) = \Delta t \cdot \|\psi^{(n)}\|_\infty,$$

which tends to zero, as $\Delta t \to 0$, uniformly in $s_1, \dots, s_n$. □

10. Normally Ordered White Noise Equations in R^d

Given a notion of stochastic integral, one can study the problem of the meaning, existence, uniqueness, and unitarity of the corresponding integral equations. We will study integral equations of the form

$$\begin{aligned} Y_t &= Y_0 + \int_{\mathbf{R}^d} L_{01}(s,t)Y_s a_s \, \mathrm{d}s + \int_{\mathbf{R}^d} L_{10}(s,t)a_s^+ Y_s \, \mathrm{d}s + \\ &+ \int_{\mathbf{R}^d} L_{11}(s,t)a_s^+ Y_s a_s \, \mathrm{d}s + \int_{\mathbf{R}^d} L_{00}(s,t)Y_s \, \mathrm{d}s, \end{aligned} \tag{10.1}$$

where the coefficients $L_{\varepsilon,\varepsilon'}(s,t)$ $(\varepsilon, \varepsilon' = 0, 1)$ are linear operators acting on $\mathcal{H}_S$ such that,

(i) for any $(\varepsilon, \varepsilon' = 0, 1)$ and $s, t \in \mathbf{R}^d$, the operator $L_{\varepsilon,\varepsilon'}(s,t)$ is bounded;
(ii) defining

$$\max_{\varepsilon,\varepsilon'=0,1} \|L_{\varepsilon,\varepsilon'}(s,t)\| =: l(s,t), \tag{10.2}$$

then for any bounded set $B \subseteq \mathbf{R}^d$, the functions

$$s \in \mathbf{R}^d \mapsto l(s,t) \tag{10.3}$$

are integrable for each $t \in B$ and the set of integrals, as a function of $t \in B$, is bounded;
(iii) for any bounded set $B \subseteq \mathbf{R}^d$, there exists a constant $L \geqslant 0$ such that, for any natural integer k one has

$$\int_{\mathbf{R}^d} \cdots \int_{\mathbf{R}^d} \mathrm{d}s_1 \ldots \mathrm{d}s_k l(s_1,t) l(s_2,s_1) \ldots l(s_k,s_{k-1}) \leqslant \frac{L^k}{k!} \tag{10.4}$$

uniformly in $t \in B$. (In Section 12, we shall give examples of coefficients $L_{\varepsilon,\varepsilon'}(s,t)$ which satisfy this condition.)

We shall write Equation (10.1) in the notation

$$Y_t = Y_0 + \int_{\mathbf{R}^d} L_{\varepsilon,\varepsilon'}(s,t) \, \mathrm{d}\Lambda_\varepsilon^{\varepsilon'}(s) Y_s, \tag{10.5}$$

where summation is understood in the indices $\varepsilon, \varepsilon' \in \{0, 1\}$.

In this notation we define the kth iterated approximation solution of Equation (10.5) by

$$Y_t^{(0)} = Y_0, \tag{10.6}$$

$$Y_t^{(k+1)} = \int_{\mathbf{R}^d} L_{\varepsilon\varepsilon'}(s,t) \, \mathrm{d}\Lambda_\varepsilon^{\varepsilon'}(s) Y_s^{(k)}. \tag{10.7}$$

The iterated series, associated to Equation (10.1) is

$$\sum_{k=0}^{\infty} Y_t^{(k)}. \tag{10.8}$$

In this section we shall fix the set

$$\mathcal{S}_0 = \{f \in L^2(\mathbf{R}^d) : \max\{\|f\|_\infty, \|f\|_2\} \leqslant 1\} \tag{10.9}$$

and we denote $\mathcal{E}(\mathcal{S}_0)$ as the corresponding set of exponential vectors. It is known that $\mathcal{S}_0$ is a total set in $\mathcal{F}$.

THEOREM 1. *Suppose that the coefficients of Equation* (10.1) *satisfy conditions* (i), (ii), (iii) *and, moreover,*

$$\|L\| < \frac{1}{16e}. \tag{10.10}$$

Then the iterated series (10.8) *converges, strongly in norm on* $\mathcal{E}(\mathcal{S}_0)$ *to a solution of this equation uniformly in bounded subsets of* $\mathbf{R}^d$.

For the proof of Theorem 1 we shall use several lemmata.

LEMMA 2. *Let* $L_{\varepsilon,\varepsilon'}(s,t)$ $(\varepsilon, \varepsilon' = 0, 1)$ *be linear operators on* $\mathcal{H}_S$ *satisfying the conditions* (i), (ii) *and* (iii), *then for any* $t \in B \subseteq \mathbf{R}^d$, *a bounded set*, $n \in \mathbf{N}$ *and* $f \in \mathcal{S}_0$ *one has*

$$\begin{aligned} &\left\| P_n \int L_{\varepsilon\varepsilon'}(s,t) Y_s \, \mathrm{d}\Lambda_\varepsilon^{\varepsilon'}(s) \psi_f \right\|^2 \\ &\quad \leqslant 8n \int l(s,t) \, \mathrm{d}s (\|P_{n-1} Y_s \psi_f\|^2 + \|P_n Y_s \psi_f\|^2). \end{aligned} \tag{10.11}$$

In particular, $Y_t^{(k)}$ *defined by* (10.7) *verifies that*

$$\begin{aligned} &\|P_n Y_t^{(k+1)} \psi_f\|^2 \\ &\quad \leqslant 8n \int \mathrm{d}s \, l(s,t) (\|P_{n-1} Y_s^{(k)} \psi_f\|^2 + \|P_n Y_s^{(k)} \psi_f\|^2). \end{aligned} \tag{10.12}$$

Proof. First of all,

$$\begin{aligned} &\left\| P_n \int L_{\varepsilon,\varepsilon'}(s,t) Y_s \, \mathrm{d}\Lambda_\varepsilon^{\varepsilon'}(s) \psi_f \right\|^2 \\ &\quad \leqslant \left(\sum_{\varepsilon,\varepsilon'=0,1} \left\| P_n \int_{\mathbf{R}^d} L_{\varepsilon,\varepsilon'}(s,t) \, \mathrm{d}\Lambda_\varepsilon^{\varepsilon'}(s) Y_s \psi_f \right\| \right)^2. \end{aligned}$$

The Schwarz inequality

$$\left(\sum_{j=1}^{M} a_j\right)^2 \leqslant M \sum_{j=1}^{M} a_j^2$$

(M is an integer and M and a_j are real numbers), implies that, for any $n \in \mathbf{N}$,

$$\begin{aligned} &\left\| P_n \int L_{\varepsilon,\varepsilon'}(s,t) Y_s \, \mathrm{d}\Lambda_\varepsilon^{\varepsilon'}(s) \psi_f \right\|^2 \\ &\quad \leqslant 4 \sum_{\varepsilon,\varepsilon'=0,1} \left\| P_n \int_{\mathbf{R}^d} L_{\varepsilon,\varepsilon'} Y_s \, \mathrm{d}\Lambda_\varepsilon^{\varepsilon'}(s) \psi_f \right\|^2 . \end{aligned} \tag{10.13}$$

Now we investigate the quantity in the right-hand side of (10.13) term by term according to the values of ε, ε'.

By letting a_s act on the exponential vector, we deduce

$$\begin{aligned} &\left\| P_n \int L_{01}(s,t) Y_s a_s \psi_f \, \mathrm{d}s \right\|^2 \\ &\quad \leqslant \|f\|_2^2 \int \|P_n L_{01}(s,t) Y_s \psi_f\|^2 \, \mathrm{d}s \\ &\quad \leqslant \|f\|_2^2 \int \mathrm{d}s l(s,t) \|P_n Y_s \psi_f\|^2 \, \mathrm{d}s. \end{aligned} \tag{10.14}$$

From formula (6.4), one has

$$\left\| P_n \int \mathrm{d}s \; L_{10}(s,t) a_s^+ Y_s \psi_f \right\|^2 \leqslant n \int \mathrm{d}s \; l(s,t) \|P_{n-1} Y_s \psi_f\|^2 \tag{10.15}$$

and from formula (7.3),

$$\begin{aligned} &\left\| P_n \int \mathrm{d}s \; L_{11}(s,t) a_s^+ Y_s a_s \psi_f \right\|^2 \\ &\quad \leqslant \|f\|_\infty^2 n \int \mathrm{d}s \; l(s,t) \|P_{n-1} Y_s \psi_f\|^2 . \end{aligned} \tag{10.16}$$

Finally, the usual properties of Bochner's integral imply that

$$\left\| P_n \int \mathrm{d}s L_{00}(s,t) Y_s \psi_f \right\|^2 \leqslant \int \mathrm{d}s \; l(s,t) \|P_n Y_s \psi_f\|^2 . \tag{10.17}$$

Because of our assumption (10.9) on f, the sum of the left-hand sides of (10.14), (10.15), (10.16), (10.17) is less than or equal to

$$\begin{aligned} &2 \int \mathrm{d}s \; l(s,t) \|P_n Y_s \psi_f\|^2 + 2n \int \mathrm{d}s \; l(s,t) \|P_{n-1} Y_s \psi_f\|^2 \\ &\quad \leqslant 2n \int \mathrm{d}s \; l(s,t) (\|P_{n-1} Y_s \psi_f\|^2 + \|P_n Y_s \psi_f\|^2) \end{aligned}$$

and this is (10.11). To deduce (10.12), one simply applies (10.11) to the definition of $Y_t^{(k+1)}$. □

LEMMA 3. *If the series*

$$\sum_{k=0}^{\infty} \|Y_t^{(k)}\psi_f\| \tag{10.18}$$

converges uniformly on a bounded set B in $\mathbf{R}^d$, then for each $t \in B$ there exists a unique operator Y_t on $\mathcal{H}_S \otimes \mathcal{E}(\mathcal{S}_0)$ such that

$$\sum_{k=0}^{\infty} Y_t^{(k)} = Y_t \tag{10.19}$$

and the series on the left-hand side of (10.19) *converges strongly in norm on $\mathcal{E}(\mathcal{S}_0)$, uniformly for $t \in B$. Moreover, the function $\mapsto Y_t$ is a solution of Equation* (10.1).

Proof. From Lemma 2 we know that there exists an operator Y_t on $\mathcal{H}_S \otimes \mathcal{E}(\mathcal{S}_0)$ such that (10.19) hol ds. And the convergence estimates also imply that the stochastic integrals of Y_t for the basic integrators exist. To prove that Y_t satisfies Equation (10.1) it will be sufficient to prove that, for each $n \in \mathbf{N}$, $P_n Y_t$ satisfies Equation (10.1). To show this, we use the estimate of Lemma 2 to deduce that

$$\begin{aligned}&\left\| P_n \int_{\mathbf{R}^d} L_{\varepsilon\varepsilon'}(s,t) Y_s \,\mathrm{d}\Lambda_\varepsilon^{\varepsilon'}(s)\psi_f - P_n \int_{\mathbf{R}^d} \sum_{k=1}^{N} L_{\varepsilon,\varepsilon'}(s,t) Y_s^{(k)} \,\mathrm{d}\Lambda_\varepsilon^{\varepsilon'}(s)\psi_f \right\| \\ &\quad \leqslant 8n \int_{\mathbf{R}^d} \sum_{k=N+1}^{\infty} \|P_n Y_s^{(k)}\psi_f\| l(s,t)\,\mathrm{d}s.\end{aligned} \tag{10.20}$$

By assumption, for each $t \in B$, the function $s \mapsto l(s,t)$ is integrable. Therefore, the right-hand side of (10.20) tends to zero by dominated convergence as $N \to \infty$.

Letting $N \to \infty$ in (10.20), we see that Y satisfies Equation (10.1) and this completes the proof. □

LEMMA 4. *Let $I_{n,k}$ ($n, k \in \mathbf{N}$) be positive numbers satisfying the inequality*

$$I_{n,k+1} \leqslant cn(I_{n,k} + I_{n-1,k}), \tag{10.21}$$

where $c > 0$ is a constant, then

$$I_{n,k+1} \leqslant (2cn)^k \sum_{m=n-k}^{n} I_{m,0}. \tag{10.22}$$

Proof. By iterating the inequality (10.21) we see that the right-hand side is equal to

$$\begin{aligned}
&cn(cnI_{n,k-1} + cnI_{n-1,k-1} + c(n-1)I_{n-1,k-1} + c(n-1)I_{n-2,k-1}) \\
&\quad \leqslant (cn)^2(I_{n,k-1} + 2I_{n-1,k-1} + I_{n-2,k-1}) \\
&\quad \leqslant (cn)^3(I_{n,k-2} + 3I_{n-1,k-2} + 3I_{n-2,k-2} + I_{n-3,k-2}) \ldots \\
&\quad \leqslant (cn)^k(I_{n,0} + h_1 I_{n-1,0} + h_2 I_{n-2,0} + \cdots + h_k I_{n-k,0}),
\end{aligned}$$

where the coefficients h_α satisfy $h_\alpha \leqslant 2^k$ and (10.22) immediately follows from this. □

Proof of Theorem 1. Introducing the notation

$$I_{n,k+1}(s) := \| P_n Y_t^{(k+1)} \psi_f \|^2,$$

we have from Lemma 2

$$I_{n,k+1}(t) \leqslant \int \mathrm{d}s\, l(s,t) 8n (I_{n,k}(s) + I_{n-1,k}(s)),$$

therefore, arguing as in Lemma 4

$$\begin{aligned}
I_{n,k+1}(t) \;\leqslant\; & 16^k n^k \sum_{m=n-k}^{n} I_{m,0}(s_k) \int \cdots \int \mathrm{d}s_1 \ldots \mathrm{d}s_k l(s_1,t) \times \\
& \times l(s_2,s_1) \ldots l(s_k,s_{k-1}).
\end{aligned} \tag{10.23}$$

But for any $s_k \in \mathbf{R}^d$

$$I_{m,0}(s_k) = \| P_m Y_0 \psi_f \|^2 = \| Y_0 \|^2 \frac{\| f \|^{2m}}{m!}$$

and, without loss of generality, we can assume that

$$\| Y_0 \| = 1. \tag{10.24}$$

Moreover, according to assumption (iii), the multiple integral in (10.23) is dominated by $L^k/k!$. In conclusion

$$\| P_n Y_t^{(k+1)} \psi_f \|^2 \leqslant \frac{(16L)^k}{k!} n^k \sum_{m=n-k}^{n} \frac{\| f \|^{2m}}{m!}. \tag{10.25}$$

Since for large m the sequence $\| f \|^{2m}/m!$ is decreasing the sum in (24) is majorized by

$$k \frac{\| f \|^{2(n-k)}}{(n-k)!}.$$

Therefore

$$\|P_n Y_t^{(k+1)}\psi_f\|^2 \leqslant \frac{(16L)^k}{(k-1)!}\frac{n^k\|f\|^{2(n-k)}}{(n-k)!}.$$

So in order to estimate

$$\|Y_t^{(k+1)}\psi_f\|^2,$$

we are lead to estimate the series

$$\begin{aligned}\sum_{n\geqslant k}\frac{n^k\|f\|^{2(n-k)}}{(n-k)!} &= \frac{\mathrm{d}^k}{\mathrm{d}t^k}\bigg|_{t=0}\sum_{n\geqslant k}\mathrm{e}^{tn}\frac{\|f\|^{2(n-k)}}{(n-k)!} \\ &= \frac{\mathrm{d}^k}{\mathrm{d}t^k}\bigg|_{t=0}\mathrm{e}^{tk}\mathrm{e}^{\|f\|^2\mathrm{e}^t}. \end{aligned} \tag{10.26}$$

Moreover, because of our assumption (10.9) on the test functions f, we can restrict our attention to the case in which $\|f\| = 1$ in (10.26). (We could have put $\|f\| = 1$ directly in (10.25), but it is convenient to leave it to show the opportunity of introducing *Bell numbers depending on a parameter*.) In this case by Leibnitz rule the expression (10.26) is

$$\sum_{h=0}^{k}\binom{k}{h}k^h B_2(k-h), \tag{10.27}$$

where $B_2(k-h)$ are the Bell numbers of order 2 as defined in [9].

Under this assumption denoting

$$c := 16L, \tag{10.28}$$

we have

$$\begin{aligned}\|P_n Y_t^{(k+1)}\psi_f\|^2 &\leqslant \frac{c^k}{(k-1)!}\sum_{h=0}^{k}\binom{k}{h}k^h B_2(k-h) \\ &= (kc^k)\sum_{h=0}^{k}\frac{k^h}{h!}\frac{B_2(k-h)}{(k-h)!}. \end{aligned} \tag{10.29}$$

Now, since all the terms involved are positive, clearly

$$\sum_{h=0}^{k}\frac{k^h}{h!}\frac{B_2(k-h)}{(k-h)!} \leqslant \left(\sum_{h=0}^{k}\frac{k^h}{h!}\right)\left(\sum_{h'=0}^{k}\frac{B_2(k-h)}{(k-h)!}\right)$$

and, from [9] we know that this is

$$\leqslant \mathrm{e}^k G_2(1)/2,$$

where G_2 is an analytic function. Therefore

$$\|Y_t^{(k+1)}\psi_f\|^2 \leqslant G_2(1)k(ce)^k/2. \tag{10.30}$$

But if $ce < 1$ or, equivalently due to (10.28), if

$$L < \frac{1}{16e}$$

the series on the right-hand side of (10.30) is convergent. □

11. An Example

In this section we produce an example of coefficients which satisfy condition (iii) of Equation (10.11). Let, for $s, t \in \mathbf{R}^d$

$$L_{\varepsilon,\varepsilon'}(s,t) = L_{\varepsilon,\varepsilon'}\psi(|s|)\chi_{[0,|t|)}(|s|)\varphi(\hat{s},\hat{t}), \tag{11.1}$$

where $L_{\varepsilon,\varepsilon'} \in \mathcal{B}(\mathcal{H}_S)$ $(\varepsilon, \varepsilon' = 0, 1)$,

$$\chi_I(x) = \begin{cases} 0, & \text{if } x \notin I \subseteq \mathbf{R}, \\ 1, & \text{if } x \in I. \end{cases} \tag{11.2}$$

$\psi\colon \mathbf{R}_+ \to \mathbf{C}$ and $\varphi\colon S^{(d)} \times S^{(d)} \to \mathbf{C}$ are continuous functions ($S^{(d)}$ is the unit sphere in $\mathbf{R}^d$) and

$$t = |t|\hat{t} \in \mathbf{R}^d; \quad |t| \in \mathbf{R}_+; \ \hat{t} \in S^{(d)} \quad \text{(unit sphere in } \mathbf{R}^d) \tag{11.3}$$

is the polar decomposition of $t \in \mathbf{R}^d$. Then

$$\begin{aligned}
&\int_{\mathbf{R}^d} \cdots \int_{\mathbf{R}^d} \mathrm{d}s_1 \ldots \mathrm{d}s_k l(s_1,t) l(s_1,s_2) \ldots l(s_k,s_{k-1}) \\
&\quad = \int \cdots \int \rho_1^{d-1}\,\mathrm{d}\rho_1\,\mathrm{d}\hat{s}_1 \ldots \rho_k^{d-1}\,\mathrm{d}\rho_k\,\mathrm{d}\hat{s}_k\,\chi_{[0,t]}(\rho_1)\chi_{[0,\rho_1)}(\rho_2)\ldots\chi_{[0,\rho_{n-1})}(\rho_n)\times \\
&\qquad \times \varphi(\hat{s}_1,\hat{t})\varphi(\hat{s}_2,\hat{s}_1)\ldots\varphi(s_k,s_{k-1})\psi(\rho_1)\ldots\psi(\rho_k) \\
&\quad \leqslant (|t|^{d-1})^k \|\varphi\|_\infty^k \sigma_d^k \cdots \int_0^{|t|} \mathrm{d}\rho_1 \int_0^{\rho_1} \mathrm{d}\rho_2 \ldots \int_0^{\rho_{k-1}} \mathrm{d}\rho_k \psi(\rho_1)\ldots\psi(\rho_k) \\
&\quad = (|t|^{d-1})^k \|\varphi\|_\infty^k \sigma_d^k \frac{(\int_0^{|t|} \psi(s)\,\mathrm{d}\rho)^k}{k!}.
\end{aligned}$$

Therefore, if $B \subseteq \mathbf{R}^d$ is a bounded set and $t \in B$, condition (iii) is satisfied.

References

1. Accardi, L., Frigerio, A. and Lu, Y. G.: On the weak coupling limit problem, In: *Quantum Probability and Applications IV*, Lecture Notes in Math. 1396, Springer, New York, 1987, pp. 20–58.

2. Accardi, L., Frigerio, A. and Lu, Y. G.: The weak coupling limit as a quantum functional central limit, *Comm. Math. Phys.* **131** (1990), 537–570.
3. Accardi, L., Lu, Y. G. and Volovich, I. V.: The stochastic sector of quantum field theory, Volterra Preprint 138, 1993; *Mat. Zam.* (1994).
4. Accardi, L., Lu, Y. G. and Volovich, I. V.: Non-commutative (quantum) probability, master fields and stochastic bosonization, Volterra Preprint CVV-198-94, hep-th/9412241.
5. Accardi, L. and Mohari, A.: Stochastic flows and imprimitivity systems, In: *Quantum Probability and Related Topics*, World Scientific, Singapore, 1994, pp. 43–65. Volterra Preprint, 1994.
6. Accardi, L., Lu, Y. G. and Volovich, I. V.: Nonlinear extensions of classical and quantum stochastic calculus and essentially infinite dimensional analysis, In: L. Accardi and Chris Heyde (eds), *Probability Towards 2000*, Lecture Notes in Statist. 128, Springer, New York, 1998, pp. 1–33.
7. Accardi, L. and Volovich, I. V.: Quantum white noise with singular non-linear interaction, Volterra Preprint 278, 1997.
8. Accardi, L., Lu, Y. G. and Volovich, I. V.: *A White Noise Approach to Classical and Quantum Stochastic Calculus*, Volterra Preprint 375, Rome, July 1999, World Scientific, Singapore, 2000.
9. Cochran, W. G, Kuo, H.-H. and Sengupta, A.: A new class of white noise generalized functions, *Infin. Dimens. Anal. Quantum Probab. Relat. Top.* **1** (1998), 43–67.
10. Hida, T.: *Analysis of Brownian Functionals*, Carleton Mathematical Lecture Notes 13, 1975.
11. Hida, T.: The impact of classical functional analysis on white noise calculus, Volterra Preprints 90, 1992.
12. Hida, T., Kuo, H.-H., Potthoff, J. and Streit L.: *White Noise. An Infinite Dimensional Calculus*, Kluwer Acad. Publ., Dordrecht, 1993, pp. 185–231.
13. Hida, T., Potthoff, J. and Streit, L.: *White Noise Analysis and Applications*, Kluwer Acad. Publ., Dordrecht, 1994.
14. Hudson, R. L. and Parthasarathy, K. R.: Construction of quantum diffusions, In: L. Accardi, A. Frigerio and V. Gorini (eds), *Quantum Probability and Applications to the Quantum Theory of Irreversible Processes*, Lecture Notes in Math. 1055, Springer, New York, 1984.
15. Hudson, R. L. and Parthasarathy, K. R.: Quantum Itô's formula and stochastic evolutions, *Comm. Math. Phys.* **93** (1984), 301–323.
16. Kuo, H.-H.: *White Noise Distribution Theory*, CRC Press, Boca Raton, 1996.
17. Parthasarathy, K. R.: Quantum stochastic calculus, Preprint, 1995.
18. Obata, N.: Coherent state representation and unitarity condition in white noise calculus, *J. Korean Math. Soc.* (2000), to appear; White noise operator theory: fundamental concepts and developing applications, to appear in *Proc. Volterra Internat. School*, World Scientific, Singapore, 2000.

Acta Applicandae Mathematicae 63: 27–40, 2000.

Stochastic Dynamics of Compact Spins: Ergodicity and Irreducibility

SERGIO ALBEVERIO[1], ALEXEI DALETSKII[1], YURI KONDRATIEV[1] and MICHAEL RÖCKNER[2]
[1]*Institut für Angewandte Mathematik, Universität Bonn, D-53115 Bonn, Germany*
[2]*Fakultät für Mathematik, Bielefeld Universität, D-33615 Bielefeld, Germany*

(Received: 5 May 1999)

Abstract. Stochastic dynamics associated with Gibbs measures on $M^{\mathbf{Z}^d}$, where M is a compact Riemannian manifold and $\mathbf{Z}^d$ is an integer lattice, is considered. Equivalence of its L^2-ergodicity and the extremality of the corresponding Gibbs measure is proved.

Mathematics Subject Classifications (2000): 58B90, 58J65, 58Z05.

Key words: Dirichlet form, Gibbs measure, ergodic semigroup.

1. Introduction

This paper is devoted to the investigation of L^2-ergodic properties of the stochastic dynamics associated with Gibbs measures of classical lattice systems with a compact Riemannian manifold M as the spin space. That is, we study the semigroup $T_t = e^{-tH_\mu}$ in $L^2(\mu)$, where μ is a fixed Gibbs measure on $M^{\mathbf{Z}^d}$ and H_μ is the corresponding Dirichlet operator. It is well known that the Gibbs measure given by a fixed set of potentials $\mathcal{U}$ or, in other words, by a fixed logarithmic derivative, is in general not unique. Such measures form a convex set denoted by $\mathcal{G}(\mathcal{U})$. The main result of this work is the characterization of the extreme elements of $\mathcal{G}(\mathcal{U})$ in terms of ergodicity of the associated stochastic dynamics. This fact was proved for Gibbs states of classical and quantum lattice systems with flat spin spaces in [8], and for some models of Euclidean quantum field theory in [9]. Similar results for classical continuous systems were obtained in [10]. Compactness of the spin space in our case gives us the possibility to consider a quite general class of potentials, more general than in the case of noncompact spin spaces, including interactions of infinite range.

The structure of the paper is the following. In Section 2 we introduce the main notations and conditions on the family of potentials $\mathcal{U}$ which ensure existence of the corresponding Gibbs measure μ (cond. (U1), see, e.g., [13, 12]). Extending the ideas of the works [17] and [2], we prove that under the conditions (U1) and (U2) the Gibbs measure μ is differentiable and can be characterized in terms of its logarithmic derivative (Theorem 1). Then we formulate the main result of the paper: equivalence of irreducibility of the Dirichlet form $\mathcal{E}_\mu$ associated with μ and extremality of μ under the conditions (U1)–(U3) (Theorem 2).

In Section 3 we prove the uniqueness of the stochastic dynamics associated with μ under the condition (U3), i.e. the essential self-adjointness of the operator H_μ on the set of cylinder (finitely based) smooth functions in $L_2(\mu)$ (Theorem 3). This proof extends the scheme of the work [6]. A probabilistic proof of this fact under stronger conditions on the potentials is given in [4].

In Section 4 we prove Theorem 2. First we consider the following general situation. Given a differentiable measure μ on $M^{\mathbf{Z}^d}$, we construct a bilinear form $\mathcal{E}_\mu^{\max}$ which is an extension of the Dirichlet form $\mathcal{E}_\mu$ and in general may be not a Dirichlet form itself. Following the scheme proposed in [10], we prove the equivalence of irreducibility of $\mathcal{E}_\mu^{\max}$ and extremality of μ in the convex set of all differentiable measures possessing the same logarithmic derivative (Theorem 4). In the case of Gibbs measures satisfying conditions (U1)–(U3), Theorem 3 implies that the forms $\mathcal{E}_\mu^{\max}$ and $\mathcal{E}_\mu$ coincide, from which the proof of Theorem 2 follows.

DEDICATION. It's a great pleasure for the authors to dedicate this work to Professor Takeyuki Hida on the occasion of his 70th birthday. He has always been a source of inspiration for us.

2. Setting and the Main Result

Let M be a smooth compact N-dimensional Riemannian manifold. We will denote by $(\cdot,\cdot)_x$ the corresponding scalar product in the tangent space of T_xM and omit the index x if possible. We will use the notation $d_X u$ for the derivative (resp. Levi-Civita covariant derivative) of the function (resp. of the vector field) u along the vector field X. The corresponding gradient associated with the Riemannian structure $(\cdot,\cdot)$ will be denoted by ∇u.

Let us consider the integer lattice $\mathbf{Z}^d$, $d \geqslant 1$, and define the space $\mathbf{M}$, which is an infinite product of manifolds $M_k = M$:

$$\mathbf{M} \equiv M^{\mathbf{Z}^d} = \times_{k\in\mathbf{Z}^d} M_k \ni x = (x_k)_{k\in\mathbf{Z}^d}. \tag{1}$$

$\mathbf{M}$ is endowed with the product topology. Given $\Lambda \subset \mathbf{Z}^d$

$$\mathbf{M} \ni x \mapsto x_\Lambda = (x_k)_{k\in\Lambda} \in M^\Lambda \tag{2}$$

denotes the natural projection of $\mathbf{M}$ onto M^Λ.

Let $\mathcal{A}$ be the family of all finite subsets of $\mathbf{Z}^d$. We will denote by $\mathcal{F}C^m(\mathbf{M})$ the space of m-times continuously differentiable real-valued cylinder functions on $\mathbf{M}$,

$$\mathcal{F}C^m(\mathbf{M}) = \bigcup_{\Lambda\in\mathcal{A}} C^m(M^\Lambda). \tag{3}$$

For such functions the symbols ∇_k resp. Δ_k will denote the gradient, resp. the Laplace–Beltrami operator, with respect to the variable x_k. We will use the notation $\overline{\nabla} u = (\nabla_k u)_{k \in \mathbf{Z}^d}$.

We also define the space $\mathcal{F}C^m(\mathbf{M} \to T\mathbf{M})$ of m-times differentiable cylinder vector fields on $\mathbf{M}$ with both domain and range consisting of cylinder elements:

$$\mathcal{F}C^m(\mathbf{M} \to T\mathbf{M}) \ni X = (X_k)_{k \in \Lambda}, \qquad \Lambda \in \mathcal{A}, X_k \in \mathcal{F}C^m(\mathbf{M} \to TM_k). \tag{4}$$

Let us remark that for $u \in \mathcal{F}C^m(\mathbf{M})$ we have $\overline{\nabla} u \in \mathcal{F}C^m(\mathbf{M} \to T\mathbf{M})$. We will use the notations

$$\langle X, Y \rangle = \sum_{k \in \mathbf{Z}^d} (X_k, Y_k), \qquad \mathbf{div} X = \sum_{k \in \mathbf{Z}^d} \mathrm{div}_k X_k, \tag{5}$$

div_k meaning the divergence with respect to x_k.

Let us consider a family of potentials $\mathcal{U} = (U_\Lambda)_{\Lambda \in \mathcal{A}}$, $U_\Lambda \in C(M^\Lambda)$. Let $\mathcal{A}(k)$ be the family of all sets $\Lambda \in \mathcal{A}$ containing the point $k \in \mathbf{Z}^d$. We will assume the following:

$$\text{(U1)} \qquad \sum_{\Lambda \in \mathcal{A}(k)} \sup_{x \in \mathbf{M}} |U_\Lambda(x)| < \infty \tag{6}$$

for any $k \in \mathbf{Z}^d$. Let us recall the definition of a Gibbs measure on the Borel σ-algebra $\mathcal{B}(\mathbf{M})$, associated with $\mathcal{U}$. For any $\Lambda \in \mathcal{A}$ we introduce the energy of the interaction in the volume Λ with fixed boundary condition $\xi \in \mathbf{M}$ as

$$V_\Lambda(x_\Lambda | \xi) = \sum_{\Lambda' \cap \Lambda \neq \emptyset} U_{\Lambda'}(y), \tag{7}$$

where $y = (x_\Lambda, \xi_{\Lambda^c}) \in \mathbf{M}$, $\Lambda^c = \mathbf{Z}^d \backslash \Lambda$. We define the corresponding Gibbs measure in the volume Λ with boundary condition ξ as the measure on $\mathcal{B}(M^\Lambda)$

$$\mathrm{d}\mu_\Lambda(x_\Lambda | \xi) = \frac{1}{Z_\Lambda(\xi)} \mathrm{e}^{-V_\Lambda(x_\Lambda | \xi)} \mathrm{d}x_\Lambda, \tag{8}$$

where $\mathrm{d}x = \bigotimes_{k \in \Lambda} \mathrm{d}x_k$ is the product of the Riemannian volume measures $\mathrm{d}x_k$ on M_k and

$$Z_\Lambda(\xi) = \int_{M^\Lambda} \mathrm{e}^{-V_\Lambda(x_\Lambda | \xi)} \mathrm{d}x_\Lambda. \tag{9}$$

These measures are well-defined for any finite volume Λ and all boundary conditions $\xi \in \mathbf{M}$.

For any $f \in \mathcal{F}C(\mathbf{M})$ we put

$$(\mathsf{E}_\Lambda f)(\xi) = \int f(x_\Lambda, \xi_{\Lambda^c}) \, \mathrm{d}\mu_\Lambda(x_\Lambda | \xi). \tag{10}$$

DEFINITION 1. A probability measure μ on $\mathcal{B}(\mathbf{M})$ is called a Gibbs measure (for given $\mathcal{U}$) if

$$\int \mathsf{E}_\Lambda f \, \mathrm{d}\mu = \int f \, \mathrm{d}\mu \tag{11}$$

for each $\Lambda \in \mathcal{A}$ and any $f \in \mathcal{F}C(\mathbf{M})$.

Remark 1. Condition (11) is equivalent to the assumption that $\mu_\Lambda(\cdot|\xi)$ is the conditional measure associated with μ under the condition ξ_{Λ^c}.

Remark 2. Heuristically μ can be given by the expression

$$\mathrm{d}\mu(x) = \frac{1}{Z} \mathrm{e}^{E(x)} \mathrm{d}x, \quad E(x) = \sum_{\Lambda \in \mathcal{A}} U_\Lambda(x), \tag{12}$$

where $\mathrm{d}x = \bigotimes_k \mathrm{d}x_k$ is the product of the volume measures on M_k.

Let $\mathcal{G}(\mathcal{U})$ be the family of all such Gibbs measures. The family $\mathcal{G}(\mathcal{U})$ is non-empty under the condition (U1), see, e.g., [13, 12].

DEFINITION 2. A measure ν on $\mathbf{M}$ is called differentiable if the following integration by parts formula holds true: for any $u \in \mathcal{F}C^1(\mathbf{M})$, and any vector field $X \in \mathcal{F}C^\infty(\mathbf{M} \to T\mathbf{M})$

$$\int \langle \overline{\nabla} u(x), X(x) \rangle \, \mathrm{d}\nu(x) = - \int \beta_X^\nu u(x) \, \mathrm{d}\nu(x), \tag{13}$$

with some $\beta_X^\nu \in L^2(\mathbf{M}, \nu)$. β_X^ν is called the logarithmic derivative of ν in the direction X.

Let us now assume that the family of potentials $\mathcal{U}$ satisfies (in addition to (U1)) the following condition:

$$\text{(U2)} \quad \text{each } U_\Lambda \in C^1(M^\Lambda) \text{ and } \sup_{k \in \mathbf{Z}^d} \sum_{\Lambda \in \mathcal{A}} |||\nabla_k U_\Lambda|||_{TM} < \infty, \tag{14}$$

where $|||\nabla_k U_\Lambda|||_{TM} := \sup_{x \in \mathbf{M}} \|\nabla_k U_\Lambda(x)\|_{T_{x_k} M}$.

The following statement shows that the Gibbs measure $\nu \in \mathcal{G}(\mathcal{U})$ can be completely characterized by its logarithmic derivative. It was proved first in the special situation of path measures in [17]. The case of compact spins with finite range of interactions was considered in [2], the case of linear spin spaces was treated in [8]. In order to make the paper more self-contained, we give here a sketch of the proof.

THEOREM 1. *The following conditions are equivalent:*

(i) *the measure ν belongs to the class $\mathcal{G}(\mathcal{U})$;*

(ii) *the measure* ν *is differentiable and its logarithmic derivative is given by*

$$\beta_X^\nu(x) = \sum_{k\in\mathbf{Z}^d} ((b_k(x), X_k(x)) + \operatorname{div} X_k(x)), \tag{15}$$

where $b_k(x) = -\nabla_k V_k(x)$, $V_k(x) = \sum_{\Lambda\in\mathcal{A}(k)} U_\Lambda(x)$.

Sketch of the proof. (i) $\Rightarrow$ (ii): this follows in the straighforward manner from the integration by parts formula for the measures μ_Λ and condition (11).

(ii) $\Rightarrow$ (i): Let ν_k^y, $k \in \mathbf{Z}^d$, $y \in \mathbf{M}^{\mathbf{Z}^d\setminus\{k\}}$, be the conditional measure of ν on M_k under the condition y. It follows easily from (13) and (15) that ν_k^y is differentiable and the following integration by parts formula holds for any smooth vector field Y on M and any $u \in C^1(M)$:

$$\int (\nabla u(x_k), Y(x_k))\,\mathrm{d}\nu_k^y(x_k)$$
$$= -\int ((\nabla V_k^y(x_k), Y(x_k))_k + \operatorname{div} Y(x_k))u(x_k)\,\mathrm{d}\nu_k^y(x_k),$$

where

$$V_k^y(x_k) := V_k((x_k, y)) \tag{16}$$

or

$$\int \operatorname{div} Y(x_k)u(x_k)\,\mathrm{d}\nu_k^y(x_k)$$
$$= -\int ((\nabla u(x_k), Y(x_k)) - (\nabla V_k^y(x_k), Y(x_k))u(x_k))\,\mathrm{d}\nu_k^y(x_k). \tag{17}$$

Let us remark that

$$(\nabla u(x_k), Y(x_k)) - (\nabla V_k^y(x_k), Y(x_k))u(x_k)$$
$$= \mathrm{e}^{V_k^y(x_k)}(\mathrm{d}_Y \mathrm{e}^{-V_k^y} u(x_k))(x_k) \tag{18}$$

and therefore

$$\int \operatorname{div} Y(x_k)u(x_k)\,\mathrm{d}\nu_k^y(x_k)$$
$$= -\int \mathrm{e}^{V_k^y(x_k)}(\mathrm{d}_Y \mathrm{e}^{-V_k^y} u(x_k))(x_k)\,\mathrm{d}\nu_k^y(x_k). \tag{19}$$

Setting $g = \mathrm{e}^{-V_k^y}u$, $\widetilde{\nu}_k = \mathrm{e}^{V_k^y}\cdot\nu_k^y$, we obtain the identity

$$\int \operatorname{div} Y(x_k)g(x_k)\,\mathrm{d}\widetilde{\nu}_k(x_k) = -\int (\mathrm{d}_Y g)(x_k)\,\mathrm{d}\widetilde{\nu}_k(x_k), \tag{20}$$

which completely characterizes the Riemannian volume on M_k [11]. Therefore

$$\mathrm{d}\widetilde{\nu}_k(x_k) = C(y)\,\mathrm{d}x_k \tag{21}$$

and

$$\mathrm{d}\nu_k^y(x_k) = \mathrm{e}^{-V_k^y(x_k)} C(y)\,\mathrm{d}x_k. \tag{22}$$

As ν_k^y is a probability measure, we have

$$C(y) = \left(\int \mathrm{e}^{-V_k^y(x_k)}\,\mathrm{d}x_k \right)^{-1} \tag{23}$$

and

$$\mathrm{d}\nu_k^y(x_k) = \mathrm{d}\mu_{\{k\}}(x_{\{k\}}|\xi), \quad \xi = (x_k, y). \tag{24}$$

To complete the proof, it is enough to remark that a Gibbs measure is completely characterized by its one-point conditional measures [13]. □

Let us recall that $\mathcal{G}(\mathcal{U})$ is a convex set [13]. Denote by $\mathcal{G}_{\mathrm{ext}}(\mathcal{U})$ the set of extreme elements of $\mathcal{G}(\mathcal{U})$. That is, $\mu \in \mathcal{G}_{\mathrm{ext}}(\mathcal{U})$ iff the equality $\mu = \tau\mu_1 + (1-\tau)\mu_2$, $\mu_1, \mu_2 \in \mathcal{G}(\mathcal{U})$ implies $\tau = 0$ or $\tau = 1$.

Let $\mu \in \mathcal{G}(\mathcal{U})$ be fixed. We introduce also the set $\mathcal{G}_\mu(\mathcal{U})$ of elements of $\mathcal{G}(\mathcal{U})$ which are absolutely continuous with respect to μ.

Let us consider the pre-Dirichlet form $\mathcal{E}_\mu$ on the space $\mathcal{F}C^2(\mathbf{M})$:

$$\mathcal{E}_\mu(u, v) = \frac{1}{2} \sum_k \int (\nabla_k u(x), \nabla_k v(x))\,\mathrm{d}\mu(x), \quad u, v \in \mathcal{F}C^2(\mathbf{M}). \tag{25}$$

Obviously $\mathcal{E}_\mu$ is associated with the operator H_μ defined on $\mathcal{F}C^2(\mathbf{M})$,

$$H_\mu u(x) = -\frac{1}{2} \sum_k \Delta_k u(x) - \frac{1}{2} \sum_k (b_k(x), \nabla_k u(x)), \tag{26}$$

in the sense that

$$\mathcal{E}_\mu(u, v) = \int H_\mu u(x) \cdot v(x)\,\mathrm{d}\mu(x), \quad u, v \in \mathcal{F}C^2(\mathbf{M}).$$

This implies that $\mathcal{E}_\mu$ is closable. We will denote its closure also by $\mathcal{E}_\mu$. We preserve the notation H_μ for the generator of the Dirichlet form $\mathcal{E}_\mu$ and denote by $D(H_\mu)$ its domain.

The operator H_μ generates the semigroup $T_t = \mathrm{e}^{-tH_\mu}$ in $L^2(\mathbf{M}, \mu)$, which defines the stochastic dynamics associated with μ.

Let us recall that a Dirichlet form $\mathcal{E}$ is called irreducible if $\mathcal{E}(u, u) = 0$ implies $u = 0$. We also recall the following known result characterizing the irreducibility of $\mathcal{E}_\mu$.

PROPOSITION 1. *The following assertions are equivalent:*

(i) $\mathcal{E}_\mu$ *is irreducible;*

(ii) *the semigroup* T_t *is irreducible, i.e. if* $G \in L^2(\mathbf{M},\mu)$ *such that*

$$T_t(GF) = GT_t(F) \tag{27}$$

for all bounded measurable F *and any* $t > 0$, *then* $G = \text{const}$;

(iii) *if* $G \in L^2(\mathbf{M},\mu)$ *such that* $T_t(G) = G$ *for any* $t > 0$, *then* $G = \text{const}$;

(iv) *the semigroup* T_t *is ergodic, i.e.*

$$\int \left(T_t F - \int F \, \mathrm{d}\mu \right)^2 \mathrm{d}\mu \to 0, \quad t \to \infty, \tag{28}$$

for all $F \in L^2(\mathbf{M}, \mu)$;

(v) *if* $F \in D(H_\mu)$ *and* $H_\mu F = 0$, *then* $F = \text{const}$ (*'uniqueness of ground state'*).

Proof. The proof is completely analogous to the one of [8]. □

Let us now suppose that the family of potentials $\mathcal{U}$ satisfies the following condition:

$$\text{(U3)} \quad \text{each } U_\Lambda \in C^2(\Lambda) \text{ and } \sup_{k \in \mathbf{Z}^d} \sum_{j \in \mathbf{Z}^d} \sum_{\Lambda \in \mathcal{A}} |||\nabla_j \nabla_k U_\Lambda|||_{TM \otimes TM} < \infty, \tag{29}$$

where

$$|||\nabla_j \nabla_k U_\Lambda|||_{TM \otimes TM} := \sup_{x \in \mathbf{M}} ||\nabla_j \nabla_k U_\Lambda(x)||_{T_{x_j} M \otimes T_{x_k} M}.$$

We can now formulate the main result of this paper.

THEOREM 2. *Let* $\mu \in \mathcal{G}(\mathcal{U})$, *where the family of potentials* $\mathcal{U}$ *satisfies the conditions* (U1), (U2), (U3). *Then the following assertions are equivalent:*

(i) $\mu \in \mathcal{G}_{\text{ext}}(\mathcal{U})$;

(ii) $\mathcal{G}_\mu(\mathcal{U}) = \{\mu\}$;

(iii) *the form* $\mathcal{E}_\mu$ *is irreducible.*

Proof. The proof of the theorem will be given in the Section 3 in a more general framework. □

3. Uniqueness of the Dynamics

The aim of this section is to prove the uniqueness of the stochastic dynamics associated with the fixed Gibbs measure $\mu \in \mathcal{G}(\mathcal{U})$.

THEOREM 3. *For any family* $\mathcal{U}$ *of potentials which satisfy the assumptions* (U1), (U2), (U3), *and any Gibbs measure* $\mu \in \mathcal{G}(\mathcal{U})$ *the pre-Dirichlet operator* H_μ *defined on* $\mathcal{F}C^2(\mathbf{M})$ *is an essentially self-adjoint operator in* $L^2(\mathbf{M}, \mu)$.

Remark 3. The essential self-adjointness of H_μ was proved in [6] in the case of interactions of a finite range and in [4] in the case interactions of infinite range (under stronger conditions on the potentials).

Proof. We will follow essentially the scheme of [5, 6]. Let us approximate the potentials U_Λ by smooth functions $U_\Lambda^n \in C^\infty(M^\Lambda)$, $n \in \mathbf{N}$, such that

$$\|U_\Lambda^n - U_\Lambda\|_{C^2(M^\Lambda)} \leqslant \alpha_\Lambda \mathrm{e}^{-n} \tag{30}$$

for some sequence $(\alpha_\Lambda)_{\Lambda\in\mathcal{A}}$ so that $\sum \alpha_\Lambda < \infty$. It is easy to see that the potentials U_Λ^n satisfy the conditions (U2) and (U3) uniformly in n,

$$\sup_{n\in\mathbf{N}} \sup_{k\in\mathbf{Z}^d} \sum_{\Lambda\in\mathcal{A}} |||\nabla_k U_\Lambda^n|||_{TM} < \infty, \tag{31}$$

$$\sup_{n\in\mathbf{N}} \sup_{k\in\mathbf{Z}^d} \sum_{j\in\mathbf{Z}^d} \sum_{\Lambda\in\mathcal{A}} |||\nabla_j \nabla_k U_\Lambda^n|||_{TM\otimes TM} < \infty. \tag{32}$$

Let us set

$$V_k^n(x) = \sum_{\Lambda\in\mathcal{A}(k), d(\Lambda)\leqslant n} U_\Lambda^n(x), \tag{33}$$

where $\mathrm{d}(\Lambda) = \max_{k\in\Lambda} |k|$, and

$$b_k^n(x) = \nabla_k V_k^n(x). \tag{34}$$

Let us remark that $V_k^n \in C^\infty(M^{\Lambda_n})$, where $\Lambda_n = \{k \in \mathbf{Z}^d : |k| \leqslant n\}$, and therefore $b_k^n = 0$ for $|k| > n$.

For any $n \in \mathbf{N}$ we define the differential operator H_n on the domain $\mathcal{F}C^2(\mathbf{M}) \subset L^2(\mathbf{M}, \mu)$ by the formula

$$H_n u(x) = -\frac{1}{2} \sum_{k\in\mathbf{Z}^d} \Delta_k u(x) - \frac{1}{2} \sum_{k\in\mathbf{Z}^d} (b_k^n(x), \nabla_k u(x)). \tag{35}$$

We will use the parabolic criterium of essential self-adjointness [20]. Let us consider the following Cauchy problems:

$$\frac{\mathrm{d}}{\mathrm{d}t} u_n(t,x) + H_n u_n(t,x) = 0, \qquad u_n(0,x) = f(x), \quad t \in [0,1], \tag{36}$$

for arbitrary $f \in \mathcal{F}C^2(\mathbf{M})$. If we can prove the existence of strong solutions

$$u_n \colon [0,1] \to L^2(\mathbf{M}, \mu) \tag{37}$$

to (36) such that

$$u_n(t) \in D(H_\mu) \tag{38}$$

for any $n \in \mathbf{N}$ and $t \in [0, 1]$, where $D(H_\mu)$ is the domain of H_μ, and

$$\int_0^1 \|(H_\mu - H_n)u_n(t)\|_{L^2(\mathbf{M},\mu)}\,\mathrm{d}t \to 0, \quad n \to \infty, \tag{39}$$

then the operator H_μ is essentially self-adjoint in $L^2(\mathbf{M}, \mu)$.

Let us remark first that the Cauchy problem (36) with fixed $f \in \mathcal{F}C^2(\mathbf{M})$ is finite-dimensional. Hence the classical solution u_n to (36) exists and, moreover, $u_n(t) \in \mathcal{F}C^2(\mathbf{M})$ and is a C^1-function in t. Therefore the conditions (37) and (38) are fulfilled.

In order to prove the condition (39), let us choose some weight sequence $p = (p_k)_{k\in\mathbf{Z}^d}$ of positive numbers, $p \in l_1$. We introduce the norm $\|X\|_p$ resp. $\|X\|_{p^{-1}}$ of the vector field $X = (X_k)_{k\in\mathbf{Z}^d}$ over $\mathbf{M}$, $X_k(x) \in T_{x_k}M$:

$$\|X\|_p^2 = \sum_{k\in\mathbf{Z}^d} p_k |||X_k|||_{TM}^2, \tag{40}$$

$$\|X\|_{p^{-1}}^2 = \sum_{k\in\mathbf{Z}^d} p_k^{-1} |||X_k|||_{TM}^2 \tag{41}$$

(which may be in general infinite). Let us remark that for any function $g \in \mathcal{F}C^1(\mathbf{M})$ and any fixed $x \in \mathbf{M}$ its gradient $\overline{\nabla} g(x) = (\nabla_k g(x))_{k\in\mathbf{Z}^d}$ has a finite number of nonzero components. Therefore for any p we have

$$\|\overline{\nabla} g\|_p < \infty, \qquad \|\overline{\nabla} g\|_{p^{-1}} < \infty. \tag{42}$$

LEMMA 1. *There exists a weight sequence $p \in l_1$ such that*

$$\|\overline{\nabla} u_n(t)\|_{p^{-1}} \leqslant C_1\,\mathrm{e}^{C_2 t/2} \|\overline{\nabla} f\|_{p^{-1}} \tag{43}$$

for some constants C_1, C_2 uniformly in n.

The proof of the lemma will be given at the end of this section.

Let us fix a weight sequence p from the lemma. Now we can check the condition (39). We have

$$(H_\mu - H_n)u_n(x) = \sum_k (b_k(x) - b_k^n(x), \nabla_k u_n(x)), \tag{44}$$

and, because of (43)

$$\sup_{x\in\mathbf{M}} |(H_\mu - H_n)u_n(x, t)| \leqslant C_1 \mathrm{e}^{C_2 t/2} \|b - b^n\|_p \|\overline{\nabla} f\|_{p^{-1}} \tag{45}$$

uniformly in n. Hence

$$\int_0^1 \|(H_\mu - H_n)u_n(t)\|_{L^2(\mathbf{M},\mu)}\,\mathrm{d}t \leqslant C_3 \|b - b^n\|_p. \tag{46}$$

By definition,

$$\begin{aligned}\|b_k(x) - b_k^n(x)\|_{T_{x_k}M} &\leqslant \Big\| \sum_{\Lambda\in\mathcal{A}} \nabla_k U_\Lambda(x) - \sum_{\substack{\Lambda\in\mathcal{A}\\ d(\Lambda)\leqslant n}} \nabla_k U_\Lambda^n(x) \Big\|_{T_{x_k}M} \\ &\leqslant \sum_{\Lambda\in\mathcal{A}} \|\nabla_k U_\Lambda(x) - \nabla_k U_\Lambda^n(x)\|_{T_{x_k}M} + \sum_{\substack{\Lambda\in\mathcal{A}\\ d(\Lambda)>n}} \|\nabla_k U_\Lambda(x)\|_{T_{x_k}M}. \end{aligned} \tag{47}$$

Then, because of (30),

$$|||b_k - b_k^n|||_{TM} \leqslant \mathrm{e}^{-n} \sum_{\substack{\Lambda\in\mathcal{A}\\ d(\Lambda)>n}} \alpha_\Lambda |||\nabla_k U_\Lambda|||_{TM}, \tag{48}$$

and

$$\|b - b^n\|_p^2 \leqslant C_4 \mathrm{e}^{-2n} \sum_{k\in\mathbf{Z}^d} p_k + \sum_{\substack{\Lambda\in\mathcal{A}\\ d(\Lambda)>n}} \sum_{k\in\mathbf{Z}^d} p_k |||\nabla_k U_\Lambda|||_{TM}^2 \to 0, \tag{49}$$

$n \to \infty$. □

Proof of Lemma 1. We introduce $\mathrm{Hes}_{jk}(g)$ for $g \in \mathcal{F}C^2(\mathbf{M})$ and $j, k \in \mathbf{Z}^d$ to be the (j, k) block of the Hessian tensor of g and denote by Ric the Ricci curvature tensor for M. It is known [18] that the following estimation holds:

$$\begin{aligned}\frac{\partial}{\partial t}|\nabla_k u_n(t,x)|^2 &\leqslant -H_n |\nabla_k u_n(t,x)|^2 - \mathrm{Ric}(\nabla_k u_n(t,x), \nabla_k u_n(t,x)) - \\ &\quad - \sum_j \mathrm{Hes}_{jk}(V_k^n)(\nabla_j u_n(t,x), \nabla_k u_n(t,x)). \end{aligned} \tag{50}$$

Obviously for any vector fields X, Y on M $\mathrm{Hes}_{jk}(V_k^n)(X(x), Y(x)) = (R_{jk}(x)X(x), Y(x))$, where $R_{jk}(x) = \nabla_k \nabla_j \sum_{\Lambda\in\mathcal{A}} U_\Lambda(x)$. Let us consider the matrix r with elements $r_{jk} = \sup_{y\in\mathbf{M}} \|R_{jk}(y)\|$. Because of conditions (31) and (32) the sum $\sum_k r_{jk}$ is bounded uniformly in j. This is enough for the existence of a positive sequence $p = (p_k) \in l_1$ such that $\sum_j r_{jk} p_j < C p_k$ for some constant C [16]. Let us introduce the Hilbert space $l_{2,p}(\mathbf{Z}^d \to \mathbf{R}^1) \subset \left(\mathbf{R}^1\right)^{\mathbf{Z}^d}$ with the norm $\|\cdot\|$ given by the expression $\|x\|^2 = \sum_{k\in\mathbf{Z}^d} p_k |x_k|^2$. By Schur's test the matrix r generates a bounded operator in $l_{2,p}(\mathbf{Z}^d \to \mathbf{R}^1)$ with the norm less than C (see, e.g., [14]). Moreover, the matrix r is symmetric and therefore also generates a bounded operator with norm less than C in the dual space $l_{2,p}(\mathbf{Z}^d \to \mathbf{R}^1)'$ which can obviously be identified with the space $l_{2,p^{-1}}(\mathbf{Z}^d \to \mathbf{R}^1)$, where $p^{-1} = (p_k^{-1})$.

Let us remark that the operator H_n generates a positivity preserving contractive semigroup in the space $C(M^{\Lambda_n})$ for some $\Lambda_n \in \mathcal{A}$. Then:

$$\|\overline{\nabla} u_n(t)\|_{p^{-1}}^2$$

$$= \sum_k p_k^{-1} |||\nabla_k u_n(t)|||_{TM}^2 \leqslant \sum_k p_k^{-1} |||\nabla_k f|||_{TM}^2 +$$

$$+ a_1 \int_0^t \Big(\sum_{j,k} r_{jk} p_k^{-1} |||\nabla_k u_n(s)|||_{TM} |||\nabla_j u_n(s)|||_{TM} + a_2 \|\overline{\nabla} u_n(s)\|_{p^{-1}}^2 \Big) \mathrm{d}s$$

$$\leqslant \|\overline{\nabla} f\|_{p^{-1}}^2 + a_3 \int_0^t \|\overline{\nabla} u_n(s)\|_{p^{-1}}^2 \mathrm{d}s. \tag{51}$$

An application of Gronwall's inequality completes the proof. □

4. Extremality of Gibbs Measures and Irreducibility of Associated Dirichlet Forms

Let μ be a differentiable measure on $\mathbf{M}$. For any $X \in \mathcal{F}C^\infty(\mathbf{M} \to T\mathbf{M})$ we fix a μ-version β_X^μ of its logarithmic derivative. The set of all differentiable measures ν on $\mathbf{M}$ such that for any $X \in \mathcal{F}C^\infty(\mathbf{M} \to T\mathbf{M})$ we have $\beta_X^\nu = \beta_X^\mu$ μ-a.e. will be denoted by $\mathcal{G}^{\beta^\mu}$. We introduce also the set $\mathcal{G}_{ac}^{\beta^\mu} \subset \mathcal{G}^{\beta^\mu}$ of elements of $\mathcal{G}^{\beta^\mu}$ which are absolutely continuous with respect to μ with bounded densities and the set $\mathcal{G}_{\mathrm{ext}}^{\beta^\mu} \subset \mathcal{G}^{\beta^\mu}$ of extreme elements of $\mathcal{G}^{\beta^\mu}$.

We define the divergence $\mathbf{div}_\mu X \in L^2(\mathbf{M}, \mu)$ of the vector field $X \in \mathcal{F}C^\infty (\mathbf{M} \to T\mathbf{M})$ by $\mathbf{div}_\mu X := \beta_X^\mu + \mathbf{div}\, X$. Then we have the integration by parts formula:

$$\int \langle \overline{\nabla} u(x), X(x) \rangle \mathrm{d}\mu(x)$$

$$= - \int u(x) \mathbf{div}_\mu X(x)\, \mathrm{d}\mu(x), \quad u \in \mathcal{F}C^1(\mathbf{M}). \tag{52}$$

Let us introduce the operator

$$(d^\mu, D(d^\mu))\colon L^2(\mathbf{M}, \mu) \to L^2(\mathbf{M} \to T\mathbf{M}, \mu), \tag{53}$$

where $L^2(\mathbf{M} \to T\mathbf{M}, \mu)$ is the completion of $\mathcal{F}C^\infty(\mathbf{M} \to T\mathbf{M})$ in the norm $\int \langle X, X \rangle\, \mathrm{d}\mu$, as the adjoint of $(-\mathbf{div}_\mu, \mathcal{F}C^\infty(\mathbf{M} \to T\mathbf{M}))$. By definition, $u \in L^2(\mathbf{M}, \mu)$ belongs to $D(d^\mu)$ iff there exists $V_u \in L^2(\mathbf{M} \to T\mathbf{M}, \mu)$ such that

$$\int \langle V_u(x), X(x) \rangle\, \mathrm{d}\mu(x) = - \int u(x) \mathbf{div}_\mu X(x)\, \mathrm{d}\mu(x) \tag{54}$$

for all $X \in \mathcal{F}C^\infty(\mathbf{M} \to T\mathbf{M})$. Then $d^\mu u = V_u$.

Let us also define the positive symmetric bilinear form

$$\mathcal{E}_\mu^{\max}(u, v) = \frac{1}{2} \int \langle d^\mu u(x), d^\mu v(x) \rangle\, \mathrm{d}\mu(x) \tag{55}$$

with domain of definition $D(d^\mu)$.

Remark 4. The form $\mathcal{E}_\mu^{\max}$ is an extension of the form $\mathcal{E}_\mu$ and may be in general not a Dirichlet form.

The following theorem reflects a quite general fact. For example, an analogous result has been proved for continuous systems in [9]. The proof given below is an adaptation of the latter.

THEOREM 4. *The following assertions are equivalent:*

(i) $\mu \in \mathcal{G}_{\mathrm{ext}}^{\beta_\mu}$;
(ii) $\mathcal{G}_{ac}^{\beta_\mu} = \{\mu\}$;
(iii) $\mathcal{E}_\mu^{\max}(u,u) = 0$ *implies that* $u = 0$ *for any bounded* $u \in D(d^\mu)$.

Proof. (i) $\Rightarrow$ (ii): Assume that (i) holds. Let $\nu \in \mathcal{G}_{ac}^{\beta_\mu}$, $\nu = \rho \cdot \mu$ for a bounded function $\rho \in L^1(\mathbf{M}, \mu)$. Let $C = \sup_{x \in \mathbf{M}} \rho(x)$. We set

$$\mu_1 = \frac{C - \rho}{C - 1}\mu. \tag{56}$$

Then obviously $\beta^{\mu_1} = \beta^\nu$, $\mu_1 \in \mathcal{G}_{ac}^{\beta_\mu}$ and we have

$$\mu = \frac{C-1}{C}\mu_1 + \frac{1}{C}\nu. \tag{57}$$

By assumption (i) it follows that $\mu_1 = \nu$ which implies $\rho = 1$.

(ii) $\Rightarrow$ (i): Assume that (ii) holds. Let $\mu = t\mu_1 + (1-t)\mu_2$, where $\mu_1, \mu_2 \in \mathcal{G}^{\beta_\mu}$ and $t \in (0,1)$. Then both μ_1, μ_2 are absolutely continuous with respect to μ with bounded densities, and $\mu_1 = \mu_2 = \mu$ by assumption (ii). Consequently, $\mu \in \mathcal{G}_{\mathrm{ext}}^{\beta_\mu}$.

(ii) $\Rightarrow$ (iii): Assume that (ii) holds. Let $G \in D(d^\mu)$ be bounded and $\mathcal{E}_\mu^{\max}(G,G) = 0$. Since $d^\mu 1 = 0$, replacing G by G-essinf G, we may assume that $G \geqslant 0$ and $\int G\,\mathrm{d}\mu = 1$. Then $d^\mu G = 0$ and $d^\mu(FG)(x) = d^\mu F(x)G(x) = \overline{\nabla}F(x)G(x)$ for any $F \in \mathcal{F}C^\infty(\mathbf{M})$. Let us introduce the measure $\nu = G \cdot \mu$. Then

$$\begin{aligned}
&\int F(x)\beta_X^\mu(x)\,\mathrm{d}\nu(x) \\
&\quad = -\int \langle d^\mu(FG)(x), X(x)\rangle\,\mathrm{d}\mu(x) - \\
&\qquad - \int \langle d^\mu F(x), X(x)\rangle G(x)\,\mathrm{d}\mu(x) \\
&\quad = -\int \langle \overline{\nabla}F(x), X(x)\rangle\,\mathrm{d}\nu(x)
\end{aligned} \tag{58}$$

for any $X \in \mathcal{F}C^\infty(\mathbf{M} \to T\mathbf{M})$ and $F \in \mathcal{F}C^\infty(\mathbf{M})$. This implies that $\nu \in \mathcal{G}_{ac}^{\beta_\mu}$ and by assumption (ii) $G \equiv 1$.

(iii) $\Rightarrow$ (ii): Assume that (iii) holds. Let $\nu \in \mathcal{G}_{ac}^{\beta_\mu}$, $\nu = \rho \cdot \mu$ for a bounded function $\rho \in L^1(\mathbf{M}, \mu)$. Then

$$0 = \int \beta_X^\mu(x)\,\mathrm{d}\nu(x) = \int \beta_X^\mu(x)\rho(x)\,\mathrm{d}\mu(x)$$

$$= -\int \langle d^{\mu}\rho(x), X(x)\rangle \, \mathrm{d}\mu(x) \tag{59}$$

for all $X \in \mathcal{F}C^{\infty}(\mathbf{M} \to T\mathbf{M})$. It follows then that $d^{\mu}\rho = 0$, and (iii) implies that $\rho = \text{const}$. □

Let us now consider the case of $\mu \in \mathcal{G}(\mathcal{U})$. We assume that the family of potentials $\mathcal{U}$ satisfy conditions (U1), (U2), (U3). Because of the essential self-adjointness of the generator H_{μ}, we have in this case

$$\mathcal{E}_{\mu} = \mathcal{E}_{\mu}^{\max}. \tag{60}$$

Remark 5. Since $\mathcal{E}_{\mu}^{\max}$ is now a Dirichlet form, condition (iii) implies its irreducibility (see [9], Lemma 6.1).

Remark 6. It is easy to see that the irreducibility of the form $\mathcal{E}_{\mu}^{\max}$ implies that the density ρ in the proof of (iii) ⇒ (ii) can be taken to be unbounded.

Proof of Theorem 2. Follows from Theorem 4 and Remark 4. □

Acknowledgements

We are very grateful to K. D. Elworthy for fruitful and stimulating discussions. Financial support of DFG Research Project AL 214/9-2 and DFG Schwerpunkt 'Interacting particle systems' is gratefully acknowledged.

References

1. Antonjuk, A. Val. and Antonjuk, A. Vic.: Smoothness properties of semigroups for Dirichlet operators of Gibbs measures, *J. Funct. Anal.* **127**(2) (1995), 390–430.
2. Albeverio, S., Antonjuk, A. Val., Antonjuk, A. Vic. and Kondratiev, Yu.: Stochastic dynamics in some lattice spin systems, *Methods Funct. Anal. Topology* **1**(1) (1995), 3–28.
3. Albeverio, S., Daletskii, A. and Kondratiev, Yu.: A stochastic differential equation approach to some lattice spin models on compact Lie groups, *Random Oper. Stochastic Equations* **4**(3) (1996), 227–236.
4. Albeverio, S., Daletskii, A. and Kondratiev, Yu.: Infinite systems of stochastic differential equations and some lattice models on compact Riemannian manifolds, *Ukr. Math. J.* **49**(3) (1997), 326–337.
5. Albeverio, S., Kondratiev, Yu. and Röckner, M.: Dirichlet operators via stochastic analysis, *J. Funct. Anal.* **128**(1) (1995), 102–138.
6. Albeverio, S., Kondratiev, Yu. and Röckner, M.: Uniqueness of the stochastic dynamics for continuous spin systems on a lattice, *J. Funct. Anal.* **133**(1) (1995), 10–20.
7. Albeverio, S., Kondratiev, Yu. and Röckner, M.: Quantum fields, Markov fields and stochastic quantization, In: *Stochastic Analysis*, Mathematics and Physics, Nato ASI, Academic Press, New York, 1995.

8. Albeverio, S., Kondratiev, Yu. and Röckner, M.: Ergodicity of L^2-semigroups and extremality of Gibbs states, *J. Funct. Anal.* **144** (1997), 394–423.
9. Albeverio, S., Kondratiev, Yu. and Röckner, M.: Ergodicity for the stochastic dynamics of quasi-invariant measures with applications to Gibbs states, *J. Funct. Anal.* **149** (1997), 415–469.
10. Albeverio, S., Kondratiev, Yu. and Röckner, M.: Analysis and geometry on conguration spaces: The Gibbsian case, *J. Funct. Anal.* **157** (1998), 242–291.
11. Chavel, I.: *Eigenvalues in Riemannian Geometry*, Academic Press, New York, 1984.
12. Enter, V., Fernandez, R. and Sokal, D.: Regularity properties and Patologies of Position-Space renormalization-group transformations, *J. Statist. Phys.* **2**(5/6) (1993), 879–1168.
13. Georgii, H. O.: *Gibbs Measures and Phase Transitions*, Studies in Math. 9, de Gruyter, Berlin, 1988.
14. Halmos, P. R.: *A Hilbert Space Problem Book*, Springer, New York, 1982.
15. Holley, R. and Stroock, D.: Diffusions on the infinite dimensional torus, *J. Funct. Anal.* **42** (1981), 29–63.
16. Leha, G. and Ritter, G.: On solutions of stochastic differential equations with discontinuous drift in Hilbert space, *Math. Ann.* **270**, 109–123.
17. Roelly, S. and Zessin, H.: A characterization of Gibbs measures on $C[0,1]^{\mathbf{Z}^d}$ by the stochastic calculus of variations, *Ann. Inst. H. Poincaré* **29** (1993), 327–338.
18. Stroock, D. and Zegarlinski, B.: The equivalence of the logarithmic Sobolev inequality and Dobrushin–Shlosman mixing condition, *Comm. Math. Phys.* **144** (1992), 303–323.
19. Stroock, D. and Zegarlinski, B.: The logarithmic Sobolev inequality for continuous spin systems on a lattice, *J. Funct. Anal.* **104** (1992), 299–326.
20. Beresansky, Yu. M. and Kondratiev, Yu. G.: *Spectral Methods in Infinite Dimensional Analysis*, Kluwer Acad. Publ., Dordrecht, 1995.

Acta Applicandae Mathematicae **63:** 41–78, 2000.

Infinite-Dimensional Analysis and Analytic Number Theory

Dedicated to Professor Takeyuki Hida on the occasion of his 70th birthday

ASAO ARAI
Department of Mathematics, Hokkaido University, Sapporo 060-0810, Japan.
e-mail: arai@math.sci.hokudai.ac.jp

(Received: 19 January 1999)

Abstract. We consider arithmetical aspects of analysis on Fock spaces (Boson Fock space, Fermion Fock space, and Boson–Fermion Fock space) with applications to analytic number theory.

Mathematics Subject Classifications (2000): 81T99, 11A67.

Key words: Fock space, infinite dimensional analysis, analytic number theory, statistical mechanics, quantum field, supersymmetry.

1. Introduction

In recent years, connections between number theory and physics have been noted and discussed (e.g., [25, 33]). From an arithmetic point of view, 'statistical mechanics' of numbers may be particularly interesting, because it is related in a direct way to the Riemann zeta function and may give a key to solve the Riemann hypothesis ([18, 19, 21, 23, 24, 29–31] and references therein). Spector [30] pointed out intriguing relationships between analytic number theory and a free supersymmetric quantum field theory, and further discussed these aspects with notions of partial supersymmetry and 'duality' [31].

On the other hand, general mathematical structures and aspects of some models in supersymmetric quantum field theory have been studied as a subject of infinite-dimensional analysis ([3–12, 15–17]) (cf. also [22]). Motivated by the work of Spector mentioned above, we are interested in developing analytic number theory *as a field of infinite-dimensional analysis*. In this paper we start this program with reviewing fundamental aspects of relationships between analytic number theory and analysis on Fock spaces (Boson Fock space, Fermion Fock space and Boson–Fermion Fock space).

The present paper, which is intended to be of review nature, is organized as follows. In Section 2, we discuss relationships between analysis on Boson Fock space and analytic number theory. Statistical mechanical partition functions are defined in an abstract level and their arithmetical structures are analyzed. One of new aspects here is an introduction of a graded partition function which is associated with a graded structure of the abstract Boson Fock space. We apply the abstract results to a concrete case where arithmetical functions (the Riemann zeta function, the Liouville function, Dirichlet series, etc.) are described in terms of Fock space langauge. In Section 3 analysis similar to that of Section 2 is made on Fermion Fock spaces. We prove duality relations between bosonic and fermionic partition functions. One of the relevant arithmetical functions on a Fermion Fock space is the Möbius function. As applications of the abstract results, we rederive some known formulas on the Möbius function and completely multiplicative functions. We also derive a Fock space expression of Jordan's totient function. Section 4 is devoted to a brief review of fundamental aspects of analysis on the abstract Boson–Fermion Fock space developed by the present author [7]. The main objects in this analysis are infinite-dimensional Dirac and Laplace–Beltrami operators. In the last section, we discuss arithmetical aspects of the analysis of Boson–Fermion Fock spaces, generalizing the ideas in [30, 31].

2. Boson Fock Spaces and Arithmetical Functions

2.1. PARTITION FUNCTIONS AND CORRELATION FUNCTIONS

Let $\mathcal{H}$ be a separable infinite-dimensional Hilbert space with inner product $(\cdot, \cdot)_{\mathcal{H}}$ (complex linear in the second variable) and $\bigotimes_{\mathrm{s}}^n \mathcal{H}$ be the n-fold symmetric tensor product Hilbert space of $\mathcal{H}$ ($n = 0, 1, 2, \ldots$; $\bigotimes_{\mathrm{s}}^0 \mathcal{H} := \mathbf{C}$). Then the Boson Fock space over $\mathcal{H}$ is defined by

$$\mathcal{F}_{\mathrm{B}}(\mathcal{H}) := \bigoplus_{n=0}^{\infty} \bigotimes_{\mathrm{s}}^{n} \mathcal{H} \tag{2.1}$$

(e.g., [20, §5.2], [26, §II.4]).

We denote by $a_{\mathcal{H}}(f)$ ($f \in \mathcal{H}$) the annihilation operator on $\mathcal{F}_{\mathrm{B}}(\mathcal{H})$ (e.g., [20, §5.2], [27, §X.7]) ($a_{\mathcal{H}}(f)$ is antilinear in f). The set $\{a_{\mathcal{H}}(f), a_{\mathcal{H}}(f)^* \mid f \in \mathcal{H}\}$ satisfies the canonical commutation relations

$$\begin{aligned} &[a_{\mathcal{H}}(f), a_{\mathcal{H}}(g)^*] = (f, g)_{\mathcal{H}}, \qquad [a_{\mathcal{H}}(f), a_{\mathcal{H}}(g)] = 0, \\ &[a_{\mathcal{H}}(f)^*, a_{\mathcal{H}}(g)^*] = 0, \quad f, g \in \mathcal{H}, \end{aligned} \tag{2.2}$$

on

$$\mathcal{F}_{\mathrm{B},0}(\mathcal{H}) := \{\Psi = \{\Psi_n\}_{n=0}^{\infty} \in \mathcal{F}_{\mathrm{B}}(\mathcal{H}) \mid \Psi_n = 0 \text{ for all but finitely many } n\text{'s}\}, \tag{2.3}$$

the space of finite particle vectors in $\mathcal{F}_{\mathrm{B}}(\mathcal{H})$. We denote by $\Omega_{\mathcal{H}} := \{1, 0, 0, \ldots\}$ the Fock vacuum in $\mathcal{F}_{\mathrm{B}}(\mathcal{H})$. We have

$$a_{\mathcal{H}}(f)\Omega_{\mathcal{H}} = 0, \quad f \in \mathcal{H}. \tag{2.4}$$

Let A be a nonnegative self-adjoint operator on $\mathcal{H}$ and $d\Gamma_B(A)$ the second quantization of A on $\mathcal{F}_B(\mathcal{H})$:

$$d\Gamma_B(A) := \bigoplus_{n=0}^{\infty} d\Gamma_B^{(n)}(A) \tag{2.5}$$

with $d\Gamma_B^{(0)}(A) := 0$ and

$$d\Gamma_B^{(n)}(A) := \sum_{j=1}^{n} I \otimes \cdots \otimes I \otimes \overset{j\text{th}}{\breve{A}} \otimes I \otimes \cdots \otimes I,$$

where I denotes identity operator (e.g., [20, §5.2], [27, p. 302, Example 2]). Then $d\Gamma_B(A)$ is nonegative and self-adjoint. We set

$$H_B(A) := d\Gamma_B(A). \tag{2.6}$$

Remark 2.1. In quantum field theory, $H_B(A)$ describes a *free Hamiltonian* of a quantized Bose field with A being its one particle Hamiltonian.

Let C be a densely defined closed linear operator on $\mathcal{H}$ and $\bigotimes^n C$ the n-fold tensor product of C on the n-fold tensor product Hilbert space $\bigotimes^n \mathcal{H}$ of $\mathcal{H}$ ($\bigotimes^0 C := 1$). Then

$$\Gamma(C) := \bigoplus_{n=0}^{\infty} \bigotimes^n C \tag{2.7}$$

is a densely defined closed linear operator on the full Fock space

$$\mathcal{F}_{\mathrm{full}}(\mathcal{H}) := \bigoplus_{n=0}^{\infty} \bigotimes^n \mathcal{H} \tag{2.8}$$

over $\mathcal{H}$. It is well known (or easy to see) that, if C is a contraction operator, then so is $\Gamma(C)$. In particular, if C is unitary, then so is $\Gamma(C)$ (cf. [27, §X.7]).

The operator $\Gamma(C)$ is reduced by $\mathcal{F}_B(\mathcal{H})$. We denote its reduced part by $\Gamma_B(C)$. Then we have

$$\Gamma_B(e^{itA}) = e^{itH_B(A)}, \quad t \in \mathbf{R}. \tag{2.9}$$

For a self-adjoint operator T, we denote by $\sigma(T)$ (resp. $\sigma_d(T)$) the spectrum (resp. the discrete spectrum) of T.

We denote by $\mathbf{N} := \{1, 2, \ldots, \}$ the set of natural numbers.

The following lemma is easily proven (e.g., [14, Lemma 3.25]).

LEMMA 2.1. *Suppose that A is strictly positive and the spectrum of A is purely discrete with*

$$\sigma(A) = \sigma_{\mathrm{d}}(A) = \{E_n(A)\}_{n=1}^{\infty}, \tag{2.10}$$

$0 < E_1(A) \leqslant E_2(A) \leqslant \dots,\ E_n(A) \to \infty\ (n \to \infty)$, *counted with algebraic multiplicity. Then the spectrum of $H_{\mathrm{B}}(A)$ is purely discrete with*

$$\sigma(H_{\mathrm{B}}(A)) = \sigma_{\mathrm{d}}(H_{\mathrm{B}}(A)) = \bigcup_{n=1}^{\infty}\left\{\sum_{j=1}^{n} k_j E_j(A) \mid k_j \in \{0\} \cup \mathbf{N}\right\}. \tag{2.11}$$

We denote by N_{B} the number operator on $\mathcal{F}_{\mathrm{B}}(\mathcal{H})$:

$$N_{\mathrm{B}} := \mathrm{d}\Gamma_{\mathrm{B}}(I). \tag{2.12}$$

The Boson Fock space $\mathcal{F}_{\mathrm{B}}(\mathcal{H})$ is $\mathbf{Z}^2$-graded with

$$\mathcal{F}_{\mathrm{B}}(\mathcal{H}) = \mathcal{F}_{\mathrm{B},+}(\mathcal{H}) \oplus \mathcal{F}_{\mathrm{B},-}(\mathcal{H}), \tag{2.13}$$

where

$$\mathcal{F}_{\mathrm{B},+}(\mathcal{H}) := \bigoplus_{n=0}^{\infty} \bigotimes_{\mathrm{s}}^{2n} \mathcal{H}, \qquad \mathcal{F}_{\mathrm{B},-}(\mathcal{H}) := \bigoplus_{n=0}^{\infty} \bigotimes_{\mathrm{s}}^{2n+1} \mathcal{H}. \tag{2.14}$$

The self-adjoint operator $(-1)^{N_{\mathrm{B}}}$ is the grading operator of this gradation.

For $s > 0$, we define

$$Z_{\mathrm{B}}(s; A) := \mathrm{Tr}\, \mathrm{e}^{-sH_{\mathrm{B}}(A)}, \tag{2.15}$$

$$\widetilde{Z}_{\mathrm{B}}(s; A) := \mathrm{Tr}\big\{(-1)^{N_{\mathrm{B}}}\, \mathrm{e}^{-sH_{\mathrm{B}}(A)}\big\}, \tag{2.16}$$

provided that $\mathrm{e}^{-sH_{\mathrm{B}}(A)}$ is trace class on $\mathcal{F}_{\mathrm{B}}(\mathcal{H})$, where Tr denotes trace.

Remark 2.2. In statistical mechanics of quantum fields, $Z_{\mathrm{B}}(s; A)$ is called the *partition function* of the Hamiltonian $H_{\mathrm{B}}(A)$ at temperature $1/s$ (physically s denotes an *inverse temperature*). The function $\widetilde{Z}_{\mathrm{B}}(s; A)$ is not so standard. We call it the *graded partition function* of the Hamiltonian $H_{\mathrm{B}}(A)$ at temperature $1/s$. This type of partition function was considered in a concrete case by Spector [31].

To treat the partition functions in a unified way, we introduce a more general partition function

$$Z_{\mathrm{B}}(s, z; A) := \mathrm{Tr}\big(\Gamma_{\mathrm{B}}(z)\, \mathrm{e}^{-sH_{\mathrm{B}}(A)}\big) \tag{2.17}$$

with

$$z \in D := \{w \in \mathbf{C} \mid |w| \leqslant 1\}, \tag{2.18}$$

provided that $e^{-sH_B(A)}$ is trace class on $\mathcal{F}_B(\mathcal{H})$. Since $\Gamma_B(1) = I$ and $\Gamma_B(-1) = (-1)^{N_B}$, we have

$$Z_B(s, 1; A) = Z_B(s; A), \qquad Z_B(s, -1; A) = \widetilde{Z}_B(s; A). \tag{2.19}$$

For a linear operator T on a Hilbert space, we denote its domain by $D(T)$. For each $z \in \mathbf{C}$, we can define an operator z^{N_B} on $\mathcal{F}_B(\mathcal{H})$ by

$$D(z^{N_B}) := \left\{\Psi = \{\Psi^{(n)}\}_{n=0}^{\infty} \in \mathcal{F}_B(\mathcal{H}) \mid \sum_{n=0}^{\infty} |z|^{2n} \|\Psi^{(n)}\|^2 < \infty\right\}, \tag{2.20}$$

$$(z^{N_B}\Psi)^{(n)} := z^n \Psi^{(n)}, \quad \Psi \in D(z^{N_B}),\ n \geqslant 0. \tag{2.21}$$

It is easy to see that

$$\Gamma_B(z) = z^{N_B}. \tag{2.22}$$

If T is a trace class operator on a Hilbert space, then one can define $\det(I + T)$, the determinant for $I + T$, in an intrinsic way [28, §XIII.17]. It follows that

$$\det(I + T) = \prod_{n=1}^{N(T)} (1 + E_n(T)), \tag{2.23}$$

where $\{E_n(T)\}_{n=1}^{N(T)}$ are the eigenvalues of T counted with algebraic multiplicity [28, Theorem XIII.106]. We set $\det I := 1$.

In what follows, we assume the following.

(A) *The operator A is strictly positive, self-adjoint and, for some $s > 0$, e^{-sA} is trace class on $\mathcal{H}$.*

Remark 2.3. Under Assumption (A), e^{-tA} is trace class on $\mathcal{H}$ for all $t > s$.

THEOREM 2.2. *Let $z \in D$. Then the operator $\Gamma_B(z)\, e^{-sH_B(A)}$ is trace class on $\mathcal{F}_B(\mathcal{H})$ and*

$$Z_B(s, z; A) = \frac{1}{\det(1 - z\, e^{-sA})}. \tag{2.24}$$

In particular,

$$Z_B(s; A) = \frac{1}{\det(I - e^{-sA})}, \tag{2.25}$$

$$\widetilde{Z}_B(s; A) = \frac{1}{\det(I + e^{-sA})}. \tag{2.26}$$

Proof. By the assumption, e^{-sA} is compact. Hence, by the Hilbert–Schmidt theorem, the spectrum of A is purely discrete, satisfying the assumption of Lemma 2.1. Therefore we have

$$\sigma_{\mathrm{d}}\big(\Gamma_{\mathrm{B}}(z)\,\mathrm{e}^{-sH_{\mathrm{B}}(A)}\big) = \bigcup_{n=1}^{\infty}\left\{\prod_{j=1}^{n} z^{k_j}\,\mathrm{e}^{-sk_jE_j(A)} \mid k_j \in \{0\}\cup\mathbf{N}\right\},$$

from which the desired assertion follows. See [14, Theorem 3.26] (cf. also [20, Proposition 5.2.27]). □

Using Theorem 2.2 and the product law of the determinant $\det(\cdot)$, we can derive relations of partition functions at different temperatures:

THEOREM 2.3. *For all $n \in \mathbf{N}$ and $z \in D$,*

$$Z_{\mathrm{B}}(s, z; A) = \det\left(\sum_{k=0}^{n-1} z^k\,\mathrm{e}^{-ksA}\right) Z_{\mathrm{B}}(ns, z^n; A) \tag{2.27}$$

and

$$Z_{\mathrm{B}}(s, z; A) Z_{\mathrm{B}}(s, -z; A) = Z_{\mathrm{B}}(2s, z^2; A). \tag{2.28}$$

In particular,

$$Z_{\mathrm{B}}(s; A) = \det\left(\sum_{k=0}^{n-1} \mathrm{e}^{-ksA}\right) Z_{\mathrm{B}}(ns; A), \tag{2.29}$$

$$Z_{\mathrm{B}}(ns; A)\widetilde{Z}_{\mathrm{B}}(ns; A) = Z_{\mathrm{B}}(2ns; A). \tag{2.30}$$

Proof. By (2.24),

$$\frac{Z_{\mathrm{B}}(s, z; A)}{Z_{\mathrm{B}}(ns, z^n; A)} = \frac{\det(1 - z^n\,\mathrm{e}^{-nsA})}{\det(1 - z\,\mathrm{e}^{-sA})}.$$

Note that

$$1 - z^n\,\mathrm{e}^{-nsA} = (1 - z\,\mathrm{e}^{-sA})\left(\sum_{k=0}^{n-1} z^k\,\mathrm{e}^{-ksA}\right).$$

Hence, by the product law of $\det(\cdot)$ [28, Theorem XIII.105],

$$\det(1 - z^n\,\mathrm{e}^{-nsA}) = \det(1 - z\,\mathrm{e}^{-sA})\det\left(\sum_{k=0}^{n-1} z^k\,\mathrm{e}^{-ksA}\right).$$

Thus (2.27) follows. □

Remark 2.4. In general, relationships among theories at different coupling constants are referred to as 'duality' [31]. Equation (2.30) is a duality relation, where the coupling constant is the inverse temperature.

In statistical mechanics, *correlation functions* are also important objects. Let $f \in D(A^{-1/2})$ and $g \in D(A^{-1/2}) \cap D(A)$. Then, using the well known estimates

$$\|a(f)\Psi\|_{\mathcal{F}_B(\mathcal{H})} \leqslant \|A^{-1/2}f\|_{\mathcal{H}}\|H_B(A)^{1/2}\Psi\|_{\mathcal{F}_B(\mathcal{H})}, \tag{2.31}$$

$$\|a(f)^*\Psi\|_{\mathcal{F}_B(\mathcal{H})} \leqslant \|A^{-1/2}f\|_{\mathcal{H}}\|H_B(A)^{1/2}\Psi\|_{\mathcal{F}_B(\mathcal{H})} + \|f\|_{\mathcal{H}}\|\Psi\|_{\mathcal{F}_B(\mathcal{H})} \tag{2.32}$$

and commutation properties of the annihilation and creation operators with $H_B(A)$, one can show that $a(f)^*a(g)$ is $H_B(A)$-bounded (cf. [5, §II]). Hence, for all $t > 0$, $a(f)^*a(g)\,e^{-tH_B(A)}$ is bounded. Let $t > s$. Then, by the identity

$$a(f)^*a(g)\,e^{-tH_B(A)} = a(f)^*a(g)\,e^{-(t-s)H_B(A)}\,e^{-sH_B(A)},$$

$a(f)^*a(g)\,e^{-tH_B(A)}$ is trace class. Hence we can define

$$R_B(t,z;f,g;A) := \frac{\mathrm{Tr}(\Gamma_B(z)a_{\mathcal{H}}(f)^*a_{\mathcal{H}}(g)\,e^{-tH_B(A)})}{Z_B(t,z;A)}, \quad z \in D. \tag{2.33}$$

This is called a *two-point correlation function.* In the same manner as in [20, Proposition 5.2.28], we can show that

$$R_B(t,z;f,g;A) = (g, z\,e^{-tA}(1 - z\,e^{-tA})^{-1}f)_{\mathcal{H}}. \tag{2.34}$$

Remark 2.5. Correlation functions of the form

$$\mathrm{Tr}\big(\Gamma_B(z)a_{\mathcal{H}}(f_1)^* \ldots a_{\mathcal{H}}(f_n)^*a_{\mathcal{H}}(g_1)\ldots a_{\mathcal{H}}(g_m)\,e^{-tH_B(A)}\big)/Z_B(t,z;A)$$

can be defined for f_j, g_k in a dense subspace in $\mathcal{H}$. These are computed in terms of two-point correlation functions [20, §5.2]. Moreover, we can consider a perturbation of $H_B(A)$ by a symmetric operator V on $\mathcal{F}_B(\mathcal{H})$ such that $H := H_B(A) + V$ defines a self-adjoint operator in the sense of sequilinear form and derive trace formulas for the heat semi-group e^{-sH} $(s > 0)$ [13].

2.2. ARITHMETICAL ASPECTS

Let the spectrum of A be as in (2.10) and denote by ϕ_n a normalized eigenvector of A with eigenvalue $E_n(A)$:

$$A\phi_n = E_n(A)\phi_n, \tag{2.35}$$

such that $\{\phi_n\}_{n=1}^{\infty}$ is an orthonormal system of $\mathcal{H}$. Then $\{\phi_n\}_{n=1}^{\infty}$ is complete in $\mathcal{H}$. We set

$$a_n := a_{\mathcal{H}}(\phi_n). \tag{2.36}$$

Then we have

$$[a_n, a_m^*] = \delta_{mn}, \qquad [a_n, a_m] = 0, \qquad [a_n^*, a_m^*] = 0, \quad n, m \geqslant 1, \tag{2.37}$$

on $\mathcal{F}_{\mathrm{B},0}(\mathcal{H})$.

We denote by

$$\mathcal{P} := \{p_n\}_{n=1}^{\infty} \tag{2.38}$$

the set of all prime numbers with $p_n < p_{n+1}$, $n \geqslant 1$ ($p_1 = 2$, $p_2 = 3$, $p_3 = 5$, $p_4 = 7$, $p_5 = 11, \ldots$).

By definition, an arithmetical function is a complex-valued function on $\mathbf{N}$. An arithmetical function f is called *completely multiplicative* if it satisfies

$$f(1) = 1, \qquad f(mn) = f(m)f(n), \quad m, n \in \mathbf{N}.$$

Let $N \geqslant 2$ be a natural number. Then, by the fundamental theorem of arithmetic, there exists a unique set $\{i_1, \ldots, i_n, \alpha_1, \ldots, \alpha_n\} \subset \mathbf{N}$ ($i_1 < \cdots < i_n$) such that

$$N = (p_{i_1})^{\alpha_1} \ldots (p_{i_n})^{\alpha_n}. \tag{2.39}$$

Then we define an arithmetical function $\gamma(N)$ by

$$\gamma(N) := \sum_{k=1}^{n} \alpha_k \tag{2.40}$$

and $\gamma(1) := 0$.

The arithmetical function defined by $\lambda(1) := 1$ and

$$\lambda(N) := (-1)^{\gamma(N)} \tag{2.41}$$

is called the *Liouville function* [1, §2.12]. This function is completely multiplicative.

Using the representation (2.39) of N, we can define a vector $\Psi_N \in \mathcal{F}_{\mathrm{B}}(\mathcal{H})$ by

$$\Psi_N := C_N (a_{i_1}^*)^{\alpha_1} \ldots (a_{i_n}^*)^{\alpha_n} \Omega_{\mathcal{H}}, \tag{2.42}$$

where

$$C_N := \frac{1}{\sqrt{\alpha_1! \ldots \alpha_n!}}$$

is a normalization constant so that $\|\Psi_N\| = 1$. We set

$$\Psi_1 := \Omega_{\mathcal{H}}. \tag{2.43}$$

LEMMA 2.4. *The set* $\{\Psi_N\}_{N=1}^{\infty}$ *is a complete orthonormal system (CONS) of* $\mathcal{F}_{\mathrm{B}}(\mathcal{H})$.

Proof. This is due to the fact that $\mathcal{D} := \{\Omega_{\mathcal{H}},\ a_{j_1}^* \cdots a_{j_n}^* \Omega_{\mathcal{H}} \mid n \geqslant 1,\ j_1 \leqslant j_2 \leqslant \dots \leqslant j_n,\ j_k \geqslant 1,\ k = 1, \dots, n\}$ is a complete orthogonal system of $\mathcal{F}_{\mathrm{B}}(\mathcal{H})$. Note that $\{\Psi_N\}_{N=1}^{\infty}$ is just obtained by relabeling and normalizing vectors in $\mathcal{D}$. □

Remark 2.6. The set $\{\Psi_N\}_{n=1}^{\infty}$ of vectors was introduced in [30].

LEMMA 2.5. *For all $N \in \mathbf{N}$, Ψ_N is a unique eigenvector (up to constant multiples) of $\Gamma_{\mathrm{B}}(z)$ with eigenvalue $z^{\gamma(N)}$.*

Proof. Let N be as in (2.39). Then $N_{\mathrm{B}}\Psi_N = \gamma(N)\Psi_N$, which, combined with (2.22) implies the desired assertion. □

We introduce a function $F_A\colon \mathbf{N} \to (0, \infty)$ as follows: $F_A(1) := 1$ and if $N \geqslant 2$ is represented as (2.39), then

$$F_A(N) := \prod_{k=1}^{n} \mathrm{e}^{\alpha_k E_{i_k}(A)}. \tag{2.44}$$

It is easy to see that F_A is completely multiplicative.

LEMMA 2.6. *For all $N \in \mathbf{N}$, Ψ_N is a unique eigenvector (up to constant multiples) of $H_{\mathrm{B}}(A)$ with eigenvalue* $\log F_A(N)$.

Proof. Let N be given by (2.39). Then

$$H_{\mathrm{B}}(A)\Psi_N = \left(\sum_{k=1}^{n} \alpha_k E_{i_k}(A)\right)\Psi_N = (\log F_A(N))\Psi_N. \tag{2.45}$$

Hence, the desired assertion follows. □

By Lemmas 2.5 and 2.6, we have

$$Z_{\mathrm{B}}(s, z; A) = \sum_{N=1}^{\infty} \frac{z^{\gamma(N)}}{F_A(N)^s}, \quad z \in D. \tag{2.46}$$

THEOREM 2.7. *For all $z \in D$,*

$$\sum_{N=1}^{\infty} \frac{z^{\gamma(N)}}{F_A(N)^s} = \frac{1}{\prod_{n=1}^{\infty}(1 - z\,\mathrm{e}^{-sE_n(A)})}. \tag{2.47}$$

In particular,

$$\sum_{N=1}^{\infty} \frac{1}{F_A(N)^s} = \frac{1}{\prod_{n=1}^{\infty}(1 - \mathrm{e}^{-sE_n(A)})}, \tag{2.48}$$

$$\sum_{N=1}^{\infty} \frac{\lambda(N)}{F_A(N)^s} = \frac{1}{\prod_{n=1}^{\infty}(1 + \mathrm{e}^{-sE_n(A)})}. \tag{2.49}$$

Proof. By (2.24), we have

$$Z_{\mathrm{B}}(s,z;A)=\frac{1}{\prod_{n=1}^{\infty}(1-z\,\mathrm{e}^{-sE_n(A)})},$$

which, combined with (2.46), implies (2.47). □

Remark 2.7. Formulas (2.47) may be regarded as a general form unifying arithmetical formulas known under the name of *Euler products* [1, Chapter 11]. See Section 2.3 below.

We introduce a function $\varrho(N,m)$: $\mathbf{N}\times\mathbf{N}\to\{0\}\cup\mathbf{N}$ by

$$\varrho(1,m):=0, \tag{2.50}$$

$$\varrho(N,m):=\sum_{k=1}^{n}\alpha_k\delta_{i_k m} \tag{2.51}$$

if $N\geqslant 2$ is expressed as (2.39) ($N,m\in\mathbf{N}$).

THEOREM 2.8. *Let $t>s$. Then, for all $m\in\mathbf{N}$ and $z\in D$,*

$$\sum_{N=1}^{\infty}\frac{z^{\gamma(N)}\varrho(N,m)}{F_A(N)^t}=\frac{z}{\mathrm{e}^{tE_m(A)}-z}Z_{\mathrm{B}}(t,z;A). \tag{2.52}$$

Proof. Putting $f=g=\phi_m$ in (2.34), we have $R_{\mathrm{B}}(t,z;\phi_m,\phi_m;A)=z/(\mathrm{e}^{tE_m(A)}-z)$. It is easy to see that

$$\mathrm{Tr}\big(\Gamma_{\mathrm{B}}(z)a_m^*a_m\,\mathrm{e}^{-tH_{\mathrm{B}}(A)}\big)=\sum_{N=1}^{\infty}\frac{z^{\gamma(N)}\varrho(N,m)}{F_A(N)^t}.$$

Thus (2.52) follows. □

Let $N\geqslant 2$ be given as (2.39). Then, each divisor m of N is of the form

$$m=p_{i_1}^{r_1}\cdots p_{i_n}^{r_n} \tag{2.53}$$

with $0\leqslant r_j\leqslant\alpha_j$, $j=1,\ldots,n$. We define a vector $\Psi_{N,m}\in\mathcal{F}_{\mathrm{B}}(\mathcal{H})$ by

$$\Psi_{N,m}:=C_{N,m}a_{i_1}^{*\,r_1}\ldots a_{i_n}^{*\,r_n}\Omega_{\mathcal{H}}, \tag{2.54}$$

where $C_{N,m}>0$ is a normalization constant. For $m\in\mathbf{N}$ and $N\in\mathbf{N}$, we mean by $m|N$ that m is a divisor of N. The set $\{\Psi_{N,m}\}_{m|N}$ of vectors is orthonormal. We introduce

$$\mathcal{F}_{\mathrm{B}}^{(N)}(\mathcal{H}):=\mathcal{L}\{\Psi_{N,m}\}_{m|N}, \tag{2.55}$$

where $\mathcal{L}\{\cdot\}$ means the subspace spanned algebraically by the vectors in the set $\{\cdot\}$. We set $\mathcal{F}_{\mathrm{B}}^{(1)}:=\{\alpha\Omega_{\mathcal{H}}\mid\alpha\in\mathbf{C}\}$. We denote by P_N the orthogonal projection from $\mathcal{F}_{\mathrm{B}}(\mathcal{H})$ onto $\mathcal{F}_{\mathrm{B}}^{(N)}(\mathcal{H})$.

PROPOSITION 2.9. *Let $z \in D$. Then, for all N,*

$$\mathrm{Tr}\big(P_N \Gamma_{\mathrm{B}}(z)\, \mathrm{e}^{-sH_{\mathrm{B}}(A)} P_N\big) = \sum_{m|N} \frac{z^{\gamma(m)}}{F_A(m)^s}. \tag{2.56}$$

Proof. We have

$$N_{\mathrm{B}} \Psi_{N,m} = \gamma(m) \Psi_{N,m} \quad \text{and} \quad H_{\mathrm{B}}(A) \Psi_{N,m} = \log F_A(m) \Psi_{N,m}.$$

From these facts the desired formula follows. □

2.3. APPLICATIONS TO ANALYTIC NUMBER THEORY

A basic object in analytic number theory is the *Dirichlet series*

$$D(s, f) := \sum_{n=1}^{\infty} \frac{f(n)}{n^s} \tag{2.57}$$

for an arithmetical function f and $s \in \mathbf{C}$, provided that the infinite series converges. The *Riemann zeta function*

$$\zeta(s) := \sum_{n=1}^{\infty} \frac{1}{n^s}, \quad s > 1, \tag{2.58}$$

is a special case of $D(s, f)$, i.e., the case $f \equiv 1$. We first show that $\zeta(s)$ and $D(s, \lambda)$ can be represented as partition functions of $H_{\mathrm{B}}(A)$ with a suitable A. For this purpose, we consider the case where $\mathcal{H}$ is given by

$$\ell^2 := \bigoplus_{n=1}^{\infty} \mathbf{C} = \left\{ \psi = \{\psi_n\}_{n=1}^{\infty} \mid \psi_n \in \mathbf{C}, n \geqslant 1, \sum_{n=1}^{\infty} |\psi_n|^2 < \infty \right\}. \tag{2.59}$$

On this Hilbert space we define an operator $\omega_{\mathcal{P}}$ as follows:

$$D(\omega_{\mathcal{P}}) = \left\{ \psi = \{\psi_n\}_{n=1}^{\infty} \in \ell^2 \mid \sum_{n=1}^{\infty} |(\log p_n)\psi_n|^2 < \infty \right\}, \tag{2.60}$$

$$(\omega_{\mathcal{P}} \psi)_n = (\log p_n)\psi_n, \quad \psi \in D(\omega_{\mathcal{P}}),\ n \geqslant 1. \tag{2.61}$$

Then $\omega_{\mathcal{P}}$ is strictly positive and self-adjoint. Moreover, the spectrum of $\omega_{\mathcal{P}}$ is purely discrete with

$$\sigma(\omega_{\mathcal{P}}) = \sigma_{\mathrm{d}}(\omega_{\mathcal{P}}) = \{\log p_n\}_{n=1}^{\infty} \tag{2.62}$$

with the multiplicity of each eigenvalue $\log p_n$ being one. A normalized eigenvector of $\omega_{\mathcal{P}}$ with eigenvalue $\log p_n$ is given by

$$e_n := \{\delta_{nj}\}_{j=1}^{\infty} \in \ell^2. \tag{2.63}$$

THEOREM 2.10. *For all $s > 1$ and $z \in D$,*

$$Z_{\mathrm{B}}(s, z; \omega_{\mathcal{P}}) = \sum_{N=1}^{\infty} \frac{z^{\gamma(N)}}{N^s}. \tag{2.64}$$

In particular,

$$\zeta(s) = Z_{\mathrm{B}}(s; \omega_{\mathcal{P}}), \tag{2.65}$$

$$D(s, \lambda) = \widetilde{Z}_{\mathrm{B}}(s; \omega_{\mathcal{P}}). \tag{2.66}$$

Proof. One can easily show that

$$F_{\omega_{\mathcal{P}}}(N) = N. \tag{2.67}$$

Hence, by (2.46), we obtain (2.64). □

Applying Theorem 2.7 with $A = \omega_{\mathcal{P}}$, we obtain the following corollary:

COROLLARY 2.11. *For all $s > 1$ and $z \in D$,*

$$\sum_{N=1}^{\infty} \frac{z^{\gamma(N)}}{N^s} = \frac{1}{\prod_{p\in\mathcal{P}}(1 - zp^{-s})}. \tag{2.68}$$

In particular,

$$\zeta(s) = \frac{1}{\prod_{p\in\mathcal{P}}(1 - p^{-s})}, \tag{2.69}$$

$$D(s, \lambda) = \frac{1}{\prod_{p\in\mathcal{P}}(1 + p^{-s})}. \tag{2.70}$$

An application of Theorem 2.8 gives the following corollary:

COROLLARY 2.12. *For all $s > 1$, $n \in \mathbf{N}$ and $z \in D$,*

$$\sum_{N=1}^{\infty} \frac{z^{\gamma(N)}\varrho(N, n)}{N^s} = \frac{z}{p_n^s - z} Z_{\mathrm{B}}(s, z; \omega_{\mathcal{P}}). \tag{2.71}$$

In particular,

$$\sum_{N=1}^{\infty} \frac{\varrho(N, n)}{N^s} = \frac{\zeta(s)}{p_n^s - 1}, \tag{2.72}$$

$$\sum_{N=1}^{\infty} \frac{\lambda(N)\varrho(N, n)}{N^s} = -\frac{D(s, \lambda)}{p_n^s + 1}. \tag{2.73}$$

Moreover, (2.30) yields the following corollary:

COROLLARY 2.13. *Let $s > 1$. Then*

$$D(s,\lambda) = \frac{\zeta(2s)}{\zeta(s)}. \tag{2.74}$$

The operator $\omega_{\mathcal{P}}$ may be regarded as as a special case of a more general operator associated with a completely multiplicative function. Let f be a completely multiplicative function such that $0 < f(n) < 1$ for all $n \geqslant 2$ and

$$\sum_{n=1}^{\infty} f(p_n) < \infty, \tag{2.75}$$

and define an operator A_f on ℓ^2 by

$$D(A_f) = \left\{ \psi = \{\psi_n\}_{n=1}^{\infty} \mid \sum_{n=1}^{\infty} |\log f(p_n)|^2 |\psi_n|^2 < \infty \right\}, \tag{2.76}$$

$$(A_f\psi)_n = [-\log f(p_n)]\psi_n, \quad \psi \in D(A_f),\ n \geqslant 1. \tag{2.77}$$

Then A_f is a strictly positive self-adjoint operator. Since

$$\sum_{n=1}^{\infty} \mathrm{e}^{\log f(p_n)} = \sum_{n=1}^{\infty} f(p_n) < \infty,$$

the self-adjoint operator e^{-A_f} is trace class on ℓ^2. It is easy to see that

$$F_{A_f}(N) = \frac{1}{f(N)}, \quad N \in \mathbf{N}. \tag{2.78}$$

Hence, we have

$$Z_{\mathrm{B}}(1, z; A_f) = \sum_{n=1}^{\infty} z^{\gamma(n)} f(n), \quad z \in D. \tag{2.79}$$

In particular,

$$Z_{\mathrm{B}}(1; A_f) = \sum_{n=1}^{\infty} f(n), \tag{2.80}$$

$$\widetilde{Z}_{\mathrm{B}}(1; A_f) = \sum_{n=1}^{\infty} f(n)\lambda(n). \tag{2.81}$$

Applying Theorem 2.7, we obtain the following fact.

COROLLARY 2.14. *Let f be as above. Then, for all $z \in D$,*

$$\sum_{n=1}^{\infty} z^{\gamma(n)} f(n) = \frac{1}{\prod_{p\in\mathcal{P}}(1 - zf(p))}. \tag{2.82}$$

In particular,

$$\sum_{n=1}^{\infty} f(n) = \frac{1}{\prod_{p\in\mathcal{P}}(1-f(p))}, \tag{2.83}$$

$$\sum_{n=1}^{\infty} f(n)\lambda(n) = \frac{1}{\prod_{p\in\mathcal{P}}(1+f(p))}. \tag{2.84}$$

Theorem 2.8 gives the following corollary:

COROLLARY 2.15. *Let f be as above. Then, for all $n \in \mathbf{N}$ and $z \in D$,*

$$\sum_{N=1}^{\infty} z^{\gamma(N)}\varrho(N,n)f(N) = \frac{zf(p_n)}{1-zf(p_n)}Z_{\mathrm{B}}(1,z;A_f). \tag{2.85}$$

In particular,

$$\sum_{N=1}^{\infty} \varrho(N,n)f(N) = \frac{f(p_n)}{1-f(p_n)}Z_{\mathrm{B}}(1;A_f), \tag{2.86}$$

$$\sum_{N=1}^{\infty} \varrho(N,n)\lambda(N)f(N) = -\frac{f(p_n)}{1+f(p_n)}\widetilde{Z}_{\mathrm{B}}(1;A_f). \tag{2.87}$$

Applying Proposition 2.9, we have for all $s > 1$

$$\mathrm{Tr}\big(P_N z^{N_{\mathrm{B}}}\,\mathrm{e}^{-sH_{\mathrm{B}}(\omega_{\mathcal{P}})}P_N\big) = \sum_{m|N}\frac{z^{\gamma(m)}}{m^s}, \qquad z \in D. \tag{2.88}$$

In particular,

$$\mathrm{Tr}\big(P_N\,\mathrm{e}^{-sH_{\mathrm{B}}(\omega_{\mathcal{P}})}P_N\big) = \sum_{m|N}\frac{1}{m^s}, \tag{2.89}$$

$$\mathrm{Tr}\big(P_N(-1)^{N_{\mathrm{B}}}\,\mathrm{e}^{-sH_{\mathrm{B}}(\omega_{\mathcal{P}})}P_N\big) = \sum_{m|N}\frac{\lambda(m)}{m^s}. \tag{2.90}$$

3. Fermion Fock Spaces and Arithmetical Functions

3.1. PARTITION FUNCTIONS AND CORRELATION FUNCTIONS

Let $\mathcal{K}$ be a separable infinite-dimensional Hilbert space and $\bigotimes_{\mathrm{as}}^n \mathcal{K}$ be the n-fold antisymmetric tensor product Hilbert space of $\mathcal{K}$ ($n = 0, 1, 2, \ldots, \bigotimes_{\mathrm{as}}^0 \mathcal{K} := \mathbf{C}$). Then the Fermion Fock space over $\mathcal{K}$ is defined by

$$\mathcal{F}_{\mathrm{F}}(\mathcal{K}) := \bigoplus_{n=0}^{\infty}\bigotimes_{\mathrm{as}}^{n}\mathcal{K} \tag{3.1}$$

(e.g., [20, §5.2], [26, §II.4]).

We denote by $b_{\mathcal{K}}(u)$ ($u \in \mathcal{K}$) the annihilation operator on $\mathcal{F}_{\mathrm{F}}(\mathcal{K})$ (e.g., [20, §5.2]). $b_{\mathcal{K}}(u)$ is antilinear in u and bounded with $\|b(u)\| = \|u\|_{\mathcal{K}}$. The set $\{b_{\mathcal{K}}(u), b_{\mathcal{K}}(u)^* \mid u \in \mathcal{K}\}$ satisfies the canonical anti-commutation relations

$$\{b_{\mathcal{K}}(u), b_{\mathcal{K}}(v)^*\} = (u, v)_{\mathcal{K}}, \tag{3.2}$$

$$\{b_{\mathcal{K}}(u), b_{\mathcal{K}}(v)\} = 0, \qquad \{b_{\mathcal{K}}(u)^*, b_{\mathcal{K}}(v)^*\} = 0, \quad u, v \in \mathcal{K}, \tag{3.3}$$

where $\{X, Y\} := XY + YX$. We denote by $\Omega_{\mathcal{K}} := \{1, 0, 0, \ldots\}$ the Fock vacuum in $\mathcal{F}_{\mathrm{F}}(\mathcal{K})$. We have

$$b_{\mathcal{K}}(u)\Omega_{\mathcal{K}} = 0, \quad u \in \mathcal{K}. \tag{3.4}$$

Let T be a nonnegative self-adjoint operator on $\mathcal{K}$ and $\mathrm{d}\Gamma_{\mathrm{F}}(T)$ the second quantization of T in $\mathcal{F}_{\mathrm{F}}(\mathcal{K})$:

$$\mathrm{d}\Gamma_{\mathrm{F}}(T) := \bigoplus_{n=0}^{\infty} \mathrm{d}\Gamma_{\mathrm{F}}^{(n)}(T) \tag{3.5}$$

with $\mathrm{d}\Gamma_{\mathrm{F}}^{(0)}(T) := 0$ and

$$\mathrm{d}\Gamma_{\mathrm{F}}^{(n)}(T) := \sum_{j=1}^{n} I \otimes \cdots \otimes I \otimes \underbrace{T}_{j\text{th}} \otimes I \otimes \cdots \otimes I.$$

Then $\mathrm{d}\Gamma_{\mathrm{F}}(T)$ is nonnegative and self-adjoint.

We set

$$H_{\mathrm{F}}(T) := \mathrm{d}\Gamma_{\mathrm{F}}(T). \tag{3.6}$$

LEMMA 3.1. *Suppose that T is strictly positive and the spectrum of T is purely discrete with*

$$\sigma(T) = \sigma_{\mathrm{d}}(T) = \{E_n(T)\}_{n=1}^{\infty}, \tag{3.7}$$

$0 < E_1(T) \leqslant E_2(T) \leqslant \cdots$, $E_n(T) \to \infty$ $(n \to \infty)$, *counted with algebraic multiplicity. Then the spectrum of $H_{\mathrm{F}}(T)$ is purely discrete with*

$$\sigma(H_{\mathrm{F}}(T)) = \sigma_{\mathrm{d}}(H_{\mathrm{F}}(T)) = \bigcup_{n=1}^{\infty} \left\{ \sum_{j=1}^{n} k_j E_j(T) \mid k_j = 0, 1 \right\}. \tag{3.8}$$

Proof. See, e.g., [14, Lemma 4.3]. □

Remark 3.1. Note the difference between the spectral property of $H_{\mathrm{B}}(A)$ (Lemma 2.1) and that of $H_{\mathrm{F}}(T)$.

We set

$$N_{\mathrm{F}} := \mathrm{d}\Gamma_{\mathrm{F}}(I), \tag{3.9}$$

the number operator on $\mathcal{F}_{\mathrm{F}}(\mathcal{K})$.

As in the Boson Fock space, $\mathcal{F}_{\mathrm{F}}(\mathcal{K})$ is $\mathbf{Z}_2$-graded with

$$\mathcal{F}_{\mathrm{F}}(\mathcal{K}) = \mathcal{F}_{\mathrm{F},+}(\mathcal{K}) \oplus \mathcal{F}_{\mathrm{F},-}(\mathcal{K}), \tag{3.10}$$

where

$$\mathcal{F}_{\mathrm{F},+}(\mathcal{K}) := \bigoplus_{n=0}^{\infty} \bigotimes_{\mathrm{as}}^{2n} \mathcal{K}, \qquad \mathcal{F}_{\mathrm{F},-}(\mathcal{K}) := \bigoplus_{n=0}^{\infty} \bigotimes_{\mathrm{as}}^{2n+1} \mathcal{K}. \tag{3.11}$$

The self-adjoint operator $(-1)^{N_{\mathrm{F}}}$ is the grading operator of this gradation.

For all $z \in D$, the operator $\Gamma(z)$ on the full Fock space $\mathcal{F}_{\mathrm{full}}(\mathcal{K})$ over $\mathcal{K}$ is reduced by $\mathcal{F}_{\mathrm{F}}(\mathcal{K})$. We denote by $\Gamma_{\mathrm{F}}(z)$ its reduced part. As in the case of the Boson Fock space, we have

$$\Gamma_{\mathrm{F}}(z) = z^{N_{\mathrm{F}}}. \tag{3.12}$$

Let $s > 0$, $z \in D$ and

$$Z_{\mathrm{F}}(s, z; T) := \mathrm{Tr}\big(\Gamma_{\mathrm{F}}(z)\, \mathrm{e}^{-sH_{\mathrm{F}}(T)}\big), \tag{3.13}$$

provided that $\mathrm{e}^{-sH_{\mathrm{F}}(T)}$ is trace class on $\mathcal{F}_{\mathrm{F}}(\mathcal{H})$. In particular, we define

$$Z_{\mathrm{F}}(s; T) := Z_{\mathrm{F}}(s, 1; T) = \mathrm{Tr}\, \mathrm{e}^{-sH_{\mathrm{F}}(T)}, \tag{3.14}$$

$$\widetilde{Z}_{\mathrm{F}}(s; T) := Z_{\mathrm{F}}(s, -1; T) = \mathrm{Tr}\big\{(-1)^{N_{\mathrm{F}}}\, \mathrm{e}^{-sH_{\mathrm{F}}(T)}\big\}. \tag{3.15}$$

In what follows, we assume the following.

(T) *For some* $s > 0$, e^{-sT} *is trace class on* $\mathcal{K}$.

THEOREM 3.2. *For all* $z \in D$, $\Gamma_{\mathrm{F}}(z)\, \mathrm{e}^{-sH_{\mathrm{F}}(T)}$ *is trace class on* $\mathcal{F}_{\mathrm{F}}(\mathcal{K})$ *and*

$$Z_{\mathrm{F}}(s, z; T) = \det(I + z\, \mathrm{e}^{-sT}). \tag{3.16}$$

In particular,

$$Z_{\mathrm{F}}(s; T) = \det(I + \mathrm{e}^{-sT}), \tag{3.17}$$

$$\widetilde{Z}_{\mathrm{F}}(s; T) = \det(I - \mathrm{e}^{-sT}). \tag{3.18}$$

Proof. We have

$$\Gamma_{\mathrm{F}}(z)\, \mathrm{e}^{-sH_{\mathrm{F}}(T)} = \Gamma(z\, \mathrm{e}^{-sT}) | \mathcal{F}_{\mathrm{F}}(\mathcal{K}). \tag{3.19}$$

Hence, by the definition of $\det(I + \cdot)$ [28, p. 323], we obtain (3.16). □

Remark 3.2. By (3.8) and the functional calculus, we have

$$\sigma_{\mathrm{d}}\big(\Gamma_{\mathrm{F}}(z)\,\mathrm{e}^{-sH_{\mathrm{F}}(T)}\big) = \bigcup_{n=1}^{\infty}\Big\{\prod_{j=1}^{n} z^{k_j}\mathrm{e}^{-sk_jE_j(T)} \mid k_j = 0, 1\Big\}.$$

One can use this relation to prove (3.16) (e.g., [14, Theorem 4.4]).

Remark 3.3. In Theorem 3.2, we do not need assume that T is *strictly* positive.

By Theorems 2.2 and 3.2, we have interesting relations between bosonic and fermionic partition functions:

COROLLARY 3.3. *Consider the case* $\mathcal{H} = \mathcal{K}$ *and* A *be an operator on* $\mathcal{H}$ *obeying Assumption* (A) *in Section 2. Then, for all* $z \in D$,

$$Z_{\mathrm{B}}(s, -z; A) = \frac{1}{Z_{\mathrm{F}}(s, z; A)}. \tag{3.20}$$

In particular,

$$Z_{\mathrm{B}}(s; A) = \frac{1}{\widetilde{Z}_{\mathrm{F}}(s; A)}, \qquad \widetilde{Z}_{\mathrm{B}}(s; A) = \frac{1}{Z_{\mathrm{F}}(s; A)}. \tag{3.21}$$

In the same way as in Theorem 2.3, we can prove the following theorem:

THEOREM 3.4. *For all* $n \in \mathbf{N}$ *and* $z \in D$,

$$Z_{\mathrm{F}}(ns, -z^n; T) = \det\Big(\sum_{k=1}^{n-1} z^k\,\mathrm{e}^{-skT}\Big) Z_{\mathrm{F}}(s, -z; T), \tag{3.22}$$

$$Z_{\mathrm{F}}(s, -z; T) Z_{\mathrm{F}}(s, z; T) = Z_{\mathrm{F}}(2s, -z^2; T). \tag{3.23}$$

In particular,

$$\widetilde{Z}_{\mathrm{F}}(ns; T) = \det\Big(\sum_{k=0}^{n-1} \mathrm{e}^{-ksT}\Big) \widetilde{Z}_{\mathrm{F}}(s; T), \tag{3.24}$$

$$\widetilde{Z}_{\mathrm{F}}(s; T) Z_{\mathrm{F}}(s; T) = \widetilde{Z}_{\mathrm{F}}(2s; T). \tag{3.25}$$

Remark 3.4. If T is strictly positive, these relations follow from Theorem 2.3 and (3.21).

Remark 3.5. Relation (3.23) is a form of *duality* of fermionic partition functions. A special case is discussed in [31].

COROLLARY 3.5. *Consider the case* $\mathcal{H} = \mathcal{K}$ *and* A *be an operator on* $\mathcal{H}$ *obeying* (A). *Then*

$$Z_{\mathrm{B}}(2s, z^2; A) Z_{\mathrm{F}}(s, z; A) = Z_{\mathrm{B}}(s, z; A). \tag{3.26}$$

Proof. This follows from (3.23) and Corollary 3.3. □

Remark 3.6. Relation (3.26) is also a form of *duality* of fermionic and bosonic partition functions. For a special case, see [31].

Let $u, v \in \mathcal{K}$ and $z \in D$. Then a *fermionic two-point correlation function* is defined by

$$R_{\mathrm{F}}(s, z; u, v; T) := \frac{\mathrm{Tr}\big(z^{N_{\mathrm{F}}}\,\mathrm{e}^{-sH_{\mathrm{F}}(T)} b_{\mathcal{K}}(u)^* b_{\mathcal{K}}(v)\big)}{Z_{\mathrm{F}}(s, z; T)}. \tag{3.27}$$

It is easy to see (cf., e.g. [20]) that

$$R_{\mathrm{F}}(s, z; u, v; T) = (v, z\,\mathrm{e}^{-sT}(1 + z\,\mathrm{e}^{-sT})^{-1} u)_{\mathcal{K}}. \tag{3.28}$$

Remark 3.7. Correlation functions of the form

$$\mathrm{Tr}\big(z^{N_{\mathrm{F}}}\,\mathrm{e}^{-sH_{\mathrm{F}}(T)} b_{\mathcal{K}}(u_1)^* \ldots b_{\mathcal{K}}(u_n)^* b_{\mathcal{K}}(v_1) \ldots b_{\mathcal{K}}(v_m)\big)/Z_{\mathrm{F}}(s, z; T)$$

$(u_j, v_k \in \mathcal{K})$ are defined. These are computed in terms of two-point correlation functions ([20, §5.2], [3, 7]).

3.2. ARITHMETICAL ASPECTS

Let the spectrum of T be as in (3.7) and denote by u_n a normalized eigenvector of T with eigenvalue $E_n(T)$:

$$Tu_n = E_n(T)u_n \tag{3.29}$$

such that $\{u_n\}_{n=1}^{\infty}$ is an orthonormal system. Then $\{u_n\}_{n=1}^{\infty}$ is complete in $\mathcal{K}$. We set

$$b_n := b_{\mathcal{K}}(u_n). \tag{3.30}$$

Then we have

$$\{b_n, b_m^*\} = \delta_{mn}, \qquad \{b_n, b_m\} = 0, \qquad \{b_n^*, b_m^*\} = 0, \quad n, m \geqslant 1. \tag{3.31}$$

In particular, $b_n^2 = 0$, $b_n^{*2} = 0$, $n \in \mathbf{N}$.

For $N \in \mathbf{N}$ we define $\nu(N)$ by $\nu(1) := 1$ and

$$\nu(N) = n, \quad N \geqslant 2, \tag{3.32}$$

if N is represented as (2.39) [1, p. 247].

A natural number $m \geqslant 2$ is called *square free* if it is written as a product of mutually different prime numbers. As a convention, 1 is defined to be square free. We denote by $\mathcal{S}_0$ the set of square free elements in $\mathbf{N}$:

$$\mathcal{S}_0 := \{m \in \mathbf{N} \mid m \text{ is square free}\}. \tag{3.33}$$

For each $N \in \mathbf{N}$, we define a set $\mathcal{S}_0(N)$ as follows:

$$\mathcal{S}_0(1) := \{1\}, \tag{3.34}$$

$$\mathcal{S}_0(N) := \{m \in \mathcal{S}_0 \mid m \text{ is a divisor of } N\}, \quad N \geqslant 2. \tag{3.35}$$

Let $N \geqslant 2$ be given as (2.39). Then each element m of $\mathcal{S}_0(N)$ is of the form

$$m = p_{i_1}^{q_1} \cdots p_{i_n}^{q_n}, \tag{3.36}$$

where $q_j = 0$ or $q_j = 1$ $(j = 1, \ldots, n)$. Corresponding to this, we define a vector $\Phi_{N,m}$ by

$$\Phi_{N,m} := b_{i_1}^{*\,q_1} \cdots b_{i_n}^{*\,q_n} \Omega_{\mathcal{K}}. \tag{3.37}$$

Let

$$\mathcal{F}_{\mathrm{F}}^{(1)}(\mathcal{K}) := \{c\Omega_{\mathcal{K}} \mid c \in \mathbf{C}\},$$

$$\mathcal{F}_{\mathrm{F}}^{(N)}(\mathcal{K}) := \mathcal{L}\{\Phi_{N,m} \mid m \in \mathcal{S}_0(N)\}, \quad N \geqslant 2. \tag{3.38}$$

Then $\mathcal{F}_{\mathrm{F}}^{(N)}(\mathcal{K})$ is finite-dimensional with $\dim \mathcal{F}_{\mathrm{F}}^{(N)}(\mathcal{K}) = 2^{\nu(N)}$. We denote by R_N the orthogonal projection from $\mathcal{F}_{\mathrm{F}}(\mathcal{K})$ onto $\mathcal{F}_{\mathrm{F}}^{(N)}(\mathcal{K})$.

Let $N \geqslant 2$ be of the form (2.39),

$$\mathcal{K}_N := \mathcal{L}\{u_{i_k} \mid k = 1, \ldots, n\} \tag{3.39}$$

and T_N be the restriction of T to $\mathcal{K}_N$. Then we can show that

$$\mathrm{Tr}\big(R_N z^{N_{\mathrm{F}}} \mathrm{e}^{-sH_{\mathrm{F}}(T)} R_N\big) = \det(1 + z\,\mathrm{e}^{-sT_N}). \tag{3.40}$$

Let $m \in \mathcal{S}_0$, $m \geqslant 2$ and

$$m = p_{i_1} \cdots p_{i_r} \tag{3.41}$$

be its factorization in prime numbers $(i_j \neq i_k,\ j \neq k)$. Then we define a vector Φ_m in $\mathcal{F}_{\mathrm{F}}(\mathcal{K})$ by

$$\Phi_m := b_{i_1}^* \cdots b_{i_r}^* \Omega_{\mathcal{K}}. \tag{3.42}$$

For $m = 1$, we set

$$\Phi_1 := \Omega_{\mathcal{K}}. \tag{3.43}$$

For $m \notin \mathcal{S}_0$, we define

$$\Phi_m := 0. \tag{3.44}$$

LEMMA 3.6. *The set $\{\Phi_m\}_{m \in \mathcal{S}_0}$ is a CONS of $\mathcal{F}_{\mathrm{F}}(\mathcal{K})$.*

Proof. By using (3.31) and (3.4), one easily sees that $\{\Phi_m\}_{m \in \mathcal{S}_0}$ is an orthonormal system of $\mathcal{F}_{\mathrm{F}}(\mathcal{K})$. It is well known that $\mathcal{F}_{\mathrm{F}}(\mathcal{K})$ is generated by $\Omega_{\mathcal{K}}$ and the vectors of the form $b_{i_1}^* \ldots b_{i_r}^* \Omega_{\mathcal{K}}$ with $r \in \mathbf{N}, i_j \in \mathbf{N},\ i_j \neq i_k,\ j \neq k$. Thus the assertion follows. □

Remark 3.8. The set $\{\Phi_m\}_m$ was introduced in [30].

The *Möbius function* μ: $\mathbf{N} \to \{0, \pm 1\}$ is defined as follows: $\mu(1) := 1$, $\mu(m) := 0$ if $m \notin \mathcal{S}_0$ and $\mu(m) := (-1)^r$ if m is written as the product of mutually different r prime numbers. We have

$$\mu(m) = (-1)^{\gamma(m)}, \quad m \in \mathcal{S}_0. \tag{3.45}$$

LEMMA 3.7. *For all $m \in \mathcal{S}_0$, Φ_m is an eigenvector of N_{F} with eigenvalue $\gamma(m)$. In particular, for all $m \in \mathcal{S}_0$ and $z \in D$,*

$$z^{N_{\mathrm{F}}}\Phi_m = z^{\gamma(m)}\Phi_m, \qquad (-1)^{N_{\mathrm{F}}}\Phi_m = \mu(m)\Phi_m. \tag{3.46}$$

Proof. Let $m \in \mathcal{S}_0$ be as in (3.41). Then $N_{\mathrm{F}}\Phi_m = r\Phi_m = \gamma(m)\Phi_m$, which implies that $z^{N_{\mathrm{F}}}\Phi_m = z^{\gamma(m)}\Phi_m$. □

LEMMA 3.8. *For all $m \in \mathcal{S}_0$, Φ_m is an eigenvector of $H_{\mathrm{F}}(T)$ with eigenvalue $\log F_T(m)$, where F_T is defined by* (2.44) *with $A = T$.*

Proof. Let $m \in \mathcal{S}_0$ be expressed as (3.41). Then

$$H_{\mathrm{F}}(T)\Phi_m = (E_{i_1}(T) + \cdots + E_{i_r}(T))\Phi_m = (\log F_T(m))\Phi_m.$$

Hence, the desired assertion follows. □

It follows from Lemmas 3.7 and 3.8 that

$$Z_{\mathrm{F}}(s, z; T) = \sum_{m=1}^{\infty} \frac{z^{\gamma(m)}|\mu(m)|}{F_T(m)^s}, \quad z \in D, \tag{3.47}$$

where we have used that $\mu(m) = 0$ for all $m \notin \mathcal{S}_0$ and $|\mu(m)| = 1$ for all $m \in \mathcal{S}_0$. In particular,

$$Z_{\mathrm{F}}(s; T) = \sum_{m=1}^{\infty} \frac{|\mu(m)|}{F_T(m)^s}, \tag{3.48}$$

$$\widetilde{Z}_{\mathrm{F}}(s; T) = \sum_{m=1}^{\infty} \frac{\mu(m)}{F_T(m)^s}. \tag{3.49}$$

By (3.47) and Theorem 3.2, we obtain the following theorem:

THEOREM 3.9. *Let $z \in D$. Then*

$$\sum_{m=1}^{\infty} \frac{z^{\gamma(m)}|\mu(m)|}{F_T(m)^s} = \prod_{n=1}^{\infty}(1 + z\,\mathrm{e}^{-sE_n(T)}). \tag{3.50}$$

In particular,

$$\sum_{m=1}^{\infty} \frac{|\mu(m)|}{F_T(m)^s} = \prod_{n=1}^{\infty}(1 + \mathrm{e}^{-sE_n(T)}), \tag{3.51}$$

$$\sum_{m=1}^{\infty} \frac{\mu(m)}{F_T(m)^s} = \prod_{n=1}^{\infty}(1 - \mathrm{e}^{-sE_n(T)}). \tag{3.52}$$

Theorems 3.9 and 2.7 imply the following corollary:

COROLLARY 3.10. *Let $z \in D$. Then,*

$$\sum_{m=1}^{\infty} \frac{z^{\gamma(m)}|\mu(m)|}{F_T(m)^s} = \frac{1}{\sum_{n=1}^{\infty} \frac{(-z)^{\gamma(n)}}{F_T(n)^s}}. \tag{3.53}$$

In particular,

$$\sum_{m=1}^{\infty} \frac{|\mu(m)|}{F_T(m)^s} = \frac{1}{\sum_{n=1}^{\infty} \frac{\lambda(n)}{F_T(n)^s}}, \tag{3.54}$$

$$\sum_{m=1}^{\infty} \frac{\mu(m)}{F_T(m)^s} = \frac{1}{\sum_{n=1}^{\infty} \frac{1}{F_T(n)^s}}. \tag{3.55}$$

We introduce a function η on $\mathbf{N} \times \mathbf{N}$ by

$$\eta(1, n) := 0, \tag{3.56}$$

$$\eta(m, n) := \sum_{k=1}^{r} (-1)^{k-1} \delta_{i_k n} \tag{3.57}$$

if $m \in \mathcal{S}_0$ is expressed as (3.41). If $m \notin \mathcal{S}_0$, then $\eta(m, n) := 0$ for all $n \in \mathbf{N}$.

THEOREM 3.11. *Let $z \in D$ and $n \in \mathbf{N}$. Then*

$$\sum_{m=1}^{\infty} \frac{z^{\gamma(m)}\eta(m, n)}{F_T(m)^s} = \frac{z}{\mathrm{e}^{sE_n(T)} + z} Z_{\mathrm{F}}(s, z; T). \tag{3.58}$$

In particular,

$$\sum_{m=1}^{\infty} \frac{\eta(m, n)}{F_T(m)^s} = \frac{Z_{\mathrm{F}}(s; T)}{\mathrm{e}^{sE_n(T)} + 1}, \tag{3.59}$$

$$\sum_{m=1}^{\infty} \frac{\mu(m)\eta(m, n)}{F_T(m)^s} = -\frac{\widetilde{Z}_{\mathrm{F}}(s; T)}{\mathrm{e}^{sE_n(T)} - 1}. \tag{3.60}$$

Proof. Similar to the proof of Theorem 2.8 [use (3.28)]. □

The left-hand side of (3.40) is equal to $\sum_{m\in\mathcal{S}_0(N)} z^{\gamma(m)}/F_T(m)^s$. Hence, we obtain

$$\sum_{m|N} \frac{z^{\gamma(m)}|\mu(m)|}{F_T(m)^s} = \det(1 + z\,\mathrm{e}^{-sT_N}). \tag{3.61}$$

3.3. APPLICATIONS TO ANALYTIC NUMBER THEORY

Consider the case where $\mathcal{H} = \ell^2$ and $T = \omega_{\mathcal{P}}$. Let $z \in D$ and $s > 1$. Then, by (2.67), we have

$$Z_{\mathrm{F}}(s, z; \omega_{\mathcal{P}}) = \sum_{m=1}^{\infty} \frac{z^{\gamma(m)}|\mu(m)|}{m^s}. \tag{3.62}$$

In particular,

$$Z_{\mathrm{F}}(s; \omega_{\mathcal{P}}) = \sum_{m=1}^{\infty} \frac{|\mu(m)|}{m^s}, \tag{3.63}$$

$$\widetilde{Z}_{\mathrm{F}}(s; \omega_{\mathcal{P}}) = \sum_{m=1}^{\infty} \frac{\mu(m)}{m^s}. \tag{3.64}$$

Let f be a completely multiplicative function as in Section 2.3 and $z \in D$. Then, by (2.78), we have

$$Z_{\mathrm{F}}(1, z; A_f) = \sum_{m=1}^{\infty} z^{\gamma(m)}|\mu(m)| f(m). \tag{3.65}$$

In particular,

$$Z_{\mathrm{F}}(1; A_f) = \sum_{m=1}^{\infty} |\mu(m)| f(m), \tag{3.66}$$

$$\widetilde{Z}_{\mathrm{F}}(1; A_f) = \sum_{m=1}^{\infty} \mu(m) f(m). \tag{3.67}$$

By Theorem 3.9, we obtain the following.

COROLLARY 3.12. *For all $z \in D$,*

$$\sum_{m=1}^{\infty} z^{\gamma(m)}|\mu(m)| f(m) = \prod_{p\in\mathcal{P}} (1 + z f(p)). \tag{3.68}$$

In particular,

$$\sum_{m=1}^{\infty} |\mu(m)| f(m) = \prod_{p \in \mathcal{P}} (1 + f(p)), \tag{3.69}$$

$$\sum_{m=1}^{\infty} \mu(m) f(m) = \prod_{p \in \mathcal{P}} (1 - f(p)). \tag{3.70}$$

Theorem 3.11 gives the following.

COROLLARY 3.13. *For all $n \in \mathbf{N}$ and $z \in D$,*

$$\sum_{m=1}^{\infty} z^{\gamma(m)} \eta(m,n) f(m) = \frac{z f(p_n)}{1 + z f(p_n)} Z_{\mathrm{F}}(1, z; A_f). \tag{3.71}$$

In particular,

$$\sum_{m=1}^{\infty} \eta(m,n) f(m) = \frac{f(p_n)}{f(p_n) + 1} Z_{\mathrm{F}}(1; A_f), \tag{3.72}$$

$$\sum_{m=1}^{\infty} \mu(m) \eta(m,n) f(m) = \frac{f(p_n)}{f(p_n) - 1} \widetilde{Z}_{\mathrm{F}}(1; A_f). \tag{3.73}$$

Jordan's totient function $J_s(N)$ ($s \geqslant 0$, $N \in \mathbf{N}$) is defined by $J_s(1) := 1$ and, for $N \geqslant 2$,

$$J_s(N) = N^s \prod_{p|N;\, p \in \mathcal{P}} \left(1 - \frac{1}{p^s}\right) \tag{3.74}$$

[1, p. 48]. The special case

$$\varphi(N) = J_1(N) \tag{3.75}$$

is *Euler's totient function* [1, pp. 25, 27]. We have

$$\det(1 - \mathrm{e}^{-s(\omega_{\mathcal{P}})_N}) = \prod_{p|N;\, p \in \mathcal{P}} \left(1 - \frac{1}{p^s}\right), \quad s \geqslant 0,\ N \geqslant 2. \tag{3.76}$$

Hence we obtain

$$J_s(N) = N^s \det(1 - \mathrm{e}^{-s(\omega_{\mathcal{P}})_N}), \quad s \geqslant 0,\ N \geqslant 2, \tag{3.77}$$

which, together with (3.40), implies that

$$J_s(N) = N^s \operatorname{Tr}(R_N (-1)^{N_{\mathrm{F}}} \mathrm{e}^{-s H_{\mathrm{F}}(\omega_{\mathcal{P}})} R_N), \quad s \geqslant 0,\ N \in \mathbf{N}. \tag{3.78}$$

This gives an expression of Jordan's totient function in terms of Fock space objects. Formula (3.61) implies the well-known identity [1, p. 48]:

$$J_s(N) = \sum_{m|N} \mu(m) \left(\frac{N}{m}\right)^s, \quad s \geqslant 0, \ N \in \mathbf{N}. \tag{3.79}$$

4. Dirac and Laplace–Beltrami Operators on the Abstract Boson–Fermion Fock Space

In this section we review the basic results of the analysis of the abstract Boson–Fermion Fock space [3, 6, 7, 16].

4.1. EXTERIOR DIFFERENTIAL OPERATORS

Let $\mathcal{H}$ and $\mathcal{K}$ be Hilbert spaces. Then the Boson–Fermion Fock space associated with the pair $\langle \mathcal{H}, \mathcal{K} \rangle$ is defined by the tensor product Hilbert space

$$\mathcal{F}_{\mathrm{BF}}(\mathcal{H}, \mathcal{K}) := \mathcal{F}_{\mathrm{B}}(\mathcal{H}) \otimes \mathcal{F}_{\mathrm{F}}(\mathcal{K}). \tag{4.1}$$

We define

$$\Omega := \Omega_{\mathcal{H}} \otimes \Omega_{\mathcal{K}} \in \mathcal{F}_{\mathrm{BF}}(\mathcal{H}, \mathcal{K}), \tag{4.2}$$

the *vacuum* in $\mathcal{F}_{\mathrm{BF}}(\mathcal{H}, \mathcal{K})$.

The annihilation operators $a_{\mathcal{H}}(f)$ and $b_{\mathcal{K}}(u)$ ($f \in \mathcal{H}, u \in \mathcal{K}$) can be extended to operators on $\mathcal{F}_{\mathrm{BF}}(\mathcal{H}, \mathcal{K})$ as

$$a(f) := a_{\mathcal{H}}(f) \otimes I, \qquad b(u) := I \otimes b_{\mathcal{K}}(u). \tag{4.3}$$

We denote by $\mathsf{C}(\mathcal{H}, \mathcal{K})$ the set of densely defined closed linear operators from $\mathcal{H}$ to $\mathcal{K}$.

For each $S \in \mathsf{C}(\mathcal{H}, \mathcal{K})$, we define a subspace

$$\begin{aligned} \mathcal{D}_S \ := \ & \mathcal{L}\big\{a(f_1)^* \ldots a(f_n)^* b(u_1)^* \ldots b(u_m)^* \Omega \mid n, m \geqslant 0, \ f_j \in D(S), \\ & j = 1, \ldots, n, \ u_k \in D(S^*), \ k = 1, \ldots, m\big\}. \end{aligned} \tag{4.4}$$

Since $D(S)$ and $D(S^*)$ are dense in $\mathcal{H}$ and $\mathcal{K}$, respectively, it follows that $\mathcal{D}_S$ is dense in $\mathcal{F}_{\mathrm{BF}}(\mathcal{H}, \mathcal{K})$.

PROPOSITION 4.1 ([7]). *For each $S \in \mathsf{C}(\mathcal{H}, \mathcal{K})$, there exists a unique densely defined closed linear operator d_S on $\mathcal{F}_{\mathrm{BF}}(\mathcal{H}, \mathcal{K})$ with the following properties:*

(i) *$\mathcal{D}_S$ is a core of d_S.*

(ii) *For each vector* $\Psi \in \mathcal{D}_S$ *of the form*

$$\Psi = a(f_1)^* \dots a(f_n)^* b(u_1)^* \dots b(u_m)^* \Omega, \tag{4.5}$$

d_S *acts as*

$$d_S\Psi = 0 \quad \text{for } n = 0,$$

$$d_S\Psi = \sum_{j=1}^{n} a(f_1)^* \dots \widehat{a(f_j)}^* \dots a(f_n)^* b(Sf_j)^* b(u_1)^* \dots b(u_m)^* \Omega \quad \text{for } n \geqslant 1,$$

where $\widehat{T}$ *indicates the omission of* T.

We call the operator d_S an *exterior differential operator* on $\mathcal{F}_{\mathrm{BF}}(\mathcal{H}, \mathcal{K})$. Fundamental properties of the operator d_S are given in the following proposition.

PROPOSITION 4.2 ([7]). *Let* $S \in \mathsf{C}(\mathcal{H}, \mathcal{K})$. *Then the following* (i)–(iv) *hold.*

(i) $d_S^2 = 0$.

(ii) *For each CONS* $\{u_n\}_{n=1}^\infty$ *of* $\mathcal{K}$ *with* $u_n \in D(S^*)$,

$$d_S\Psi = \sum_{n=1}^{\infty} a(S^*u_n) b(u_n)^* \Psi, \quad \Psi \in \mathcal{D}_S,$$

where the convergence is taken in the strong topology of $\mathcal{F}_{\mathrm{FB}}(\mathcal{H}, \mathcal{K})$.

(iii) *For each CONS* $\{\phi_n\}_{n=1}^\infty$ *of* $\mathcal{H}$ *with* $\phi_n \in D(S)$, *we have*

$$(\Phi, d_S\Psi) = \lim_{N\to\infty} \left(\Phi, \sum_{n=1}^{N} a(\phi_n) b(S\phi_n)^* \Psi\right), \quad \Phi, \Psi \in \mathcal{D}_S.$$

(iv) $\mathcal{D}_S \subset D(d_S^*)$ *and, for all vectors* Ψ *of the form* (4.5) *with* $m \geqslant 1$,

$$d_S^*\Psi = \sum_{k=1}^{m} (-1)^{k-1} a(S^*u_k)^* a(f_1)^* \dots a(f_n)^* b(u_1)^* \dots \widehat{b(u_k)}^* \dots b(u_m)^* \Omega.$$

If $m = 0$, *then* $d_S^*\Psi = 0$.

The Boson–Fermion Fock space $\mathcal{F}_{\mathrm{BF}}(\mathcal{H}, \mathcal{K})$ is $\mathbf{Z}_2$-graded:

$$\mathcal{F}_{\mathrm{BF}}(\mathcal{H}, \mathcal{K}) = \mathcal{F}_+(\mathcal{H}, \mathcal{K}) \oplus \mathcal{F}_-(\mathcal{H}, \mathcal{K}) \tag{4.6}$$

with

$$\mathcal{F}_\pm(\mathcal{H}, \mathcal{K}) := \mathcal{F}_{\mathrm{B}}(\mathcal{H}) \otimes \mathcal{F}_{\mathrm{F},\pm}(\mathcal{K}). \tag{4.7}$$

The grading operator of this $\mathbf{Z}_2$-gradation is given by

$$\Gamma_{\mathrm{F}} := (-1)^{I \otimes N_{\mathrm{F}}} = I \otimes (-1)^{N_{\mathrm{F}}}. \tag{4.8}$$

Let

$$\mathcal{F}_n(\mathcal{H}, \mathcal{K}) := \mathcal{F}_{\mathrm{B}}(\mathcal{H}) \otimes \left(\bigotimes_{\mathrm{as}}^{n} \mathcal{K} \right). \tag{4.9}$$

Then

$$\mathcal{F}_{\mathrm{BF}}(\mathcal{H}, \mathcal{K}) = \bigoplus_{n=0}^{\infty} \mathcal{F}_n(\mathcal{H}, \mathcal{K}), \tag{4.10}$$

$$\mathcal{F}_{+}(\mathcal{H}, \mathcal{K}) = \bigoplus_{n=0}^{\infty} \mathcal{F}_{2n}(\mathcal{H}, \mathcal{K}), \qquad \mathcal{F}_{-}(\mathcal{H}, \mathcal{K}) = \bigoplus_{n=0}^{\infty} \mathcal{F}_{2n+1}(\mathcal{H}, \mathcal{K}). \tag{4.11}$$

We denote by Π_n the orthogonal projection from $\mathcal{F}_{\mathrm{BF}}(\mathcal{H}, \mathcal{K})$ onto $\mathcal{F}_n(\mathcal{H}, \mathcal{K})$. The following fact is easily proven.

PROPOSITION 4.3. *For all $n \geqslant 0$, $\Pi_{n+1} d_S \subset d_S \Pi_n$. In particular, d_S maps $D(d_S) \cap \mathcal{F}_n(\mathcal{H}, \mathcal{K})$ to $\mathcal{F}_{n+1}(\mathcal{H}, \mathcal{K})$.*

By this proposition, we can define, for each $n \in \mathbf{N}$, a densely define closed linear operator $d_{S,n}$ from $\mathcal{F}_n(\mathcal{H}, \mathcal{K})$ to $\mathcal{F}_{n+1}(\mathcal{H}, \mathcal{K})$ by

$$D(d_{S,n}) := D(d_S) \cap \mathcal{F}_n(\mathcal{H}, \mathcal{K}), \quad d_{S,n}\Psi := d_S\Psi, \ \Psi \in D(d_{S,n}). \tag{4.12}$$

We have

$$d_{S,n+1} d_{S,n} = 0, \quad n \geqslant 0. \tag{4.13}$$

4.2. DIRAC AND LAPLACE–BELTRAMI OPERATORS

In what follows, we fix $S \in \mathsf{C}(\mathcal{H}, \mathcal{K})$. We define

$$Q_S := d_S + d_S^* \tag{4.14}$$

with $D(Q_S) = D(d_S) \cap D(d_S^*)$. We call it a *Dirac operator* on $\mathcal{F}_{\mathrm{BF}}(\mathcal{H}, \mathcal{K})$.

Since S^*S and SS^* are nonnegative self-adjoint operators on $\mathcal{H}$ and $\mathcal{K}$, respectively, we can define

$$L_S := H_{\mathrm{B}}(S^*S) \otimes I + I \otimes H_{\mathrm{F}}(SS^*) \tag{4.15}$$

acting in $\mathcal{F}_{\mathrm{BF}}(\mathcal{H}, \mathcal{K})$, which is nonnegative and self-adjoint (cf. [26, §VIII.10, Corollary]).

For a linear operator A on a Hilbert space, we set

$$C^{\infty}(A) := \bigcap_{n=1}^{\infty} D(A^n).$$

Since $C^{\infty}(S^*S)$ and $C^{\infty}(SS^*)$ are dense in $\mathcal{H}$ and $\mathcal{K}$, respectively, the set

$$\mathcal{D}_S^{\infty} := \mathcal{L}\{a(f_1)^* \dots a(f_n)^* b(u_1)^* \dots b(u_m)^* \Omega \mid n, m \geqslant 0,\ f_j \in C^{\infty}(S^*S),\ j = 1, \dots, n,\ u_k \in C^{\infty}(SS^*),\ k = 1, \dots, m\} \tag{4.16}$$

is dense in $\mathcal{F}_{\mathrm{BF}}(\mathcal{H}, \mathcal{K})$.

THEOREM 4.4 ([7]). (i) *The operator Q_S is self-adjoint, and essentially self-adjoint on every core of L_S. In particular, Q_S is essentially self-adjoint on $\mathcal{D}_S^{\infty}$.*

(ii) *The operator Γ_{F} leaves $D(Q_S)$ invariant and $\Gamma_{\mathrm{F}} Q_S + Q_S \Gamma_{\mathrm{F}} = 0$ on $D(Q_S)$.*

(iii) *The following operator equations hold*:

$$L_S = Q_S^2 = d_S^* d_S + d_S d_S^*.$$

Remark 4.1. (i) The operators d_S and d_S^* leave $\mathcal{D}_S^{\infty}$ invariant and so does Q_S.

(ii) Part (ii) of Theorem 4.4 justifies calling Q_S a Dirac type operator (cf. [32, Chapter 5]).

(iii) Because of part (iii) of Theorem 4.4, we call the operator L_S the *Laplace–Beltrami operator* associated with the exterior differential operator d_S.

(iv) For all $z \in \mathbf{C}$ with $|z| = 1$, $L_{zS} = L_S$. Hence $L_S = Q_{zS}^2$.

4.3. ANALYTICAL INDICES OF DIRAC OPERATORS AND PARTITION FUNCTIONS

By Theorem 4.4(i) and (ii), there exists a unique densely defined closed linear operator $Q_{S,+}$ from $\mathcal{F}_+(\mathcal{H}, \mathcal{K})$ to $\mathcal{F}_-(\mathcal{H}, \mathcal{K})$ such that

$$Q_S = \begin{pmatrix} 0 & Q_{S,+}^* \\ Q_{S,+} & 0 \end{pmatrix}, \tag{4.17}$$

where the matrix representation is relative to the orthogonal decomposition (4.6).

In general, for a densely defined closed linear operator T from a Hilbert space to a Hilbert space, its analytical index is defined by

$$\operatorname{ind}(T) := \dim\ker T - \dim\ker T^*$$

provided that at least one of $\dim\ker T$ and $\dim\ker T^*$ is finite.

The following theorem is concerned with Fredholm property of $Q_{S,+}$.

THEOREM 4.5 [7]. (i) *If S is Fredholm with* $\ker S = \{0\}$, *then* $Q_{S,+}$ *is Fredholm with*

$$\operatorname{ind}(Q_{S,+}) = \delta_{0,\dim\ker S^*}.$$

(ii) *If S is semi-Fredholm with* $\dim\ker S \geqslant 1$ *and* $\ker S^* = \{0\}$, *then* $Q_{S,+}$ *is semi-Fredholm with*

$$\operatorname{ind}(Q_{S,+}) = \dim\ker Q_{S,+} = +\infty.$$

Remark 4.2. Let V be a symmetric operator on $\mathcal{F}_{\mathrm{BF}}(\mathcal{H}, \mathcal{K})$ and $Q_S(V) := Q_S + V$. For a class of V such that $\{\Gamma_{\mathrm{F}}, Q_S(V)\} = 0$ on $D(Q_S(V))$, an index formula of $Q_S(V)_+ := Q_S(V)|\mathcal{F}_+(\mathcal{H}, \mathcal{K})$ is established in terms of a functional integral representation [3, 7].

As for the heat semi-group e^{-sL_S} $(s > 0)$, the following theorem holds.

THEOREM 4.6 ([7]). *Let $s > 0$. Suppose that* e^{-sS^*S} *and* e^{-sSS^*} *are trace class on* $\mathcal{H}$ *and* $\mathcal{K}$, *respectively, with* $\ker S = \{0\}$. *Then* e^{-sL_S} *is trace class on* $\mathcal{F}_{\mathrm{BF}}(\mathcal{H}, \mathcal{K})$ *and*

$$\operatorname{Tr}\left(\Gamma_{\mathrm{F}}\,\mathrm{e}^{-sL_S}\right) = \operatorname{ind}(Q_{S,+}) = \delta_{0,\dim\ker S^*}. \tag{4.18}$$

We have another orthogonal decomposition

$$\mathcal{F}_{\mathrm{BF}}(\mathcal{H}, \mathcal{K}) = \mathcal{G}_+(\mathcal{H}, \mathcal{K}) \oplus \mathcal{G}_-(\mathcal{H}, \mathcal{K}) \tag{4.19}$$

with

$$\mathcal{G}_\pm(\mathcal{H}, \mathcal{K}) := \mathcal{F}_{\mathrm{B},\pm}(\mathcal{H}) \otimes \mathcal{F}_{\mathrm{F}}(\mathcal{K}). \tag{4.20}$$

The grading operator of this gradation is given by

$$\Gamma_{\mathrm{B}} := (-1)^{N_{\mathrm{B}}\otimes I} = (-1)^{N_{\mathrm{B}}} \otimes I. \tag{4.21}$$

It is easy to see that

$$\Gamma_{\mathrm{B}} Q_S \subset -Q_S \Gamma_{\mathrm{B}}. \tag{4.22}$$

Hence, as in the case of the decomposition (4.6), there exists a unique densely defined closed linear operator $\widetilde{Q}_{S,+}$ from $\mathcal{G}_+(\mathcal{H}, \mathcal{K})$ to $\mathcal{G}_-(\mathcal{H}, \mathcal{K})$ such that

$$Q_S = \begin{pmatrix} 0 & \widetilde{Q}_{S,+}^* \\ \widetilde{Q}_{S,+} & 0 \end{pmatrix}, \tag{4.23}$$

where the matrix representation is relative to the orthogonal decomposition (4.19).

THEOREM 4.7. (i) *If S is Fredholm with* $\ker S = \{0\}$ *and* $\dim\ker S^* < \infty$, *then* $\widetilde{Q}_{S,+}$ *is Fredholm with*

$$\mathrm{ind}(\widetilde{Q}_{S,+}) = \dim\ker \widetilde{Q}_{S,+} = 2^{\dim\ker S^*}.$$

(ii) *If S is semi-Fredholm with* $\dim\ker S \geqslant 1$ *and* $\ker S^* = \{0\}$, *then* $\widetilde{Q}_{S,+}$ *is semi-Fredholm with*

$$\mathrm{ind}(\widetilde{Q}_{S,+}) = \dim\ker \widetilde{Q}_{S,+} = +\infty.$$

(iii) *If S is semi-Fredholm with* $\ker S = \{0\}$ *and* $\dim\ker S^* = +\infty$, *then* $\widetilde{Q}_{S,+}$ *is semi-Fredholm with* $\mathrm{ind}(\widetilde{Q}_{S,+}) = \dim\ker \widetilde{Q}_{S,+} = +\infty$.

Proof. Similar to the proof of Theorem 4.5. □

THEOREM 4.8. *Under the same assumption as in Theorem* 4.6, e^{-sL_S} *is trace class on* $\mathcal{F}_{\mathrm{BF}}(\mathcal{H}, \mathcal{K})$ *and*

$$\mathrm{Tr}\big(\Gamma_{\mathrm{B}}\, \mathrm{e}^{-sL_S}\big) = \mathrm{ind}(\widetilde{Q}_{S,+}) = 2^{\dim\ker S^*}. \tag{4.24}$$

Proof. Similar to the proof of Theorem 4.6. □

4.4. CONNECTION WITH SUPERSYMMETRY

Let $\mathcal{X}$ be a Hilbert space and H, $\{Q_j\}_{j=1}^N$, τ, be self-adjoint operators on $\mathcal{X}$. Then the quadruple $\{\mathcal{X}, H, \{Q_j\}_{j=1}^N, \tau\}$ is called a *supersymmetric quantum mechanics* (SSQM) with N-supersymmetry if the following (S.1)–(S.4) are satisfied:

(S.1) τ is bounded and $\tau^2 = I$.
(S.2) For all $j = 1, \ldots, N$, $H = Q_j^2$.
(S.3) For each $j = 1, \ldots, N$, the operator τ leaves $D(Q_j)$ invariant and $\{\tau, Q_j\} = 0$ on $D(Q_j)$.
(S.4) For all $j, k = 1, \ldots, N$ with $j \neq k$,

$$(Q_j\Psi, Q_k\Phi) + (Q_k\Psi, Q_j\Phi) = 0, \quad \Psi, \Phi \in D(Q_j) \cap D(Q_k).$$

The operators Q_j and H are called a *supercharge* and a *supersymmetric Hamiltonian*, respectively.

For mathematical discussions of SSQM, see, e.g., [2, 22, 32].

It is easy to see that, for all $z \in \mathbf{C}$ with $|z| = 1$, the quadruple $\{\mathcal{F}_{\mathrm{BF}}(\mathcal{H}, \mathcal{K}), L_S, \{Q_{zS}, Q_{izS}\}, \Gamma_{\mathrm{F}}\}$ is a SSQM. This SSQM produces various supersymmetric quantum field models in concrete realizations [7, 12]. For mathematical analysis of models in supersymmetric quantum field theory, see [4, 8, 9, 12, 22].

The quadruple $\{\mathcal{F}_{\mathrm{BF}}(\mathcal{H}, \mathcal{K}), L_S, \{Q_{zS}, Q_{izS}\}, \Gamma_{\mathrm{B}}\}$ ($|z| = 1$) also is a SSQM.

4.5. OTHER ASPECTS

Decomposition theorems of De Rham–Hodge–Kodaira type on the exterior differential operators $d_{S,n}$ are established in [7, 16]. Fundamental spaces associated with the Laplace–Beltrami operator L_S are introduced in [17] and their structures are investigated. The self-adjointness of a perturbed Dirac operator $Q_S(V) = Q_S + V$ with some symmetric operator V is discussed in [10]. The strong anticommutativity of two Dirac operators Q_S and Q_T $(S, T \in \mathrm{C}(\mathcal{H}, \mathcal{K}))$ and its applications to representations of a supersymmetry algebra are studied in [15] (cf. also [11]).

5. Arithmetical Aspects of Boson–Fermion Fock Spaces

5.1. SOME GENERAL ASPECTS

Let $\mathcal{H}$ and $\mathcal{K}$ be Hilbert spaces, and A and T be nonnegative self-adjoint operators on $\mathcal{H}$ and $\mathcal{K}$ respectively. Then the operator

$$H(A, T) := H_{\mathrm{B}}(A) \otimes I + I \otimes H_{\mathrm{F}}(T) \tag{5.1}$$

on $\mathcal{F}_{\mathrm{BF}}(\mathcal{H}, \mathcal{K})$ is nonnegative and self-adjoint.

We assume the following.

(AT) *The operators A and T satisfy* (A) *in Section* 2 *and* (T) *in Section* 3 *respectively.*

Under this assumption, $\mathrm{e}^{-sH(A,T)}$ is trace class and we can define a partition function

$$Z(s, z, w; A, T) := \mathrm{Tr}\big(\Gamma_{\mathrm{B}}(z) \otimes \Gamma_{\mathrm{F}}(w)\,\mathrm{e}^{-sH(A,T)}\big), \quad z, w \in D. \tag{5.2}$$

Let

$$N_{\mathrm{BF}} := N_{\mathrm{B}} \otimes I + I \otimes N_{\mathrm{F}}, \tag{5.3}$$

the number operator on $\mathcal{F}_{\mathrm{BF}}(\mathcal{H}, \mathcal{K})$, and set

$$\Gamma_{\mathrm{BF}} := (-1)^{N_{\mathrm{BF}}} = \Gamma_{\mathrm{B}}\Gamma_{\mathrm{F}} = \Gamma_{\mathrm{F}}\Gamma_{\mathrm{B}}. \tag{5.4}$$

As special cases of $Z(s, z, w; A, T)$, we define the following partition functions:

$$Z(s; A, T) := Z(s, 1, 1; A, T) = \mathrm{Tr}\,\mathrm{e}^{-sH(A,T)}, \tag{5.5}$$
$$\widetilde{Z}(s; A, T) := Z(s, -1, -1; A, T) = \mathrm{Tr}\big(\Gamma_{\mathrm{BF}}\,\mathrm{e}^{-sH(A,T)}\big), \tag{5.6}$$
$$\widetilde{Z}_{\mathrm{B}}(s; A, T) := Z(s, -1, 1; A, T) = \mathrm{Tr}\big(\Gamma_{\mathrm{B}}\,\mathrm{e}^{-sH(A,T)}\big), \tag{5.7}$$
$$\widetilde{Z}_{\mathrm{F}}(s; A, T) := Z(s, 1, -1; A, T) = \mathrm{Tr}\big(\Gamma_{\mathrm{F}}\,\mathrm{e}^{-sH(A,T)}\big). \tag{5.8}$$

We have

$$Z(s, z, w; A, T) = Z_{\mathrm{B}}(s, z; A)Z_{\mathrm{F}}(s, w; T), \quad z, w \in D. \tag{5.9}$$

In particular,

$$Z(s;A,T)=Z_{\mathrm{B}}(s;A)Z_{\mathrm{F}}(s;T), \tag{5.10}$$
$$\widetilde{Z}(s;A,T):=\widetilde{Z}_{\mathrm{B}}(s;A)\widetilde{Z}_{\mathrm{F}}(s;T), \tag{5.11}$$
$$\widetilde{Z}_{\mathrm{B}}(s;A,T):=\widetilde{Z}_{\mathrm{B}}(s;A)Z_{\mathrm{F}}(s;T), \tag{5.12}$$
$$\widetilde{Z}_{\mathrm{F}}(s;A,T):=Z_{\mathrm{B}}(s;A)\widetilde{Z}_{\mathrm{F}}(s;T). \tag{5.13}$$

Hence properties of the partition functions of $H(A,T)$ introduced above are reduced to those of $H_{\mathrm{B}}(A)$ and $H_{\mathrm{F}}(T)$. However, if one can represent the left-hand sides on (5.9)–(5.13) in various ways, (5.9)–(5.13) may produce nontrivial arithmetical relations for eigenvalues of A and T. Moreover, different expressions of $\mathrm{Tr}(X\,\mathrm{e}^{-sH(A,T)})$ with X an operator on $\mathcal{F}_{\mathrm{BF}}(\mathcal{H},\mathcal{K})$ may yield interesting arithmetical relations. Here we present only an outline of investigations along this line.

We carry over the notation in the preceding sections. Let $N\geqslant 2$ be of the form (2.39) and $m\in\mathcal{S}_0(N)$. Then we can write

$$m=(p_{i_1})^{q_1}(p_{i_2})^{q_2}\dots(p_{i_n})^{q_n}, \tag{5.14}$$

where $q_j=0$ or $q_j=1$. Based on these factorizations, we define a vector

$$\Omega_{N,m}:=C_{N,m}\big[(a_{i_1}^*)^{\alpha_1-q_1}\dots(a_{i_n}^*)^{\alpha_n-q_n}\Omega_{\mathcal{H}}\big]\otimes\big[(b_{i_1}^*)^{q_1}\dots(b_{i_n}^*)^{q_n}\Omega_{\mathcal{K}}\big], \tag{5.15}$$

where $C_{N,m}>0$ is a normalization constant. For $N=1$ and $m=1$, we set

$$\Omega_{1,1}:=\Omega.$$

LEMMA 5.1. *The set* $\{\Omega_{N,m}\mid N\geqslant 1,\ m\in\mathcal{S}_0(N)\}$ *is a CONS of* $\mathcal{F}_{\mathrm{BF}}(\mathcal{H},\mathcal{K})$.

Proof. The subset of vectors of the form

$$(a_{j_1}^*\dots a_{j_n}^*\Omega_{\mathcal{H}})\otimes(b_{k_1}^*\dots b_{k_\ell}^*\Omega_{\mathcal{K}}),$$

$j_1\leqslant j_2\leqslant\dots;\ k_1<k_2<\dots\ (j_m,k_i\in\mathbf{N})$, is a complete orthogonal system of $\mathcal{F}_{\mathrm{BF}}(\mathcal{H},\mathcal{K})$. As is easily seen, vectors of this form is a constant multiple of $\Omega_{N,m}$ for some $\langle N,m\rangle$. □

Remark 5.1. The CONS $\{\Omega_{N,m}\}$ was introduced in [30].

For each $N\in\mathbf{N}$, the subspace

$$\mathcal{F}_{\mathrm{BF}}^{(N)}:=\mathcal{L}\{\Omega_{N,m}\mid m\in\mathcal{S}_0(N)\} \tag{5.16}$$

is finite-dimensional with

$$\dim\mathcal{F}_{\mathrm{BF}}^{(N)}=2^{\nu(N)}. \tag{5.17}$$

By Lemma 5.1, we have

$$\mathcal{F}_{\mathrm{BF}}(\mathcal{H},\mathcal{K})=\bigoplus_{N=1}^{\infty}\mathcal{F}_{\mathrm{BF}}^{(N)}. \tag{5.18}$$

The following fact is easily proven.

LEMMA 5.2. *Let $N \in \mathbf{N}$, $m \in \mathcal{S}_0(N)$ and $z, w \in D$. Then $\Omega_{N,m}$ is an eigenvector of $\Gamma_{\mathrm{B}}(z) \otimes \Gamma_{\mathrm{F}}(w)$ with eigenvalue $z^{\gamma(N)-\gamma(m)} w^{\gamma(m)}$.*

For each $N \in \mathbf{N}$, we define a function $Y_{A,T}(N, \cdot)$ on $\mathcal{S}_0(N)$ by

$$Y_{A,T}(N, m) := \prod_{k=1}^{n} \mathrm{e}^{(\alpha_k - q_k) E_{i_k}(A) + q_k E_{i_k}(T)}, \quad m \in \mathcal{S}_0(N), \tag{5.19}$$

when N and m are represented as (2.39) and (5.14) respectively. Note that

$$Y_{A,T}(N, m) = F_A\left(\frac{N}{m}\right) F_T(m). \tag{5.20}$$

LEMMA 5.3. *Let $N \in \mathbf{N}$ and $m \in \mathcal{S}_0(N)$. Then $\Omega_{N,m}$ is an eigenvector of $H(A, T)$ with eigenvalue* $\log Y_{A,T}(N, m)$.

Proof. Let N and m be as in (2.39) and (5.14) respectively. Then

$$\begin{aligned} H(A, T)\Omega_{N,m} &= \left(\sum_{k=1}^{n} \{(\alpha_k - q_k) E_{i_k}(A) + q_k E_{i_k}(T)\} \right) \Omega_{N,m} \\ &= (\log Y_{A,T}(N, m))\Omega_{N,m}. \end{aligned}$$

Hence, the desired assertion follows. □

THEOREM 5.4. *Let $z, w \in D$. Then*

$$Z(s, z, w; A, T) = \sum_{N=1}^{\infty} \sum_{m|N} \frac{z^{\gamma(N)-\gamma(m)} w^{\gamma(m)} |\mu(m)|}{Y_{A,T}(N, m)^s}. \tag{5.21}$$

In particular,

$$Z(s; A, T) = \sum_{N=1}^{\infty} \sum_{m|N} \frac{|\mu(m)|}{Y_{A,T}(N, m)^s}, \tag{5.22}$$

$$\widetilde{Z}(s; A, T) := \sum_{N=1}^{\infty} \lambda(N) \sum_{m|N} \frac{|\mu(m)|}{Y_{A,T}(N, m)^s}, \tag{5.23}$$

$$\widetilde{Z}_{\mathrm{B}}(s; A, T) := \sum_{N=1}^{\infty} \lambda(N) \sum_{m|N} \frac{\mu(m)}{Y_{A,T}(N, m)^s}, \tag{5.24}$$

$$\widetilde{Z}_{\mathrm{F}}(s; A, T) := \sum_{N=1}^{\infty} \sum_{m|N} \frac{\mu(m)}{Y_{A,T}(N, m)^s}. \tag{5.25}$$

Proof. For a trace class operator X on $\mathcal{F}_{\mathrm{BF}}(\mathcal{H}, \mathcal{K})$, we have

$$\mathrm{Tr}\, X = \sum_{N=1}^{\infty} \sum_{m \in \mathcal{S}_0(N)} (\Omega_{N,m}, X\Omega_{N,m}). \tag{5.26}$$

We need only apply this formula, using Lemmas 5.2 and 5.3. □

By Theorem 5.4 and (5.10)–(5.13), we obtain the following formulas:

COROLLARY 5.5. *Let $z, w \in D$. Then*

$$\sum_{N=1}^{\infty}\sum_{m|N}\frac{z^{\gamma(N)-\gamma(m)}w^{\gamma(m)}|\mu(m)|}{Y_{A,T}(N,m)^s} = Z_{\mathrm{B}}(s,z;A)Z_{\mathrm{F}}(s,w;T). \tag{5.27}$$

In particular,

$$\sum_{N=1}^{\infty}\sum_{m|N}\frac{|\mu(m)|}{Y_{A,T}(N,m)^s} = Z_{\mathrm{B}}(s;A)Z_{\mathrm{F}}(s;T), \tag{5.28}$$

$$\sum_{N=1}^{\infty}\lambda(N)\sum_{m|N}\frac{|\mu(m)|}{Y_{A,T}(N,m)^s} = \widetilde{Z}_{\mathrm{B}}(s;A)\widetilde{Z}_{\mathrm{F}}(s;T), \tag{5.29}$$

$$\sum_{N=1}^{\infty}\lambda(N)\sum_{m|N}\frac{\mu(m)}{Y_{A,T}(N,m)^s} = \widetilde{Z}_{\mathrm{B}}(s;A)Z_{\mathrm{F}}(s;T), \tag{5.30}$$

$$\sum_{N=1}^{\infty}\sum_{m|N}\frac{\mu(m)}{Y_{A,T}(N,m)^s} = Z_{\mathrm{B}}(s;A)\widetilde{Z}_{\mathrm{F}}(s;T). \tag{5.31}$$

Remark 5.2. If we put into the right-hand sides of (5.27)–(5.31) the formulas established in Sections 2 and 3, then we obtain explicit formulas, which are nontrivial.

Remark 5.3. By rescaling as $T \to tT/s$ ($t > 0$) in (5.27)–(5.31), we can obtain relations at different temperatures $1/s$ and $1/t$. Hence, (5.27)–(5.31) include 'duality relations'. See Section 5.3 below.

PROPOSITION 5.6. *Let $N \geqslant 2$. Then*

$$\mathrm{Tr}_{\mathcal{F}_{\mathrm{BF}}^{(N)}}\Gamma_{\mathrm{F}} = 0, \tag{5.32}$$

where $\mathrm{Tr}_{\mathcal{F}_{\mathrm{BF}}^{(N)}}$ means trace restricted to the subspace $\mathcal{F}_{\mathrm{BF}}^{(N)}$.

Proof. Let $N \geqslant 2$ and m be as in (2.39) and (5.14), respectively. Then the cardinal number of the set $\mathcal{S}_0(N)$ is 2^n, including 1. In the expression (5.14) of m, if there is a j such that $q_j = 1$, then $m' = m/p_{i_j} \in \mathcal{S}_0(N)$. If $q_1 = q_2 = \cdots = q_n = 0$ (i.e., $m = 1$), then, $m' := p_{i_1}$ is in $\mathcal{S}_0(N)$. In this case, if $\Gamma_{\mathrm{F}}\Omega_{N,m} = \pm\Omega_{N,m}$, then $\Gamma_{\mathrm{F}}\Omega_{N,m'} = \mp\Omega_{N,m'}$. Hence, $\mathcal{S}_0(N)$ consists of 2^{n-1} pairs (m,m') that satisfy this relation. Using this fact, we obtain (5.32). □

COROLLARY 5.7. *Let $N \geqslant 2$. Then*

$$\sum_{m|N}\mu(m) = 0. \tag{5.33}$$

Proof. We have by (5.32)

$$0 = \mathrm{Tr}_{\mathcal{F}_{\mathrm{BF}}^{(N)}} \Gamma_{\mathrm{F}} = \sum_{m \in \mathcal{S}_0(N)} (\Omega_{N,m}, \Gamma_{\mathrm{F}} \Omega_{N,m}) = \sum_{m \in \mathcal{S}_0(N)} \mu(m).$$

Since $\mu(m) = 0$ if $m \notin \mathcal{S}_0$, we obtain (5.33). □

Remark 5.4. Equation (5.33) is a well-known formula of the Möbius function [1, p. 25].

5.2. PARTIAL SUPERSYMMETRY

A notion of *partial supersymmetry* was introduced in [31] in a physical way. In the context of our general formulation of supersymmetric quantum field theory presented in Section 4, a partial supersymmetry means to consider a 'distorted' supersymmetric Hamiltonian

$$\begin{aligned} H_{\mathrm{PS}}(a, b; S) &:= L_S + aH_{\mathrm{B}}(S^*S) \otimes I + bI \otimes H_{\mathrm{F}}(SS^*) \\ &= L_S + H(aS^*S, bSS^*) \end{aligned} \tag{5.34}$$

on the Boson–Fermion Fock space $\mathcal{F}_{\mathrm{BF}}(\mathcal{H}, \mathcal{K})$ with parameters $a, b \geqslant 0$, where $S \in \mathsf{C}(\mathcal{H}, \mathcal{K})$. Note that

$$H_{\mathrm{PS}}(a, b; S) = H((a+1)S^*S, (b+1)SS^*). \tag{5.35}$$

Hence, in the caes $a = b$, supersymmetry recovers with $H_{\mathrm{PS}}(a, a; S) = (a+1)L_S$.

Let $\beta > 0$ and suppose that $\mathrm{e}^{-(a+1)\beta S^*S}$ and $\mathrm{e}^{-(b+1)\beta SS^*}$ are trace class. Then $\mathrm{e}^{-\beta H_{\mathrm{PS}}(a,b;S)}$, $\mathrm{e}^{-(a+1)\beta H_{\mathrm{B}}(S^*S)}$ and $\mathrm{e}^{-(b+1)\beta H_{\mathrm{F}}(SS^*)}$ are trace class and the following formula holds:

$$\begin{aligned} &\mathrm{Tr}\big(\Gamma_{\mathrm{B}}(z) \otimes \Gamma_{\mathrm{F}}(w)\, \mathrm{e}^{-\beta H_{\mathrm{PS}}(a,b;S)}\big) \\ &\quad = Z_{\mathrm{B}}((a+1)\beta, z; S^*S) Z_{\mathrm{F}}((b+1)\beta, w; SS^*), \quad z, w \in D. \end{aligned} \tag{5.36}$$

In particular,

$$\mathrm{Tr}\, \mathrm{e}^{-\beta H_{\mathrm{PS}}(a,b;S)} = Z_{\mathrm{B}}((a+1)\beta; S^*S) Z_{\mathrm{F}}((b+1)\beta; SS^*), \tag{5.37}$$

$$\mathrm{Tr}\big(\Gamma_{\mathrm{BF}}\, \mathrm{e}^{-\beta H_{\mathrm{PS}}(a,b;S)}\big) = \widetilde{Z}_{\mathrm{B}}((a+1)\beta; S^*S) \widetilde{Z}_{\mathrm{F}}((b+1)\beta; SS^*), \tag{5.38}$$

$$\mathrm{Tr}\big(\Gamma_{\mathrm{F}}\, \mathrm{e}^{-\beta H_{\mathrm{PS}}(a,b;S)}\big) = Z_{\mathrm{B}}((a+1)\beta; S^*S) \widetilde{Z}_{\mathrm{F}}((b+1)\beta; SS^*), \tag{5.39}$$

$$\mathrm{Tr}\big(\Gamma_{\mathrm{B}}\, \mathrm{e}^{-\beta H_{\mathrm{PS}}(a,b;S)}\big) = \widetilde{Z}_{\mathrm{B}}((a+1)\beta; S^*S) Z_{\mathrm{F}}((b+1)\beta; SS^*). \tag{5.40}$$

Note that the left-hand sides of these equations may be regarded as correlation functions in the SSQM described by the supersymmetric Hamiltonian L_S:

$$\mathrm{Tr}\big(\Gamma_{\mathrm{B}}(z) \otimes \Gamma_{\mathrm{F}}(w)\, \mathrm{e}^{-\beta H_{\mathrm{PS}}(a,b;S)}\big) = \big\langle \Gamma_{\mathrm{B}}(z) \otimes \Gamma_{\mathrm{F}}(w)\, \mathrm{e}^{-\beta H(aS^*S, bSS^*)} \big\rangle_\beta, \tag{5.41}$$

where

$$\langle X \rangle_\beta := \mathrm{Tr}\big(X\, \mathrm{e}^{-\beta L_S}\big). \tag{5.42}$$

In particular,

$$\mathrm{Tr}\, \mathrm{e}^{-\beta H_{\mathrm{PS}}(a,b;S)} = \big\langle \mathrm{e}^{-\beta H(aS^*S,bSS^*)} \big\rangle_\beta, \tag{5.43}$$

$$\mathrm{Tr}\big(\Gamma_{\mathrm{BF}}\, \mathrm{e}^{-\beta H_{\mathrm{PS}}(a,b;S)}\big) = \big\langle \Gamma_{\mathrm{BF}}\, \mathrm{e}^{-\beta H(aS^*S,bSS^*)} \big\rangle_\beta, \tag{5.44}$$

$$\mathrm{Tr}\big(\Gamma_{\mathrm{F}}\, \mathrm{e}^{-\beta H_{\mathrm{PS}}(a,b;S)}\big) = \big\langle \Gamma_{\mathrm{F}}\, \mathrm{e}^{-\beta H(aS^*S,bSS^*)} \big\rangle_\beta, \tag{5.45}$$

$$\mathrm{Tr}\big(\Gamma_{\mathrm{B}}\, \mathrm{e}^{-\beta H_{\mathrm{PS}}(a,b;S)}\big) = \big\langle \Gamma_{\mathrm{B}}\, \mathrm{e}^{-\beta H(aS^*S,bSS^*)} \big\rangle_\beta. \tag{5.46}$$

By computing these correlation functions in various ways, we would obtain from (5.37)–(5.40) nontrivial duality relations. But here we omit the details.

5.3. APPLICATIONS TO ANALYTIC NUMBER THEORY

We consider the case where $\mathcal{H} = \mathcal{K} = \ell^2$ and $A = T = \omega_{\mathcal{P}}$. Then we have

$$Y_{\omega_{\mathcal{P}},\omega_{\mathcal{P}}}(N, m) = N. \tag{5.47}$$

Hence Corollary 5.5 gives

$$\sum_{N=1}^{\infty} \sum_{m|N} \frac{z^{\gamma(N)-\gamma(m)} w^{\gamma(m)} |\mu(m)|}{N^s} = Z_{\mathrm{B}}(s, z; \omega_{\mathcal{P}}) Z_{\mathrm{F}}(s, w; \omega_{\mathcal{P}}), \quad s > 1. \tag{5.48}$$

In particular, for all $s > 1$,

$$\sum_{N=1}^{\infty} \frac{2^{\nu(N)}}{N^s} = \frac{\zeta(s)}{D(s,\lambda)}, \tag{5.49}$$

$$\sum_{N=1}^{\infty} \frac{\lambda(N) 2^{\nu(N)}}{N^s} = \frac{D(s,\lambda)}{\zeta(s)}. \tag{5.50}$$

Remark 5.5. In the present case, (5.30) and (5.31) imply Corollary 3.3, since Proposition 5.6 holds.

Let f be the completely multiplicative function considered in Section 2.3. Then, taking $S = \sqrt{A_f}$ we have

$$H := H(A_f, A_f) = L_{\sqrt{A_f}}.$$

Hence, by Theorems 4.6 and 4.8, for all $s > 1$,

$$\mathrm{Tr}\big(\Gamma_{\mathrm{F}}\, \mathrm{e}^{-sH}\big) = 1, \tag{5.51}$$

$$\mathrm{Tr}\big(\Gamma_{\mathrm{B}}\, \mathrm{e}^{-sH}\big) = 1. \tag{5.52}$$

These are supersymmetric identities. It is shown [30] that (5.51) implies that

$$\sum_{m=1}^{\infty} \mu(m) f(m) = \frac{1}{\sum_{n=1}^{\infty} f(n)}. \tag{5.53}$$

In the same manner as in the derivation of (5.53) [30], we can show that (5.52) implies that

$$\sum_{m=1}^{\infty} |\mu(m)| f(m) = \frac{1}{\sum_{n=1}^{\infty} \lambda(n) f(n)}. \tag{5.54}$$

We have for all $s, t > 1$

$$Y_{s\omega_{\mathcal{P}}, t\omega_{\mathcal{P}}}(N, m) = N^s m^{t-s}. \tag{5.55}$$

Hence, by Corollary 5.5 with rescaling $T \to tT/s$, we obtain

$$\begin{aligned} &\sum_{N=1}^{\infty} \sum_{m|N} \frac{z^{\gamma(N)-\gamma(m)} w^{\gamma(m)} |\mu(m)|}{N^s m^{t-s}} \\ &\quad = Z_{\mathrm{B}}(s, z; \omega_{\mathcal{P}}) Z_{\mathrm{F}}(t, w; \omega_{\mathcal{P}}), \quad t > s > 1. \end{aligned} \tag{5.56}$$

In particular,

$$\sum_{N=1}^{\infty} \frac{1}{N^s} \sum_{m|N} \frac{|\mu(m)|}{m^{t-s}} = \frac{\zeta(s)}{D(t, \lambda)}, \tag{5.57}$$

$$\sum_{N=1}^{\infty} \frac{\lambda(N)}{N^s} \sum_{m|N} \frac{|\mu(m)|}{m^{t-s}} = \frac{D(s, \lambda)}{\zeta(t)}, \tag{5.58}$$

$$\sum_{N=1}^{\infty} \frac{\lambda(N)}{N^s} \sum_{m|N} \frac{\mu(m)}{m^{t-s}} = \frac{D(s, \lambda)}{D(t, \lambda)}, \tag{5.59}$$

$$\sum_{N=1}^{\infty} \frac{1}{N^s} \sum_{m|N} \frac{\mu(m)}{m^{t-s}} = \frac{\zeta(s)}{\zeta(t)}, \quad t \geqslant s > 1. \tag{5.60}$$

These may be regarded as 'duality relations' for the Riemann zeta function $\zeta(s)$ and the Dirichlet series $D(s, \lambda)$.

Using (3.79) to rewrite the left-hand sides of (5.59) and (5.60), we obtain

$$\sum_{N=1}^{\infty} \frac{\lambda(N)}{N^t} J_{t-s}(N) = \frac{D(s, \lambda)}{D(t, \lambda)}, \tag{5.61}$$

$$\sum_{N=1}^{\infty} \frac{J_{t-s}(N)}{N^t} = \frac{\zeta(s)}{\zeta(t)}, \quad t \geqslant s > 1. \tag{5.62}$$

In particular,

$$\sum_{N=1}^{\infty} \frac{\lambda(N)\varphi(N)}{N^s} = \frac{D(s-1,\lambda)}{D(s,\lambda)}, \tag{5.63}$$

$$\sum_{N=1}^{\infty} \frac{\varphi(N)}{N^s} = \frac{\zeta(s-1)}{\zeta(s)}, \quad s > 2. \tag{5.64}$$

References

1. Apostol, T. M.: *Introduction to Analytic Number Theory*, Springer-Verlag, New York, 1976.
2. Arai, A.: Supersymmetry and singular perturbations, *J. Funct. Anal.* **60** (1985), 378–393.
3. Arai, A.: Path integral representation of the index of Kähler–Dirac operators on an infinite dimensional manifold, *J. Funct. Anal.* **82** (1989), 330–369.
4. Arai, A.: Supersymmetric embedding of a model of a quantum harmonic oscillator interacting with infinitely many bosons, *J. Math. Phys.* **30** (1989), 512–520.
5. Arai, A.: Perturbation of embedded eigenvalues: A general class of exactly soluble models in Fock spaces, *Hokkaido Math. J.* **19** (1990), 1–34.
6. Arai, A.: A general class of infinite dimensional Dirac operators and related aspects, In: S. Koshi (ed.), *Functional Analysis and Related Topics*, World Scientific, Singapore, 1991.
7. Arai, A.: A general class of infinite dimensional Dirac operators and path integral representation of their index, *J. Funct. Anal.* **105** (1992), 342–408.
8. Arai, A.: Dirac operators in Boson–Fermion Fock spaces and supersymmetric quantum field theory, *J. Geom. Phys.* **11** (1993), 465–490.
9. Arai, A.: Supersymmetric extension of quantum scalar field theories, In: H. Araki *et al.* (eds), *Quantum and Non-Commutative Analysis*, Kluwer Acad. Publ., Dordrecht, 1993.
10. Arai, A.: On self-adjointness of Dirac operators in Boson–Fermion Fock spaces, *Hokkaido Math. J.* **23** (1994), 319–353.
11. Arai, A.: Operator-theoretical analysis of a representation of a supersymmetry algebra in Hilbert space, *J. Math. Phys.* **36** (1995), 613–621.
12. Arai, A.: Supersymmetric quantum field theory and infinite dimensional analysis, *Sugaku Expos.* **9** (1996), 87–98.
13. Arai, A.: Trace formulas, a Golden–Thompson inequality and classical limit in Boson Fock space, *J. Funct. Anal.* **136** (1996), 510–547.
14. Arai, A.: *Introduction to Mathematical Methods of Quantum Field Theory* (in Japanese), Lecture Note Ser. in Math. 5, Osaka University, Osaka Mathematical Publications, Osaka, 1997.
15. Arai, A.: Strong anticommutativity of Dirac operators on Boson–Fermion Fock spaces and representations of a supersymmetry algebra, *Math. Nachr.* **207** (1999), 61–77.
16. Arai, A. and Mitoma, I.: De Rham–Hodge–Kodaira decomposition in ∞-dimensions, *Math. Ann.* **291** (1991), 51–73.
17. Arai, A. and Mitoma, I.: Comparison and nuclearity of spaces of differential forms on topological vector spaces, *J. Funct. Anal.* **111** (1993), 278–294.
18. Bakas, I. and Bowick, M. J.: Curiosities of arithmetic gases, *J. Math. Phys.* **32** (1991), 1881–1884.
19. Bost, J.-B. and Connes, A.: Hecke algebras, type III factors and phase transitions with spontaneous symmetry breaking in number theory, *Selecta Math. (N.S.)* **1** (1995), 411–457.
20. Bratteli, O. and Robinson, D. W.: *Operator Algebras and Quantum Statistical Mechanics 2*, 2nd edn, Springer, Berlin, 1997.

21. Contucci, P. and Knauf, A.: The low activity phase of some Dirichlet series, *J. Math. Phys.* **37** (1996), 5458–5475.
22. Jaffe, A. and Lesniewski, A.: Supersymmetric quantum fields and infinite dimensional analysis, In: G.'t Hooft *et al.* (eds), *Nonperturbative Quantum Field Theory*, Plenum, New York, 1988.
23. Julia, B.: On the 'statistics' of primes, *J. Phys. France* **50** (1989), 1371–1375.
24. Julia, B.: Statistical theory of numbers, In: J.-M. Luck, P. Moussa and M. Waldschmidt (eds), *Number Theory and Physics*, Springer, Berlin, 1990.
25. Luck, J.-M., Moussa, P. and Waldschmidt, M. (eds), *Number Theory and Physics*, Springer, Berlin, 1990.
26. Reed, M. and Simon, B.: *Methods of Modern Mathematical Physics Vol. I: Functional Analysis*, Academic Press, New York, 1972.
27. Reed, M. and Simon, B.: *Methods of Modern Mathematical Physics Vol. II: Fourier Analysis, Self-adjointness*, Academic Press, New York, 1975.
28. Reed, M. and Simon, B.: *Methods of Modern Mathematical Physics Vol. IV: Analysis of Operators*, Academic Press, New York, 1978.
29. Spector, D.: Multiplicative functions, Dirichlet convolution, and quantum systems, *Phys. Lett. A* **140** (1989), 311–316.
30. Spector, D.: Supersymmetry and the Möbius inversion function, *Comm. Math. Phys.* **127** (1990), 239–252.
31. Spector, D.: Duality, partial supersymmetry, and arithmetic number theory, *J. Math. Phys.* **39** (1998), 1919–1927.
32. Thaller, B.: *The Dirac Equation*, Springer, Berlin, 1992.
33. Waldschmidt, M., Moussa, P., Luck, J.-M. and Itzykson, C. (eds): *From Number Theory to Physics*, Springer, Berlin, 1992.

Acta Applicandae Mathematicae **63:** 79–87, 2000.

Bell Numbers, Log-Concavity, and Log-Convexity ⋆

NOBUHIRO ASAI[1], IZUMI KUBO[2] and HUI-HSIUNG KUO[3]
[1]*Graduate School of Mathematics, Nagoya University, Nagoya 464-8602, Japan*⋆⋆
[2]*Department of Mathematics, Graduate School of Science, Hiroshima University, Higashi-Hiroshima 739-8526, Japan*
[3]*Department of Mathematics, Louisiana State University, Baton Rouge, LA 70803, U.S.A.*

(Received: 8 January 1999)

Abstract. Let $\{b_k(n)\}_{n=0}^{\infty}$ be the Bell numbers of order k. It is proved that the sequence $\{b_k(n)/n!\}_{n=0}^{\infty}$ is log-concave and the sequence $\{b_k(n)\}_{n=0}^{\infty}$ is log-convex, or equivalently, the following inequalities hold for all $n \geqslant 0$,

$$1 \leqslant \frac{b_k(n+2)b_k(n)}{b_k(n+1)^2} \leqslant \frac{n+2}{n+1}.$$

Let $\{\alpha(n)\}_{n=0}^{\infty}$ be a sequence of positive numbers with $\alpha(0) = 1$. We show that if $\{\alpha(n)\}_{n=0}^{\infty}$ is log-convex, then

$$\alpha(n)\alpha(m) \leqslant \alpha(n+m), \quad \forall n, m \geqslant 0.$$

On the other hand, if $\{\alpha(n)/n!\}_{n=0}^{\infty}$ is log-concave, then

$$\alpha(n+m) \leqslant \binom{n+m}{n}\alpha(n)\alpha(m), \quad \forall n, m \geqslant 0.$$

In particular, we have the following inequalities for the Bell numbers

$$b_k(n)b_k(m) \leqslant b_k(n+m) \leqslant \binom{n+m}{n} b_k(n)b_k(m), \quad \forall n, m \geqslant 0.$$

Then we apply these results to characterization theorems for CKS-space in white noise distribution theory.

Mathematics Subject Classifications (2000): 60H40, 11B73, 05A10, 05A15.

Key words: Bell numbers, log-concavity, log-convexity, CKS-space, characterization theorem, white noise distribution theory.

⋆ Research supported by the Daiko Foundation 1998 and the Kamiyama Foundation 1999 (N.A.), U.S. Army Research Office grant #DAAH04-94-G-0249, Academic Frontier in Science of Meijo University, and the National Science Council of Taiwan (H.-H.K.)
⋆⋆ Current address: International Institute for Advanced Studies, 9-3, Kizugawadai, Kizu, Kyoto 619-0225, Japan

1. The Main Theorems

For an integer $k \geqslant 2$, let $\exp_k(x)$ denote the k-times iterated exponential function

$$\exp_k(x) = \underbrace{\exp(\exp \cdots (\exp(x)))}_{k\text{-times}}.$$

Let $\{B_k(n)\}_{n=0}^{\infty}$ be the sequence of numbers given in the power series of $\exp_k(x)$

$$\exp_k(x) = \sum_{n=0}^{\infty} \frac{B_k(n)}{n!} x^n. \tag{1}$$

The *Bell numbers* $\{b_k(n)\}_{n=0}^{\infty}$ *of order* k are defined by

$$b_k(n) = \frac{B_k(n)}{\exp_k(0)}, \qquad n \geqslant 0.$$

The numbers $b_2(n)$, $n \geqslant 0$, with $k = 2$ are usually known as the *Bell numbers*. The first few terms of these numbers are 1, 1, 2, 5, 15, 52, 203. Note that $\exp_2(0) = e$ and so we have

$$e^{e^x - 1} = \sum_{n=0}^{\infty} \frac{b_2(n)}{n!} x^n. \tag{2}$$

A sequence $\{\delta(n)\}_{n=0}^{\infty}$ of nonnegative real numbers is called *log-concave* if

$$\delta(n)\delta(n+2) \leqslant \delta(n+1)^2, \quad \forall n \geqslant 0.$$

It is called *log-convex* if

$$\delta(n)\delta(n+2) \geqslant \delta(n+1)^2, \quad \forall n \geqslant 0.$$

The main purpose of this paper is to prove the following theorems.

THEOREM 1. *Let* $\{b_k(n)\}_{n=0}^{\infty}$ *be the Bell numbers of order k. Then the sequence* $\{b_k(n)/n!\}_{n=0}^{\infty}$ *is log-concave and the sequence* $\{b_k(n)\}_{n=0}^{\infty}$ *is log-convex.*

Note that the conclusion of this theorem is equivalent to the inequalities

$$1 \leqslant \frac{b_k(n)b_k(n+2)}{b_k(n+1)^2} \leqslant \frac{n+2}{n+1}, \qquad \forall n \geqslant 0.$$

A different proof of the log-convexity of $\{b_2(n)\}_{n=0}^{\infty}$ has been given earlier by Engel [5]. In [3] Canfield showed that the log-concavity of $\{b_2(n)/n!\}_{n=0}^{\infty}$ holds asymptotically. In a recent paper [4], Cochran *et al.* used the log-concavity of certain sequences to study characterization theorems. However, they did not show whether the sequence $\{b_k(n)/n!\}_{n=0}^{\infty}$ is log-concave. Thus our Theorem 1 fills up this gap (for details, see Section 3).

THEOREM 2. *Let* $\{\alpha(n)\}_{n=0}^{\infty}$ *be a sequence of positive numbers with* $\alpha(0) = 1$.

(a) *If* $\{\alpha(n)\}_{n=0}^{\infty}$ *is log-convex, then*

$$\alpha(n)\alpha(m) \leqslant \alpha(n+m), \quad \forall n, m \geqslant 0. \tag{3}$$

(b) *If* $\{\alpha(n)/n!\}_{n=0}^{\infty}$ *is log-concave, then*

$$\alpha(n+m) \leqslant \binom{n+m}{n}\alpha(n)\alpha(m), \quad \forall n, m \geqslant 0. \tag{4}$$

We will prove Theorems 1 and 2 in Section 2. The next theorem is an immediate consequence of these two theorems.

THEOREM 3. *The Bell numbers* $\{b_k(n)\}_{n=0}^{\infty}$ *of order* k *satisfy the inequalities*

$$b_k(n)b_k(m) \leqslant b_k(n+m) \leqslant \binom{n+m}{n}b_k(n)b_k(m), \quad \forall n, m \geqslant 0. \tag{5}$$

In a recent paper [7] it is shown that for any $k \geqslant 2$ there exist constants c_2 and c_3, depending on k, such that for all $n, m \geqslant 0$,

$$b_k(n+m) \leqslant c_2^{n+m} b_k(n)b_k(m), \qquad b_k(n)b_k(m) \leqslant c_3^{n+m} b_k(n+m).$$

Observe that from Equation (5) we get

$$b_k(n)b_k(m) \leqslant b_k(n+m) \leqslant 2^{n+m} b_k(n)b_k(m).$$

Thus in fact we can take $c_2 = 2$ and $c_3 = 1$ for the Bell numbers of any order k.

2. Proofs of Theorems 1 and 2

For the proof of Theorem 1 we prepare two lemmas and state the Bender–Canfield theorem [2].

LEMMA 1. *If* $\{\beta(n)/n!\}_{n=0}^{\infty}$ *is a log-concave sequence and* r *is a nonnegative real number such that* $\beta(2) \leqslant r\beta(1)^2$, *then the sequence* 1, $r\beta(n)/(n-1)!$, $n \geqslant 1$, *is log-concave.*

Proof. By assumption we have

$$\frac{\beta(n)}{n!}\frac{\beta(n+2)}{(n+2)!} \leqslant \left(\frac{\beta(n+1)}{(n+1)!}\right)^2.$$

When $n \geqslant 1$ this inequality is equivalent to

$$\frac{(n+1)^2}{n(n+2)}\left(\frac{\beta(n)}{(n-1)!}\frac{\beta(n+2)}{(n+1)!}\right) \leqslant \left(\frac{\beta(n+1)}{n!}\right)^2.$$

Note that $(n+1)^2 \geqslant n(n+2)$. Hence for $n \geqslant 1$,

$$\frac{\beta(n)}{(n-1)!}\frac{\beta(n+2)}{(n+1)!} \leqslant \left(\frac{\beta(n+1)}{n!}\right)^2.$$

Thus for any constant r we have

$$\frac{r\beta(n)}{(n-1)!}\frac{r\beta(n+2)}{(n+1)!} \leqslant \left(\frac{r\beta(n+1)}{n!}\right)^2, \quad \forall n \geqslant 1.$$

Moreover, the assumption $\beta(2) \leqslant r\beta(1)^2$ implies that $1 \cdot (r\beta(2)) \leqslant (r\beta(1))^2$. Thus the sequence $1,\ r\beta(n)/(n-1)!,\ n \geqslant 1$, is log-concave. □

BENDER–CANFIELD THEOREM ([2]). *Let $1, Z_1, Z_2, \ldots$ be a log-concave sequence of nonnegative real numbers and define the sequence $\{a(n)\}_{n=0}^{\infty}$ by*

$$\sum_{n=0}^{\infty} \frac{a(n)}{n!} x^n = \exp\left(\sum_{j=1}^{\infty} \frac{Z_j}{j} x^j\right).$$

Then the sequence $\{a(n)/n!\}_{n=0}^{\infty}$ is log-concave and the sequence $\{a(n)\}_{n=0}^{\infty}$ is log-convex.

LEMMA 2. *The sequence $\{b_2(n)/n!\}_{n=0}^{\infty}$ is log-concave and the sequence $\{b_2(n)\}_{n=0}^{\infty}$ is log-convex.*

Proof. Note that $e^{e^x-1} = \exp\left(\sum_{j=1}^{\infty} \frac{1}{j!} x^j\right)$. Hence, by Equation (2) we have

$$\exp\left(\sum_{j=1}^{\infty} \frac{1}{j!} x^j\right) = \sum_{n=0}^{\infty} \frac{b_2(n)}{n!} x^n.$$

Let

$$Z_j = \frac{1}{(j-1)!} \quad \text{for } j \geqslant 1.$$

It is easy to check that the sequence $1, Z_1, Z_2, \ldots$ is log-concave. Thus this lemma follows from the above Bender–Canfield theorem. □

Proof of Theorem 1. We prove the theorem by mathematical induction. By Lemma 2 the theorem is true for $k = 2$. Assume the theorem is true for k. Note that $\exp_{k+1}(x)/\exp_{k+1}(0) = \exp(\exp_k(x) - \exp_k(0))$. Hence

$$\exp\left(\exp_k(x) - \exp_k(0)\right) = \sum_{n=0}^{\infty} \frac{b_{k+1}(n)}{n!} x^n.$$

But $\exp_k(x) - \exp_k(0) = \sum_{j=1}^{\infty} \frac{B_k(j)}{j!} x^j$. Thus we get

$$\exp\left(\sum_{j=1}^{\infty} \frac{B_k(j)}{j!} x^j\right) = \sum_{n=0}^{\infty} \frac{b_{k+1}(n)}{n!} x^n.$$

Let

$$Z_j = \frac{B_k(j)}{(j-1)!}, \quad j \geqslant 1.$$

Then the above equation becomes

$$\exp\left(\sum_{j=1}^{\infty} \frac{Z_j}{j} x^j\right) = \sum_{n=0}^{\infty} \frac{b_{k+1}(n)}{n!} x^n. \tag{6}$$

By the induction assumption, the sequence $\{b_k(n)/n!\}_{n=0}^{\infty}$ is log-concave. This implies in particular that $b_k(0)b_k(2)/2 \leqslant b_k(1)^2$. But $b_k(0) = 1$ and $\exp_k(0) > 2$. Hence

$$b_k(2) \leqslant 2b_k(1)^2 < \exp_k(0)\, b_k(1)^2.$$

Thus we can apply Lemma 1 with $\beta(n) = b_k(n)$ and $r = \exp_k(0)$ to conclude that the sequence

$$1, \exp_k(0)\frac{b_k(n)}{(n-1)!}, \quad n \geqslant 1,$$

is log-concave. Note that for $n \geqslant 1$,

$$\exp_k(0)\frac{b_k(n)}{(n-1)!} = \frac{B_k(n)}{(n-1)!} = Z_n.$$

Hence, the sequence $1, Z_1, Z_2, \ldots$ is log-concave. Upon applying the Bender–Canfield theorem, we see from Equation (6) that the sequence $\{b_{k+1}(n)/n!\}_{n=0}^{\infty}$ is log-concave and the sequence $\{b_{k+1}(n)\}_{n=0}^{\infty}$ is log-convex. □

Proof of Theorem 2. To prove (a), let $\{\alpha(n)\}_{n=0}^{\infty}$ be log-convex. Then $\alpha(n)\alpha(n+2) \geqslant \alpha(n+1)^2$. Hence $\alpha(n+1)/\alpha(n) \leqslant \alpha(n+2)/\alpha(n+1)$ and this implies that for any $n \geqslant 0$ and $m \geqslant 1$,

$$\frac{\alpha(1)}{\alpha(0)} \leqslant \frac{\alpha(2)}{\alpha(1)} \leqslant \cdots \leqslant \frac{\alpha(n+m)}{\alpha(n+m-1)}.$$

Therefore, for any $n \geqslant 0$ and $m \geqslant 1$,

$$\frac{\alpha(1)}{\alpha(0)}\frac{\alpha(2)}{\alpha(1)}\cdots\frac{\alpha(m)}{\alpha(m-1)} \leqslant \frac{\alpha(n+1)}{\alpha(n)}\frac{\alpha(n+2)}{\alpha(n+1)}\cdots\frac{\alpha(n+m)}{\alpha(n+m-1)}.$$

After the cancellation we get $\alpha(n)\alpha(m) \leqslant \alpha(0)\alpha(n+m)$. But $\alpha(0) = 1$ and so Equation (3) is true when $n \geqslant 0$ and $m \geqslant 1$. When $m = 0$, Equation (3) obviously holds for any $n \geqslant 0$. Hence we have proved assertion (a).

For the proof of (b), first note that $\{\alpha(n)/n!\}_{n=0}^{\infty}$ is log-concave if and only if for all $n \geqslant 0$,

$$\frac{\alpha(n+1)}{\alpha(n)} \geqslant \frac{n+1}{n+2}\frac{\alpha(n+2)}{\alpha(n+1)}.$$

By using this inequality repeatedly, we get the following inequalities for any $n \geqslant 0$ and $m \geqslant 1$,

$$\frac{\alpha(1)}{\alpha(0)} \geqslant \frac{1}{2}\frac{\alpha(2)}{\alpha(1)} \geqslant \frac{1}{3}\frac{\alpha(3)}{\alpha(2)} \geqslant \cdots \geqslant \frac{1}{n+m}\frac{\alpha(n+m)}{\alpha(n+m-1)}.$$

Hence for any $0 \leqslant j \leqslant m-1$,

$$\frac{\alpha(j+1)}{\alpha(j)} \geqslant \frac{j+1}{n+m}\frac{\alpha(n+m)}{\alpha(n+m-1)}.$$

Therefore,

$$\frac{\alpha(1)}{\alpha(0)}\frac{\alpha(2)}{\alpha(1)}\cdots\frac{\alpha(m)}{\alpha(m-1)} \geqslant \left(\frac{1}{n+1}\frac{\alpha(n+1)}{\alpha(n)}\right)\left(\frac{2}{n+2}\frac{\alpha(n+2)}{\alpha(n+1)}\right)\cdots\left(\frac{m}{n+m}\frac{\alpha(n+m)}{\alpha(n+m-1)}\right).$$

After the cancellation we get

$$\frac{\alpha(m)}{\alpha(0)} \geqslant \frac{n!m!}{(n+m)!}\frac{\alpha(n+m)}{\alpha(n)}.$$

But $\alpha(0) = 1$. Hence we have proved that for any $n \geqslant 0, m \geqslant 1$,

$$\alpha(n+m) \leqslant \binom{n+m}{n}\alpha(n)\alpha(m).$$

Note that when $m = 0$, Equation (4) obviously holds for any $n \geqslant 0$. Thus assertion (b) is proved. □

3. Application to White Noise Distribution Theory

3.1. CHARACTERIZATION OF TEST AND GENERALIZED FUNCTIONS

The Bell numbers $\{b_k(n)\}_{n=0}^{\infty}$ for $k \geqslant 2$ provide important examples in white noise distribution theory [8]. In a recent paper [4] Cochran *et al.* have constructed a space $[\mathcal{V}]_\alpha$ of test functions and its dual space $[\mathcal{V}]_\alpha^*$ of generalized functions (CKS-space for short) from a nuclear space $\mathcal{V}$ and a sequence $\{\alpha(n)\}_{n=0}^{\infty}$ of positive numbers satisfying the following conditions:

(1) $\alpha(0) = 1$.
(2) $\inf_{n \geqslant 0} \alpha(n) > 0$.
(3) $\lim_{n \to \infty} \left(\frac{\alpha(n)}{n!}\right)^{1/n} = 0$.

For the characterization of generalized functions in $[\mathcal{V}]_\alpha^*$ (Theorem 6.4 in [4]) they assume the following condition

$$\limsup_{n \to \infty} \left(\frac{n!}{\alpha(n)} \inf_{x>0} \frac{G_\alpha(x)}{x^n} \right)^{1/n} < \infty, \tag{7}$$

where $G_\alpha(x) = \sum_{n=0}^{\infty} \alpha(n)/n!\, x^n$ is the exponential generating function of the sequence $\{\alpha(n)\}_{n=0}^{\infty}$. Furthermore, by Corollary 4.4 in [4], if the sequence $\{\alpha(n)/n!\}_{n=0}^{\infty}$ is log-concave, then the condition in Equation (7) is satisfied.

For the case $\alpha(n) = b_k(n)$, Cochran *et al.* showed in Proposition 7.4 in [4] that the condition in Equation (7) is satisfied. However, they did not show whether the sequence $\{b_k(n)/n!\}_{n=0}^{\infty}$ is log-concave. Our Theorem 1 shows that this is indeed the case.

The other conclusion in Theorem 1, i.e., $\{b_k(n)\}_{n=0}^{\infty}$ being log-convex, can be used to characterize the test functions. First we point out the following fact which can be easily checked.

FACT. *If $\{\beta(n)\}_{n=0}^{\infty}$ is log-convex, then $\{1/\beta(n)n!\}_{n=0}^{\infty}$ is log-concave.*

Recall from Theorem 1 that the sequence $\{b_k(n)\}_{n=0}^{\infty}$ is log-convex. Hence by the above fact the sequence $\{1/(b_k(n)n!)\}_{n=0}^{\infty}$ is log-concave.

In [4], Cochran *et al.* did not study the characterization of test functions in $[\mathcal{V}]_\alpha$. In recent papers [6, 1] and references therein, several theorems on the characterization of test functions and related results have been obtained. For test functions, we need to assume the following condition

$$\limsup_{n \to \infty} \left(\alpha(n) n! \inf_{x>0} \frac{G_{1/\alpha}(x)}{x^n} \right)^{1/n} < \infty, \tag{8}$$

where

$$G_{1/\alpha}(x) = \sum_{n=0}^{\infty} \frac{1}{\alpha(n)n!} x^n$$

is the exponential generating function of the sequence $\{1/\alpha(n)\}_{n=0}^{\infty}$. The same argument as in the proof of Corollary 4.4 in [4] can be used to show that if $\{1/(\alpha(n)n!)\}_{n=0}^{\infty}$ is log-concave, then the condition in Equation (8) is satisfied.

In particular, when $\alpha(n) = b_k(n)$, we know from Theorem 1 that the sequence $\{1/(b_k(n)n!)\}_{n=0}^{\infty}$ is log-concave. Thus the condition in Equation (8) is satisfied.

3.2. INEQUALITY CONDITIONS ON THE SEQUENCE $\{\alpha(n)\}_{n=0}^{\infty}$

In order to carry out the white noise distribution theory for the spaces $[\mathcal{V}]_\alpha$ and $[\mathcal{V}]_\alpha^*$ the following three conditions have been imposed on $\{\alpha(n)\}_{n=0}^{\infty}$ in [7]:

(c-1) There exists a constant c_1 such that for any $n \leqslant m$,

$$\alpha(n) \leqslant c_1^m \alpha(m).$$

(c-2) There exists a constant c_2 such that for any n and m,

$$\alpha(n+m) \leqslant c_2^{n+m} \alpha(n)\alpha(m).$$

(c-3) There exists a constant c_3 such that for any n and m,

$$\alpha(n)\alpha(m) \leqslant c_3^{n+m} \alpha(n+m).$$

Note that $c_i \geqslant 1$ for all $i = 1, 2, 3$ since $\alpha(0) = 1$. As shown in Section 3 in [7], condition (c-3) implies condition (c-1). Moreover, it has been proved in Theorem 4.8 in [7] that the Bell numbers $\{b_k(n)\}_{n=0}^{\infty}$ satisfy conditions (c-1), (c-2), and (c-3). Below we give further comments on the constants c_1, c_2, and c_3.

Obviously, if a sequence $\{\alpha(n)\}_{n=0}^{\infty}$ is nondecreasing, then condition (c-1) is satisfied and $c_1 = 1$ is the best constant satisfying condition (c-1).

From Equation (7.5) in [4] we have the formula for the sequence $\{B_k(n)\}_{n=0}^{\infty}$ defined in Equation (1) for $k \geqslant 2$:

$$B_k(n) = \sum_{j=0}^{\infty} \frac{B_{k-1}(j)}{j!} j^n, \tag{9}$$

where $B_1(n) = 1$ for all n. On the other hand, we can differentiate both sides of Equation (1) and then compare the coefficients of x^n to get the formula:

$$B_k(n+1) = \sum_{j_1+\cdots+j_k=n} \frac{n!}{j_1! \cdots j_k!} B_1(j_1) \cdots B_k(j_k). \tag{10}$$

We see from either Equation (9) or (10) that the sequence $\{B_k(n)\}_{n=0}^{\infty}$ is increasing. But $b_k(n) = B_k(n)/\exp_k(0)$ and so the sequence $\{b_k(n)\}_{n=0}^{\infty}$ is also increasing. Hence the Bell numbers satisfy condition (c-1) and the best constant for c_1 is $c_1 = 1$.

As mentioned at the end of Section 1, the Bell numbers of any order $k \geqslant 2$ satisfy the inequalities:

$$b_k(n)b_k(m) \leqslant b_k(n+m) \leqslant 2^{n+m} b_k(n)b_k(m).$$

Hence the Bell numbers satisfy conditions (c-2) and (c-3) with $c_2 = 2$ and $c_3 = 1$. Obviously, $c_3 = 1$ is the best constant for condition (c-3). As for the best constant for c_2 we have the following

CONJECTURE. *The best constant c_2 in the condition* (c-2) *for the Bell numbers $\{b_k(n)\}_{n=0}^{\infty}$ of any order $k \geqslant 2$ is $c_2 = 2$.*

Here we prove that the conjecture is true for $k = 2$. It follows from Theorem 3 that $b_2(n+m) \leqslant 2^{n+m} b_2(n) b_2(m)$. Hence, the best constant c_2 must be $c_2 \leqslant 2$. On the other hand, by Theorem 4.3 in [7],

$$\log b_2(n) = n \log n - n \log\log n - n + \mathrm{o}(n).$$

From this equality we obtain that

$$\log b_2(2n) - 2\log b_2(n) = 2n \log 2 - 2n\big(\log\log(2n) - \log\log n\big) + \mathrm{o}(n).$$

Then we get the following limit

$$\lim_{n\to\infty} \frac{1}{2n} \log \frac{b_2(2n)}{b_2(n)^2} = \log 2. \tag{11}$$

Now, put $m = n$ in condition (c-2) to get $b_2(2n) \leqslant c_2^{2n} b_2(n)^2$. This inequality implies that for all $n \geqslant 1$,

$$\frac{1}{2n} \log \frac{b_2(2n)}{b_2(n)^2} \leqslant \log c_2. \tag{12}$$

Obviously, Equations (11) and (12) show that $\log 2 \leqslant \log c_2$. Hence $c_2 \geqslant 2$. But we already noted above that $c_2 \leqslant 2$. Therefore, $c_2 = 2$.

References

1. Asai, N., Kubo, I. and Kuo, H.-H.: General characterization theorems and intrinsic topologies in white noise analysis, Preprint, 1998.
2. Bender, E. A. and Canfield, E. R.: Log-concavity and related properties of the cycle index polynomials, *J. Combin. Theory A* **74** (1996), 57–70.
3. Canfield, E. R.: Engel's inequality for Bell numbers, *J. Combin. Theory A* **72** (1995), 184–187.
4. Cochran, W. G., Kuo, H.-H. and Sengupta, A.: A new class of white noise generalized functions, *Infin. Dimens. Anal. Quantum Probab. Relat. Top.* **1** (1998), 43–67.
5. Engel, K.: On the average rank of an element in a filter of the partition lattice, *J. Combin. Theory A* **65** (1994), 67–78.
6. Kubo, I.: On characterization theorems for CKS-spaces in white noise analysis, Preprint, 1998.
7. Kubo, I., Kuo, H.-H. and Sengupta, A.: White noise analysis on a new space of Hida distributions, *Infin. Dimens. Anal. Quantum Probab. Relat. Top.* **2** (1999), 315–335.
8. Kuo, H.-H.: *White Noise Distribution Theory*, CRC Press, Boca Raton, 1996.

Acta Applicandae Mathematicae **63:** 89–100, 2000.

Poisson Equations Associated with Differential Second Quantization Operators in White Noise Analysis ⋆

Dedicated to Professor Takeyuki Hida on the Occasion of his 70th birthday

DONG MYUNG CHUNG[1] and UN CIG JI[2]
[1]*Department of Mathematics, Sogang University, Seoul 121–742, Korea*
[2]*Global Analysis Research Center, Department of Mathematics, Seoul National University, Seoul 151–742, Korea*

(Received: 24 December 1998)

Abstract. In this paper we shall show the heredity of a differentiable one-parameter semigroup under the second quantization and then discuss the resolvent of the differential second quantization operator and the potentials of test white noise functionals. As an application, we shall investigate the existence of solutions of the Poisson-type equations associated with differential second quantization operators as well as operators similar to differential second quantization operators.

Mathematics Subject Classifications (2000): 60H40, 46F25, 47D06.

Key words: white noise, differential second quantization operator, one-parameter semigroup, potential, Poisson equation.

1. Introduction

Gross [10] introduced the Gross Laplacian Δ_G on abstract Wiener space as a natural infinite-dimensional analogue of the finite-dimensional Laplacian and studied the Poisson equation associated with Δ_G. There are some related results in Carmona [1]. Kuo [18] studied the Poisson equation associated with the number operator N on abstract Wiener space.

Within a white noise setting, the Gross Laplacian Δ_G and the number operator N has been formulated by Kuo in [19, 20] as continuous linear operators acting on test white noise functions and the Cauchy problem associated with Δ_G has been studied in [20]. Kang [15] studied the heat and Poisson equations associated with the number operator N. In [5–7], the authors investigated the existence and uniqueness of solutions to the Cauchy problems associated with the linear combination of Δ_G and N, and their powers and generalized Gross Laplacian. Moreover, the Poisson equation associated with a linear combination of integral kernel operations

⋆ This work is supported by Korea Research Foundation (KRF-2000-015-DP0016).
Research supported by Global Analysis Research Center.

containing only annihilations and the number operator N has been studied in [4]. In [2], the authors discussed the heredity of a differentiable one-parameter group and the cosine family under differential second quantization.

In this paper, motivated by the results in [4, 15, 18], we shall study the Poisson equations associated with differential second quantization operators as well as operators similar to differential second quantization operators.

The paper is organized as follows. In Section 2 we briefly recall well-known results in white noise analysis. In Section 3 we shall prove the heredity of a differentiable one-parameter semigroup under the second quantization. In Section 4 we shall discuss the resolvent of the differential second quantization operators and the potentials of test white noise functions, and then apply these concepts to investigate the Poisson equations associated with differential second quantization operators as well as operators similar to differential second quantization operators. This extends the results of [4] and [15].

2. Preliminaries on White Noise Analysis

In this section we shall briefly recall some of the concepts, notation and known results in white noise analysis [14, 21, 24]. Let $H = L^2(\mathbb{R}, \mathrm{d}t)$ be the real Hilbert space of square integrable functions on $\mathbb{R}$ with norm $|\cdot|_0$. For any $p \in \mathbb{R}$ define a norm by

$$|\xi|_p = |A^p\xi|_0, \quad A = 1 + t^2 - \frac{\mathrm{d}^2}{\mathrm{d}t^2}.$$

For $p \geqslant 0$, let E_p be the Hilbert space consisting of $\xi \in H$ with $|\xi|_p < \infty$, and let E_{-p} be the completion of H with respect to the norm $|\cdot|_{-p}$. As is well known, the Schwartz space of rapidly decreasing C^∞-functions and its dual space are obtained by

$$E = \mathcal{S}(\mathbb{R}) = \operatorname*{proj\,lim}_{p\to\infty} E_p, \qquad E^* = \mathcal{S}'(\mathbb{R}) = \operatorname*{ind\,lim}_{p\to\infty} E_{-p}.$$

Thus we have a Gelfand triple

$$E \subset H \subset E^*. \tag{2.1}$$

From now on, we use the constant numbers

$$\rho = \|A^{-1}\|_{\mathrm{OP}} = \tfrac{1}{2}, \qquad \delta^2 = \|A^{-1}\|_{\mathrm{HS}}^2 = \sum_{i=0}^{\infty} \frac{1}{(2i+2)^2} < \infty.$$

Let μ be the Gaussian measure on E^* whose characteristic functional is given by

$$\int_{E^*} \exp\{i\langle x, \xi\rangle\}\mu(\mathrm{d}x) = \exp\left\{-\frac{1}{2}|\xi|_0^2\right\}, \quad \xi \in E,$$

where $\langle \cdot, \cdot \rangle$ is the canonical bilinear form on $E^* \times E$. We denote by $(L^2) \equiv L^2(E^*, \mu)$ the complex Hilbert space of μ-square integrable functions on E^* with norm $\| \cdot \|_0$. By the Wiener–Itô decomposition theorem, we have the following unitary isomorphism between (L^2) and the Boson Fock space $\Gamma(H_{\mathbb{C}})$:

$$(L^2) \ni \phi(x) = \sum_{n=0}^{\infty} \langle :x^{\otimes n}:, f_n \rangle \longleftrightarrow (f_n) \sim \phi \in \Gamma(H_{\mathbb{C}}), \quad f_n \in H_{\mathbb{C}}^{\widehat{\otimes} n}, \tag{2.2}$$

where $:x^{\otimes n}:$ denotes the Wick ordering of $x^{\otimes n}$ and $H_{\mathbb{C}}^{\widehat{\otimes} n}$ is the n-fold symmetric tensor product of the complexification of H. Note that the (L^2)-norm $\|\phi\|_0$ of ϕ is given by

$$\|\phi\|_0 = \left(\sum_{n=0}^{\infty} n! |f_n|_0^2 \right)^{1/2},$$

where $| \cdot |_0$ denotes the $H_{\mathbb{C}}^{\otimes n}$-norm for any n.

Let $\Gamma(A)$ denote the second quantization operator of A defined by

$$\Gamma(A)\phi \sim (A^{\otimes n} f_n), \quad (f_n) \sim \phi \in (L^2).$$

Then $\Gamma(A)$ becomes a positive self-adjoint operator on (L^2) with $\|\Gamma(A)^{-1}\|_{\mathrm{OP}} < 1$ and $\|\Gamma(A)^{-1}\|_{\mathrm{HS}} < \infty$. From (L^2) and $\Gamma(A)$, a complex Gelfand triple $(E) \subset (L^2) \subset (E)^*$ is constructed in the standard manner (see [14, 21, 24]). We note that (E) is a nuclear space equipped with the Hilbertian norms

$$\|\phi\|_p = \|\Gamma(A)^p \phi\|_0, \quad \phi \in (E), \ p \in \mathbb{R}.$$

An element in (E) (and in $(E)^*$) is called a *test* (*and generalized, respectively*) *white noise function*. We denote by $\langle\langle \cdot, \cdot \rangle\rangle$ the canonical $\mathbb{C}$-bilinear form on $(E)^* \times (E)$. For each $\Phi \in (E)^*$, there exists a unique sequence $\{F_n\}_{n=0}^{\infty}$, $F_n \in (E_{\mathbb{C}}^{\otimes n})^*_{\mathrm{sym}}$ such that

$$\langle\langle \Phi, \phi \rangle\rangle = \sum_{n=0}^{\infty} n! \langle F_n, f_n \rangle, \quad (f_n) \sim \phi \in (E). \tag{2.3}$$

For each $\xi \in E_{\mathbb{C}}$, the exponential vector ϕ_ξ associated with $\xi \in E_{\mathbb{C}}$ is defined by

$$\left(1, \xi, \frac{\xi^{\otimes 2}}{2!}, \dots, \frac{\xi^{\otimes n}}{n!}, \dots \right) \sim \phi_\xi \in (E).$$

It is well known that $\{\phi_\xi : \xi \in E_{\mathbb{C}}\}$ spans a dense subspace of (E). The S-transform of $\Phi \in (E)^*$ is a function on $E_{\mathbb{C}}$ defined by $S\Phi(\xi) = \langle\langle \Phi, \phi_\xi \rangle\rangle$. Then $\Phi \in (E)^*$ is uniquely determined by its S-transform $S\Phi$. Moreover, the analytic characterization theorem for S-transform has been proved by Potthoff and Streit in [26].

Let $\mathcal{L}(\mathfrak{X}, \mathfrak{Y})$ denote the space of all continuous linear operators from a locally convex space $\mathfrak{X}$ into another locally convex space $\mathfrak{Y}$. Now, we recall the analytic characterization theorem for symbols of operators on white noise functionals due to Obata [23]. For the proof, we refer the reader to [3]. For any $\Xi \in \mathcal{L}((E), (E)^*)$, a function $\widehat{\Xi}$ on $E_{\mathbb{C}} \times E_{\mathbb{C}}$ defined by

$$\widehat{\Xi}(\xi, \eta) = \langle\langle \Xi\phi_\xi, \phi_\eta \rangle\rangle, \quad \xi, \eta \in E_{\mathbb{C}}$$

is called the *symbol* of Ξ. Then $\Xi \in \mathcal{L}((E), (E)^*)$ is uniquely determined by its symbol $\widehat{\Xi}$.

THEOREM 2.1. *Let Θ be a function on $E_{\mathbb{C}} \times E_{\mathbb{C}}$ with values in $\mathbb{C}$. Then there exists a continuous operator $\Xi \in \mathcal{L}((E), (E)^*)$ such that $\Theta = \widehat{\Xi}$ if and only if*

(S1) *for any $\xi, \xi_1, \eta, \eta_1 \in E_{\mathbb{C}}$, the function $\Theta(z\xi + \xi_1, w\eta + \eta_1)$ is an entire holomorphic function of $(z, w) \in \mathbb{C} \times \mathbb{C}$;*

(S2) *there exist constant numbers $K > 0$, $a > 0$ and $p \geqslant 0$ such that*

$$|\Theta(\xi, \eta)| \leqslant K \exp\left\{a\left(|\xi|_p^2 + |\eta|_p^2\right)\right\}, \quad \xi, \eta \in E_{\mathbb{C}}.$$

Moreover, Θ is the symbol of an operator $\Xi \in \mathcal{L}((E), (E))$ if and only if Θ satisfies the condition (S1) *and*

(S2′) *for any $p \geqslant 0$ and $\epsilon > 0$, there exist constants $C \geqslant 0$ and $q \geqslant 0$ such that*

$$|\Theta(\xi, \eta)| \leqslant C \exp\left\{\epsilon\left(|\xi|_{p+q}^2 + |\eta|_{-p}^2\right)\right\}, \quad \xi, \eta \in E_{\mathbb{C}}.$$

In this case,

$$\|\Xi\phi\|_{p-1} \leqslant CM(\epsilon, q, r)\|\phi\|_{p+q+r+1}, \quad \phi \in (E),$$

where $M(\epsilon, q, r)$ is a (finite) constant for $\epsilon < (2\mathrm{e}^3\delta^2)^{-1}$, $r \geqslant r_0(q) \geqslant 0$.

Let $l, m \geqslant 0$ be integers. For each $\kappa \in (E_{\mathbb{C}}^{\otimes(l+m)})^*$, there exists a unique $\Xi_{l,m}(\kappa) \in \mathcal{L}((E), (E)^*)$ such that

$$\widehat{\Xi}_{l,m}(\kappa)(\xi, \eta) = \langle \kappa, \eta^{\otimes l} \otimes \xi^{\otimes m} \rangle \mathrm{e}^{\langle \xi, \eta \rangle}, \quad \xi, \eta \in \mathcal{E}_{\mathbb{C}}.$$

Then $\Xi_{l,m}(\kappa)$ is called the *integral kernel operator* with kernel distribution κ (see [8, 12, 24]). We sometimes use a formal integral expression:

$$\Xi_{l,m}(\kappa) = \int_{\mathbb{R}^{l+m}} \kappa(s_1, \ldots, s_l, t_1, \ldots, t_m) a_{s_1}^* \ldots a_{s_l}^* a_{t_1} \ldots a_{t_m} \, \mathrm{d}s_1 \ldots \mathrm{d}s_l \mathrm{d}t_1 \ldots \mathrm{d}t_m,$$

where $a_t = \Xi_{0,1}(\delta_t)$ and $a_t^* = \Xi_{1,0}(\delta_t)$ are the annihilation operator and the creation operator at each point $t \in \mathbb{R}$, respectively. It is well known that $\Xi_{l,m}(\kappa) \in$

$\mathcal{L}((E),(E))$ if and only if $\kappa \in E_{\mathbb{C}}^{\otimes l} \otimes (E_{\mathbb{C}}^{\otimes m})^*$. The integral kernel operators with trace τ as the kernel distribution

$$\Delta_{\mathrm{G}} = \int_{\mathbb{R}^2} \tau(s,t) a_s a_t \, \mathrm{d}s \, \mathrm{d}t, \qquad N = \int_{\mathbb{R}^2} \tau(s,t) a_s^* a_t \, \mathrm{d}s \, \mathrm{d}t,$$

are called the *Gross Laplacian* and the *number operator*, respectively, where the trace $\tau \in E_{\mathbb{C}} \otimes E_{\mathbb{C}}^*$ is defined by

$$\langle \tau, \eta \otimes \xi \rangle = \langle \eta, \xi \rangle, \quad \eta, \xi \in E_{\mathbb{C}}.$$

3. Differentiable One-Parameter Semigroups

In this section we shall prove the heredity of differentiable one-parameter semigroup under the second quantization. Let us consider a one-parameter semigroup $\{\Omega_\theta\}_{\theta \geqslant 0} \subset \mathcal{L}((E),(E))$, i.e., $\Omega_\theta \in \mathcal{L}((E),(E))$ for any $\theta \geqslant 0$ and

$$\Omega_{\theta_1+\theta_2} = \Omega_{\theta_1}\Omega_{\theta_2}, \quad \theta_1, \theta_2 \geqslant 0; \quad \Omega_0 = I \text{ (identity)}.$$

A one-parameter semigroup $\{\Omega_\theta\}_{\theta \geqslant 0} \subset \mathcal{L}((E),(E))$ is called *equicontinuous* if for any $p \geqslant 0$ there exist $M \geqslant 0$ and $q \geqslant 0$ such that

$$\|\Omega_\theta \phi\|_p \leqslant M \|\phi\|_q, \quad \phi \in (E), \ \theta \geqslant 0.$$

If, in particular, $M \leqslant 1$, then $\{\Omega_\theta\}_{\theta \geqslant 0}$ is called a *contraction semigroup*.

DEFINITION 3.1. A one-parameter semigroup $\{\Omega_\theta\}_{\theta \geqslant 0} \subset \mathcal{L}((E),(E))$ is said to be *differentiable* if there exists $\Xi \in \mathcal{L}((E),(E))$ such that

$$\lim_{\theta \downarrow 0} \left\| \frac{\Omega_\theta \phi - \phi}{\theta} - \Xi \phi \right\|_p = 0, \quad \phi \in (E), \ p \geqslant 0.$$

In that case Ξ is unique and is called the *infinitesimal generator* of the one-parameter semigroup $\{\Omega_\theta\}_{\theta \geqslant 0}$.

Note that if $\{\Omega_\theta\}_{\theta \geqslant 0}$ is a differentiable one-parameter semigroup with the infinitesimal generator $\Xi \in \mathcal{L}((E),(E))$, then it is easily shown that $\{\Omega_\theta\}_{\theta \geqslant 0}$ is infinitely many differentiable and

$$\frac{\mathrm{d}^n \Omega_\theta}{\mathrm{d}\theta^n} \phi = \Omega_\theta \Xi^n \phi = \Xi^n \Omega_\theta \phi, \quad \phi \in (E), \ \theta \geqslant 0.$$

Let $K \in \mathcal{L}(E_{\mathbb{C}}, E_{\mathbb{C}})$. Then by the kernel theorem, there exists a unique $\lambda_K \in E_{\mathbb{C}} \otimes E_{\mathbb{C}}^*$ such that

$$\langle \lambda_K, \eta \otimes \xi \rangle = \langle K\xi, \eta \rangle, \quad \xi, \eta \in E_{\mathbb{C}}. \tag{3.1}$$

In fact, for any $\eta \in E_{\mathbb{C}}$, $K\eta$ equals to the right contraction $\lambda_K \otimes_1 \eta$. For each $K \in \mathcal{L}(E_{\mathbb{C}}, E_{\mathbb{C}})$, the *differential second quantization operator* $\mathrm{d}\Gamma(K)$ is defined as

$$\langle\langle \mathrm{d}\Gamma(K)\phi_\xi, \phi_\eta \rangle\rangle = \langle K\xi, \eta \rangle \mathrm{e}^{\langle \xi, \eta \rangle}, \quad \xi, \eta \in E_{\mathbb{C}}.$$

In fact, $\mathrm{d}\Gamma(K) = \Xi_{1,1}(\lambda_K)$.

THEOREM 3.2. *Let $K \in \mathcal{L}(E_{\mathbb{C}}, E_{\mathbb{C}})$. Then K is the infinitesimal generator of a differentiable one-parameter semigroup $\{\Omega_\theta\}_{\theta \geqslant 0}$ if and only if its differential second quantization operator $\mathrm{d}\Gamma(K)$ is the infinitesimal generator of a differentiable one-parameter semigroup $\{\Gamma(\Omega_\theta)\}_{\theta\geqslant 0}$.*

Proof. It is obvious that $\{\Gamma(\Omega_\theta)\}_{\theta\geqslant 0}$ is a one-parameter semigroup. We now shall prove that $\{\Gamma(\Omega_\theta)\}_{\theta\geqslant 0}$ is a differentiable semigroup of which the infinitesimal generator is $\mathrm{d}\Gamma(K)$. Put

$$f(\theta) = \langle\langle \Gamma(\Omega_\theta)\phi_\xi, \phi_\eta \rangle\rangle = \mathrm{e}^{\langle \Omega_\theta \xi, \eta\rangle}, \quad \theta \geqslant 0,\ \xi, \eta \in E_{\mathbb{C}}.$$

Then we have

$$f'(\theta) = \langle K\Omega_\theta \xi, \eta\rangle \mathrm{e}^{\langle \Omega_\theta \xi, \eta\rangle}$$

and

$$f''(\theta) = \left(\langle K\Omega_\theta \xi, \eta\rangle^2 + \langle K^2\Omega_\theta \xi, \eta\rangle\right)\mathrm{e}^{\langle \Omega_\theta \xi, \eta\rangle}.$$

Let $\theta_0 > 0$ be fixed. Then by applying Baire category theorem, for any $r \geqslant 0$ there exist $r' \geqslant 0$ and $C_1 \geqslant 0$ such that

$$\max_{0\leqslant\theta\leqslant\theta_0} |K\Omega_\theta\xi|_r \leqslant C_1|\xi|_{r+r'}, \qquad \max_{0\leqslant\theta\leqslant\theta_0} |K^2\Omega_\theta\xi|_r \leqslant C_1|\xi|_{r+r'}$$

since Ω_θ is continuous in $\theta \geqslant 0$. Therefore, we can easily show that for any $p \geqslant 0$ and $\epsilon > 0$, there exist $C = C(K, \theta_0) \geqslant 0$ and $q \geqslant 0$ such that

$$\max_{0\leqslant\theta\leqslant\theta_0} |f''(\theta)| \leqslant C \exp\left\{\epsilon\left(|\xi|_{p+q}^2 + |\eta|_{-p}^2\right)\right\}, \quad \xi, \eta \in E_{\mathbb{C}}.$$

Now we put

$$g_\theta(\xi, \eta) = f(\theta) - f(0) - f'(0)\theta, \quad \xi, \eta \in E_{\mathbb{C}},\ \theta > 0.$$

Then by the Taylor theorem, we have, for $0 < \theta < \theta_0$,

$$\begin{aligned} |g_\theta(\xi, \eta)| &\leqslant \frac{\theta^2}{2} \max_{0\leqslant\theta\leqslant\theta_0} |f''(\theta)| \\ &\leqslant \frac{\theta^2}{2} C \exp\left\{\epsilon\left(|\xi|_{p+q}^2 + |\eta|_{-p}^2\right)\right\}, \quad \xi, \eta \in E_{\mathbb{C}}. \end{aligned}$$

It then follows from Theorem 2.1 that there exists $\Xi_\theta \in \mathcal{L}((E), (E))$ such that $\widehat{\Xi}_\theta = g_\theta$ and

$$\|\Xi_\theta \phi\|_{p-1} \leqslant \frac{\theta^2}{2} CM(\epsilon, q, r)\|\phi\|_{p+q+r+1}, \quad 0 < \theta < \theta_0,\ \phi \in (E) \tag{3.2}$$

for some constants $M(\epsilon, q, r) \geqslant 0$ and $r \geqslant 0$. On the other hand,

$$f'(0) = \langle K\xi, \eta\rangle \mathrm{e}^{\langle \xi, \eta\rangle}, \quad \xi, \eta \in E_{\mathbb{C}}.$$

Hence we have, for $0 < \theta < \theta_0$,

$$\Xi_\theta = \Gamma(\Omega_\theta) - I - \theta \, d\Gamma(K).$$

It follows from (3.2) that for any $\phi \in (E)$ and $p \geqslant 0$,

$$\lim_{\theta \downarrow 0} \left\| \frac{\Gamma(\Omega_\theta)\phi - \phi}{\theta} - d\Gamma(K)\phi \right\|_p = 0.$$

The proof of the converse is obvious since

$$\Gamma(\Omega_\theta)(0, \xi, 0, \ldots) = (0, \Omega_\theta \xi, 0, \ldots),$$
$$d\Gamma(K)(0, \xi, 0, \ldots) = (0, K\xi, 0, \ldots), \quad \xi \in E_{\mathbb{C}}$$

and $\|(0, \xi, 0, \ldots)\|_p = |\xi|_p$ for all $\xi \in E_{\mathbb{C}}$ and $p \geqslant 0$.

Let $\{\Omega_\theta\}_{\theta \geqslant 0} \subset \mathcal{L}(E_{\mathbb{C}}, E_{\mathbb{C}})$ be an equicontinuous one-parameter semigroup. Then for any $p \geqslant 0$ there exists $C \geqslant 0$ and $q \geqslant 0$ such that

$$|\Omega_\theta \xi|_p \leqslant C|\xi|_{p+q} \leqslant |\xi|_{p+q+r}, \quad \xi \in E_{\mathbb{C}}, \quad \theta \geqslant 0$$

for some $r \geqslant 0$ with $C\rho^r \leqslant 1$. Therefore $\{\Omega_\theta\}_{\theta \geqslant 0}$ is an equicontinuous one-parameter semigroup if and only if $\{\Omega_\theta\}_{\theta \geqslant 0}$ is a contraction semigroup. Moreover, for any $\phi \sim (f_n) \in (E)$, we obtain that

$$\|\Gamma(\Omega_\theta)\phi\|_p^2 = \sum_{n=0}^{\infty} n! |\Omega_\theta^{\otimes n} f_n|_p^2 \leqslant \sum_{n=0}^{\infty} n! |f_n|_{p+q+r}^2 = \|\phi\|_{p+q+r}^2$$

for some $q \geqslant 0$ and $r \geqslant 0$. Hence, $\{\Gamma(\Omega_\theta)\}_{\theta \geqslant 0} \subset \mathcal{L}((E), (E))$ is a contraction semigroup. Also, the converse is true, i.e., if $\{\Gamma(\Omega_\theta)\}_{\theta \geqslant 0}$ is a contraction semigroup, then $\{\Omega_\theta\}_{\theta \geqslant 0}$ is a contraction semigroup. Thus by Theorem 3.2, the following theorem is obvious.

THEOREM 3.3. *Let $K \in \mathcal{L}(E_{\mathbb{C}}, E_{\mathbb{C}})$. Then K is the infinitesimal generator of a differentiable contraction one-parameter semigroup $\{\Omega_\theta\}_{\theta \geqslant 0} \subset \mathcal{L}(E_{\mathbb{C}}, E_{\mathbb{C}})$ if and only if its differential second quantization operator* $d\Gamma(K)$ *is the infinitesimal generator of a differentiable contraction one-parameter semigroup $\{\Gamma(\Omega_\theta)\}_{\theta \geqslant 0} \subset \mathcal{L}((E), (E))$.*

4. Poisson Equations Associated with Differential Second Quantization Operators

In this section we shall investigate the existence of the solutions to the Poisson equations associated with differential second quantization operators as well as operators similar to differential second quantization operators.

PROPOSITION 4.1. *Let $K \in \mathcal{L}(E_{\mathbb{C}}, E_{\mathbb{C}})$ be the infinitesimal generator of a differentiable contraction one-parameter semigroup $\{\Omega_\theta\}_{\theta \geqslant 0} \subset \mathcal{L}(E_{\mathbb{C}}, E_{\mathbb{C}})$. Then the set $\{\lambda \in \mathbb{C}; \mathrm{Re}(\lambda) > 0\}$ is contained in the resolvent set $\rho(\mathrm{d}\Gamma(K))$ of $\mathrm{d}\Gamma(K)$, and we have*

$$\begin{aligned} R(\lambda; \mathrm{d}\Gamma(K))\phi &= (\lambda I - \mathrm{d}\Gamma(K))^{-1}\phi \\ &= \int_0^\infty \mathrm{e}^{-\lambda s}\Gamma(\Omega_s)\phi \, \mathrm{d}s, \quad \mathrm{Re}(\lambda) > 0, \ \phi \in (E). \end{aligned}$$

In this case, for any $\lambda \in \mathbb{C}$ with $\mathrm{Re}(\lambda) > 0$, $R(\lambda; \mathrm{d}\Gamma(K)) \in \mathcal{L}((E), (E))$.

Proof. By Theorem 3.3 $\{\Gamma(\Omega_\theta)\}_{\theta \geqslant 0} \subset \mathcal{L}((E), (E))$ is a differentiable contraction one-parameter semigroup with the infinitesimal generator $\mathrm{d}\Gamma(K)$. Hence, the proof follows from Theorem 1 in Chapter IX, 4 [28]. □

Let K be the infinitesimal generator of a differentiable contraction one-parameter semigroup $\{\Omega_\theta\}_{\theta \geqslant 0} \subset \mathcal{L}(E_{\mathbb{C}}, E_{\mathbb{C}})$. Then it follows from Proposition 4.1 that for any $\phi \in (E)$

$$\begin{aligned} \mathrm{d}\Gamma(K)R(\lambda; \mathrm{d}\Gamma(K))\phi &= \mathrm{d}\Gamma(K)(\lambda I - \mathrm{d}\Gamma(K))^{-1}\phi \\ &= -\phi + \lambda R(\lambda; \mathrm{d}\Gamma(K))\phi. \end{aligned}$$

An element $\psi \in (E)$ is called a *potential* of $\phi \in (E)$ if

$$\psi = \lim_{\lambda \downarrow 0}(\lambda I - \mathrm{d}\Gamma(K))^{-1}\phi. \tag{4.1}$$

Then it is easy to see that for a given $\phi \in (E)$ if $\psi \in (E)$ is the potential of ϕ, then ψ is a solution of the following Poisson equation:

$$\mathrm{d}\Gamma(K)u = -\phi. \tag{4.2}$$

Note that if $\phi \in (E)$ with $E(\phi) \neq 0$, where $E(\phi)$ is the expectation of ϕ, then the limit in (4.1) does not exists.

PROPOSITION 4.2. *Let $K \in \mathcal{L}(E_{\mathbb{C}}, E_{\mathbb{C}})$. If there exists $\omega > 0$ such that $\omega I + K$ is the infinitesimal generator of a differentiable contraction one-parameter semigroup $\{\Upsilon_\theta\}_{\theta \geqslant 0} \subset \mathcal{L}(E_{\mathbb{C}}, E_{\mathbb{C}})$, then the limit in (4.1) exists for any $\phi - E(\phi) \in (E)$. Moreover,*

$$\lim_{\lambda \downarrow 0}(\lambda I - \mathrm{d}\Gamma(K))^{-1}(\phi - E(\phi)) = \int_0^\infty \Gamma(\Omega_s)(\phi - E(\phi)) \, \mathrm{d}s, \tag{4.3}$$

where $\Omega_\theta = \mathrm{e}^{-\omega\theta}\Upsilon_\theta$ for any $\theta \geqslant 0$.

Proof. Let $\Omega_\theta = \mathrm{e}^{-\omega\theta}\Upsilon_\theta$ for any $\theta \geqslant 0$. Then $\{\Omega_\theta\}_{\theta \geqslant 0} \subset \mathcal{L}(E_{\mathbb{C}}, E_{\mathbb{C}})$ is a differentiable contraction one-parameter semigroup with the infinitesimal generator K. Moreover, for any $p \geqslant 0$ there exists $q \geqslant 0$ such that

$$\|\Omega_\theta \xi\|_p \leqslant \mathrm{e}^{-\omega\theta}\|\xi\|_{p+q}, \quad \xi \in E_{\mathbb{C}}.$$

Therefore, by Proposition 4.1, for any $\lambda > 0$ the resolvent operator $(\lambda I - \mathrm{d}\Gamma(K))^{-1}$ exists. On the other hand, by direct computation, we obtain that

$$\|\Gamma(\Omega_\theta)(\phi - E(\phi))\|_p \leqslant \mathrm{e}^{-\omega\theta}\|\phi - E(\phi)\|_{p+q}, \quad \phi \in (E).$$

Thus the integral in (4.3) exists and we have

$$\begin{aligned}
&\lim_{\lambda\downarrow 0}\left\| \int_0^\infty (\mathrm{e}^{-\lambda s} - 1)\Gamma(\Omega_s)(\phi - E(\phi))\,\mathrm{d}s \right\|_p \\
&\quad \leqslant \left[\lim_{\lambda\downarrow 0}\int_0^\infty \left(\mathrm{e}^{-\omega s} - \mathrm{e}^{-(\lambda+\omega)s}\right)\mathrm{d}s\right]\|\phi - E(\phi)\|_{p+q} \\
&\quad = \lim_{\lambda\downarrow 0}\left(\frac{1}{\omega} - \frac{1}{\lambda+\omega}\right)\|\phi - E(\phi)\|_{p+q} \\
&\quad = 0.
\end{aligned}$$

This completes the proof. □

From the above results, the following theorem is obvious.

THEOREM 4.3. *Let $K \in \mathcal{L}(E_{\mathbb{C}}, E_{\mathbb{C}})$. If there exists $\omega > 0$ such that $\omega I + K$ is the infinitesimal generator of a differentiable contraction one-parameter semigroup $\{\Upsilon_\theta\}_{\theta\geqslant 0} \subset \mathcal{L}(E_{\mathbb{C}}, E_{\mathbb{C}})$, then the following Poisson equation:*

$$\mathrm{d}\Gamma(K)u = -(\phi - E(\phi)), \quad \phi \in (E)$$

has a solution $u \in (E)$ and, moreover, $E(u) = 0$ and u is given by

$$u = \int_0^\infty \Gamma(\Omega_s)(\phi - E(\phi))\,\mathrm{d}s,$$

where $\Omega_\theta = \mathrm{e}^{-\omega\theta}\Upsilon_\theta$ for any $\theta \geqslant 0$.

EXAMPLE 4.4. For any $0 < \omega < 1$, $(\omega - 1)I$ is the infinitesimal generator of a differentiable contraction one-parameter semigroup $\{\mathrm{e}^{(\omega-1)\theta}\}_{\theta\geqslant 0}$. Hence, by Theorem 4.3,

$$u = \int_0^\infty \Gamma(\mathrm{e}^{-s})(\phi - E(\phi))\,\mathrm{d}s \tag{4.4}$$

is a solution of the following Poisson equation:

$$\mathrm{d}\Gamma(I)u = \phi - E(\phi), \quad \phi \in (E).$$

Note that $\mathrm{d}\Gamma(I)$ equals to the number operator N. Hence, $u \in (E)$ given in (4.4) is a solution of the Poisson equation asociated with the number operator N.

For a locally convex topological space $\mathfrak{X}$, let $\mathrm{GL}(\mathfrak{X})$ denote the set of all linear homeomorphisms on $\mathfrak{X}$. An operator $\Xi_1 \in \mathcal{L}((E),(E))$ is said to be *similar* to $\Xi_2 \in \mathcal{L}((E),(E))$ if there exists $\mathcal{G} \in \mathrm{GL}((E))$ such that $\mathcal{G}\Xi_1\mathcal{G}^{-1} = \Xi_2$. Let $\Xi_1, \Xi_2 \in \mathcal{L}((E),(E))$. If Ξ_1 is similar to Ξ_2 and $\{\Upsilon_\theta\}_{\theta\geqslant 0} \subset \mathcal{L}((E),(E))$ is a differentiable contraction semigroup with the infinitesimal generator Ξ_1, then Ξ_2 is the infinitesimal generator of a differentiable contraction semigroup $\{\mathcal{G}\Upsilon_\theta\mathcal{G}^{-1}\}_{\theta\geqslant 0}$, where $\mathcal{G} \in \mathrm{GL}((E))$ such that $\mathcal{G}\Xi_1\mathcal{G}^{-1} = \Xi_2$. Therefore, the set $\{\lambda \in \mathbb{C};\, \mathrm{Re}(\lambda) > 0\}$ is contained in the resolvent set $\rho(\Xi_2)$ of Ξ_2, and we have

$$\begin{aligned} R(\lambda;\Xi_2)\phi &= (\lambda I - \Xi_2)^{-1}\phi \\ &= \int_0^\infty \mathrm{e}^{-\lambda s}\mathcal{G}\Upsilon_s\mathcal{G}^{-1}\phi\,\mathrm{d}s, \quad \mathrm{Re}(\lambda) > 0,\ \phi \in (E). \end{aligned}$$

Note that

$$\int_0^\infty \mathrm{e}^{-\lambda s}\mathcal{G}\Upsilon_s\mathcal{G}^{-1}\phi\,\mathrm{d}s = \mathcal{G}\int_0^\infty \mathrm{e}^{-\lambda s}\Upsilon_s\mathcal{G}^{-1}\phi\,\mathrm{d}s, \quad \mathrm{Re}(\lambda) > 0,\ \phi \in (E).$$

Let $K \in \mathcal{L}(E_{\mathbb{C}}, E_{\mathbb{C}})$ and $\Xi \in \mathcal{L}((E),(E))$. If there exists $\omega > 0$ such that $\omega I + K$ is the infinitesimal generator of a differentiable contraction one-parameter semigroup and Ξ is similar to $\mathrm{d}\Gamma(K)$, then for any $\phi \in (E)$ we have

$$\begin{aligned} R_\Xi(\phi) &\equiv \lim_{\lambda\downarrow 0}(\lambda I - \Xi)^{-1}\big(\phi - \mathcal{G}E(\mathcal{G}^{-1}\phi)\big) \\ &= \int_0^\infty \mathcal{G}\Gamma(\Omega_s)(\mathcal{G}^{-1}\phi - E(\mathcal{G}^{-1}\phi))\,\mathrm{d}s, \end{aligned}$$

where $\{\Omega_\theta\}_{\theta\geqslant 0}$ is a differentiable contraction one-parameter semigroup with the infinitesimal generator K and $\mathcal{G} \in \mathrm{GL}((E))$ with $\Xi = \mathcal{G}\,\mathrm{d}\Gamma(K)\mathcal{G}^{-1}$. Therefore, $R_\Xi(\phi)$ is a solution of the following Poisson equation:

$$\Xi u = -(\phi - \mathcal{G}E(\mathcal{G}^{-1}\phi)), \quad \phi \in (E). \tag{4.5}$$

Let $\kappa \in (E_{\mathbb{C}}^{\otimes 2})^*$ and $\alpha \in \mathbb{C}$ with $\alpha \neq 0$. Then by Theorem 2.1, there exists a unique $\mathcal{G}_{\kappa,\alpha} \in \mathcal{L}((E),(E))$ such that

$$\mathcal{G}_{\kappa,\alpha}\phi_\xi = \exp\left\{\frac{1}{2\alpha}\langle\kappa, \xi^{\otimes 2}\rangle\right\}\phi_\xi, \quad \xi \in E_{\mathbb{C}}.$$

In fact, we can easily show that the function $\Theta(\xi,\eta) = \langle\langle \mathcal{G}_{\kappa,\alpha}\phi_\xi, \phi_\eta\rangle\rangle$ satisfies the conditions (S1) and (S2′) in Theorem 2.1. In this case, $\mathcal{G}_{\kappa,\alpha} \in \mathrm{GL}((E))$ and $\mathcal{G}_{\kappa,\alpha}^{-1} = \mathcal{G}_{-\kappa,\alpha}$. Note that $\mathcal{G}_{\kappa,\alpha}(E(\phi)) = E(\phi)$.

We now state the following result without proof (see [2] for the proof).

THEOREM 4.5. *Let $\kappa \in (E_{\mathbb{C}}^{\otimes 2})^*_{\mathrm{sym}}$ and $K \in \mathcal{L}(E_{\mathbb{C}}, E_{\mathbb{C}})$. If $(I \otimes K^*)\kappa = \alpha\kappa$, $\alpha \in \mathbb{C}$ with $\alpha \neq 0$, then we have*

$$(\Xi_{0,2}(\kappa) + \mathrm{d}\Gamma(K)) = \mathcal{G}_{\kappa,\alpha}\mathrm{d}\Gamma(K)\mathcal{G}_{\kappa,\alpha}^{-1}.$$

THEOREM 4.6. *Let* $\kappa \in (E_{\mathbb{C}}^{\otimes 2})_{\mathrm{sym}}^{*}$ *and* $K \in \mathcal{L}(E_{\mathbb{C}}, E_{\mathbb{C}})$ *satisfying* $(I \otimes K^{*})\kappa = \alpha\kappa$, $\alpha \in \mathbb{C}$ *with* $\alpha \neq 0$. *If there exists* $\omega > 0$ *such that* $\omega I + K$ *is the infinitesimal generator of a differentiable contraction one-parameter semigroup* $\{\Upsilon_{\theta}\}_{\theta \geqslant 0} \subset \mathcal{L}(E_{\mathbb{C}}, E_{\mathbb{C}})$, *then we have*

$$(\Xi_{0,2}(\kappa) + \mathrm{d}\Gamma(K))\mathcal{G}_{\kappa,\alpha} G \mathcal{G}_{\kappa,\alpha}^{-1}\phi = -(\phi - E(\mathcal{G}_{\kappa,\alpha}^{-1}\phi)), \quad \phi \in (E),$$

where

$$G\phi = \int_{0}^{\infty} \Gamma(\Omega_{s})(\phi - E(\phi))\,\mathrm{d}s, \quad \phi \in (E)$$

and $\Omega_{\theta} = \mathrm{e}^{-\omega\theta}\Upsilon_{\theta}$ *for any* $\theta \geqslant 0$.

The proof is obvious from Theorem 4.5 and (4.5).

References

1. Carmona, R.: Potentials on abstract Wiener space, *J. Funct. Anal.* **26** (1977), 215–230.
2. Chung, C.-H., Chung, D. M. and Ji, U. C.: One-parameter groups and cosine families of operators on white noise functions, *J. Korean Math. Soc.* **37** (2000), 687–705.
3. Chung, D. M., Chung, T. S. and Ji, U. C.: A simple proof of analytic characterization theorem for operator symbols, *Bull. Korean Math. Soc.* **34** (1997), 421–436.
4. Chung, D. M., Chung, T. S. and Ji, U. C.: Products of white noise functionals and associated derivations, *J. Korean Math. Soc.* **35** (1998), 559–574.
5. Chung, D. M. and Ji, U. C.: Transformation groups on white noise functionals and their applications, *Appl. Math. Optim.* **37** (1998), 205–223.
6. Chung, D. M. and Ji, U. C.: Some Cauchy problems in white noise analysis and associated semigroups of operators, *Stochastic Anal. Appl.* **17** (1999), 1–22.
7. Chung, D. M. and Ji, U. C.: Transformations on white noise functionals with their applications to Cauchy problems, *Nagoya Math. J.* **147** (1997), 1–23.
8. Chung, D. M., Ji, U. C. and Obata, N.: Higher powers of quantum white noises in terms of integral kernel operators, *Infin. Dimens. Anal. Quantum Probab. Relat. Top.* **1** (1998), 533–559.
9. Cochran, W. G., Kuo, H.-H. and Sengupta, A.: A new class of white noise generalized functions, *Infin. Dimens. Anal. Quantum Probab. Relat. Top.* **1** (1998), 43–67.
10. Gross, L.: Potential theory on Hilbert space, *J. Funct. Anal.* **1** (1967), 123–181.
11. Hida, T.: *Analysis of Brownian Functionals*, Carleton Math. Lecture Notes 13, Carleton University, Ottawa, 1975.
12. Hida, T., Obata, N. and Saitô, K.: Infinite dimensional rotations and Laplacians in terms of white noise calculus, *Nagoya Math. J.* **128** (1992), 65–93.
13. Hida, T., Kuo, H.-H. and Obata, N.: Transformations for white noise functionals, *J. Funct. Anal.* **111** (1993), 259–277.
14. Hida, T., Kuo, H.-H., Potthoff, J. and Streit, L. (eds): *White Noise: An Infinite Dimensional Calculus*, Kluwer Acad. Publ., Dordrecht, 1993.
15. Kang, S. J.: Heat and Poisson equations associated with number operator in white noise analysis, *Soochow J. Math.* **20** (1994), 45–55.
16. Kondratiev, Yu. G. and Streit, L.: Spaces of white noise distributions: Constructions, descriptions, applications I, *Rep. Math. Phys.* **33** (1993), 341–366.
17. Kubo, I. and Takenaka, S: Calculus on Gaussian white noise I–IV, *Proc. Japan Acad. A* **56** (1980), 376–380; 411–416; **57** (1981), 433–437; **58** (1982), 186–189.

18. Kuo, H.-H.: Potential theory associated with Uhlenbeck–Ornstein process, *J. Funct. Anal.* **21** (1976), 63–75.
19. Kuo, H.-H.: On Laplacian operators of generalized Brownian functionals, In: K. Itô and T. Hida (eds), *Stochastic Processes and Applications*, Lecture Notes in Math. 1203, Springer, New York, 1986, pp. 119–128.
20. Kuo, H.-H.: Stochastic differential equations of generalized Brownian functionals, Lecture Notes in Math. 1390, Springer, New York, 1989, pp. 138–146.
21. Kuo, H.-H.: *White Noise Distribution Theory*, CRC Press, Boca Raton, 1996.
22. Lee, Y.-J.: Applications of the Fourier–Wiener transform to differential equations on infinite dimensinal space, I, *Trans. Amer. Math. Soc.* **262** (1980), 259–283.
23. Obata, N.: An analytic characterization of symbols of operators on white noise functionals, *J. Math. Soc. Japan* **45** (1993), 421–445.
24. Obata, N.: *White Noise Calculus and Fock Space*, Lecture Notes in Math. 1577, Springer, New York, 1994.
25. Piech, M. A.: Parabolic equations associated with the number operator, *Trans. Amer. Math. Soc.* **194** (1974), 213–222.
26. Potthoff, J. and Streit, L.: A characterization of Hida distributions, *J. Funct. Anal.* **101** (1991), 212–229.
27. Saitô, K.: A C_0-group generated by the Lévy Laplacian II, *Infin. Dimens. Anal. Quantum Probab. Relat. Top.* **1** (1998), 425–437.
28. Yosida, K.: *Functional Analysis*, 6th edn, Springer, New York, 1980.

Acta Applicandae Mathematicae **63:** 101–117, 2000.

Exponential Moments of Solutions for Nonlinear Equations with Catalytic Noise and Large Deviation ⋆

ISAMU DÔKU
Department of Mathematics, Saitama University, Urawa 338-8570, Japan

(Received: 23 March 1999)

Abstract. Nonlinear equation with catalytic noise is considered. We discuss the existence of catalytic superprocess associated with the equation and derive the exponential moment formula. Moreover, we prove the large deviation principle for catalytic superprocesses.

Mathematics Subject Classifications (2000): 60G57, 60J80, 60F10.

Key words: nonlinear reaction diffusion equation, catalyst process, catalytic medium, super-Brownian motion, superprocess, branching rate functional, branching measure-valued process, collosion local time, cumulant equation, large deviation.

1. Introduction and Notations

We consider the nonlinear differential equation with catalytic noise. This is a rather new type of equation where the coefficient function of the nonlinear term is given by the so-called catalyst process [Dk99c]. The purpose of this paper is to discuss the existence of the catalytic superprocess asociated with the equation and establish the exponential moment formula (cf. Equation (9) in Section 5). Based on this, we give a probabilistic interpretation of solutions for the nonlinear equation with catalytic noise. Furthermore, as its application, we derive the large deviation principle for superprocesses. More precisely, we consider the following nonlinear reaction diffusion equation with catalytic noise:

$$\partial_s u(s, y) + (\kappa/2)\Delta u(s, y) + \psi(s, y) = X_t(\omega)u^2(s, y).$$

Here we mean by the catalytic noise X_t a super-Brownian motion (or Dawson–Watanabe superprocess) [D93, W68] with a simple branching rate functional which is given by the Lebesgue measure multiplied by a constant (cf. Remark 1). So that, it turns out to be that the system governed by the above equation describes its time evolution in a catalytic medium [Dk99b]. In Section 3 we treat the nonlinear

⋆ Research supported in part by JMESC Grant-in-Aids SR(C) 07640280 and also by JMESC Grant-in-Aids CR(A) 09304022, CR(A) 10304006.

differential equation with catalytic noise and discuss the existence and uniqueness of its solutions (cf. Theorems 1 and 2). Next we construct the Brownian collision local time (BCLT) $L = L_{[W,\rho]}$ (Proposition 3) and study some basic properties that the BCLT should satisfy (Proposition 4). The catalytic superprocess is constructed in Section 5 that is associated with the nonlinear equation with catalytic noise (Theorem 5). In other words, the integral of the nonlinear term of the equation relative to the BCLT L provides with a rigorous expression of the corresponding log-Laplace equation (cf. Equation (10)). This enables us to establish the Laplace functional formalism for the catalytic superprocess. On this account, we can derive the probabilistic representation of solutions for the nonlinear equation with catalytic noise (Theorem 6). As its application, we discuss in Section 6 the large deviation and prove the large deviation principle (LDP) for the catalytic superprocess. This is the main result in the paper (cf. Theorem 12). The proof of weak large deviation principle, which is the key result for LDP, will be given in Section 7. For other related results on LDP, see [DeR98, FK94, IL93, S97] (see also [DmZ93] for LD techniques).

Let p be a positive number such that $p > d$ where d is the space dimension. Define the reference function φ_p by $\varphi_p(y) := (1 + |y|^2)^{-p/2}$ for $y \in R^d$. We denote by $\mathcal{B}^p \equiv \mathcal{B}^p(R^d)$ the space of real-valued Borel measurable functions f on R^d such that $|f(x)| \leqslant C(f) \cdot \varphi_p(x)$ holds for every x in R^d, for some positive constant $C(f)$ depending on the function f. The space $\mathcal{B}_+^p$ consists of all positive elements in $\mathcal{B}^p$. For a time interval I in R_+, $\mathcal{B}^{p,I}$ denotes the space of the functions $f = f(s,x)$ in $\mathcal{B}(I \times R^d)$ such that there is a positive constant $C(f)$ depending on f satisfying $|f(s,\cdot)| \leqslant C(f) \cdot \varphi_p$ for $s \in I$. $\mathcal{C}^p \equiv \mathcal{C}^p(R^d)$ is the space of real-valued continuous functions f on R^d such that $|f(x)| \leqslant C(f) \cdot \varphi_p(x), \forall x \in R^d$ for some positive constant $C(f)$ depending on f. Both $\mathcal{B}^p$ and $\mathcal{C}^p$ are equipped with the norm $\|f\| := \|f/\varphi_p\|_\infty$, where $\|\cdot\|_\infty$ is the uniform norm. Likewise, $\mathcal{C}_+^p$ (resp. $\mathcal{C}^{p,I}$) is the counterpart of $\mathcal{B}_+^p$ (resp. $\mathcal{B}^{p,I}$) for continuous functions.

Let $\mathcal{M}_p \equiv \mathcal{M}_p(R^d)$ be the set of all locally finite nonnegative measures μ on R^d, such that

$$\|\mu\|_p := \langle \mu, \varphi_p \rangle = \int_{R^d} \varphi_p(y)\mu(\mathrm{d}y) < \infty.$$

$\mathcal{M}_p$ is the set of tempered measures on R^d with the p-vague topology. While, $M_F = M_F(R^d)$ is the set of all finite measures on R^d. We denote by

$$W^\kappa := \{W_s^\kappa, \Pi_{s,a}^\kappa, s \geqslant 0, a \in R^d\}$$

the d-dimensional Brownian motion with generator $\kappa\Delta/2$ and the canonical path space $\Omega = C(R_+; R^d)$. We write $W_s^\kappa(\omega) = \omega(s)$, $\omega \in \Omega$ for its canonical realization. Especially when $\kappa = 1$, we call it the standard Brownian motion and suppress the parameter $\kappa = 1$ in its notation. For convention in the theory of measure-valued processes, we would rather use the notation $P[X]$ for the mathematical expectation than the usual $E[X] = E^P[X] = \int_\Omega X\,\mathrm{d}P$.

2. Preliminaries

According to Dawson: [D93], we may use the Brownian motion (BM) with generator $\frac{1}{2}\Delta$ (as an underlying process) to define the super-Brownian motion (SBM) (or Dawson–Watanabe superprocess) in terms of the martingale problem formulation ([Dk97]). That is to say, for each initial measure μ in M_F, there exists a probability measure P_μ on $(\Omega', \mathcal{F}')$ with $\Omega' = C(R_+; M_F)$ such that $X_0 = \mu$, P_μ-a.s.,

$$M_t(\psi) := \langle X_t, \psi\rangle - \langle \mu, \psi\rangle - \int_0^t \left\langle X_s, \left(\frac{1}{2}\Delta\right)\psi\right\rangle \mathrm{d}s, \quad \forall t > 0,\ \psi \in \mathrm{Dom}(\Delta/2),$$

is a continuous $(\mathcal{F}_t')$-martingale under P_μ, where the quadratic variation process $\langle M(\psi)\rangle_t$ is given by

$$\langle M_\cdot(\psi)\rangle_t = 2\gamma \int_0^t \int_\Omega \psi(\eta)^2 X_s(\mathrm{d}\eta)\, \mathrm{d}s, \qquad \forall t > 0,\ P_\mu\text{-a.s.}$$

In connection with application to the succeeding section, we shall introduce an alternative formulation of SBM. We begin with the nonlinear parabolic equation and determine the superprocess by making use of its solution via the Laplace functional. In fact, we consider the following nonlinear reaction diffusion equation in backward formulation (of convenience for later discussion) with the terminal condition:

$$-\frac{\partial v}{\partial s} = \frac{1}{2}\Delta v - \gamma \cdot v^2, \qquad v|_{s=t} = \varphi \in \mathcal{C}_+^p, \tag{1}$$

where γ is a positive constant. It is well known ([Dy94]) that the solution $v \equiv v(\cdot, t, \cdot)$ of the log-Laplace equation

$$v(s,t,a) = \Pi_{s,a}\left[\varphi(W_t) - \int_s^t v^2(r,t,W_r)\gamma\, \mathrm{d}r\right], \qquad 0 \leqslant s \leqslant t,\ a \in R^d \tag{2}$$

uniquely solves Equation (1) in the backward formulation.

Remark 1. It is interesting to note that the second term at the right-hand side in Equation (2) can be regarded as an integral with respect to the branching rate functional K ([DF97, §2.2], [DkK99, §4]) of the special case, namely, $K(\mathrm{d}r) = \gamma\, \mathrm{d}r$. Heuristically, this simply corresponds to the event that each corresponding X-particle branches with the constant rate $\gamma > 0$, on a phenomenal basis.

Now we introduce the super-Brownian motion (SBM) as a catalyst process, which is used to describe the catalytic medium in the next section. According to Dynkin's approach [Dy94], there exists an $\mathcal{M}_p$-valued critical SBM (or Dawson–Watanabe superprocess) $X = X^K$ with branching rate functional $K(\mathrm{d}r) = \gamma\, \mathrm{d}r$ with the Laplace transition functional

$$P_{s,\mu} \exp\langle X_t^K, -\varphi\rangle = \exp\langle \mu, -v^{[\varphi]}(s,t,\cdot)\rangle, \qquad 0 \leqslant s \leqslant t, \tag{3}$$

for μ in $\mathcal{M}_p$ and $\varphi \in \mathcal{B}_+^p$, where $v^{[\varphi]}(s,t,\cdot)$ is a solution of the nonlinear reaction diffusion equation (1). In fact, $X = X^K = X^{\gamma \mathrm{d}r}$ is a time-homogeneous Markov process. From now on, we would rather use the notation ρ^γ than X^K with $K(\mathrm{d}r) = \gamma\,\mathrm{d}r$. We call it the catalyst process, as the naming is originally due to Dawson and Fleischmann [DF97].

Consequently, the solution $v(s,t,\cdot) \equiv v^{[\varphi]}(s,t,\cdot)$ of Equation (1) may be expressed by

$$v(s,t,a) = -\log P_{s,\delta_a}[\exp\langle \rho_t^\gamma, -\varphi\rangle], \quad t \geqslant s,\ a \in R^d,\ \varphi \in \mathcal{C}_+^p. \tag{4}$$

This is nothing but a probabilistic representation of solutions of the nonlinear reaction diffusion equation. This probabilistic interpretation is, for instance, due to E. B. Dynkin ([Dy91, Dk99a]).

3. Nonlinear Differential Equation with Catalytic Noise

Let $\rho^\gamma = \{\rho_t^\gamma; t \geqslant s\}$ be the catalyst process defined in the previous section. The principal object of this section is the following nonlinear reaction diffusion equation with catalytic noise:

$$\mathcal{L}u \equiv \frac{\partial u}{\partial s} + \frac{\kappa}{2}\Delta u + \psi - \rho_s^\gamma u^2 = 0, \quad u|_{s=t} = \varphi. \tag{5}$$

Here we can regard the noise term ρ_s^γ as an $\mathcal{M}_p$-valued continuous path since we have a modification $\tilde{X}^K$ of the SBM X^K with continuous paths ([DF97, Dk99c]). The existence of solutions to Equation (5) can be attributed to the problem of a generalized cumulant equation, which is actually associated with a more general equation than (5).

Before discussing the cumulant equation, we need to introduce some notations. Let $\{T_t^\kappa; t \geqslant 0\}$ be the semigroup with generator $\kappa\Delta/2$, and set

$$(U^\kappa f)(s,\cdot) := \int_s^t T_{r-s}^\kappa f(r,\cdot)\,\mathrm{d}r, \quad \text{for } f \in \mathcal{C}^{p,I},\ s \in I = [L,T].$$

DEFINITION 1. We say that a continuous additive functional (CAF) $A = A_{[W^\kappa]}$ of the Brownian motion W^κ belongs to the class $\mathcal{K}$ if A is locally admissible, i.e., if

$$\sup_{a \in R^d} \Pi_{s,a}^\kappa \int_s^t \varphi_p(W_r^\kappa) A_{[W^\kappa]}(\mathrm{d}r)$$

vanishes as s, t tends to r_0 for some positive r_0.

For $f \in \mathcal{C}^{p,I}$ and $s \in I = [L,T]$, define

$$(Z^\kappa[A]f)(s,\cdot) := \Pi_{s,a}^\kappa \int_s^t f(r, W_r^\kappa) A_{[W^\kappa]}(\mathrm{d}r), \quad A \in \mathcal{K}.$$

Conventionally, we write it formally as $\int_{[s,t]} T^{\kappa}_{r-s}(f(r,\cdot))A_{[W^{\kappa}]}(\mathrm{d}r)$. We introduce the functional

$$F(\kappa, \varphi, \psi, \rho^{\gamma}, u) := u - T^{\kappa}_{t-\cdot}\varphi - U^{\kappa}\psi + Z^{\kappa}[A](u^2) \tag{6}$$

defined for

$$\{\kappa, \varphi, \psi, \rho^{\gamma}, u\} \in R_+ \times \mathcal{C}^p \times \mathcal{C}^{p,I} \times C(R_+; \mathcal{M}_p) \times \mathcal{C}^{p,I}.$$

We will study the following generalized cumulant equation:

$$F(\kappa, \varphi, \psi, \rho^{\gamma}, u) = 0 \tag{7}$$

which covers Equation (5). The purpose of this section is to solve Equation (7). Some routine work for the functional equation theory allows us to obtain the uniqueness of solutions to (7).

THEOREM 1 (Uniqueness). *Let $A \in \mathcal{K}$ be locally bounded characteristic. For each initial data $\{\kappa, \varphi, \psi, \rho^{\gamma}\}$ in $R_+ \times \mathcal{C}^p \times \mathcal{C}^{p,I} \times C(R_+; \mathcal{M}_p)$, there exists at most one element $u \in \mathcal{C}^{p,I}$ in the sense of $\|\cdot\|$ which solves the generalized cumulant equation $F(\kappa, \varphi, \psi, \rho^{\gamma}, u) = 0$.*

Proof. Assume that there are two elements $u, v \in \mathcal{C}^{p,I}$ such that

$$F(\kappa, \varphi, \psi, \rho^{\gamma}, u) = F(\kappa, \varphi, \psi, \rho^{\gamma}, v) = 0$$

for

$$\{\kappa, \varphi, \psi, \rho^{\gamma}\} \in R_+ \times \mathcal{C}^p \times \mathcal{C}^{p,I} \times C(R_+; \mathcal{M}_p).$$

From (6) we readily obtain

$$\|u(s) - v(s)\| \leqslant \left\|Z^{\kappa}[A](u^2)(s) - Z^{\kappa}[A](v^2)(s)\right\|_{\mathcal{C}^p}. \tag{$*1$}$$

Recall that $\mathcal{C}^p$ and $\mathcal{C}^{p,I}$ are Banach algebras with respect to the pointwise product of functions. Moreover, Equation ($*1$) may be estimated from

$$\sup_{s\in I}\|u(s) + v(s)\| \left\| \int_s^t T^{\kappa}_{r-s}(u(r) - v(r))\varphi_p(\cdot)A_{[W^{\kappa}]}(\mathrm{d}r)\right\|$$

$$\leqslant C_I \left\|\int_s^t T^{\kappa}_{r-s}\varphi_p A_{[W^{\kappa}]}(\mathrm{d}r)\right\| \tag{$*2$}$$

because linear operators $\{T^{\kappa}_t\}$ acting in $\mathcal{C}^p$ are uniformly bounded over the bounded region of t, κ, and we have only to pay attention to the finiteness of the upper bound $\leqslant C_{\kappa,I}\|\varphi_p(\cdot)\|_{\infty}$ of Equation ($*1$). By local admissibility, we can choose small $\varepsilon > 0$ which gives the upper estimate of Equation ($*2$), and the required result is obtained from *reductio ad absurdum* together with the above estimate. □

Moreover, resorting to functional analysis, we can prove the existence of solutions to Equation (7) by employing the implicit function theorem and the standard iteration scheme. Now we state the assertion with the proof divided into two parts, which will be given separately below and in the Appendix.

THEOREM 2 (Existence). *Suppose the same assumptions on A as in Theorem* 1. *For each initial data* $\{\kappa, \varphi, \psi, \rho^\gamma\}$ *in* $R_+ \times \mathcal{C}^p \times \mathcal{C}^{p,I} \times C(R_+; \mathcal{M}_p)$, *there exists a solution u in $\mathcal{C}^{p,I}$ of Equation* (7).

Proof. Note that we can choose an approximating sequence $\{A^{(n)}\}_n$ of $A \in \mathcal{K}$ such that with probability one, $A^{(n)}(J) \nearrow A(J)$ as $n \to \infty$ for all open intervals J of R_+ (cf. Remark 1 in [DF97, p. 223]). Assume that there exists a unique nonnegative bounded solution u_n of Equation (7) with $A_{[W^\kappa]}$ replaced by $A^{(n)}_{[W^\kappa]}$ for each n. Then we readily obtain

$$0 \leqslant u_n(s,a) \leqslant C_{\kappa,I} \cdot \varphi_p(a), \quad s \in I,\ a \in R^d,$$

because both $T^\kappa \varphi$ and $U^\kappa \psi$ satisfy the locally bounded characteristic property so that the solution $\{u_n\}$ are uniformly dominated. Then the pointwise limit u ($\geqslant 0$) of u_n as $n \to \infty$ is also dominated. The passage to the limit $n \to \infty$ of $F(\kappa, \varphi, \psi, \rho^\gamma, u_n) = 0$ leads to the required result if we can show that

$$Z^\kappa[A^{(n)}](u_n^2)(s,a) \to Z^\kappa[A](u^2)(s,a) \quad (\text{as } n \to \infty) \tag{$*$}$$

for each $(s,a) \in I \times R^d$. In fact, we get

(i) $|Z^\kappa[A^{(n)}](u_n^2) - Z^\kappa[A^{(n)}](u^2)| \leqslant \sup_r \|u_n(r)+u(r)\| \cdot \|u_n - u\| C_{\kappa,I}\varphi_p(a) \to 0$

(as $n \to \infty$) and also

(ii) $|Z^\kappa[A^{(n)}](u^2) - Z^\kappa[A](u^2)| \leqslant \Pi^\kappa_{s,a}|\langle (A^{(n)} - A)_{[W^\kappa]}, u^2(\cdot, W_r^\kappa)\rangle| \to 0$

(as $n \to \infty$). To get $(*)$, we have only to combine the above estimates (i), (ii). Therefore, it remains to show the existence of solutions u_n of $F(\kappa, \varphi, \psi, \rho^\gamma, u)[A^{(n)}] = 0$. This will be proved in the Appendix. □

Remark 2. Dawson and Fleischmann [DF97] have treated only the special case $\kappa = 1$ and $\psi \equiv 0$ of Equation (5). So our result is a generalization of their existence and uniqueness theorem. Moreover, the method we have adopted here is based upon nonlinear functional analysis and is a more general approach to nonlinear equations than what they used in [DF97].

4. Regular Path and Collision Local Time

First we introduce a certain class of measure-valued continuous paths which is suitable for defining the corresponding collision local time (see Remark 7 below).

DEFINITION 2. The path η is said to be an element of the Regular Path Class $\mathcal{R}_p$ if $\eta \in C(R_+; \mathcal{M}_p)$ and for all $N > 0$,

$$\sup_{\substack{0 \leqslant s \leqslant N \\ a \in R^d}} \int_s^{s+\varepsilon} \langle \eta_r, \varphi_p \cdot p^\kappa(r-s, a, \cdot)\rangle \, \mathrm{d}r \to 0 \quad (\text{as } \varepsilon \to 0),$$

where p^κ is the transition density function of the BM with generator $\kappa\Delta/2$.

Roughly speaking, that η is regular means that ε-accumulated densities of the finite measure-valued path $\varphi_p \cdot \eta$ vanishes uniformly on $[0, N] \times R^d (\forall N > 0)$ as ε tends to zero.

Remark 3. The original idea of a branching rate functional is due to Dynkin's additive functional approach ([Dy94]). However, the theory is not directly applicable to the catalytic reaction diffusion equations. So Dawson and Fleischmann [DF97] extended it to cover the catalytic case ([DkK99]). Of course, our definition of regular paths and the following results are extensions of their work ([Dk99c]).

Let $\varepsilon \in (0, 1]$. Suggested by [DF97], we define $L^\varepsilon = L^\varepsilon_{[W^\kappa, \rho^\gamma]}$ by

$$L^\varepsilon_{[W^\kappa, \rho^\gamma]}(\mathrm{d}r) := \langle \rho^\gamma_r, p^\kappa(\varepsilon, W^\kappa_r, \cdot) \rangle \, \mathrm{d}r. \tag{8}$$

Then, L^ε is a continuous additive functional (CAF) of the Brownian motion W^κ in the sense of Dynkin [Dy94]. Here L^ε is also the collision local time (CLT) of $\rho^\gamma \equiv X^K$ with the ε-vicinity of the Brownian path W^κ in the sense of Barlow, Evans and Perkins [BeP91]. It is not difficult to show the following proposition. All through this section we may consider only the special case $\kappa = 1$ for the proofs of the propositions without loss of generality as far as the existence of collision local time is concerned.

PROPOSITION 3. *Let $d \leqslant 3$. If $\eta \in \mathcal{R}_p$, then there exists an additive functional $L = L_{[W^\kappa, \eta]}$ of the Brownian path W^κ such that, for every $\psi \in \mathcal{C}^{p,[0,N]}_+$ $(N > 0)$,*

$$\sup_{\substack{0 \leqslant s \leqslant N \\ a \in R^d}} \Pi^\kappa_{s,a} \sup_{s \leqslant t \leqslant N} \left| \int_s^t \psi(r, W^\kappa_r) L^\varepsilon_{[W^\kappa, \eta]}(\mathrm{d}r) - \int_s^t \psi(r, W^\kappa_r) L_{[W^\kappa, \eta]}(\mathrm{d}r) \right|^2$$

vanishes as ε approaches to zero.

Proof. Set $A^\varepsilon(\mathrm{d}r) = \langle \psi(r, \cdot)\eta_r, p(\varepsilon, W_r, \cdot) \rangle \, \mathrm{d}r$. Then note that

$$\sup_{\substack{s \leqslant N \\ a \in R^d}} \int_s^{s+\varepsilon} \int \psi(r, b) p(r - s, a, b) \eta_r(\mathrm{d}b) \, \mathrm{d}r \to 0 \quad (\text{as } \varepsilon \downarrow 0), \; N > 0.$$

By virtue of the regularity of η, the assertion immediately follows from a slight modification of the proof of Proposition 6(a) of [DF97, p. 258] and Theorem 4.1 (on the general convergence of CAF's) [EP94, p. 144], together with the aforementioned fact. Actually, L is given by the limit of A^ε divided by ψ. □

Since it is true that the path $\rho^\gamma(\omega) \equiv X^K(\omega)$ is contained in $\mathcal{R}_p$ with probability one, we can apply Proposition 3 for the case $\eta = \rho^\gamma$. Consequently, there is the so-called Brownian collision local time $L \equiv L_{[W^\kappa, \rho^\gamma]}$ of the catalyst process $\rho^\gamma = \rho(\gamma)$. It is easy to show the following properties that this limit L possesses.

PROPOSITION 4. *Let $\delta > 0$, $d \leqslant 3$ and $\xi \in (0, 1/4)$. The Brownian collision local time $L \equiv L_\delta = L_{[W^\kappa, \rho^\gamma_{\delta+(\cdot)}]}$ satisfies the following properties:*

(a) *L_δ is continuous, i.e., it does not carry mass at any single point;*
(b) *L_δ is locally admissible, i.e., as $s, t \to r_0$, ($r_0 \geqslant 0$)*

$$\sup_{a \in R^d} \Pi^\kappa_{s,a} \int_s^t \varphi_p(W^\kappa_r) L_{[W^\kappa, \rho^\gamma_{\delta+(\cdot)}]}(\mathrm{d}r) \to 0;$$

(c) *for each $N > 0$, there exists a positive constant $C_{\kappa,N}$ such that*

$$\Pi^\kappa_{s,a} \int_s^t \varphi_p(W^\kappa_r)^2 L_{[W^\kappa, \rho^\gamma_{\delta+(\cdot)}]}(\mathrm{d}r) \leqslant C_{\kappa,N} |t-s|^\xi \varphi_p(a)$$

for $0 \leqslant s \leqslant t \leqslant N$ and $a \in R^d$.

Proof. Only the property (c) matters. The same discussion as in the second step of the proof of Theorem 4 [DF97, p. 260] together with the domination property of the heat flow (e.g. (2.10) in [DF97, p. 225]) provides us with an estimate $\Pi_{s,a} \int_s^{s+\varepsilon} \varphi_p^2(W_r) L_{[W,\rho^\gamma]}(\mathrm{d}r) \leqslant C\varepsilon^\xi \varphi_p(a)$. This concludes the assertion. □

From (a) in the above, we know that the BCLT L_δ lies in the class $\mathcal{K}$. We say that the continuous additive functional A belongs to the class $\mathcal{K}^\xi$ if A additionally satisfies the conditions (b), (c) in Proposition 4. Notice that $\mathcal{K}^\xi \subset \mathcal{K}$.

Remark 4. We assume tacitly that the parameter δ is positive in the above proposition. However, if $\delta = 0$, then some proper additional condition is required to let it make sense.

Indeed, we need to pose, for instance, the following condition for the case $\delta = 0$: The mapping $R_+ \times R^d \ni (r, z) \mapsto \rho^\gamma_0 * q^\kappa(0, r, z)$ is locally ξ-Hölder continuous with Hölder constant proportional to $\|\rho^\gamma_0\|_p$, where the convolution is given by

$$\rho^\gamma_0 * q^\kappa(0, r, z) = \int_0^r \int_{R^d} p^\kappa(s, a, z) \rho^\gamma_0(\mathrm{d}a)\, \mathrm{d}s$$

with $P_{0,\mu}$-probability one.

Remark 5. Since ρ^γ is an $\mathcal{M}_p$-valued SBM, in the above-mentioned condition the quantity $\|\rho^\gamma_0\|_p := \langle \rho^\gamma_0, \varphi_p \rangle = \int \varphi_p \rho^\gamma_0(\mathrm{d}a)$ is finite.

Remark 6. Heuristically, the collision local time L between a Brownian particle with path $W^\kappa(\omega)$ and the catalytic medium $\rho^\gamma(\omega)$ is given [DF97, §1.2] by

$$L_{[W^\kappa, \rho^\gamma]}((s,t)) = \int_s^t \int \delta_b(W^\kappa_r) \rho^\gamma_r(\mathrm{d}b)\, \mathrm{d}r.$$

Here $\rho^\gamma_r(b)$ can be considered as the amount of catalyst present at time r at b which is encountered by a reactant particle with path W^κ, when ρ^γ_r is a singular measure (cf. [Dk99b, Dk98]). Hence, the term $\delta_r(W^\kappa_r)\ \rho^\gamma_r(\mathrm{d}b)$ allows us to realize the intuitive description at the microscopic level that a tagged Brownian particle with path W^κ enjoys branching according to a clock given by the additive functional $L = L_{[W^\kappa, \rho^\gamma]}$ ([DkK99]).

Remark 7. For the one-dimensional case with $\kappa = 1$, it is well known ([D93]] or [Dk99b]) that the continuous SBM lies in the nicer space of absolutely continuous measures. Therefore, there exists the Radon–Nikodym derivative $\rho_t(b) \equiv \rho_t(\mathrm{d}b)/\mathrm{d}b$ with respect to the Lebesgue measure $\mathrm{d}b$ which is even a jointly continuous density field $\{\rho_t(b); t > 0, b \in R\}$ (e.g. [KS88]). So that

$$L_{[W,\rho]}(\mathrm{d}r) := \left\{\int \delta_b(W_r)\,\mathrm{d}b\right\}\rho_r(W_r)\mathrm{d}r$$

defines a continuous additive functional of BM. For dimensions $d \geqslant 4$, the Brownian collision local time $L = L_{[W,\rho]}$ degenerates to 0 because the closure of the graph of the SBM ρ never intersects with the graph of W (cf. [BP94], Proposition 1.3, p. 1275). On the other hand, the BCLT L exists nontrivially in dimensions $d = 2$ and $d = 3$ [DF97, p. 218], although the random measures $\rho_t(\mathrm{d}b)$ are singular [DH79] and we cannot use the above-mentioned expression for the definition of BCLT. By virtue of Theorem 4.1 in [EP94, p. 144] together with Proposition 4.7 in [EP94, p. 150], we can also show a similar result for our case $L = L_{[W^\kappa, \rho^\gamma]}$. Therefore, the restriction $d \leqslant 3$ is essential for our theory.

5. Catalytic Superprocess and Exponential Moment Formula

In Section 2, according to Dynkin's approach ([Dy94]), we have introduced the Laplace functional formalism with the log-Laplace equation, whereby the Dawson–Watanabe superprocess (a finite measure-valued Markov process) can be determined. The purpose of this section consists in reconstructing the Laplace functional formalism with the log-Laplace equation being adjusted for the catalytic superprocess associated with the nonlinear equation (5) with catalytic noise. We now introduce some additional notation. In analogy to $\mathcal{M}_p$, let $\mathcal{M}_p^I$ be the set of all measures ν on $I \times R^d$ such that

$$\langle \nu, \psi \rangle_I := \iint\limits_{I \times R^d} \psi(r, b)\nu(\mathrm{d}[r, b]) < +\infty, \quad \forall \psi \in \mathcal{B}^{p,I},$$

where we furnish $\mathcal{M}_p^I$ with the weakest topology such that the maps $\nu \mapsto \langle \nu, \psi \rangle_I$ are continuous for all $\psi \in \mathcal{C}^{p,I}$. Here we use the same notation for extended functions, even in $C(I \times R_c^d)$ by the one point compactification if necessary.

Due to Dynkin's AF approach, with a slight modification of Theorem 1.1 in [Dy91] and several results in [Dy91] with minor changes, we can derive the following formulation. Here using $L^\rho = L(\rho) = L(\rho^\gamma)$, we quote the BCLT L from Section 4 instead of $L_{[W^\kappa, \rho^\gamma]}$. Now we are ready to introduce one of the principal results in this paper, which includes the exponential moment formula (9) of solutions for nonlinear differential equation with catalytic noise.

THEOREM 5 (cf. [Dk99c]). *Let $d \leqslant 3$. If the branching rate functional K is given $P_{s,\nu}$-a.s. by the BCLT $L(\rho^\gamma)$ of the SBM ρ^γ, there exist a time inhomogeneous $\mathcal{M}_p$-valued Markov process*

$$X^L \equiv X^{L(\rho^\gamma)} = \{X_t^L = X_t^{L(\rho^\gamma)}, \mathbf{P}_{s,\mu}^{\kappa,\rho^\gamma}, \ s > 0, \ \kappa \in R_+, \ \mu \in \mathcal{M}_p\}$$

and an $\mathcal{M}_p^{[s,t]}$-valued weighted occupation measure process $Y^\rho \equiv \{Y_t^\rho\}$ defined by

$$\langle Y_t^\rho, \psi \rangle_{[s,t]} := \int_s^t \langle X_r^{L(\rho^\gamma)}, \psi(r, \cdot) \rangle \, \mathrm{d}r, \quad \forall \psi \in \mathcal{C}_+^I$$

with the Laplace transition functional

$$\mathbf{P}_{s,\mu}^{\kappa,\rho^\gamma} \exp\{-(\langle X_t^{L(\rho^\gamma)}, \varphi \rangle + \langle Y_y^\rho, \psi \rangle_{[s,t]})\} = \exp\{-\langle \mu, u(s,t,\cdot) \rangle\} \tag{9}$$

$P_{s,\nu}$-a.s. *for $0 \leqslant s \leqslant t$, $\mu \in \mathcal{M}_p$, $\varphi \in \mathcal{B}_t^p$ and $\psi \in \mathcal{B}_t^{p,I}$. Here the function $u(\cdot, t\cdot) \in \mathcal{C}_+^{p,I}$ is the unique solution of the log-Laplace equation*

$$\begin{aligned} &u(s,t,a) \\ &\quad = \Pi_{s,a}^\kappa \left[\varphi(W_t^\kappa) + \int_s^t \psi(r, W_r^\kappa) \, \mathrm{d}r - \int_s^t u^2(r,t,W_r^\kappa) L(\rho^\gamma)(\mathrm{d}r) \right]. \end{aligned} \tag{10}$$

Proof. For $\mathcal{K} \ni L(\rho^\gamma)$, there exists an approximating sequence $\{L^n\}_n$ belonging to Dynkin's Admissible Class $\mathcal{K}_0$ ($\subset \mathcal{K}$) (§3.3.3 of [Dy94, pp. 49–50]). In fact, we can take L^n ($\in \mathcal{K}_0$) as

$$L^n := \int_{[\cdot]} (1 \wedge n\varphi_p)(W_r^\kappa) L(\rho^\gamma)(\mathrm{d}r), \quad n \geqslant 1.$$

Then the existence of a time-inhomogeneous $M_F \times M_F$-valued Markov process $Z^n = \{X^{L^n}, Y^{L^n}\}$ with $\mathbf{P}_{s,\mu}$, $s \geqslant 0$, $\mu \in M_F$ satisfying

$$\mathbf{P}_{s,\mu} \exp\{-\langle X_t^{L^n}, \theta\varphi \rangle - \langle Y_t^{L^n}, \zeta\psi \rangle_{[s,t]}\} = \exp\langle \mu, -v(s,t,\cdot) \rangle, \tag{$*1$}$$

for

$$0 \leqslant s \leqslant t, \mu \in M_F, (\varphi, \psi) \in \mathcal{B}_+^p \times \mathcal{B}_+^{p,[0,t]}, \theta, \zeta \geqslant 0,$$

and

$$v(s,t,a) = \theta \cdot f(s,a) + \zeta \cdot g(s,a) - \Pi_{s,a}^\kappa \int_s^t v^2(r,t,W_r^\kappa) L^n(\mathrm{d}r) \tag{$*2$}$$

for fixed $t > 0$, $f, g \in \mathcal{B}_+^{p,[0,t]}$ is guaranteed by Dynkin's additive functional theory (Theorem 1.4, [Dy91a]). Indeed, ($*1$), ($*2$) are the particular cases of (1.36), (1.37) in [Dy93, p. 1198]. The domination property

$$\sup_{s \in I = [S,L]} T_{L-s}^\kappa \varphi(a) \leqslant C_0 \|\varphi\| \varphi_p(a), \quad \varphi \in \mathcal{B}_+^p$$

allows us to extend it to the $\mathcal{M}_p$-valued Markov process version $\tilde{Z}^n$. The fact that $L^n \nearrow L(\rho^\gamma)$ and the monotone convergence $v_n \searrow v$ in Equation (∗2) verify the convergence of the corresponding (∗1) with v replaced by v_n, for fixed s, t, μ. Note that the limit functional is continuous in (φ, ψ) and also that $v(s, t, a)$ vanishes if (φ, ψ) tends to null functions. Hence, the limiting functional is the Laplace functional of a random measure $\tilde{Z} = \{X^{L(\rho)}, Y^{L(\rho)}\}$ given by (∗1). Moreover, we can construct the laws of vectors $\{Z_{t_1}, \ldots, Z_{t_k}\}$ by taking advantage of the semi-group structure of the solutions for (∗2). On this account, the Markov process $\tilde{Z}$ in question is determined by these compatible finite-dimensional distributions. Note that the Markov process $\tilde{Z}$ is independent of the choice of $\{L^n\}$. On the other hand, it follows from Kolmogorov's moment method that $X^{L(\rho^\gamma)}$ has continuous paths, since the corresponding branching rate functional (= BCLT) $L(\rho^\gamma)$ is contained in $\mathcal{K}^\xi$. Thus, we attain the pathwise definition of Y^ρ as a measure on $[s, t] \times R^d$ as indicated in the statement of Theorem 5. With the help of the expectation formula for Y^ρ, it is routine work to verify that $\tilde{Z} = \{X^{L(\rho)}, Y^\rho\}$ satisfies (9), (10), namely, $f = T^\kappa\varphi$ and $g = U^\kappa\psi$, see Theorem 3.1 in [Dy93, p. 1226] (also see [Dy91, p. 90]). Clearly v solves the cumulant equation $F[A] = 0$ with $A_{[W^\kappa]}$ replaced by $L(\rho)$. □

In the above theorem, $\mathbf{P}^{\kappa,\rho^\gamma}_{s,\mu}$, $\kappa \in R_+, s \geqslant 0, \mu \in \mathcal{M}_p$ is the quenched distribution of X^L given ρ^γ. The existence and uniqueness of the solution to the log-Laplace equation (10) are guaranteed by Theorems 1 and 2 in Section 3. We call $X^L = X^{L(\rho^\gamma)}$ the catalytic superprocess in the catalytic medium ρ^γ. By the same argument as in [Dy91] we readily obtain the probabilistic representation of solutions for Equation (5) which realizes a probabilistic interpretation of the nonlinear equation with catalytic noise in terms of catalytic superprocess.

THEOREM 6 ([Dk99c]). *Let u be the unique solution to* (5). *Then it may be expressed by*

$$\begin{aligned} u(s, y) &\equiv u^{[\kappa,\varphi,\psi,\rho^\gamma]}(s, y) = V[\varphi, \psi] \\ &:= -\log \mathbf{P}^{\kappa,\rho^\gamma}_{s,\delta_y} \exp\bigl\{-\bigl(\langle X^L_t, \varphi\rangle + \langle Y^\rho_t.\psi\rangle_{[s,t]}\bigr)\bigr\}, \quad P_{s,\nu}\text{-a.s.} \qquad (11) \end{aligned}$$

for $(s, y) \in I \times R^d$, $I = [L, T]$, $L < T$, *and* $(\varphi, \psi) \in \mathcal{C}^p_+ \times \mathcal{C}^{p,I}_+$.

Proof. To get (11) from (10), it suffices to choose, in particular, $\mu = \delta_y$. □

Remark 8. The study to derive the probabilistic representation of the log type like (11) from the exponential moments is originally due to Dynkin [Dy91]. For other similar expressions for nonlinear elliptic equations and distinct probabilistic representations in terms of path-valued process, see [Dk99a].

6. Large Deviation Principle for Catalytic Processes

Recall that $\mathcal{C}_p$ is a separable Banach space. Let $(\mathcal{C}^*_p, \|\cdot\|_*)$ denote the dual Banach space relative to $(\mathcal{C}_p, \|\cdot\|)$. Then note that we can regard $\mathcal{M}_p$ as a convex subset of

$\mathcal{C}_p^*$ equipped with the weak-* topology. The p-vague topology in $\mathcal{M}_p$ is equivalent to the induced topology in $\mathcal{M}_p$ by the weak-* topology in $\mathcal{C}_p^*$. We have

LEMMA 7. *The duality pairing $\langle\langle\cdot,\cdot\rangle\rangle$ between $\mathcal{M}_p$ and $\mathcal{C}^p$ is continuous in both components, and in particular*

$$\|\mu\|_* = |\mu|_p \ (:= \langle\mu, \varphi_p\rangle), \quad \mu \in \mathcal{M}_p. \tag{12}$$

Proof. It suffices to note that the estimate

$$|\langle\langle\mu,\varphi\rangle\rangle| \leqslant \|\varphi\|\langle\mu,\varphi_p\rangle, \quad \varphi \in \mathcal{C}^p, \mu \in \mathcal{M}_p \tag{13}$$

holds. The continuity follows immediately, from (13). The equality (12) yields from (13), too, by the definition of $\|\mu\|_*$. □

Now we introduce a metric d_p in $\mathcal{M}_p$. Denote by $p\mathcal{C}_K$ the space of all positive elements f in $C_K(R^d; R)$ with compact support, equipped with uniform convergence topology.

DEFINITION 3. The meric d_p in $\mathcal{M}_p$ is defined by

$$d_p(\mu,\nu) := \sum_{n=0}^{\infty} \frac{1}{2^{n+1}}\left(1 - \exp\left\{-\frac{|\langle\langle\mu, f_n\rangle\rangle - \langle\langle\nu, f_n\rangle\rangle|}{\|f_n\|}\right\}\right)$$

for $\mu, \nu \in \mathcal{M}_p$ and some sequence $\{f_n\}_n \subset \mathcal{C}_+^p$.

Then the following are well-known results in connection with this metric.

LEMMA 8 ([Ka83]). (a) *There exists a sequence $\{f_n\}_{n\geqslant 1}$ of functions in $p\mathcal{C}_K$ such that, with $f_0 = \varphi_p$, $d_p(\cdot,\cdot)$, is a translation-invariant metric on $\mathcal{M}_p$, which generates the p-vague topology.*

(b) *$\mathcal{M}_p$ is a separable metric space with respect to the metric d_p.*

We define the open ball $B(\nu, r)$ in $\mathcal{M}_p$ with center ν and radius r by

$$B^{d_p}(\nu; r) := \{\mu \in \mathcal{M}_p; d_p(\mu,\nu) < r\}, \quad \nu \in \mathcal{M}_p, \ r > 0.$$

LEMMA 9 ([FK94]). *Each open ball $B^{d_p}(\nu; r)$, $\nu \in \mathcal{M}_p$, $r > 0$ is a convex subset of $\mathcal{M}_p$.*

Set $R_c^d := R^d \cup \{\infty\}$ with an isolated point ∞. φ_p^* denotes the extension of φ_p to R_c^d by setting $\varphi_p^*(\infty) := 1$. Define $\mathcal{M}_p^*$ as the set of all measures μ on R_c^d satisfying $\langle\mu,\varphi_p^*\rangle < \infty$ so that we can define the p-vague topology in $\mathcal{M}_p^*$ as in the case $\mathcal{M}_p$. Then the following lemma gives a criterion for relative compactness in $\mathcal{M}_p^*$:

LEMMA 10 ([I86] or [D93]). *Let A be a subset of $\mathcal{M}_p^*$. Then A is relatively compact if and only if there is some $k \in \mathbf{N}$ such that $A \subset \{\mu \in \mathcal{M}_p^*;\ \langle\mu,\varphi_p^*\rangle \leqslant k\}$.*

Let $\mathcal{U}$ be the system of all those nonempty subsets of $\mathcal{M}_p$ which are open and convex. Define

$$S_{\mu,t}(A) := - \lim_{\Lambda\to\infty} \frac{1}{\Lambda} \log \mathbf{P}^{\kappa,\rho^\gamma}_{0,\Lambda\mu}(X_t^{L(\rho^\gamma)}/\Lambda \in A), \quad P_{0,\nu}\text{-a.s.} \tag{14}$$

for all $A \in \mathcal{U}$, $\mu \in \mathcal{M}_p$, $t \geqslant 0$. We set

$$I_{\mu,t}(\nu) := \lim_{r\downarrow 0} S_{\mu,t}(B^{d_p}(\nu; r)) \quad (\nu \in \mathcal{M}_p), \tag{15}$$

for $\mu \in \mathcal{M}_p$ and $t \geqslant 0$. We are now in a position to state one of the main results in this paper, which provides the weak large deviation principle (LDP) ([Dk99c]) for catalytic superprocesses given in the previous section.

THEOREM 11 (Weak LDP). *The family* $\{1/\Lambda \log \mathbf{P}^{\kappa,\rho^\gamma}_{0,\Lambda\mu}(X_t^L/\Lambda \in (\cdot));\ \Lambda > 0\}$ *satisfies a weak large deviation principle with convex rate functional* $I_{\mu,t}$. *In other words, the following two types of estimates hold: for every* $t \geqslant 0$ *and* $\mu \in \mathcal{M}_p$,

$$\text{(I)}\ \liminf_{\Lambda\to\infty} \frac{1}{\Lambda} \log \mathbf{P}^{\kappa,\rho^\gamma}_{0,\Lambda\mu}(X_t^{L(\rho^\gamma)}/\Lambda \in G) \geqslant - \inf_{\lambda\in G} I_{\mu,t}(\lambda), \quad P_{0,\nu}\text{-a.s.},$$

for $\forall$ *open* $G \subset \mathcal{M}_p$ *and*

$$\text{(II)}\ \limsup_{\Lambda\to\infty} \frac{1}{\Lambda} \log \mathbf{P}^{\kappa,\rho^\gamma}_{0,\Lambda\mu}(X_t^{L(\rho^\gamma)}/\Lambda \in C) \leqslant - \inf_{\lambda\in C} I_{\mu,t}(\lambda), \quad P_{0,\nu}\text{-a.s.},$$

for $\forall$ *compact* $C \subset \mathcal{M}_p$.

This is the key result for derivation of the full large deviation principle (Theorem 12 below), which is one of the main results in this paper. The proof of Theorem 11 will be given in the next section. For $\mu \in \mathcal{M}_p$, $t > 0$, we denote by $\mathcal{C}_t^p[\mu]$ the largest open set of all those functions $\varphi \in \mathcal{C}^p$ such that $\Xi(\mu, t, \varphi) := \log \mathbf{P}^{\kappa,\rho^\gamma}_{0,\mu} \exp\langle X_t^{L(\rho)}, \varphi\rangle < +\infty$.

THEOREM 12 (Full LDP). *The family* $\{1/\Lambda \log \mathbf{P}^{\kappa,\rho^\gamma}_{0,\Lambda\mu}(X_t^L/\Lambda \in (\cdot));\ \Lambda > 0\}$ *satisfies the full large deviation principle with a good rate functional* $I_{\mu,t}$.

Proof. Take $\mu \in \mathcal{M}_p$, $t > 0$ and $\varphi \in \mathcal{C}_t^p[\mu]$ and fix them. Since $\mathcal{C}_t^p[\mu]$ is open, we can find $\theta > 0$ such that $(1+\theta)\varphi \in \mathcal{C}_t^p[\mu]$. Set the event $A := \{\langle X_t^{L(\rho)}/\Lambda, \varphi\rangle > N\} \subset \Omega$. A simple calculation reads

$$(*1)\ \Lambda^{-1} \log \mathbf{P}^{\kappa,\rho^\gamma}_{0,\Lambda\mu}[\exp\langle X_t^{L(\rho)}, \varphi\rangle;\ A] \leqslant \Xi(\Lambda\mu, t, (1+\theta)\varphi)/\Lambda - \theta N,$$

$\Lambda, N > 0$. The branching property [Dy94, p. 9] implies that $\Xi(\Lambda\mu, t, \varphi) = \Lambda \cdot \Xi(\mu, t, \varphi)$, hence the right-hand side in $(*1)$ proves to be finite. First letting $\Lambda \to \infty$ and then $N \to \infty$, we obtain

$$(*2)\ \lim_{N\to\infty} \lim_{\Lambda\to\infty} \log \mathbf{P}^{\kappa,\rho^\gamma}_{0,\Lambda\mu}[\exp\langle X_t^{L(\rho)}, \varphi\rangle; A] = -\infty.$$

For $\theta > 0$ small enough, we can deduce from $(*2)$ that

(∗3) $\lim_{N\to\infty}\lim_{\Lambda\to\infty}\Lambda^{-1}\log\mathbf{P}^{\kappa,\rho^\gamma}_{0,\Lambda\mu}(\langle X^{L(\rho)}_t/\Lambda,\theta\varphi_p\rangle\geqslant N)=-\infty$.

By taking (12) and Lemma 10 into consideration, for each $M>0$ we can find a compact subset C_M of $\mathcal{M}_p$ such that

(∗4) $\limsup_{\Lambda\to\infty}\Lambda^{-1}\log\mathbf{P}^{\kappa,\rho^\gamma}_{0,\Lambda\mu}(X^{L(\rho)}_t/\Lambda\in(C_M)^c)\leqslant -M$,

as far as we reinterpret the measure and distribution as those on R^d_c (or on $\mathcal{M}^*_p$) respectively. Equation (∗4) implies that $\{X^{L(\rho)}_t/\Lambda;\ \Lambda>0\}\subset\mathcal{M}^*_p$ is exponentially tight [DmZ93]. Hence, the assertion is immediately obtained from Lemma 2.1.5 of [DeS89, p. 40] together with Weak LDP (Theorem 11). □

7. Proof of Weak Large Deviation Principle

For $t>0$, $\mu\in\mathcal{M}_p$, and a convex subset $A\in\mathcal{B}(\mathcal{M}_p)$ (fixed), we define the function

$$F(\Lambda):=\mathbf{P}^{\kappa,\rho^\gamma}_{0,\Lambda\mu}(X^{L(\rho)}_t/\Lambda\in A),\qquad \Lambda>0. \tag{16}$$

Then we have:

LEMMA 13 (Supermultiplicativity). *$F(\Lambda+\Lambda')\geqslant F(\Lambda)F(\Lambda')$ holds for any $\Lambda,\Lambda'>0$.*

Proof. Let $\Lambda,\Lambda'>0$ and fixed. Suppose that $\{X^{L,1}_t,X^{L,2}_t\}$ is distributed according to the product measure $\mathbf{P}^{\kappa,\rho}_{0,\Lambda\mu}\times\mathbf{P}^{\kappa,\rho}_{0,\Lambda'\mu}$. Then it follows that

$$F(\Lambda)F(\Lambda')=(\mathbf{P}^{\kappa,\rho}_{0,\Lambda\mu}\times\mathbf{P}^{\kappa,\rho}_{0,\Lambda'\mu})(X^{L,1}_t/\Lambda\in A,\ X^{L,2}_t/\Lambda'\in A).$$

Recall that A is convex in $\mathcal{M}_p$, so that the convex combination $(\Lambda+\Lambda')^{-1}(X^{L,1}_t+X^{L,2}_t)$ lies in A if both $X^{L,1}_t/\Lambda$ and $X^{L,2}_t/\Lambda'$ are contained in A. Therefore, we immediately get

$$F(\Lambda)F(\Lambda')\leqslant(\mathbf{P}^{\kappa,\rho}_{0,\Lambda\mu}\times\mathbf{P}^{\kappa,\rho}_{0,\Lambda'\mu})((\Lambda+\Lambda')^{-1}(X^{L,1}_t+X^{L,2}_t)\in A).$$

While, the exponential moment formula Equation (9) in Theorem 5 ([Dk99c]) directly yields the branching property. Hence, the law of the sum $X^{L,1}_t+X^{L,2}_t$ is given by $\mathbf{P}^{\kappa,\rho}_{0(\Lambda+\Lambda')\mu}$. Thus we attain the required inequality. □

For $\mu\in\mathcal{M}_p$, $t>0$, and $A\in\mathcal{U}$, we set

$$\Phi(\Lambda):=-\log\mathbf{P}^{\kappa,\rho}_{0,\Lambda\mu}(X^L_t/\Lambda\in A),\ R>0.$$

An application of supermultiplicativity (Lemma 13) concludes that the function $\Phi(\Lambda)\in R_+\cup\{\infty\}$ is subadditive. We need the following result.

PROPOSITION 14. *Suppose that A is an open convex Borel subset of $\mathcal{M}_p$. If there is some positive number Λ such that $F(\Lambda)>0$, then F is bounded away from 0 on some nontrivial open interval.*

Proof. The result follows from the branching property, the translation invariance of the metric d_p (Lemma 8) and the supermultiplicativity. In fact, the standard routine argument takes care of it with Lemma 9. The whole proof is quite lengthy but easy, hence it is omitted. □

Then, from Proposition 14 we can conclude that Φ is either bounded on some nonempty open interval, or identically $+\infty$. Consequently, we know that the subadditivity of Φ guarantees the existence of all limits $S_{\mu,t}(A) \in [0, +\infty]$, $A \in \mathcal{U}$, ($\mu \in \mathcal{M}_p, t \geqslant 0$), by repeating the discussion in Lemma 4.2.5 [DeS89, pp. 112–113]. Moreover, by virtue of monotonicity, we obtain:

LEMMA 15. *For $\pi \in \mathcal{M}_p$, the value $I_{\mu,t}(\pi)$ is equal to* $\sup\{S_{\mu,t}(A); \pi \in A \in \mathcal{U}\}$.

Since all open balls $B^{d_p}(\nu; r)$, $\nu \in \mathcal{M}_p, r > 0$, belong to $\mathcal{U}$ by Lemma 9, clearly, we have the following lemma:

LEMMA 16. *$I_{\mu,t}: \mathcal{M}_p \to [0, +\infty]$ is a lower semi-continuous functional.*

The convexity of $I_{\mu,t}$ is obvious from the inequality $I_{\mu,t}(\pi_1) + I_{\mu,t}(\pi_2) \geqslant 2I_{\mu,t}(\{\pi_1 + \pi_2\}/2)$, $\pi_1, \pi_2 \in \mathcal{M}_p$. The above inequality follows from a direct computation together with the branching property and Lemma 15. On this account, the first type estimate (I) is immediately derived from (14), (15), and Lemma 15. On the other hand, we can deduce the second type estimate (II) from compactness by employing the similar argument in [DeS89, §3.1, p. 62]. Summing up, we complete the proof of weak LDP.

Appendix

Proof of Existence (Theorem 2). To show the existence of solutions u_n of $F(\kappa, \varphi, \psi, \rho^\gamma, u)[A^{(n)}] = 0$. Let $\mathcal{H}$ be the set of all those $\{\kappa, \varphi, \psi\} \in R_+ \times \mathcal{C}^p \times \mathcal{C}^{p,I}$ such that there exists an element $u := u^{[\kappa,\varphi,\psi,\rho^\gamma]} \in \mathcal{C}^{p,I}$, with $P_{s,\nu}$-probability one, for which

$$F(\kappa, \varphi, \psi, \rho^\gamma, u)[A^{(n)}] = 0, \qquad P_{s,\nu}\text{-a.s. } A^{(n)} \in \mathcal{K}_0,$$

where $\mathcal{K}_0$ is Dynkin's Admissible Class (see the proof of Theorem 5). To assert the existence of solutions is equivalent to showing that the set $\mathcal{H}$ is open. Note that the mapping $R_+ \times \mathcal{C}^p \ni (\kappa, \varphi) \mapsto T^\kappa_{T-(\cdot)}\varphi \in \mathcal{C}^{p,I}$ is continuous. By the domination property, $\{T^\kappa_t; t > 0\}$ as $T^\kappa_t \in \mathcal{L}(\mathcal{C}^p; \mathcal{C}^p)$ are uniformly bounded over a bounded region of t, κ. Hence, by the dominated convergence theorem we conclude that the mapping: $R_+ \times \mathcal{C}^{p,I} \ni (\kappa, \psi) \mapsto U^\kappa f \in \mathcal{C}^{p,I}$ is continuous. On this account, it is easy to see that

LEMMA 17. *For $A \in \mathcal{K}_0$ given, $F(\cdot, \rho^\gamma, \cdot)[A]$ maps $R_+ \times \mathcal{C}^p \times \mathcal{C}^{p,I} \times \mathcal{C}^{p,I}$ continuously into $\mathcal{C}^{p,I}$, $P_{s,\nu}$-a.s.*

Furthermore, the Fréchet derivative of F with respect to u at each point is given by

$$D_u F(\kappa, \varphi, \psi, \rho^\gamma, u)(v) = v + 2 \cdot Z^\kappa[A^{(n)}](uv), \quad v \in \mathcal{C}^{p,I}. \tag{17}$$

PROPOSITION 18. *For each* $\{\kappa, \varphi, \psi, u\} \in R_+ \times \mathcal{C}^p \times \mathcal{C}^{p,I} \times \mathcal{C}^{p,I}$, $P_{s,\nu}$-a.s.,

(a) $D_u F(v)$ *is linear in* u;
(b) $D_u F(v)$ *is continuous in* $\{\kappa, \varphi, \psi, u\}$;
(c) *the operator* $D_u F(\cdot)\colon \mathcal{C}^{p,I} \to \mathcal{C}^{p,I}$ *is bounded linear and bijective.*

Proof. (a), (b) are trivial. A similar estimate in the proof of Theorem 1 leads to $D_u F$ being an injective operator. To complete the proof, we need to find $v \in \mathcal{C}^{p,I}$ with

$$v(s) + 2\int_s^T T^\kappa_{r-s}(u(r)v(r))A^{(n)}_{[W^\kappa]}(\mathrm{d}r) = w(s) \in \mathcal{C}^{p,I}, \quad s \in I,\ P_{s,\nu}\text{-a.s.}$$

However, the usual principal theorem for linear operator equations in Banach spaces may take care of this problem with an iteration scheme and induction argument, [Z86]. □

To go back to the proof of Theorem 2, assume that $F(\kappa_0, \varphi_0, \psi_0, \rho^\gamma, g_0) = 0$, $P_{s,\nu}$-a.s. for a fixed point $\{\kappa_0, \varphi_0, \psi_0, u_0\}$. An application of the implicit function theorem (cf. Theorem 4.B, [Z86]) with Theorem 1 and Equation (17) provides the existence of an open neighborhood U_0 of $\{\kappa_0, \varphi_0, \psi_0\}$ in $R_+\times \mathcal{C}^p\times \mathcal{C}^{p,I}$ such that there exists a unique map $\{\kappa, \varphi, \psi\} \mapsto u^{[\kappa,\varphi,\psi,\rho^\gamma]}$ defined on U_0 with $F(\kappa, \varphi, \psi, \rho^\gamma, u^{[\kappa,\varphi,\psi,\rho^\gamma]}) = 0$, $P_{s,\nu}$-a.s., because we employed Proposition 18. Thus we have attained that the nonempty set $\mathcal{H}$ previously defined is open. This concludes the required assertion.

Acknowledgements

This work was completed in part during a stay in Nagoya for the Second International Conference on Quantum Information, at Meijo University, 1–5 March, 1999. The author would like to thank the organizers for giving him a chance to talk about his results during the conference. He is also grateful to Professor T. Hida (Professor Emeritus of Nagoya University) and Professor I. V. Volovich (Steklov Mathematical Institute) for their helpful comments and useful suggestions.

References

[BeP91] Barlow, M. T., Evans, S. N. and Perkins, E. A.: Collision local times and measure-valued processes, *Canad. J. Math.* **43** (1991), 897–938.

[BP94] Barlow, M. T. and Perkins, E. A.: On the filtration of historical Brownian motion, *Ann. Probab.* **22** (1994), 1273–1294.

[D93] Dawson, D. A.: *Measure-Valued Markov Processes*, Lecture Notes in Math. 1541, Springer, New York, 1993, pp. 1–260.

[DF97] Dawson, D. A. and Fleischmann, K.: A continuous super-Brownian motion in a super-Brownian medium, *J. Theoret. Probab.* **10** (1997), 213–276.

[DH79] Dawson, D. A. and Hochberg, K. J.: The carrying dimension of a stochastic measure diffusion, *Ann. Probab.* **7** (1979), 693–703.

[DeR98] Deuschel, J.-D. and Rosen, J.: Occupation time large deviations for critical branching Brownian motion, super-Brownian motion and related processes, *Ann. Probab.* **26** (1998), 602–643.

[DeS89] Deuschel, J.-D. and Stroock, D. W.: *Large Deviations*, Academic Press, Boston, 1989.

[DmZ93] Dembo, A. and Zeitouni, O.: *Large Deviations Techniques and Applications*, Jones and Bartlett, Boston, 1993.

[Dk97] Dôku, I.: Nonlinear SPDE with a large parameter and martingale problem for the measure-valued random process with interaction, *J. Saitama Univ. Math. Nat. Sci.* **46** (1997), 1–9.

[Dk98] Dôku, I.: Elementary presentation of studies on catalytic stochastic processes in outline, *Collection of Abstracts of RIMS Workshop on Stoch. Anal. MVSP*, 4–6 Nov. 1998, p. 3.

[Dk99a] Dôku, I.: A note on characterization of solutions for nonlinear equations via regular set analysis, *J. Saitama Univ. Math. Nat. Sci.* **48** (1999), 1–14.

[Dk99b] Dôku, I.: An overview of the studies on catalytic stochastic processes, to appear in *RIMS Kokyuroku* (*Kyoto Univ.*) **1089** (1999).

[Dk99c] Dôku, I.: Weak large deviation principle for superprocesses related to nonlinear differential equations with catalytic noise, Preprint, 1999.

[DkK99] Dôku, I. and Kojima, N.: An introduction to the super-Brownian motion with catalytic medium in Dawson–Fleischmann's work, to appear in *RIMS Kokyuroku* (*Kyoto Univ.*) **1089** (1999).

[Dy91] Dynkin, E. B.: A probabilistic approach to one class of nonlinear differential equations, *Probab. Theory Related Fields* **89** (1991), 89–115.

[Dy91a] Dynkin, E. B.: Path processes and historical superprocesses, *Probab. Theory Related Fields* **90** (1991), 1–36.

[Dy93] Dynkin, E. B.: Superprocesses and partial differential equations, *Ann. Probab.* **21** (1993), 1185–1262.

[Dy94] Dynkin, E. B.: *An Introduction to Branching Measure-Valued Processes*, Amer. Math. Soc., Providence, 1994.

[EP94] Evans, S. N. and Perkins, E. A.: Measure-valued branching diffusions with singular interactions, *Canad. J. Math.* **46** (1994), 120–168.

[FK94] Fleischmann, K. and Kaj, I.: Large deviation probabilities for some rescaled superprocesses, *Ann. Inst. H. Poincaré* **30** (1994), 607–645.

[I86] Iscoe, I.: A weighted occupation time for a class of measure-valued branching processes, *Probab. Theory Related Fields* **71** (1986), 85–116.

[IL93] Iscoe, I. and Lee, T.-Y.: Large deviations for occupation times of measure-valued branching Brownian motions, *Stochastics Stochastics Rep.* **45** (1993), 177–209.

[Ka83] Kallenberg, O.: *Random Measures*, 3rd edn, Akademie-Verlag, Berlin, 1983.

[KS88] Konno, N. and Shiga, T.: Stochastic partial differential equations for some measure-valued diffusions, *Probab. Theory Related Fields* **79** (1988), 201–225.

[S97] Schied, A.: Moderate deviations and functional LIL for super-Brownian motion, *Stochastic Process. Appl.* **72** (1997), 11–25.

[W68] Watanabe, S.: A limit theorem of branching processes and continuous state branching processes, *J. Math. Kyoto Univ.* **8** (1968), 141–176.

[Z86] Zeidler, E.: *Nonlinear Functional Analysis and its Applications*, Vol. I, Springer-Verlag, New York, 1986.

Acta Applicandae Mathematicae **63:** 119–135, 2000.

Ornstein–Uhlenbeck Path Integral and Its Application *

HIROSHI EZAWA
Department of Physics, Gakushuin University, Mejiro, Toshima-ku, Tokyo 171-8588, Japan

(Received: 27 January 1999)

Abstract. Introducing a path integral for the Ornstein–Uhlenbeck process distorted by a potential $V(x)$, we find out the $T \to \infty$ limit of the probability distributions of $X[\omega] := 1/T^\nu \int_0^T V(\omega(t))\,\mathrm{d}t$ for Ornstein–Uhlenbeck process $\omega(t)$, with appropriate values of the exponent ν that depend on V. The results are compared with those for the Wiener process.

Mathematics Subject Classifications (2000): 47D08, 60K40.

Key words: Ornstein–Uhlenbeck process, Wiener process, Feynman–Kac formula, long-time average.

1. Introduction

For a given random process $\omega \in \mathcal{P}$ over a time interval $[0, T]$, and a function $V(x)$, we define a random variable

$$X[\omega] = \frac{1}{T^\nu} \int_0^T V(\omega(t))\,\mathrm{d}t, \tag{1.1}$$

and wish to study its probability distribution in the long-time limit $T \to \infty$.

$X[\omega]$ may be regarded as a long-time average of the values of the field V scanned by a wanderer undergoing the process $\mathcal{P}$, although the exponent ν we consider is not necessarily one.

In fact, the exponent ν has to be chosen appropriately for the probability distribution to be nontrivial, the choice depending on the process $\mathcal{P}$ and the function V. We consider the cases of the Ornstein–Uhlenbeck process, in which $\nu = 1$ or $1/2$ depending upon whether $\int_{-\infty}^{\infty} V(x) \exp[-x^2/(2\beta D)]\,\mathrm{d}x$ is nonvanishing or vanishing (see (2.1) for β and D). The process can be extended over the infinite time interval, so that one may talk about the $T \to \infty$ limit of (1.1) and its probability distribution.

The results will be compared with those for the Wiener process. Here, however there is no transition probability for an infinite time interval so that we consider

* It is a great pleasure to dedicate this article to Prof. Takeyuki Hida at the occasion of his seventieth birthday. We are deeply grateful to him for his constant leadership for many years in the mathematical physics circle, in particular in the study of path integrals and white noise, not only in Japan but also globally. I wish him many happy returns.

$T \to \infty$ limit of the distribution of (1.1) and not the distribution of the $T \to \infty$ limit of (1.1). The value of ν is 1/2 or 1/4 depending upon whether $\int_{-\infty}^{\infty} V(x)\,dx$ is nonvanishing or vanishing.

The contrast we find between the results for the Ornstein–Uhlenbeck and the Wiener processes is amazing.

For our purpose, we use the path integral idea for the process $\mathcal{P}$ distorted by the potential V.

2. Ornstein–Uhlenbeck Path Integral

We consider the Ornstein–Uhlenbeck (OU) Process [1–3], whose transition probability from x at time T_0 to $(y, y+\Delta y)$ at T is given by the probability density,

$$G_0(y, T|x, T_0) = \frac{1}{\sqrt{2\pi\beta D(1-\mathrm{e}^{-2\beta(T-T_0)})}} \exp\left[-\frac{(y - x\mathrm{e}^{-\beta(T-T_0)})^2}{2\beta D(1-\mathrm{e}^{-2\beta(T-T_0)})}\right], \tag{2.1}$$

times Δy with the diffusion constant D and the relaxation time $1/\beta$, both being strictly positive. The process is stationary, as is clear from (2.1), depending on time only through the difference $T - T_0$.

The probability density G_0 satisfies the backward equation

$$\left(\frac{\partial}{\partial T} + \mathcal{H}_0\right) G_0(y, T|x, T_0) = 0 \tag{2.2}$$

with the initial condition

$$\lim_{T \to T_0+} G_0(y, T|x, T_0) = \delta(y - x), \tag{2.3}$$

where

$$\mathcal{H}_0 := -\beta^2 D \frac{\partial^2}{\partial x^2} + \beta x \frac{\partial}{\partial x}. \tag{2.4}$$

We should have taken $-\partial/\partial T_0$ instead if really faithful to the choice 'backward'.

It also satisfies the normalization condition

$$\int_{-\infty}^{\infty} G_0(y, T|x, T_0)\,\mathrm{d}y = 1, \tag{2.5}$$

and the Kolmogorov equation

$$G_0(y, T|x, T_0) = \int_{-\infty}^{\infty} G_0(y, T|x_1, t_1) G_0(x_1, t_1|x, T_0)\,\mathrm{d}x_1 \tag{2.6}$$

for $T > t_1 > T_0$. Quite generally, in fact

$$G_0(x', t'|x, t) := \frac{1}{\sqrt{\pi a(\tau)}} \exp\left[-\frac{(x' - b(\tau)x)^2}{a(\tau)}\right] \quad (\tau := t' - t) \tag{2.7}$$

gives

$$\frac{1}{\sqrt{\pi^2 a(\tau_1)a(\tau_2)}}\int_{-\infty}^{\infty}\exp\left[-\frac{(y-b(\tau_2)x_1)^2}{a(\tau_2)}-\frac{(x_1-b(\tau_1)x)^2}{a(\tau_1)}\right]\mathrm{d}x_1$$
$$=\frac{1}{\sqrt{\pi(a(\tau_1)b(\tau_2)^2+a(\tau_2))}}\exp\left[-\frac{(y-b(\tau_1)b(\tau_2)x)^2}{a(\tau_1)b(\tau_2)^2+a(\tau_2)}\right], \tag{2.8}$$

implying the composition law

$$a(\tau_2+\tau_1)=a(\tau_1)b(\tau_2)^2+a(\tau_2), \qquad b(\tau_2+\tau_1)=b(\tau_1)b(\tau_2), \tag{2.9}$$

which the functions in (2.1)

$$a(\tau)=1-\mathrm{e}^{-2\beta\tau}, \qquad b(\tau)=\mathrm{e}^{-\beta\tau}, \tag{2.10}$$

satisfy.

The Kolmogorov equation (2.6) guarantees the existence of a measure for the OU paths $x=x(t)$ as an extention from the cylinder set measure (Kolmogorov extension theorem); the cylinder set measure for the OU paths is obtained in the following way:

Note to begin with that the Kolmogorov equation (2.6) gives

$$G_0(y,t+\Delta t|x,t_0)=\int_{-\infty}^{\infty}G_0(y,t+\Delta t|x_1,t)G_0(x_1,t|x,t_0)\,\mathrm{d}x_1, \tag{2.11}$$

where, for infinitesimal $\Delta t>0$,

$$G_0(y,t+\Delta t|x_1,t)$$
$$=\frac{1}{\sqrt{4\pi\beta^2 D\Delta t}}\exp\left[-\frac{(y-x_1)^2}{4\beta^2 D\Delta t}-\frac{(y-x_1)x_1}{2\beta D}-\frac{x_1^2}{4D}\Delta t\right], \tag{2.12}$$

since from (2.10)

$$a(\Delta t)=2\beta\Delta t, \qquad b(\Delta t)=1-\beta\Delta t. \tag{2.13}$$

Conversely, the composition law (2.9) gives

$$a(0)=0, \qquad b(0)=1, \tag{2.14}$$

and, when differentiated* with respect to τ_2 at $\tau_2=0$,

$$\frac{\mathrm{d}}{\mathrm{d}\tau}a(\tau)=2qa(\tau)+p, \qquad \frac{\mathrm{d}}{\mathrm{d}\tau}b(\tau)=qb(\tau), \tag{2.15}$$

where $p:=\mathrm{d}a/\mathrm{d}\tau|_{\tau=0}$, $\mathrm{d}b/\mathrm{d}\tau|_{\tau=0}:=q$.

* The derivatives with respect to τ_1 at $\tau_1=0$ do not give independent equations.

It follows from the differential equation (2.15) and the initial condition (2.14) that

$$a(\tau) = C(e^{2q\tau} - 1), \qquad b(\tau) = e^{q\tau} \quad \left(C := \frac{p}{2q}\right), \tag{2.16}$$

where one could, if desired, take q and C as free parameters in place of q and p. Indeed, (2.16) satisfies the composition law (2.9) with these parameters arbitrarily chosen. The parameters are fixed by (2.13). Thus, (2.12) recovers the right G_0 when substituted in (2.11).

Repeated use of (2.6) yields

$$\begin{aligned} &G_0(x_N, t_N | x_0, t_0) \\ &= \int \cdots \int \left\{ \prod_{k=1}^{N-1} G_0(x_{k+1}, t_{k+1} | x_k, t_k) \, \mathrm{d}x_k \right\} G(x_1, t_1 | x_0, t_0), \end{aligned}$$

defining the cylinder set measure for the paths. In particular, when $t_{k+1} - t_k \to 0$, (2.11) and (2.12) lead to the path-integral representation,

$$G_0(y, T | x, 0) = \int \mathrm{d}\mu_{\mathrm{OU}, x \to y} \tag{2.17}$$

with the path space measure being given by

$$\mathrm{d}\mu_{\mathrm{OU}, x \to y} = \exp\left[-\frac{1}{2\beta D} \int_0^T \omega(t) \, \mathrm{d}\omega(t) - \frac{1}{4D} \int_0^T \omega^2(t) \, \mathrm{d}t \right] \mathrm{d}\mu_{\mathrm{W}, x \to y}, \tag{2.18}$$

where $\mathrm{d}\mu_{\mathrm{W}, x \to y}$ is the Wiener measure for the Brownian bridge $\omega(t)$ from x to y that corresponds to the first term in the square bracket in (2.12), and $\int_0^T \omega(t) \, \mathrm{d}\omega(t)$, the Itô integral with $\mathrm{d}\omega(t)$ 'sticking out' towards future. Equation (2.5) implies that

$$\int_{-\infty}^{\infty} \mathrm{d}y \int \mathrm{d}\mu_{\mathrm{OU}, x \to y} = 1. \tag{2.19}$$

We can consider a path space measure for the paths distorted [4] by a given potential $V(x)$, such that the (relative, i.e., not normalized) transition probability density be given by

$$G_\lambda(y, T | x, 0) = \int \exp\left[-\lambda \int_0^T V(\omega(t)) \, \mathrm{d}t \right] \mathrm{d}\mu_{\mathrm{OU}, x \to y}. \tag{2.20}$$

It is the solution to the backward equation

$$\left(\frac{\partial}{\partial T} + \mathcal{H} \right) G_\lambda(y, T | x, 0) = 0, \tag{2.21}$$

with

$$\mathcal{H} = \mathcal{H}_0 + V(x) = -\beta^2 D \frac{\partial^2}{\partial x^2} + \beta x \frac{\partial}{\partial x} + V(x) \tag{2.22}$$

and with the initial condition,

$$\lim_{T\to 0+} G_\lambda(y, T|x, 0) = \delta(y - x). \tag{2.23}$$

3. Average over the Paths

We shall present our ideas by referring to the OU-process, but it can be applied to other processes as well.

3.1. FIELD SCANNING

We define

$$A(x, T; \lambda) := \int_{-\infty}^{\infty} \left\langle \exp\left[-\lambda \int_0^T V(\omega(t))\,\mathrm{d}t\right]\right\rangle_{x\to y} \mathrm{d}y \tag{3.1}$$

with the right-hand side denoting the expression (2.20) integrated over all the possible values of the terminal point y:

$$A(x, T; \lambda) = \int_{-\infty}^{\infty} G_\lambda(y, T|x, 0)\,\mathrm{d}y. \tag{3.2}$$

Since, further, the measure $\mathrm{d}\mu_{\mathrm{OU},x\to y}$ is normalized by (2.19), definition (3.1) can be regarded as giving the characteristic function for the random variable (1.1) over the OU process, if

$$\lambda = -i\eta T^{-\nu}. \tag{3.3}$$

Formula (1.1) may be called the time average of the values of the field $V(x)$ scanned by a wanderer undergoing the OU process, $x = \omega(t)$. Normally one takes $\nu = 1$ for the time average of $V(\omega(t))$. As we shall see from the examples below, we have to choose a value of ν not necessarily 1 in order to obtain a nontrivial $T \to \infty$ limit of the probability distribution for $X[\omega]$.

3.2. LONG-TIME AVERAGE

We wish to calculate the long-time limit,

$$\chi_\infty(\eta, x) := \lim_{T\to\infty} A(x, T; -i\eta T^{-\nu}), \tag{3.4}$$

using the Laplace transform

$$\hat{A}(x, s; \lambda) = \int_0^\infty A(x, T; \lambda)\mathrm{e}^{-sT}\,\mathrm{d}T. \tag{3.5}$$

The link is provided by the general formula,

$$\lim_{T\to\infty} F(T) = \lim_{b\to 0} b \int_0^\infty \mathrm{e}^{-bT} F(T)\,\mathrm{d}T. \tag{3.6}$$

We define

$$\begin{aligned} &I_b(x,\sigma_1,\sigma_2) \\ &\quad := b\int_0^\infty \left\{\hat{A}(x,b+u\sigma_1;-iu^\nu) - \hat{A}(x,b+u\sigma_2;-iu^\nu)\right\}\frac{\mathrm{d}u}{u} \end{aligned} \tag{3.7}$$

with $\mathrm{d}u/u$ to manage the scale changes to follow and the difference in the curly brackets to kill the divergence caused by $1/u$. Put definition (3.5) in (3.7) to get

$$I_b(x,\sigma_1,\sigma_2) := b\int_0^\infty \frac{\mathrm{d}u}{u}\int_0^\infty \mathrm{d}T \left(\mathrm{e}^{-(b+u\sigma_1)T} - \mathrm{e}^{-(b+u\sigma_2)T}\right)A(x,T;-iu^\nu)$$

and change the scale of u by taking $v = uT$ instead, then

$$I_b(x,\sigma_1,\sigma_2) = b\int_0^\infty \mathrm{d}v\,\frac{\mathrm{e}^{-\sigma_1 v}-\mathrm{e}^{-\sigma_2 v}}{v}\int_0^\infty \mathrm{d}T\, A\left(x,T;-i\frac{v^\nu}{T^\nu}\right)\mathrm{e}^{-bT}, \tag{3.8}$$

which relates the limit we look for,

$$\begin{aligned} \chi_\infty(v^\nu,x) &:= \lim_{T\to\infty} A\left(x,T,-i\frac{v^\nu}{T^\nu}\right) \\ &= \lim_{b\to 0} b\int_0^\infty A\left(x,T,-i\frac{v^\nu}{T^\nu}\right)\mathrm{e}^{-bT}\mathrm{d}T, \end{aligned} \tag{3.9}$$

to the Laplace transform $\hat{A}(x,s,\lambda)$ in (3.5) where λ is a fixed parameter.

One step further. In (3.7) we introduce the scale change u to bu and new notations

$$\xi_l = 1 + u\sigma_l, \qquad \eta = -iu^\nu \quad (l = 1,2) \tag{3.10}$$

and

$$\Lambda_b(x,\xi,\eta) := b\hat{A}(x,b\xi,b^\nu\eta). \tag{3.11}$$

Then, combining (3.7) and (3.8), we obtain

$$\begin{aligned} &\int_0^\infty \frac{\mathrm{e}^{-\sigma_1 v}-\mathrm{e}^{-\sigma_2 v}}{v}\chi_\infty(v^\nu,x)\,\mathrm{d}v \\ &\quad = \lim_{b\to 0}\int_0^\infty \left\{\Lambda_b(x,\xi_1,\eta_1) - \Lambda_b(x,\xi_2,\eta_2)\right\}\frac{\mathrm{d}u}{u}, \end{aligned} \tag{3.12}$$

which is our key equation.

Thus, in order to find out the $T \to \infty$ limit of the probability distribution of (1.1), our strategy is to calculate the right-hand side of (3.12) in terms of the

Laplace transform (3.5) of (3.1), and then disentangle the left-hand side to obtain (3.9).

3.3. EQUATIONS FOR THE CHARACTERISTIC FUNCTION

We shall first construct a differential equation for $A(x, T; \lambda)$ using relation (3.2) to the (relative) transition probability $G_\lambda(y, T|x, 0)$ that satisfies the differential equation (2.21). After Laplace transformation in T, the differential equation will be converted into an integral equation, which turns out to be useful for taking the limit $b \to 0$ in (3.12).

The Laplace transform of the (relative) transition probability density,

$$\hat{G}_\lambda(x', x; s) := \int_0^\infty G_\lambda(x', T|x, 0) \mathrm{e}^{-sT}\, \mathrm{d}T \tag{3.13}$$

satisfies, by (2.2),

$$(\mathcal{H}_0 + s + \lambda V)\hat{G}_\lambda(x', x; s) = \delta(x' - x). \tag{3.14}$$

Integrating both sides with respect to the terminal point y, we find

$$(\mathcal{H}_0 + s + \lambda V)\hat{A}(x, s; \lambda) = 1, \tag{3.15}$$

an equation for the Laplace transform of the characteristic function, which we now transform to an integral equation for later convenience.

The Green function for $\lambda = 0$ satisfying

$$(\mathcal{H}_0 + s)\hat{G}_0(x', x; s) = \delta(x' - x) \tag{3.16}$$

is the Laplace transform of (2.1) with $T_0 = 0$, which, after a change of the variable of integration, takes the form,

$$\hat{G}_0(x', x; s) = \frac{1}{\beta}\int_0^1 \alpha^{(s/\beta)-1}\sqrt{\frac{1}{2\pi\beta D(1-\alpha^2)}}\exp\left[-\frac{(x'-\alpha x)^2}{2\beta D(1-\alpha^2)}\right] \mathrm{d}\alpha. \tag{3.17}$$

We can use this Green function to convert (3.15) into an integral equation,

$$\hat{A}(x, s; \lambda) = \frac{1}{s} - \lambda \int_{-\infty}^{\infty} \hat{G}_0(x', x; s) V(x') \hat{A}(x', s; \lambda)\, \mathrm{d}x', \tag{3.18}$$

in which the first term on the right has come from

$$\int_{-\infty}^{\infty} \hat{G}_0(x', x; s)\, \mathrm{d}x' = \frac{1}{\beta}\int_0^1 \alpha^{s/\beta - 1}\, \mathrm{d}\alpha = \frac{1}{s}. \tag{3.19}$$

In general, a solution to the homogeneous version of (3.15) can be added to the right-hand side of (3.18), but now it cannot be, because, as $\lambda \to 0$, $G_\lambda(y, x; s)$ should reduce to $G_0(y, x; s)$ and hence $\hat{A}(x, s; \lambda)$ to (3.19).

4. The OU Scanning

We make some preparatory observations. In terms of (3.11), the integral equation (3.18) reads:

$$\Lambda_b(x,\xi,\eta) = \frac{1}{\xi} - b^\nu \eta \int_{-\infty}^{\infty} V(x')\Lambda_b(x',\xi,\eta)\hat{G}_0(x',x;b\xi)\,\mathrm{d}x'. \tag{4.1}$$

The Green function (3.17),

$$\hat{G}_0(x',x;b\xi) = \frac{1}{b\xi}\int_0^1 \frac{\mathrm{d}\alpha^{b\xi/\beta}}{\mathrm{d}\alpha}\sqrt{\frac{1}{2\pi\beta D(1-\alpha^2)}}\exp\left[-\frac{(x'-\alpha x)^2}{2\beta D(1-\alpha^2)}\right]\mathrm{d}\alpha, \tag{4.2}$$

is difficult to calculate. However, the graph $(\alpha, \alpha^{b\xi/\beta})$ for small b looks quite like the one of

$$\alpha^{b\xi/\beta} = \begin{cases} f(\alpha), & 0 \leqslant \alpha < \alpha_1, \\ 1, & \alpha_1 \leqslant \alpha \leqslant 1, \end{cases} \tag{4.3}$$

where $f(\alpha) \sim \alpha^{b\xi/\beta}$ interpolates between $f(0) = 0$ and $f(\alpha_1) = 1$. The location of α_1 may be chosen by requiring the bound,

$$0 < 1 - \alpha^{b\xi/\beta} \sim -\frac{b\xi}{\beta}\log\alpha < \delta, \quad 0 \leqslant \alpha \leqslant \alpha_1$$

with a given small constant δ, so that

$$\alpha_1 \sim \mathrm{e}^{-\beta\delta/b\xi}. \tag{4.4}$$

The contribution to the integral comes from the neighborhood of $\alpha = 0$ where the derivative $\mathrm{d}\alpha^{b\xi/\beta}/\mathrm{d}\alpha$ is large. Since the other factor of the integrand is slowly varying there, we can replace it by its value at α_*, somewhere in the interval $[0, \alpha_1]$, and integrate the derivative only, obtaining

$$\hat{G}_0(x',x;b\xi) = \frac{1}{b\xi}\sqrt{\frac{1}{2\pi\beta D(1-\alpha_*^2)}}\exp\left[-\frac{(x'-\alpha_* x)^2}{2\beta D(1-\alpha_*^2)}\right]. \tag{4.5}$$

4.1. TIME AVERAGE OF OU SCAN

With the approximate form (4.5) for the Green function, (4.1) becomes

$$\begin{aligned}\Lambda_b(x,\xi,\eta) &= \frac{1}{\xi} - \frac{b^{\nu-1}\eta}{\xi}\int_{-\infty}^{\infty} V(x')\Lambda_b(x',\xi,\eta)\sqrt{\frac{1}{2\pi\beta D(1-\alpha_*^2)}}\times \\ &\quad\times \exp\left[-\frac{(x'-\alpha_* x)^2}{2\beta D(1-\alpha_*^2)}\right]\mathrm{d}x'.\end{aligned} \tag{4.6}$$

This equation implies that $\Lambda_b(x, \xi, \eta)$ is continuous and differentiable with respect to x. Iff

$$\nu = 1, \tag{4.7}$$

(4.6) has the limit $b \to 0$,

$$\Lambda_0(x, \xi, \eta) = \frac{1}{\xi} - \frac{\eta}{\xi}\int_{-\infty}^{\infty} V(x')\Lambda_0(x', \xi, \eta)\sqrt{\frac{1}{2\pi\beta D}}\exp\left[-\frac{x'^2}{2\beta D}\right]\mathrm{d}x' \tag{4.8}$$

provided that the limit can be taken inside the integral sign. Since the right-hand side is now independent of x, so is $\Lambda_0(x, \xi, \eta)$ on the left. Therefore,

$$\Lambda_0(\cdot, \xi, \eta) = \frac{1}{\xi + \langle V\rangle_\infty \eta}, \tag{4.9}$$

where $\langle V\rangle_\infty$ is the weighted average of V,

$$\langle V\rangle_\infty = \int_{-\infty}^{\infty} V(y)\sqrt{\frac{1}{2\pi\beta D}}\exp\left[-\frac{y^2}{2\beta D}\right], \tag{4.10}$$

with the subscript ∞ indicating that the weight function is the $T \to \infty$ limit of (2.1), the transition probability density of the OU-process.

We put the result (4.9) in (3.12) to obtain, writing σ and σ' for σ_1 and σ_2,

$$\lim_{b\to 0} I_b(\cdot, \sigma, \sigma') = \int_0^{\infty}\left\{\frac{1}{1 + (\sigma - i\langle V\rangle_\infty)u} - \frac{1}{1 + (\sigma' - i\langle V\rangle_\infty)u}\right\}\frac{\mathrm{d}u}{u},$$

which determines the right-hand side of (3.12) to give

$$\int_0^{\infty} \frac{\mathrm{e}^{-\sigma v} - \mathrm{e}^{-\sigma' v}}{v}\chi_\infty(v; \cdot)\,\mathrm{d}v = \log\frac{\sigma' - i\langle V\rangle_\infty}{\sigma - i\langle V\rangle_\infty}. \tag{4.11}$$

Now put $\sigma' = \sigma + \Delta\sigma$ and let $\Delta\sigma \to 0$, then

$$\int_0^{\infty} \chi_\infty(v; \cdot)\mathrm{e}^{-\sigma v}\,\mathrm{d}v = \frac{1}{\sigma - i\langle V\rangle}$$

and hence

$$\chi_\infty(v; \cdot) = \mathrm{e}^{iv\langle V\rangle_\infty}, \tag{4.12}$$

which is independent of x as indicated by the dot. With the characteristic function given by (4.12), the probability distribution of (1.1) with $\nu = 1$, the long-time average of $V(\omega(t))$ with $\omega(t)$ undergoing the OU-process starting from $\omega(0) = x$, is given by a delta function,

$$P(X) = \delta(X - \langle V\rangle_\infty), \tag{4.13}$$

supported at the space average (4.10) with the weight (2.1) at $T \to \infty$. The distribution is independent of the starting point x, a reflection of the damping $\mathrm{e}^{-\beta T}x$ in the transition probability density (2.1) of the OU-process.

4.2. THE CASE OF VANISHING AVERAGE

Since the random variable (1.1) for $\nu = 1$ has turned out not to be random but to have the sharp distribution (4.13) at $\langle V \rangle_\infty$, we wish now to look at the difference, or rather

$$X_1[\omega] := \frac{1}{T^\nu}\int_0^T \left\{V(\omega(t)) - \langle V \rangle_\infty\right\} \mathrm{d}t \tag{4.14}$$

with some ν less than (4.7).

Let us define $V_1(x) := V(x) - \langle V \rangle_\infty$, so that

$$\langle V_1 \rangle_\infty = 0 = \int_{-\infty}^{\infty} V_1(x)\mathrm{e}^{-x^2/(2\beta D)}\,\mathrm{d}x. \tag{4.15}$$

The necessary and sufficient condition for this to happen is that there exists a function $U(x)$ vanishing at infinity and such that

$$V_1(x)\exp\left[-\frac{x^2}{2\beta D}\right] = U'(x). \tag{4.16}$$

Put this representation into (4.1), and integrate by parts, then

$$\Lambda_b(x,\xi,\eta) = \frac{1}{\xi} + \mathrm{I} + \mathrm{II}, \tag{4.17}$$

where

$$\begin{aligned} \mathrm{I} &:= b^\nu \eta \int_{-\infty}^{\infty} U(x')\Lambda_b{}'(x',\xi,\eta)\hat{H}_0(x',x,b\xi)\,\mathrm{d}x', \\ \mathrm{II} &:= b^\nu \eta \int_{-\infty}^{\infty} U(x')\Lambda_b(x',\xi,\eta)\hat{H}_0{}'(x',x,b\xi)\,\mathrm{d}x', \end{aligned} \tag{4.18}$$

with

$$\begin{aligned} &\hat{H}_0(x',x;b\xi) \\ &\quad := \hat{G}_0(x',x;b\xi)\exp\left[\frac{x'^2}{2\beta D}\right] \\ &\quad = \frac{1}{b\xi}\int_0^1 \frac{\mathrm{d}\alpha^{b\xi/\beta}}{\mathrm{d}\alpha}\sqrt{\frac{1}{2\pi\beta D(1-\alpha^2)}}\exp\left[-\frac{\alpha^2 x'^2 - 2\alpha x' x + \alpha^2 x^2}{2\beta D(1-\alpha^2)}\right]\mathrm{d}\alpha. \end{aligned}$$

In the approximation (4.5), we have

$$\begin{aligned} &\hat{H}_0(x',x;b\xi) \\ &\quad = \frac{1}{b\xi}\sqrt{\frac{1}{2\pi\beta D(1-\alpha_*^2)}}\exp\left[-\frac{\alpha_*^2 x'^2 - 2\alpha_* x' x + \alpha_*^2 x^2}{2\beta D(1-\alpha_*^2)}\right], \end{aligned} \tag{4.19}$$

while the approximation does not apply to $H_0'(x', x; b\xi)$ because the key role of $\mathrm{d}\alpha^{b\xi/\beta}/\mathrm{d}\alpha$ is suppressed by the α coming down upon differentiation by x'. The derivative $\Lambda_b'(x', \xi, \eta)$ is taken from (4.1),

$$\begin{aligned}\Lambda_b'(x, \xi; \eta) &= \frac{b^{\nu}\eta}{\sqrt{2\pi}\beta}\int_0^1 \mathrm{d}\alpha\, \alpha^{b\xi/\beta}\int_{-\infty}^{\infty}\mathrm{d}x'\, V_1(x')\Lambda_b(x', \xi, \eta)\times\\ &\quad\times\frac{x'-\alpha x}{[\beta D(1-\alpha^2)]^{3/2}}\exp\left[-\frac{(x'-\alpha x)^2}{2\beta D(1-\alpha^2)}\right],\end{aligned}$$

where again $\mathrm{d}\alpha^{b\xi/\beta}/\mathrm{d}\alpha$ is replaced by $\alpha^{b\xi/\beta}$, making that approximation inapplicable. Thus,

$$\begin{aligned}I &= \frac{b^{2\nu-1}\eta^2}{2\pi\beta\xi}\int_0^1 \mathrm{d}\alpha\alpha^{b\xi/\beta}\int_{-\infty}^{\infty}\mathrm{d}x'\int_{-\infty}^{\infty}\mathrm{d}x''\, U(x')V_1(x'')\Lambda_b(x'', \xi, \eta)\times\\ &\quad\times\frac{x''-\alpha x'}{[\beta D(1-\alpha^2)]^{3/2}}\frac{1}{[\beta D(1-\alpha_*^2)]^{1/2}}\times\\ &\quad\times\exp\left[-\frac{(x''-\alpha x')^2}{2\beta D(1-\alpha^2)}\right]\exp\left[-\frac{\alpha_*^2 x'^2 - 2\alpha_* x' x + \alpha_*^2 x^2}{2\beta D(1-\alpha_*^2)}\right],\end{aligned}$$

which has nontrivial limit as $b \to 0$, iff

$$\nu = \tfrac{1}{2}. \tag{4.20}$$

This value of ν forces II in (4.18) to vanish in the limit and, consequently, (4.17) becomes independent of x. Hence,

$$\Lambda_0(\cdot, \xi, \eta) = \frac{1}{\xi - C\eta^2}, \tag{4.21}$$

where

$$C := \frac{1}{2\pi\beta\sqrt{\beta D}}\int_{-\infty}^{\infty}\mathrm{d}x'\int_{-\infty}^{\infty}\mathrm{d}x''\, U(x')V_1(x'')\frac{\partial}{\partial x''}K(x'', x'), \tag{4.22}$$

with

$$K(x'', x') := \int_0^1 \frac{1}{\sqrt{2\pi\beta D(1-\alpha^2)}}\exp\left[-\frac{(x''-\alpha x')^2}{2\beta D(1-\alpha^2)}\right]\mathrm{d}\alpha, \tag{4.23}$$

because $\alpha^{b\xi/\beta} \to 1$ and $\alpha_* \to 0$ in the limit.

For (4.21), we obtain from (3.12)

$$\int_0^{\infty}\chi_{\infty}(v^{1/2}, \cdot)\,\mathrm{d}v = \log\left|\frac{\sigma_2 + C}{\sigma_1 + C}\right|,$$

so that

$$\int_0^{\infty}\chi_{\infty}(v^{1/2}, \cdot)\mathrm{e}^{-\sigma v}\,\mathrm{d}v = \frac{1}{\sigma + C},$$

hence

$$\chi_\infty(v^{1/2}) = \mathrm{e}^{-Cv}.$$

Then, inverse Fourier transformation gives

$$P_1(X_1) = \frac{1}{2\sqrt{\pi C}}\mathrm{e}^{-X_1^2/(4C)}, \tag{4.24}$$

since $C > 0$ as we shall show below. Thus, the random variable (4.14) with $\nu = 1/2$ has a Gaussian distribution of average 0 and variance $2C$.

Now, we show that $C > 0$. Using

$$\frac{\partial}{\partial x''}K(x'', x') = -\int_0^1 \sqrt{\frac{1}{2\pi\beta D(1-\alpha^2)}}\frac{\partial}{\partial x'}\exp\left[-\frac{(x''-\alpha x')^2}{2\beta D(1-\alpha^2)}\right]\frac{\mathrm{d}\alpha}{\alpha}, \tag{4.25}$$

we obtain by integration by parts

$$C = \frac{1}{2\pi\beta\sqrt{\beta D}}\int_0^1 \frac{\mathrm{d}\alpha}{\alpha} \times \\ \times \int_{-\infty}^{\infty}\mathrm{d}x''\int_{-\infty}^{\infty}\mathrm{d}x' U'(x')U'(x'')\exp\left[-\frac{\alpha^2 x''^2 - 2\alpha x'' x' + \alpha^2 x'^2}{2\beta D(1-\alpha^2)}\right],$$

which can be put in the form,

$$C = \frac{1}{2\pi\beta\sqrt{\beta D}}\int_0^1 \frac{\mathrm{d}\alpha}{\alpha}\frac{1}{\sqrt{2\pi\beta D(1-\alpha^2)}} \times \\ \times \int_{-\infty}^{\infty}\mathrm{d}x''\int_{-\infty}^{\infty}\mathrm{d}x'\, V_\alpha(x')\exp\left[-\frac{\alpha(x'-x'')^2}{2\beta D(1-\alpha^2)}\right]V_\alpha(x''), \tag{4.26}$$

to have manifest positivity when

$$V_\alpha(x) := U'(x)\exp\left[\frac{\alpha x^2}{2\beta D(1+\alpha)}\right] = V(x)\exp\left[-\frac{x^2}{2\beta D(1+\alpha)}\right] \tag{4.27}$$

is rewritten in terms of its Fourier transform. Some remarks are in order about the integral in (4.25), which diverges at $\alpha = 0$ and also at $\alpha = 1$ when $x'' - x' = 0$. Nevertheles, the α-integral in (4.26) does not diverge at $\alpha = 0$ because then $\int_{-\infty}^{\infty} U'(x)\,\mathrm{d}x = 0$, nor at $\alpha = 1$ any longer. Therefore, we have first to regularize (4.25) by cutting off the lower and upper ends of the range of the α-integration, and then remove the cut-off in (4.26).

5. The Brownian Scanning

It is interesting to compare the cases we have treated with those of the Wiener (W) process.

5.1. TIME AVERAGE OF BROWNIAN SCAN

We now wish to find out the $T \to \infty$ limit of the probability distribution of

$$\mathrm{x}[\omega] = \frac{1}{T^{\nu}} \int_0^T V(\omega(t))\,\mathrm{d}t \tag{5.1}$$

for $\omega(t)$ undergoing the W-process.

As one might have noticed, we denote the quantities related to the W-process by lower case letters that correspond to those of the OU-process.

For the Wiener process with the generator

$$h = -D\frac{\partial^2}{\partial x^2} \tag{5.2}$$

the Laplace transform of the Green function is given by

$$\hat{g}_0(y, x; s) = \frac{1}{\sqrt{4Ds}} \exp\left[-\sqrt{\frac{s}{D}}|y - x|\right], \tag{5.3}$$

for which Equation (4.1) becomes

$$\begin{aligned} \lambda_b(x, \xi, \eta) &= \frac{1}{\xi} - \frac{b^{\nu - 1/2}\eta}{\sqrt{4D\xi}} \int_{-\infty}^{\infty} V(x')\lambda_b(x', \xi, \eta) \times \\ &\quad \times \exp\left[-\sqrt{\frac{b\xi}{D}}|x' - x|\right]\mathrm{d}x'. \end{aligned} \tag{5.4}$$

Assume

$$\int_{-\infty}^{\infty} |V(x)|\,\mathrm{d}x < \infty. \tag{5.5}$$

Then, the limit $b \to 0$ exists iff

$$\nu = 1/2, \tag{5.6}$$

and in fact

$$\lambda_0(x, \xi, \eta) = \frac{1}{\xi} - \frac{\eta}{\sqrt{4D\xi}} \int_{-\infty}^{\infty} V(x')\lambda_b(x', \xi, \eta)\,\mathrm{d}x'. \tag{5.7}$$

Since the right-hand side of this equation is independent of x and so is the left-hand side,

$$\lambda_0(\cdot, \xi, \eta) = \frac{1}{\xi + a\sqrt{\xi}\eta} \tag{5.8}$$

with

$$a := \frac{1}{\sqrt{4D}} \int_{-\infty}^{\infty} V(x)\,\mathrm{d}x, \tag{5.9}$$

for which (3.12) give

$$\int_0^\infty \frac{e^{-\sigma v} - e^{-\sigma' v}}{v} \chi_\infty(v^{1/2}; \cdot)\, dv = 2 \log \frac{\sqrt{\sigma'} - ia}{\sqrt{\sigma} - ia} \tag{5.10}$$

with $v^{1/2}$ as the argument of χ because $\nu = 1/2$.

Put $\sigma' = \sigma + \Delta\sigma$ and let $\Delta\sigma \to 0$, then

$$\int_0^\infty \chi_\infty(v; \cdot) e^{-\sigma v}\, dv = \frac{1}{\sqrt{\sigma}\sqrt{\sigma - ia}}. \tag{5.11}$$

Inverse Laplace transform gives [5]

$$\chi_\infty(\eta) = e^{-a^2\eta^2} \cdot \frac{2}{\sqrt{\pi}} \int_{-ia\eta}^\infty e^{-t^2}\, dt, \tag{5.12}$$

which has the "erfc" function standing after the center dot. Put $t = \tau - ia\eta$, then

$$\chi_\infty(\eta) = \frac{2}{\sqrt{\pi}} \int_0^\infty e^{-\tau^2 + 2ia\eta\tau}\, d\tau. \tag{5.13}$$

This is the characteristic function of the probability density $p(x)$ we are after. The density itself is given by the inverse Fourier transform,

$$p(x) = \frac{1}{\pi^{3/2}} \int_{-\infty}^\infty d\eta\, e^{-i\eta x} \int_0^\infty d\tau\, e^{-\tau^2 + 2ia\eta\tau}.$$

Changing the order of integration, we obtain a one-sided Gaussian distribution,

$$p(x) = \frac{1}{\sqrt{\pi} a} \begin{cases} e^{-(x/2a)^2} & (ax \geqslant 0), \\ 0 & (ax < 0), \end{cases} \tag{5.14}$$

which is the probability density for the random variable (5.1) with $\nu = 1/2$, the long-time average of the Brownian scan of the field V.

The average and the variance of $\mathrm{x}[\omega]$ turn out to be

$$\begin{aligned} \langle \mathrm{x} \rangle &= \frac{2a}{\sqrt{\pi}} = \frac{1}{\sqrt{\pi D}} \int_{-\infty}^\infty V(x)\, dx, \\ \langle (\Delta \mathrm{x})^2 \rangle &= \left(\frac{\pi}{2} - 1\right) \left(\frac{1}{\sqrt{\pi D}} \int_{-\infty}^\infty V(x)\, dx \right)^2. \end{aligned} \tag{5.15}$$

5.2. CASE OF VANISHING AVERAGE

If $V(x)$ is such that

$$\int_{-\infty}^\infty V(x)\, dx = 0, \tag{5.16}$$

then (5.9) vanishes and the probability density (5.14) collapses to a delta function. One could ask if a value of ν less than (5.6) exists that makes (5.1) a bona fide random variable having a nontrivial probability distribution [6].

The necessary and sufficient condition for (5.16) is that $V(x)$ can be written as a derivative,

$$V(x) = U'(x) \tag{5.17}$$

of a function $U(x)$ vanishing at infinity. (5.4) with $V(x)$ replaced by $U'(x)$ gives, after integration by parts

$$\begin{aligned}&\lambda_b(x,\xi,\eta)\\&= \frac{1}{\xi} + \frac{b^{\nu-1/2}\eta}{\sqrt{4D\xi}}\int_{-\infty}^{\infty} U(x')\lambda_b{}'(x',\xi,\eta)\exp\left[-\sqrt{\frac{b\xi}{D}}|x'-x|\right]\mathrm{d}x' -\\&\quad - \frac{b^{\nu}\eta}{2D}\int_{-\infty}^{\infty} U(x')\lambda_b(x',\xi,\eta)\varepsilon(x'-x)\exp\left[-\sqrt{\frac{b\xi}{D}}|x'-x|\right]\mathrm{d}x'.\end{aligned}$$

Substitute the derivative of (5.4) in this equation, then

$$\begin{aligned}&\lambda_b(x,\xi,\eta)\\&= \frac{1}{\xi} + \frac{b^{2\nu-1/2}}{\sqrt{4D^3\xi}}\int_{-\infty}^{\infty}\mathrm{d}x'\int_{-\infty}^{\infty}\mathrm{d}x'' U(x')U'(x'')\lambda_b(x'',\xi,\eta)\varepsilon(x'-x'')\times\\&\quad\times\exp\left[-\sqrt{\frac{b\xi}{D}}(|x'-x|+|x''-x'|)\right]+\\&\quad+\frac{b^{\nu}\eta}{2D}\int_{-\infty}^{\infty}\mathrm{d}x'\, U(x')\lambda_b(x',\xi,\eta)\varepsilon(x-x')\exp\left[-\sqrt{\frac{b\xi}{D}}|x'-x|\right].\end{aligned} \tag{5.18}$$

Assume that

$$\int_{-\infty}^{\infty}|U(x)|\,\mathrm{d}x < \infty \quad \text{and} \quad \int_{-\infty}^{\infty}|V(x)|\,\mathrm{d}x < \infty, \tag{5.19}$$

then the limit $b \to 0$ of the right-hand side exists iff

$$\nu = 1/4, \tag{5.20}$$

which is smaller than (5.6) as expected, and $\lim_{b\to 0}\lambda_b(x,\xi,\eta)$ is independent of x, the starting point of the W-process. We get

$$\lambda_0(\cdot,\xi,\eta) = \frac{1}{\xi - c\sqrt{\xi}\eta^2}, \tag{5.21}$$

a form quite similar to (5.8) with

$$c := \frac{\|U\|^2}{\sqrt{4D^3}}, \tag{5.22}$$

after using

$$\int_{-\infty}^{\infty} \mathrm{d}x' \int_{-\infty}^{\infty} \mathrm{d}x'' \varepsilon(x'-x'')U'(x')U(x'') = \int_{-\infty}^{\infty} U^2(x)\,\mathrm{d}x =: \|U\|^2, \quad (5.23)$$

which is finite under our assumption (5.19).

Now, (3.12) and (3.12) give similarly to (5.10)

$$\int_0^{\infty} \frac{\mathrm{e}^{-\sigma v} - \mathrm{e}^{-\sigma' v}}{v} \chi_{\infty}(v^{1/4};\cdot)\,\mathrm{d}v = 2\log \frac{\sqrt{\sigma'}+c}{\sqrt{\sigma}+c},$$

so that by the limiting procedure we used before,

$$\int_0^{\infty} \chi_{\infty}(v^{1/4};\cdot)\mathrm{e}^{-\sigma v}\,\mathrm{d}c = \frac{1}{\sqrt{\sigma}(\sqrt{\sigma}+c)}.$$

Therefore, everything goes similarly to the previous subsection except of course that $\nu = 1/4$ here rather than $1/2$ there,

$$\chi_{\infty}(\eta) = \mathrm{e}^{c^2\eta^4} \frac{2}{\sqrt{\pi}} \int_{c\eta^2}^{\infty} \mathrm{e}^{-t^2}\mathrm{d}t = \frac{2}{\pi} \int_0^{\infty} \mathrm{e}^{-\tau^2-2c\eta^2\tau}\,\mathrm{d}\tau, \quad (5.24)$$

so that

$$p_1(x) = \frac{1}{2\pi} \int_{-\infty}^{\infty} \chi_{\infty}(\eta)\mathrm{e}^{-i\eta x}\,\mathrm{d}\eta = \frac{1}{\pi^{3/2}} \int_{-\infty}^{\infty} \mathrm{d}\eta\,\mathrm{e}^{-i\eta x} \int_0^{\infty} \mathrm{e}^{-\tau^2-2c\eta^2\tau}$$

is given by

$$p_1(x) = \frac{1}{\pi\sqrt{2c}} \int_0^{\infty} \exp\left[-\frac{x^2}{8c\tau} - \tau^2\right] \frac{\mathrm{d}\tau}{\sqrt{\tau}}. \quad (5.25)$$

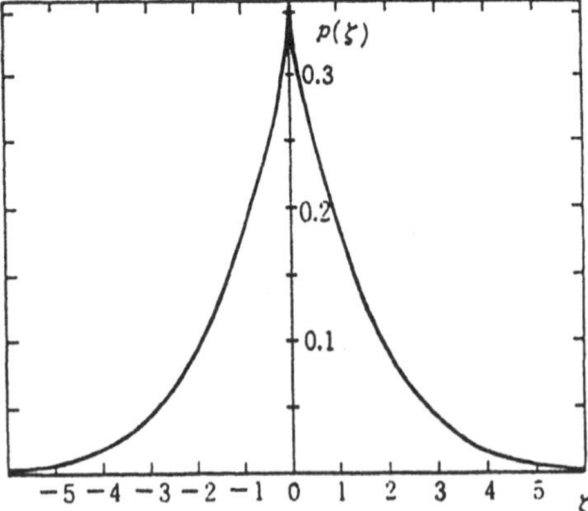

Figure 1. Probability distribution of the Brownian scan, the case of vanishing average. Note: $\zeta := \mathrm{x}[\omega]/\{\|U\|/(2D)^{3/4}\}$.

This is the $T \to \infty$ limit of the probability distribution for the Brownian scan, (5.1) with $\nu = 1/4$, for the case of V characterized by (5.16). The density is sharply peaked at the origin (Figure 1).

6. Conclusion

We have introduced a path integral (2.17) over the Ornstein–Uhlenbeck process (2.1) and its variant (2.20) over paths distorted (or, weighted in a way) by a potential V. The latter integral can be regarded as giving a characteristic function for the probability distribution of (1.1), i.e.

$$X[\omega] = \frac{1}{T^\nu} \int_0^T V(\omega(t))\, dt$$

for the Ornstein–Uhlenbeck process ω. We have studied $T \to \infty$ limit of the probability distribution. The nontrivial limit exists iff $\nu = 1$, and it has turned out to be the δ-function supported at the weighted average (4.10) of V. The deviation from the average, or in other words $X[\omega]$ for the function V of vanishing average, has the limit distribution iff $\nu = 1/2$ and it is Gaussian, (4.24). These results are obtained by calculations involving an approximation, while the following ones for the Wiener process are exact. For the Wiener process ω, the limit distribution of $X[\omega]$ exists iff $\nu = 1/2$ and is a one-sided Gaussian (5.14) with the width proportional to an average $\int_{-\infty}^{\infty} V(x)\, dx$. When this average vanishes, the distribution collapses to a delta-function, making the alternative limit exist iff $\nu = 1/4$ to have the shape (5.25) shown in Figure 1.

Acknowledgements

I thank Prof. Larry A. Shepp for his thoughtful initiatives during the work sketched in Section 5.2; I owe most of the ideas to him, and those ideas are the bases of the calculations in the other sections, too.

Thanks are also due to Profs Toru Nakamura and Keiji Watanabe for helpful discussions and for critical reading of the manuscript.

References

1. Uhlenbeck, G. E. and Ornstein, L. S.: *Phys. Rev.* **36** (1930), 823–841.
2. Nelson, E.: *Dynamical Theory of Brownian Motion*, Math. Notes, Princeton Univ. Press, Princeton, 1967.
3. Hida, T.: *Brownian Motion*, Springer, Heidelberg, 1980.
4. Ezawa, H., Klauder, J. R. and Shepp, L. A.: *Ann. Phys.* **88** (1974), 588–620.
5. Abramowitz, M. and Stegun, I. (eds): *Handbook of Mathematical Functions*, Dover, New York.
6. Shepp, L. A. and Ezawa, H.: (1974), unpublished. This work is the origin of the methods used in the present paper.

Acta Applicandae Mathematicae **63:** 137–139, 2000.

Remarks on a Noncanonical Representation for a Stationary Gaussian Process

YUJI HIBINO[1], MASUYUKI HITSUDA[2] and HIROSHI MURAOKA[3]
[1]*Faculty of Science and Engineering, Saga University, Saga, Japan*
[2]*Faculty of Science, Kumamoto University, Kumamoto, Japan*
[3]*Department of Mechanical Engineering, The University of Tokyo, Tokyo, Japan*

(Received: 5 January 1999)

Abstract. For a stationary centered Gaussian process, we construct a noncanonical representation which has an infinite-dimensional orthogonal complement that is nontrivial. The authors have already proposed a systematic method for the construction of noncanonical representation having a finite-dimensional orthogonal complement.

Mathematics Subject Classifications (2000): 60G15, 60G10.

Key words: stationary Gaussian process, canonical representation, moving average representation.

Karhunen [3] gave a complete result that the purely nondeterministic stationary Gaussian process has a canonical representation with respect to a single white noise. In addition, he succeeded in characterizing the canonical kernel as the *root* of the spectral density for the process. A little later, the general theory of canonical representation of a Gaussian process was proposed by Lévy [4].

In order to organize the consideration below, let us review Karhunen's results. Suppose that a purely nondeterministic stationary, centered Gaussian process $X = \{X(t); t \in \mathbf{R}\}$ has a moving average representation

$$X(t) = \int_{-\infty}^{t} F(t-u)\,\mathrm{d}B(u), \quad t \in \mathbf{R}. \tag{1}$$

Then the inverse Fourier transform c of F is uniquely decomposed into the five parts, namely,

$$c(\lambda) = Cc_0(\lambda)c_1(\lambda)c_2(\lambda)c_3(\lambda), \tag{2}$$

with notations

$$C = \mathrm{e}^{i\gamma},$$

$$c_0(\lambda) = \sqrt{2\pi}\exp\left\{-\frac{1}{\pi i}\int_{-\infty}^{\infty}\frac{1+w\lambda}{w-\lambda}\frac{\log f(w)}{1+w^2}\,\mathrm{d}w\right\},$$

$$c_1(\lambda) = \prod_{n=1}^{\infty}\frac{\lambda - z_n}{\lambda - \overline{z_n}}\frac{|z_n+i|}{z_n+i}\frac{|z_n-i|}{z_n-i},$$

$$c_2(\lambda) = \mathrm{e}^{i\alpha\lambda},$$
$$c_3(\lambda) = \exp\left\{\frac{1}{\pi i}\int_{-\infty}^{\infty}\frac{1+w\lambda}{w-\lambda}\,\mathrm{d}\beta(w)\right\},$$

where γ is a real constant, $\{z_n\}$ is a complex sequence satisfying $\Im(z_n) > 0$ and $\sum_{n=1}^{\infty}\Im(z_n)/(1+|z_n|^2) < \infty$, α is a nonnegative constant and $\beta(w)$ is a nondecreasing function of bounded variation whose derivative is vanished for almost all w.

The components c_0 and $c_1c_2c_3$ are called the *outer function* and the *inner function* of c, respectively. In more detail, we call c_1, c_2 and c_3 the *Blaschke part*, the *delay part* and the *singular part*, respectively.

Karhunen [3] gave a standard criterion for the canonical representation of a stationary Gaussian process. That is to say, we can find a Brownian motion $B = \{B(t); t \in \mathbf{R}\}$ such that (1) has a property

$$H_t(X) = H_t(B), \quad t \in \mathbf{R},$$

if and only if the inner part of c is trivial, namely, $c(\lambda) = Cc_0(\lambda)$. Here $H_t(X)$ (resp. $H_t(B)$) means a linear span of $X(s)$ (resp. $B(s)$), $s \leqslant t$.

This result means that the noncanonical property of the stationary process arises from three causes, since the inner function is composed of three parts. In this article, we especially attend to the type arising from the Blaschke part. The Blaschke part c_1 undertakes the zero points of c in the upper half-plane.

Before taking up the main question, we briefly note the delay part. The delay part c_2 plays a role of the time shift by α. That is to say, if c_2 in (2) is not absent, $\{B(t) - B(t-s); 0 < s \leqslant \alpha\}$ is independent of $H_t(X)$ for any $t \in \mathbf{R}$.

Up to now the role of the singular part has not been clarified. In order to concentrate our attention to the term c_1 in the decomposition (2), in the sequel of the present article we assume that the terms c_2 and c_3 are trivial.

The following theorem makes a definite relation of the zero points of c to the orthogonal complement of $H_t(X)$ in $H_t(B)$ for each $t \in \mathbf{R}$.

THEOREM 1. *Let a stationary centered Gaussian process X have a representation* (1). *The random variable $\int_{-\infty}^{t}\mathrm{e}^{-i\lambda_0 u}\,\mathrm{d}B(u)$ in $H_t(B)$ is orthogonal to $H_t(X)$ for each $t \in \mathbf{R}$, only if the point λ_0 in the upper half-plane is a zero point of c_1.*

Proof. Since the element of $H_t(X)$ is a linear combination of $X(s)$, $s \leqslant t$, or its limit, it is sufficient to prove that any $X(s)$, $s \leqslant t$, is orthogonal to the random variable $\int_{-\infty}^{t}\mathrm{e}^{-i\lambda_0 u}\,\mathrm{d}B(u)$. We obtain for any $s \leqslant t$,

$$\begin{aligned} E\left[\left(\int_{-\infty}^{t}\mathrm{e}^{-i\lambda_0 u}\,\mathrm{d}B(u)\right)X(s)\right] &= \int_{-\infty}^{s}\mathrm{e}^{-i\lambda_0 u}F(s-u)\,\mathrm{d}u \\ &= \int_{0}^{\infty}\mathrm{e}^{-i\lambda_0(s-v)}F(v)\,\mathrm{d}v \\ &= c(\lambda_0)\mathrm{e}^{-i\lambda_0 s}, \end{aligned}$$

so the statement is obvious. □

We are able to construct a noncanonical representation of $\tilde{X}$ having a nontrivial infinite-dimensional orthogonal complement in $H_t(B)$. That is to say, a new process, which has the same law of the given process, can be constructed by using the rest of the σ-field removed an infinite-dimensional subspace from the one of the original process.

THEOREM 2. *Denote by c_0 the outer part of a given stationary Gaussian process X. Let the stationary process $\tilde{X}$ have the representation* (1) *by using the Fourier transform $\tilde{F}$ of*

$$\tilde{c}(\lambda) = c_0(\lambda) \prod_{n=1}^{\infty} \frac{p_n + i\lambda}{p_n - i\lambda} \frac{|p_n - 1|}{p_n - 1}, \tag{3}$$

where the sequence $\{p_n\}$ satisfies

$$\sum_{n=1}^{\infty} \frac{p_n}{1 + p_n^2} < \infty \quad \text{and} \quad p_n > 0, \quad n \in \mathbf{N}. \tag{4}$$

Then $\tilde{X}$ is a noncanonical representation of X having infinite-dimensional orthogonal complement, namely,

$$H_t(\tilde{X}) \perp \left\{ \int_{-\infty}^{t} \mathrm{e}^{p_n u} \, \mathrm{d}B(u); n \in \mathbf{N} \right\}, \quad t \in \mathbf{R}. \tag{5}$$

Proof. The condition (4) guarantees the convergence of the infinite product of the right-hand side of (3). Hence, thanks to Theorem 1, the statement is proved. □

Remark. The linear span of the right of (5) is smaller than the closed linear manifold $H_t(B)$. Its orthogonal complement contains $H_t(\tilde{X})$. This means that the system $\{\mathrm{e}^{p_n u}; n \in \mathbf{N}\}$ satisfying (4) is not complete in $L^2((-\infty, t])$ for any $t \in \mathbf{R}$. This fact is known as the Szász theorem [5].

References

1. Hibino, Y., Hitsuda, M. and Muraoka, H.: Construction of noncanonical representations of a Brownian motion, *Hiroshima Math. J.* **27** (1997), 439–448.
2. Hida, T. and Hitsuda, M.: *Gaussian Processes, Representation and Applications,* Amer. Math. Soc., Providence, 1993.
3. Karhunen, K.: Über die Struktur stationärer zufälliger Funktionen, *Ark. Mat.* **1** (1950), 141–160.
4. Lévy, P.: Sur une classe de courbes de l'espace de Hilbert et sur une équation intégrale non linéaire, *Ann. Sci. Ecole Norm. Sup.* **73** (1956), 121–156.
5. Paley, R. E. A. C. and Wiener, N.: *Fourier Transforms in the Complex Domain,* Amer. Math. Soc., Providence, 1934.

Acta Applicandae Mathematicae **63:** 141–150, 2000.

A White Noise Approach to Stochastic Neumann Boundary-Value Problems

Dedicated to Prof. Takeyuki Hida on the occasion of his 70th birthday

HELGE HOLDEN[1] and BERNT ØKSENDAL[2]
[1] *Department of Mathematical Sciences, Norwegian University of Science and Technology, N-7491 Trondheim, Norway. e-mail: holden@math.ntnu.no, http://www.math.ntnu.no/~holden/*
[2] *Department of Mathematics, University of Oslo, N-0316 Oslo, Norway, and Norwegian School of Economics and Business Administration, Hellevn 30, N-5035 Bergen–Sandviken, Norway. e-mail: oksendal@math.uio.no*

(Received: 22 February 1999)

Abstract. We illustrate the use of white noise analysis in the solution of stochastic partial differential equations by explicitly solving the stochastic Neumann boundary-value problem

$$LU(x) - c(x)U(x) = 0, \quad x \in \mathcal{D} \subset \mathbb{R}^d,$$
$$\gamma(x) \cdot \nabla U(x) = -W(x), \quad x \in \partial\mathcal{D},$$

where L is a uniformly elliptic linear partial differential operator and $W(x)$, $x \in \mathbb{R}^d$, is d-parameter white noise.

Mathematics Subject Classifications (2000): 60H15, 60H40.

Key words: stochastic Neumann problem, white noise analysis.

1. Introduction

Since the seminal book by Hida [4] appeared in 1980, there has been a rapid development of white noise theory and its applications. In particular, white noise theory has found many spectacular applications in mathematical physics (see, e.g., [5, 7] and references therein).

In addition, white noise calculus turns out to be useful in the study of stochastic differential equations, both ordinary and partial (see [6]).

The purpose of this paper is to illustrate the application of white noise calculus to stochastic partial differential equations (SPDEs) by studying the stochastic boundary-value problem of Neumann type:

$$LU(x) - c(x)U(x) = 0 \quad \text{for } x \in \mathcal{D}, \tag{1.1}$$
$$\gamma(x) \cdot \nabla U(x) = -W(x) \quad \text{for } x \in \partial\mathcal{D}. \tag{1.2}$$

Here $\mathcal{D} \subset \mathbb{R}^d$ is a given bounded C^2 domain and L is a partial differential operator of the form

$$L = \sum_{i=1}^{d} b_i(x)\frac{\partial}{\partial x_i} + \sum_{i,j=1}^{d} a_{ij}(x)\frac{\partial^2}{\partial x_i \partial x_j}, \tag{1.3}$$

$c(x) \geqslant c > 0$ and $\gamma(x)$ are given functions (satisfying certain conditions) and $W(x) = W(x, \omega)$, $\omega \in \Omega$, is *white noise.*

One may think of U as a temperature in a medium governed by the differential equation $LU(x) - c(x)U(x) = 0$. Then the Neumann boundary condition models heat flux across the boundary which in our case is given by white noise. (See, e.g., [2].)

As pointed by Walsh [9] for similar SPDEs, it is not possible to find a solution $u(x, \omega)$ of (1.1)–(1.2) which is a regular stochastic process (random field), unless the dimension d is very low. To cover the general case it is necessary to introduce some kind of weak solution concept. One possibility is to look for solutions $u(x, \omega)$ such that

$$u(\cdot, \omega) \text{ is a distribution (in } x) \text{ for a.a. } \omega. \tag{1.4}$$

This is the approach chosen by Walsh.

Such an approach causes difficulties for nonlinear SPDEs because one will have to define nonlinear operations on distributions. However, it is possible to adapt the Colombeau nonlinear theory of distributions to some nonlinear SPDEs. (See, e.g., [8].)

The other possibility is to look for solutions $u(x, \omega)$ such that

$$u(x, \cdot) \text{ is a } \textit{stochastic} \text{ distribution (in } \omega) \text{ for all } x. \tag{1.5}$$

Such an approach fits in well with the white noise theory, where both the Hida space $(\mathcal{S})^*$ and the more general Kondratiev space $(\mathcal{S})$ of stochastic distributions are at our disposal. Moreover, in these spaces there is a natural product $\diamondsuit$ (the Wick product) and a corresponding theory for nonlinear operations on distributions.

In this paper we will use this second approach and look for solutions of the type (1.5). We will prove that, under some conditions, Equations (1.1)–(1.2) have the unique solution

$$u(x, \omega) = \widehat{E}^x\left[\int_0^\infty \exp\left(-\int_0^t c(x_s(\hat{\omega}))\,\mathrm{d}s\right) W(x_t(\hat{\omega}), \omega)\,\mathrm{d}\xi_t(\hat{\omega})\right], \tag{1.6}$$

where $(x_t, \xi_t) = (x_t(\hat{\omega}), \xi_t(\hat{\omega}))$, $\hat{\omega} \in \widehat{\Omega}$, is the solution of the Skorokhod stochastic differential equation

$$\mathrm{d}x_t = b(x_t)\,\mathrm{d}t + \sigma(x_t)\,\mathrm{d}\beta_t + \gamma(x_t)\,\mathrm{d}\xi_t, \tag{1.7}$$

where $\beta_t = \beta_t(\hat{\omega})$ is the d-dimensional Brownian motion on a filtered probability space $(\widehat{\Omega}, \widehat{\mathcal{F}}_t, \widehat{P}^x)$. The pair (x_t, ξ_t) is unique under the conditions that $x_t \in \overline{\mathcal{D}}$ for all t, ξ_t is a nondecreasing adapted process increasing only when $x \in \partial\mathcal{D}$. The process ξ_t is called *local time* of x_t at $\partial\mathcal{D}$ and the process x_t is called the *reflection* (at $\partial\mathcal{D}$, at the angle γ) of the Itô diffusion y_t given by

$$\mathrm{d}y_t = b(y_t)\,\mathrm{d}t + \sigma(y_t)\,\mathrm{d}\beta_t. \tag{1.8}$$

For a more detailed explanation, see Section 3. We refer to [1] and [3] for information about Skorokhod stochastic differential equations and the associated Neumann boundary conditions.

The process $u(x, \cdot)$ defined in (1.6) belongs to the space $(\mathcal{S})^*$ for all x.

2. A Brief Review of the White Noise Theory

For convenience, we briefly recall the concepts and terminology we will use from white noise theory. For more information, we refer to [6].

In the following, we let $(\mu, \mathcal{S}'(\mathbb{R}^d), \mathcal{B})$ be the white noise probability space. Here $\Omega = \mathcal{S}'(\mathbb{R}^d)$ is the space of tempered distributions on $\mathbb{R}^d$, $\mathcal{B}$ denotes the Borel σ-algebra on Ω and μ is defined by the property that

$$\int_{\mathcal{S}'(\mathbb{R}^d)} \mathrm{e}^{i\langle\omega,\phi\rangle}\,\mathrm{d}\mu(\omega) = \exp(-\tfrac{1}{2}\|\phi\|^2) \tag{2.1}$$

for all $\phi \in \mathcal{S}(\mathbb{R}^d)$ (the Schwartz space of rapidly decreasing smooth functions on $\mathbb{R}^d$), where $\langle\omega, \phi\rangle$ denotes the action of $\omega \in \mathcal{S}'(\mathbb{R}^d)$ on ϕ and $\|\phi\|^2 = \int_{\mathbb{R}^d} \phi(x)^2\,\mathrm{d}x$, $\mathrm{d}x$ being the Lebesgue measure on $\mathbb{R}^d$.

We let $\{\eta_k\}_{k=1}^{\infty}$ be the basis of $L^2(\mathbb{R}^d)$ consisting of tensor products of the *Hermite functions*

$$\xi_n(x) = \pi^{-1/4}((n-1)!)^{-1/2}\mathrm{e}^{-x^2/2}h_{n-1}(\sqrt{2}x), \quad n \in \mathbb{N},\ x \in \mathbb{R}, \tag{2.2}$$

where

$$h_m(y) = (-1)^m \mathrm{e}^{y^2/2}\frac{\mathrm{d}^m}{\mathrm{d}y^m}\left(\mathrm{e}^{-y^2/2}\right), \quad m = 0, 1, 2, \ldots,\ y \in \mathbb{R} \tag{2.3}$$

are the *Hermite polynomials*.

The symbol $\mathcal{I}$ denotes the set of all multi-indices $\alpha = (\alpha_1, \ldots, \alpha_n)$ where the α_i's are nonnegative integers and $n \in \mathbb{N}$. According to the chaos expansion theorem, any $f \in L^2(\mu)$ can be written uniquely

$$f(\omega) = \sum_{\alpha\in\mathcal{I}} a_\alpha H_\alpha(\omega) \tag{2.4}$$

(convergence in $L^2(\mu)$) where $a_\alpha \in \mathbb{R}$ and

$$H_\alpha(\omega) = \prod_{j=1}^{n} h_{\alpha_j}(\langle\omega, \eta_j\rangle) \quad \text{if } \alpha = (\alpha_1, \ldots, \alpha_n) \in \mathcal{I}. \tag{2.5}$$

For $-1 \leqslant \rho \leqslant 1$ and $q \in \mathbb{Z}$ we define the *Kondratiev norms* $\| \cdot \|_{\rho,q}$ by

$$\|F\|^2_{\rho,q} = \sum_\alpha a^2_\alpha (\alpha!)^{1+\rho}(2\mathbb{N})^{q\alpha} \tag{2.6}$$

if F is the (formal) expansion

$$F(\omega) = \sum_{\alpha \in \mathcal{I}} a_\alpha H_\alpha(\omega), \tag{2.7}$$

where $(2\mathbb{N})^{q\alpha} = \prod_{j=1}^n (2j)^{q\alpha_j}$ if $\alpha = (\alpha_1, \dots, \alpha_n) \in \mathcal{I}$.

The corresponding *Kondratiev Hilbert spaces* are defined by

$$(\mathcal{S})_{\rho,q} = \{F : \|F\|_{\rho,q} < \infty\}. \tag{2.8}$$

The *Kondratiev* (*stochastic*) *test function spaces* are defined by (if $0 \leqslant \rho \leqslant 1$)

$$(\mathcal{S})_\rho = \bigcap_{r=1}^\infty (\mathcal{S})_{\rho,r}, \quad \text{with the projective topology.} \tag{2.9}$$

The *Kondratiev* (*stochastic*) *distribution* spaces are defined by (if $0 \leqslant \rho \leqslant 1$)

$$(\mathcal{S})_{-\rho} = \bigcup_{r=1}^\infty (\mathcal{S})_{-\rho,-r}, \quad \text{with the inductive topology.} \tag{2.10}$$

In particular, if we choose $\rho = 0$ in (2.9) we get the *Hida test function spaces*

$$(\mathcal{S}) = (\mathcal{S})_0 \tag{2.11}$$

and if we choose $\rho = 0$ in (2.10) we get the *Hida distribution spaces*

$$(\mathcal{S})^* = (\mathcal{S})_{-0}. \tag{2.12}$$

The (*singular, d-parameter*) *white noise* is defined by

$$W(x, \omega) = \sum_{k=1}^\infty \eta_k(x) H_{\epsilon^{(k)}}(\omega), \tag{2.13}$$

where $\epsilon^{(k)} = (0, 0, \dots, 1, \dots)$ with 1 on the kth place. We easily verify that $W(x, \cdot) \in (\mathcal{S})^*$ for all x by considering

$$\begin{aligned} \|W(x, \cdot)\|^2_{0,-q} &= \sum_k \eta^2_k(x)(\epsilon^{(k)})!(2\mathbb{N})^{-q\epsilon_k} \\ &= \sum_k \eta^2_k(x)(2k)^{-q} < \infty \quad \text{for all } q > 1, \end{aligned}$$

because $\sup_k(\sup_{x\in\mathbb{R}^d} |\eta_k(x)|) < \infty$. Hence $W(x, \cdot) \in (\mathcal{S})_{0,-q}$ for all $q > 1$ and in particular $W(x, \cdot) \in (\mathcal{S})^*$.

To each $F(\omega) = \sum_{\alpha \in \mathcal{I}} a_\alpha H_\alpha(\omega) \in (\mathcal{S})_{-1}$ we can associate the expansion

$$\mathcal{H}F(z) = \sum_{\alpha \in \mathcal{I}} a_\alpha z^\alpha \quad \text{for } z = (z_1, z_2, \ldots) \in \mathbb{C}^{\mathbb{N}}, \tag{2.14}$$

where $\mathbb{C}^{\mathbb{N}}$ is the set of all sequences $(z_1, z_2, \ldots)$ of complex numbers z_j and

$$z^\alpha = z_1^{\alpha_1} z_2^{\alpha_2} \cdots \quad \text{if } \alpha = (\alpha_1, \alpha_2, \ldots).$$

The function $\mathcal{H}F(z)$ is called the *Hermite transform* of F. One can characterize the elements of $(\mathcal{S})_{-1}$ by means of their Hermite transforms as follows:

THEOREM 2.1. *A formal expansion* $F(\omega) = \sum_{\alpha \in \mathcal{I}} a_\alpha H_\alpha(\omega)$ *belongs to* $(\mathcal{S})_{-1}$ *if and only if there exists* $q < \infty$ *such that* $\mathcal{H}F(z) = \sum_{\alpha \in \mathcal{I}} a_\alpha z^\alpha$ *is a (uniformly convergent and) bounded analytic function on the infinite-dimensional ellipsoids*

$$\mathbb{K}_q(R) = \left\{ (\zeta_1, \zeta_2, \ldots) \in \mathbb{C}^{\mathbb{N}} : \sum_{\alpha \neq 0} |\zeta^\alpha|^2 (2\mathbb{N})^{q\alpha} < R^2 \right\}. \tag{2.15}$$

For a proof, see [6], Theorem 2.6.11.

The key to the solution of stochastic partial differential equations of the type (1.1)–(1.2) is the following, which is a special case of Theorem 4.1.1 in [6].

THEOREM 2.2. *Suppose we seek a solution* $U(t, x)\colon G \to (\mathcal{S})_{-1}$ *of a linear stochastic partial differential equation of the form*

$$A(t, x, \partial_t, \nabla_x, U, W(x)) = 0. \tag{2.16}$$

Suppose the Hermite transformed equation

$$A(t, x, \partial_t, \nabla_x, u, \widetilde{W}(x, z)) = 0 \tag{2.17}$$

has a (strong) solution $u = u(t, x, z)$ *for each value of the parameter* $z \in \mathbb{K}_q(R)$ *for some* $q < \infty$, $R < \infty$. *Moreover, suppose that* $u(t, x, z)$ *and all its partial derivatives involved in* (2.17) *are uniformly bounded for* $(t, x, z) \in G \times \mathbb{K}_q(R)$, *continuous with respect to* $(t, x) \in G$ *for all* $z \in \mathbb{K}_q(R)$ *and analytic with respect to* $z \in \mathbb{K}_q(R)$ *for all* $(t, x) \in G$.

Then there exists $U(t, x)\colon G \to (\mathcal{S})_{-1}$ *such that* $u(t, x, z) = \mathcal{H}U(t, x, z)$ *for all* $(t, x, z) \in G \times \mathbb{K}_q(R)$ *and* $U(t, x)$ *solves (in the strong sense in* $(\mathcal{S})_{-1}$*) Equation* (2.16).

3. Solution of the Stochastic Neumann Boundary-Value Problem

Before we state and prove the main result of this paper we briefly review some results about Skorokhod stochastic differential equations and *deterministic* Neumann boundary value problems.

In the following we let $\mathcal{D}$ be a bounded domain in $\mathbb{R}^d$ with a C^2 boundary. This means that the boundary of $\mathcal{D}$, $\partial\mathcal{D}$, is locally the graph of a C^2 function (a twice continuously differentiable function). We assume that we are given Lipschitz continuous functions $b\colon \mathbb{R}^d \to \mathbb{R}^d$ and $\sigma\colon \mathbb{R}^d \to \mathbb{R}^{d\times d}$ and a C^2 function $\gamma\colon \partial\mathcal{D} \to \mathbb{R}^d$ such that

$$\gamma(x)\cdot\nu(x) > 0 \quad \text{for all } x \in \partial\mathcal{D}, \tag{3.1}$$

where $\nu(x)$ is the inward pointing unit normal at $x \in \partial\mathcal{D}$.

Consider, as in (1.7), the following *Skorokhod* stochastic differential equation (in the unknown processes x_t, ξ_t)

$$\mathrm{d}x_t = b(x_t)\,\mathrm{d}t + \sigma(x_t)\,\mathrm{d}\beta_t + \gamma(x_t)\,\mathrm{d}\xi_t, \quad x_0 = x \in \overline{\mathcal{D}}, \tag{3.2}$$

where we require that x_t, ξ_t are adapted and

$$x_t \in \overline{\mathcal{D}} \quad \text{for all } t \geqslant 0. \tag{3.3}$$

$$\xi_t \text{ is continuous, nondecreasing and } \xi_t \text{ increases only when } x_t \in \partial\mathcal{D}. \tag{3.4}$$

The process x_t is called the *reflection at* $\partial\mathcal{D}$ of the process y_t given by

$$\mathrm{d}y_t = b(y_t)\,\mathrm{d}t + \sigma(y_t)\,\mathrm{d}\beta_t, \quad y_0 = x \in \overline{\mathcal{D}} \tag{3.5}$$

and ξ_t is called the *local time* of x_t at $\partial\mathcal{D}$.

The following result can be found in [1], Theorem 12.1:

THEOREM 3.1. *There exists a unique solution* (x_t, ξ_t) *of the Skorokhod stochastic differential equation* (3.2)–(3.4).

We proceed to consider the connection to Neumann boundary value problems: With b, σ as above define

$$Lu(x) = \sum_{i=1}^{d} b_i(x)\frac{\partial u}{\partial x_i} + \sum_{i,j=1}^{d} a_{ij}(x)\frac{\partial^2 u}{\partial x_i \partial x_j}, \tag{3.6}$$

where

$$a_{ij}(x) = \frac{1}{2}\left(\sigma\sigma^T\right)_{ij}(x) = \frac{1}{2}\sum_{k=1}^{n} \sigma_{ik}(x)\sigma_{kj}(x). \tag{3.7}$$

Assume that L is *uniformly elliptic*, i.e., there exists an $\epsilon > 0$ such that

$$\sum_{i,j=1}^{d} a_{ij}(x) y_i y_j \geqslant \epsilon |y|^2 \tag{3.8}$$

for all $y = (y_1, \ldots, y_n) \in \mathbb{R}^d$.

Let f be a C^2 function on $\partial\mathcal{D}$ and let c be a C^2 function on $\mathcal{D}$ such that

$$c(x) \geqslant c > 0 \quad \text{for all } x \in \mathcal{D}. \tag{3.9}$$

The *Neumann problem* is to find a function $u \in C^2(\overline{\mathcal{D}})$ such that

$$Lu(x) - c(x)u(x) = 0 \quad \text{for } x \in \mathcal{D} \tag{3.10}$$

and

$$\nabla u(x) \cdot \gamma(x) = -f(x) \quad \text{for } x \in \partial\mathcal{D}. \tag{3.11}$$

Recall that if $h: G \subset \mathbb{R}^d \to \mathbb{R}$ is a given function and $0 < \lambda < 1$, we define the Hölder norms

$$\|h\|_{C^\lambda(G)} = \sup_{y \in G} |h(y)| + \sup_{y,z \in G} \frac{|h(y) - h(z)|}{|y - z|^\lambda} \tag{3.12}$$

and

$$\|h\|_{C^{2+\lambda}(G)} = \sum_{0 \leqslant |\alpha| \leqslant 2} \|\partial^\alpha h\|_{C^\lambda(G)}, \tag{3.13}$$

where the sum is taken over all multi-indices $\alpha = (\alpha_1, \ldots, \alpha_d)$ with $|\alpha| = \alpha_1 + \cdots + \alpha_d \leqslant 2$ and

$$\partial^\alpha h = \frac{\partial^{\alpha_1} \cdots \partial^{\alpha_d}}{\partial x_1^{\alpha_1} \cdots \partial x_d^{\alpha_d}} h.$$

The following result can be obtained by combining Theorem 6.2 with the conclusion of Section III.7 in [1] (see also [3]):

THEOREM 3.2. *We have that*

(i) *Under the above assumptions there exists a unique solution* $u \in C^2(\overline{\mathcal{D}})$ *of the Neumann problem* (3.10)–(3.11).

(ii) *Moreover, the solution can be represented in the form*

$$u(x) = \widehat{E}^x \left[\int_0^\infty \exp\left(- \int_0^t c(x_s)\, \mathrm{d}s \right) f(x_t)\, \mathrm{d}\xi_t \right], \tag{3.14}$$

where (x_t, ξ_t), *with probability law* $\widehat{P}^x$, *is the solution of the Skorokhod equation* (3.2)–(3.4) *and* $\widehat{E}^x$ *denotes expectation with respect to* $\widehat{P}^x$.

(iii) *For all* $\lambda \in (0, 1)$ *there exists* $K < \infty$ *such that*

$$\|u\|_{C^{2+\lambda}(\overline{\mathcal{D}})} \leqslant K \, \|f\|_{C^\lambda(\partial\mathcal{D})}. \tag{3.15}$$

We can now state and prove the main result of this paper:

THEOREM 3.3. *Let $\mathcal{D}$, b, σ, γ, L and c be as above. Then the stochastic Neumann problem*

$$LU(x) - c(x)U(x) = 0 \quad \text{for } x \in \mathcal{D}, \tag{3.16}$$

$$\gamma(x) \cdot \nabla U(x) = -W(x) \quad \text{for } x \in \partial\mathcal{D}, \tag{3.17}$$

where $W(x) = W(x, \omega)$, $\omega \in \Omega$, is the d-parameter white noise, has a unique solution $U: \overline{\mathcal{D}} \to (\mathcal{S})^*$ given by

$$U(x) = U(x, \omega) = \widehat{E}^x \left[\int_0^\infty \exp\left(- \int_0^t c(x_s)\,\mathrm{d}s \right) W(x_t, \omega)\,\mathrm{d}\xi_t \right], \tag{3.18}$$

where (x_t, ξ_t) solves the Skorokhod equation (3.2)–(3.4).

Proof. The key to our method is Theorem 2.2. So we consider the Hermite transformed equation

$$Lu(x, z) - c(x)u(x, z) = 0 \quad \text{for } x \in \mathcal{D}, \tag{3.19}$$

$$\gamma(x) \cdot \nabla u(x, z) = -\widetilde{W}(x, z) \quad \text{for } x \in \partial\mathcal{D}, \tag{3.20}$$

where $z = (z_1, z_2, \ldots) \in \mathbb{C}_0^{\mathbb{N}}$ (the set of finite sequences in $\mathbb{C}^{\mathbb{N}}$). As before the differential operators L and ∇ act on the x-variable and $z = (z_1, z_2, \ldots) \in \mathbb{C}_0^{\mathbb{N}}$ is to be regarded as a parameter. Fix $z \in \mathbb{C}_0^{\mathbb{N}}$.

By considering separately the real and imaginary part of

$$\widetilde{W}(x, z) = \sum_{k=1}^{\infty} \eta_k(x) z_k,$$

we get by Theorem 3.2 that the unique solution $u(x, z)$ of (3.19)–(3.20) is given by

$$u(x, z) = \widehat{E}^x \left[\int_0^\infty \exp\left(- \int_0^t c(x_s)\,\mathrm{d}s \right) \widetilde{W}(x_t, z)\,\mathrm{d}\xi_t \right], \tag{3.21}$$

where (x_t, ξ_t) solves the Skorokhod equation (3.2)–(3.4). Moreover, $u(x, z)$ and all its partial derivatives up to order 2 are uniformly bounded for $(x, z) \in \overline{\mathcal{D}} \times \mathbb{K}_2(R)$ for all $R < \infty$. This follows by (3.15) plus the fact that

$$\begin{aligned} \|\widetilde{W}(\cdot, z)\|_{C^\lambda(\partial\mathcal{D})} &\leqslant \sup_k \|\eta_k\|_{C^\lambda(\partial\mathcal{D})} \sum_{k=1}^{\infty} |z_k| \\ &\leqslant M_1 \sum_{k=1}^{\infty} (2k)^{-2} \sum_{k=1}^{\infty} |z_k|^2 (2k)^2 \end{aligned}$$

$$
\begin{aligned}
&= M_2 \sum_{k=1}^{\infty} |z^{\epsilon^{(k)}}|^2 (2\mathbb{N})^{2\epsilon^{(k)}} \\
&\leqslant M_2 \sum_{\alpha \neq 0} |z^{\alpha}|^2 (2\mathbb{N})^{2\alpha} \\
&< M_2 R^2 \quad \text{if } z \in \mathbb{K}_2(R)
\end{aligned}
$$

and similar estimates for the partial derivatives. Similarly we see that $u(x, z)$ is analytic with respect to $z \in \mathbb{K}_2(R)$ for all $x \in \overline{\mathcal{D}}$. So by Theorem 2.2 there exists $U(x) \in (\mathcal{S})_{-1}$ such that $u(x, z) = \mathcal{H}U(x, z)$ for all $(x, z) \in \overline{\mathcal{D}} \times \mathbb{K}_2(R)$ and $U(x)$ solves (in the strong sense) the stochastic Neumann equation (3.16)–(3.17).

Finally we verify that $U(x) \in (\mathcal{S})^*$: Since $U(x)$ has the expansion

$$
U(x, \omega) = \sum_{k=1}^{\infty} g_k(x) H_{\epsilon^{(k)}}(\omega),
$$

with

$$
g_k(x) = \widehat{E}^x \left[\int_0^{\infty} \exp\left(- \int_0^t c(x_s)\, \mathrm{d}s \right) \eta_k(x_t)\, \mathrm{d}\xi_t \right]
$$

we see that

$$
\sup_k \left\{ g_k^2(x) \epsilon^{(k)}! \, (2\mathbb{N})^{-q\epsilon^{(k)}} \right\} = \sup_k \{ g_k^2(x) k^{-q} \} < \infty \quad \text{for all } q \geqslant 0.
$$

4. A Remark about the Solution

As pointed out in the introduction, our solution (3.18) is a *stochastic distribution*. This means that – just like for ordinary deterministic distributions – it acts on its corresponding test function space, which in this case is the space $(\mathcal{S})^*$ of Hida test functions. If we choose $\psi \in (\mathcal{S})^*$, then the action of U on ψ is given explicitly by

$$
\langle U, \psi \rangle = \widehat{E}^x \left[\int_0^{\infty} \exp\left(- \int_0^t c(x_s)\, \mathrm{d}s \right) \langle W(x_s, \cdot), \psi \rangle \, \mathrm{d}\xi_s \right],
$$

where

$$
\begin{aligned}
\langle W(x_s, \cdot), \psi \rangle &= \sum_{k=1}^{\infty} \eta_k(x_s) \langle H_{\epsilon^{(k)}}, \psi \rangle = \sum_{k=1}^{\infty} \eta_k(x_s) \langle \langle \cdot, \eta_k \rangle, \psi \rangle \\
&= \sum_{k=1}^{\infty} \eta_k(x_s) E_{\mu}[\langle \omega, \eta_k \rangle \psi(\omega)].
\end{aligned}
$$

In particular, if $\psi(\omega) = \psi_l(\omega) = \langle \omega, \eta_l \rangle$ then

$$
\langle U, \psi \rangle = \widehat{E}^x \left[\int_0^{\infty} \exp\left(- \int_0^t c(x_s) \mathrm{d}s \right) \eta_l(x_s)\, \mathrm{d}\xi_s \right].
$$

We could regard this as the average of U with respect to the (Gaussian) stochastic weight function $\psi_l(\omega)$.

5. Concluding Remarks

The main purpose of this paper has been to illustrate how white noise analysis can be used to solve stochastic partial differential equations by applying the method to stochastic Neumann boundary-value problems of the type (1.1)–(1.2).

By inspecting the proof we see that the same method applies to the more general equations

$$LU(x) - c(x)U(x) = -g(x) \quad \text{in } \mathcal{D},$$

$$\gamma(x) \cdot \nabla U(x) - \lambda(x)U(x) = -f(x) \quad \text{on } \partial\mathcal{D},$$

where $f: \partial\mathcal{D} \to (\mathcal{S})^*$ and $g: \mathcal{D} \to (\mathcal{S})^*$ are given $(\mathcal{S})^*$-valued functions and $\lambda(x) > 0$ is smooth. Moreover, the conditions we have assumed on L, c, γ, λ and $\mathcal{D}$ can be relaxed. In fact, we do not even need to assume uniform ellipticity if we allow ourselves to consider $(\mathcal{S})^*$-valued solutions $U(x, \cdot)$ which solve the equation in a week sense with respect to x.

References

1. Bass, R.: *Diffusions and Elliptic Operators*, Springer, Berlin, 1998.
2. Duff, G. F. and Naylor, D.: *Differential Equations of Applied Mathematics*, Wiley, New York, 1966.
3. Freidlin, M.: *Functional Integration and Partial Differential Equations*, Princeton University Press, Princeton, 1985.
4. Hida, T.: *Brownian Motion*, Springer, Berlin, 1980.
5. Hida, T., Kuo, H.-H., Potthoff, J. and Streit, L.: *White Noise Analysis*, Kluwer Acad. Publ., Dordrecht, 1993.
6. Holden, H., Øksendal, B., Ubøe, J. and Zhang, T.: *Stochastic Partial Differential Equations*, Birkhäuser, Basel, 1996.
7. Kuo, H.-H.: *White Noise Distribution Theory*, CRC Press, Boca Raton, 1996.
8. Oberguggenberger, M. and Russo, F.: Nonlinear SPDEs: Colombeau solutions and pathwise limits, In: L. Decreusefond, J. Gjerde, B. Øksendal and A. S. Üstünel (eds), *Stochastic Analysis and Related Topics VI*, Birkhäuser, Boston, 1998, pp. 319–332.
9. Walsh, J. B.: An introduction to stochastic partial differential equations, In: R. Carmona, H. Kesten and J. B. Walsh (eds), *École d'été de probabilités de Saint-Flour XIV 1984*, Lecture Notes in Math. 1180, Springer, Berlin, 1984, pp. 236–433.

Acta Applicandae Mathematicae **63:** 151–164, 2000.

Quantum Cable Equations in Terms of Generalized Operators *

ZHIYUAN HUANG[1], CAISHI WANG[1,2] and XIANGJUN WANG[1]
[1]*Department of Mathematics, Huazhong University of Science and Technology, Wuhan, Hubei 430074, P.R. China*
[2]*Department of Computer Science, Northwest Normal University, Lanzhou, Gansu 730070, P.R. China*

(Received: 24 March 1999)

Abstract. A quantum cable equation in the sense of generalized operators is introduced. The existence and uniqueness of solutions are established and the continuity and a Markov-like property of the solution are obtained.

Mathematics Subject Classification (2000): 60H40.

Key words: white noise, quantum cable equation, generalized operator.

1. Introduction

The stochastic cable equation was proposed as a model for the evolution in space and time of the electric potential on a neuron [14, 15]. It is believed that this potential evolves through a combination of diffusion and random fluctuation. Recently, it has been argued that the propagation of nerve signals might be influenced by quantum coherence effects within the microtubules in the cytoskeletons of the neurons [13]. To describe the microtubular potential as an observable that evolves in space and time through diffusion and quantum effects, Applebaum [1] has introduced the following quantized cable equation:

$$\begin{aligned} &\frac{\partial U}{\partial t} = \frac{\partial^2 U}{\partial x^2} - U + U\dot{M}, \\ &\frac{\partial U}{\partial x}(0,t) = \frac{\partial U}{\partial x}(a,t) = 0, \\ &U(x,0) = U_0(x), \end{aligned} \tag{1.1}$$

where M is a quantum stochastic spectral integral with respect to the creation, number, and annihilation processes in a Fock space, and the solution will be a family of linear operators in the same space (see [1]). Obviously Equation (1.1) is an equation in terms of Hilbert space operators. In this paper, instead of Equa-

* Project 19631030 supported by NSF of China.

tion (1.1), we will consider a partial differential equation in terms of generalized operators over a Gel'fand triple (see Equation (1.2) below).

Let $(E) \subset (L^2) \subset (E)^*$ be the framework of white noise analysis (cf. [2]). The continuous linear operators from (E) to $(E)^*$ are called generalized operators. In general, the classical products of generalized operators are meaningless. Nevertheless, the Wick products of them introduced in [7] always make sense. Moreover, all generalized operators constitute a nuclear Wick algebra which is commutative. In [3] and [11], the authors considered certain stochastic differential equations with multiplicative functional-valued noise terms. The product considered there is a Wick product of generalized functionals. Motivated by the above works, we propose the following quantum cable equation in terms of generalized operators:

$$
\begin{aligned}
&\frac{\partial V(x,t)}{\partial t} = \frac{\partial^2 V(x,t)}{\partial x^2} - V(x,t) + V(x,t) \diamondsuit K(x,t), \\
&\quad 0 < x < l,\ t > 0, \\
&\frac{\partial V}{\partial x}(0,t) = \frac{\partial V}{\partial x}(l,t) = 0, \quad t \geqslant 0, \\
&V(x,0) = J(x), \quad 0 \leqslant x \leqslant l,
\end{aligned}
\tag{1.2}
$$

where $l > 0$ is given, $\diamondsuit$ stands for the Wick product. $\{K(x,t) \mid (x,t) \in [0,l] \times \mathbf{R}_+\}$ is a given family of generalized operators which plays the role of quantum noises. The initial process $\{J(x) \mid x \in [0,l]\}$ and the solution $V(x,t)$ are all generalized operators.

In the mathematical descriptions of quantum physics, many observables and quantities are not *bona fide* Hilbert space operators, but can only be regarded as generalized operators. It has been shown in [6] that by using generalized operators, one can formulate the formal manipulations in quantum free fields at a mathematically rigorous level. From this point of view, Equation (1.2) might be a better model for the evolution in space and time through quantum effects of the microtubular potential mentioned above.

In the present paper, we will establish the existence and uniqueness of solutions to Equation (1.2) and prove the continuity as well as a Markov-like property of the solution.

2. Preliminaries

In this section, we briefly quote some concepts, notations and propositions in white noise analysis. For details, see [2] or [9]. Denote by H the Hilbert space $L^2(\mathbf{R}^2)$ with inner product $\langle \cdot, \cdot \rangle$ and norm $|\cdot|_0$. Let $E = S(\mathbf{R})^2$ be the Schwartz test function space. Then E is a countably Hilbertian nuclear space with standard norms $\{|\cdot|_p \mid p \geqslant 0\}$ and $E \subset H \subset E^*$ constitutes a Gel'fand triple. The canonical bilinear form on $E^* \times E$ is denoted by $\langle \cdot, \cdot \rangle$ which is consistent with the inner product in H.

Let $(E) \subset (L^2) \subset (E)^*$ be the canonical framework of white noise analysis associated with $E \subset H \subset E^*$. The canonical bilinear form on $(E)^* \times (E)$ is denoted by $\langle\langle\cdot,\cdot\rangle\rangle$. For $f\in E_{\mathbf{c}}$, let $\mathcal{E}(f)$ be the exponential vecter corresponding to f. Then $\{\mathcal{E}(f)\mid f\in E_{\mathbf{c}}\}$ is total in (L^2).

Let $\mathcal{L}\equiv\mathcal{L}[(E),(E)^*]$ be the space of all generalized operators. For $X\in\mathcal{L}$, its symbol $\widehat{X}$ is defined as

$$\widehat{X}(f,g)=\langle\langle X\mathcal{E}(f),\mathcal{E}(g)\rangle\rangle \mathrm{e}^{-\langle f,g\rangle},$$

$f,g\in E_{\mathbf{c}}$. If $Y,Z\in\mathcal{L}$, then their Wick product $Y\Diamond Z$ is defined as

$$\widehat{Y\Diamond Z}(f,g)=\widehat{Y}(f,g)\widehat{Z}(f,g),\quad f,g\in E_{\mathbf{c}}.$$

The following two lemmas will be used in the proofs of our main results.

LEMMA 2.1 ([2, 9]). *Let $\{X_n\}_{n\geqslant 1}\subset\mathcal{L}$ be such that*

(1) $\forall f,g\in E_{\mathbf{c}}$, *$\{\widehat{X}_n(f,g)\}_{n\geqslant 1}$ is convergent in* $\mathbf{C}$;
(2) *there exist constants $p\geqslant 0$, $k_1,k_2>0$ such that*

$$\left|\widehat{X}_n(f,g)\right|\leqslant k_1\exp\left\{k_2(|f|_p^2+|g|_p^2)\right\},$$

$\forall f,g\in E_{\mathbf{c}}, n\geqslant 1$.

Then there exists a unique $X\in\mathcal{L}$ such that $X_n\to X$ in $\mathcal{L}$.

LEMMA 2.2 ([2, 9]). *Let $(\Omega,\mathcal{F},\nu)$ be a measure space and $X:\Omega\to\mathcal{L}$ be such that*

(1) $\widehat{X(\cdot)}(f,g):\Omega\to\mathbf{C}$ *is measurable;*
(2) *there exist constants $p\geqslant 0$, $k>0$ and a positive function $\rho\in L^1(\Omega,\nu)$ such that*

$$\left|\widehat{X(\omega)}(f,g)\right|\leqslant\rho(\omega)\exp\{k(|f|_p^2+|g|_p^2)\},$$

$\forall f,g\in E_{\mathrm{c}}$ *and* ν-a.a. $\omega\in\Omega$.

Then $\int_\Omega X(\omega)\nu(\mathrm{d}\omega)$ exists as a Bochner integral in $\mathcal{L}$ and

$$\widehat{\int_\Omega X(\omega)\nu(\mathrm{d}\omega)}(f,g)=\int_\Omega\widehat{X(\omega)}(f,g)\nu(\mathrm{d}\omega),\quad f,g\in E_{\mathbf{c}}.$$

3. Formulation of the Equation

In this section, we make rigorous sense of Equation (1.2).

Let G: $[0,l]\times[0,l]\times(0,+\infty)\to\mathbf{R}_+$ be the Green's function for the classical cable equation associated with Equation (1.2). Then G has following properties:

(1) G is continuous;
(2) $\int_0^l G_s(x,y)G_t(y,z)\,\mathrm{d}y=G_{s+t}(x,z)$;

(3) $G_t(x, y) = G_t(y, x)$;
(4) for each $0 < r < 3$, there exists $k_r > 0$ such that

$$\int_0^l [G_t(x, y)]^r \, dy \leqslant k_r e^{-rt} t^{\frac{1-r}{2}}, \quad x \in [0, l], \ t > 0.$$

(For details, see [15].)

In the sequel, we put

$$\psi(x, t) \equiv \int_0^l G_t(x, y) \, dy, \quad (x, t) \in [0, l] \times \mathbf{R}_+.$$

Obviously, $0 \leqslant \psi(x, t) \leqslant k_1$, $(x, t) \in [0, l] \times \mathbf{R}_+$.

DEFINITION 3.1. Let V: $[0, l] \times \mathbf{R}_+ \to \mathcal{L}$ be an $\mathcal{L}$-valued function. We say that V is a solution to Equation (1.2) if it satisfies the following integral equations:

$$V(x, t) = \int_0^l G_t(x, y) J(y) \, dy + \int_0^t \int_0^l G_{t-s}(x, y) V(y, s) \Diamond K(y, s) \, dy \, ds, \tag{3.1}$$

$(x, t) \in [0, l] \times \mathbf{R}_+$.

Remark. The above definition is reasonable. In fact, by taking formal symbols of Equations (1.2) and (3.1), we have

$$\begin{aligned} &\frac{\partial \widehat{V(x, t)}(f, g)}{\partial t} \\ &\quad = \frac{\partial^2 \widehat{V(x, t)}(f, g)}{\partial x^2} - \widehat{V(x, t)}(f, g) + \widehat{V(x, t)}(f, g) \widehat{K(x, t)}(f, g), \\ &\quad 0 < x < l, \ t > 0, \\ &\frac{\partial \widehat{V(0, t)}(f, g)}{\partial x} = \frac{\partial \widehat{V(l, t)}(f, g)}{\partial x} = 0, \quad t \geqslant 0, \\ &\widehat{V(x, 0)}(f, g) = \widehat{J(x)}(f, g), \quad 0 \leqslant x \leqslant l \end{aligned}$$

and

$$\widehat{V(x, t)}(f, g) = \int_0^l G_t(x, y) \widehat{J(y)}(f, g) \, dy + \int_0^t \int_0^l G_{t-s}(x, y) \widehat{V(y, s)}(f, g) \widehat{K(y, s)}(f, g) \, dy \, ds,$$

$(x, t) \in [0, l] \times \mathbf{R}_+$.

By the classical theory of partial differential equations, we know that the above two complex-valued equations are equivalent under certain smoothness conditions.

4. Existence and Uniqueness of Solutions

In this section we establish the existence and uniqueness of solutions to Equation (1.2). We first prove a theorem which will facilitate the proof of the main theorem.

THEOREM 4.1. *Let* $K\colon [0,l]\times\mathbf{R}_+\to\mathcal{L}$ *and* $J\colon [0,l]\to\mathcal{L}$ *be such that*

(1) *the functions* $\widehat{K(\cdot,\cdot)}(f,g)\colon [0,l]\times\mathbf{R}_+\to\mathbf{C}$ *and* $\widehat{J(\cdot)}(f,g)\colon [0,l]\to\mathbf{C}$ *are continuous for any* $f,g\in E_{\mathbf{c}}$;

(2) *there exist constants* $p\geqslant 0$, $a,b,c,d>0$ *such that*

$$\left|\widehat{K(y,s)}(f,g)\right|\leqslant a+b(|f|_p^2+|g|_p^2),$$

$f,g\in E_{\mathbf{c}}$, $(y,s)\in[0,l]\times\mathbf{R}_+$, *and that*

$$\left|\widehat{J(x)}(f,g)\right|\leqslant c\exp\left\{d(|f|_p^2+|g|_p^2)\right\},$$

$f,g\in E_{\mathbf{c}}$, $x\in[0,l]$.

Then, there exists a sequence of $\mathcal{L}$*-valued functions* $V_n\colon [0,l]\times\mathbf{R}_+\to\mathcal{L}$, $n\geqslant 0$, *such that*

(1) *for all* $n\geqslant 0$ *and* $f,g\in E_{\mathbf{c}}$ *the function* $\widehat{V_n(\cdot,\cdot)}(f,g)\colon [0,l]\times\mathbf{R}_+\to\mathbf{C}$ *is continuous;*

(2) *for* $n\geqslant 1$, *it holds that*

$$\left|\widehat{V_n(x,t)}(f,g)\right|$$
$$\leqslant c\psi(x,t)\exp\left\{d(|f|_p^2+|g|_p^2)\right\}\sum_{i=0}^{n}\frac{1}{i!}t^i\left[a+b(|f|_p^2+|g|_p^2)\right]^i,$$

$f,g\in E_{\mathbf{c}}$, $(x,t)\in[0,l]\times\mathbf{R}_+$;

(3) $V_0(x,t)=\int_0^l G_t(x,y)J(y)\,\mathrm{d}y$, $\quad(x,t)\in[0,l]\times\mathbf{R}_+$,

$$V_n(x,t)=V_0(x,t)+\int_0^t\int_0^l G_{t-s}(x,y)V_{n-1}(y,s)\Diamond K(y,s)\,\mathrm{d}y\,\mathrm{d}s,$$

$n\geqslant 1$, $(x,t)\in[0,l]\times\mathbf{R}_+$.

Proof. It is easy to see that the integral $\int_0^l G_t(x,y)J(y)\,\mathrm{d}y$ exists and belongs to $\mathcal{L}$ for any $(x,t)\in[0,l]\times\mathbf{R}_+$.

Set

$$V_0(x,t)\equiv\int_0^l G_t(x,y)J(y)\,\mathrm{d}y,\quad (x,t)\in[0,l]\times\mathbf{R}_+.$$

Then for any $f, g \in E_{\mathbf{c}}$, the function $\widehat{V_0(\cdot,\cdot)}(f, g)$: $[0, l] \times \mathbf{R}_+ \to \mathbf{C}$ is obviously continuous.

For $x \in [0, l]$, $t > 0$, let

$$H_1(y, s) = G_{t-s}(x, y) V_0(y, s) \diamondsuit K(y, s),$$

$(y, s) \in [0, l] \times [0, t)$. Then, we have

$$\widehat{H_1(y, s)}(f, g) = G_{t-s}(x, y)\widehat{V_0(y, s)}(f, g)\widehat{K(y, s)}(f, g),$$

$(y, s) \in [0, l] \times [0, t)$, $f, g \in E_{\mathbf{c}}$. Hence, the function $\widehat{H_1(\cdot,\cdot)}(f, g)$: $[0, l] \times [0, t) \to \mathbf{C}$ is continuous. Moreover, we have

$$\begin{aligned} &\big|\widehat{H_1(y, s)}(f, g)\\ &\quad \leqslant G_{t-s}(x, y)\big[a + b(|f|_p^2 + |g|_p^2)\big]\big|\widehat{V_0(y, s)}(f, g)\big|\\ &\quad \leqslant G_{t-s}(x, y)\big[a + b(|f|_p^2 + |g|_p^2)\big]c\psi(y, s)\exp\big\{d(|f|_p^2 + |g|_p^2)\big\}\\ &\quad \leqslant ck_1 G_{t-s}(x, y)\big[a + b(|f|_p^2 + |g|_p^2)\big]\exp\big\{d(|f|_p^2 + |g|_p^2)\big\}, \end{aligned}$$

$f, g \in E_{\mathbf{c}}$, $(y, s) \in [0, l] \times [0, t)$. Therefore, by Lemma 2.2 we know that the integral

$$\int_0^t \int_0^l G_{t-s}(x, y) V_0(y, s) \diamondsuit K(y, s)\, \mathrm{d}y\, \mathrm{d}s = \int_0^t \int_0^l H_1(y, s)\, \mathrm{d}y\, \mathrm{d}s$$

exists and belongs to $\mathcal{L}$.

Now, define

$$V_1(x, t) \equiv V_0(x, t) + \int_0^t \int_0^l G_{t-s}(x, y) V_0(y, s) \diamondsuit K(y, s)\, \mathrm{d}y\, \mathrm{d}s,$$

$(x, t) \in [0, l] \times \mathbf{R}_+$. Then, we have

$$\begin{aligned} \widehat{V_1(x, t)}(f, g) &= \widehat{V_0(x, t)}(f, g) + \\ &\quad + \int_0^t \int_0^l G_{t-s}(x, y)\widehat{V_0(y, s)}(f, g)\widehat{K(y, s)}(f, g)\, \mathrm{d}y\, \mathrm{d}s, \end{aligned}$$

$f, g \in E_{\mathbf{c}}$, $(x, t) \in [0, l] \times \mathbf{R}_+$, which implies that the function $\widehat{V_1(\cdot,\cdot)}(f, g)$: $[0, l] \times \mathbf{R}_+ \to \mathbf{C}$ is continuous. Moreover, we have

$$\begin{aligned} &\big|\widehat{V_1(x, t)}(f, g)\big|\\ &\quad \leqslant \big|\widehat{V_0(x, t)}(f, g)\big| + \int_0^t \int_0^l G_{t-s}(x, y)\big|\widehat{V_0(y, s)}(f, g)\big| \times\\ &\qquad \times \big|\widehat{K(y, s)}(f, g)\big|\, \mathrm{d}y\, \mathrm{d}s\\ &\quad \leqslant \big|\widehat{V_0(x, t)}(f, g)\big| + \big[a + b(|f|_p^2 + |g|_p^2)\big] \int_0^t \int_0^l G_{t-s}(x, y) \times \end{aligned}$$

$$\times \left|\widehat{V_0(y,s)}(f,g)\right| \mathrm{d}y\,\mathrm{d}s$$
$$\leqslant c\psi(x,t)\exp\left\{d(|f|_p^2+|g|_p^2)\right\} + c\left[a+b(|f|_p^2+|g|_p^2)\right]\times$$
$$\times \exp\left\{d(|f|_p^2+|g|_p^2)\right\}\int_0^t\int_0^l G_{t-s}(x,y)\cdot\psi(y,s)\,\mathrm{d}y\,\mathrm{d}s$$
$$= c\psi(x,t)\exp\left\{d(|f|_p^2+|g|_p^2)\right\}\sum_{i=0}^{1}\frac{1}{i!}t^i\left[a+b(|f|_p^2+|g|_p^2)\right]^i,$$

$f, g \in E_{\mathbf{c}}$, $(x,t) \in [0,l]\times \mathbf{R}_+$.

Assume that for some $n \geqslant 1$, V_n: $[0,l]\times\mathbf{R}_+ \to \mathcal{L}$ has been defined and satisfies the properties below: for any $f, g \in E_{\mathbf{c}}$, the function $\widehat{V_n(\cdot,\cdot)}(f,g)$ is continuous and further

$$\left|\widehat{V_n(x,t)}(f,g)\right|$$
$$\leqslant c\psi(x,t)\exp\left\{d(|f|_p^2+|g|_p^2)\right\}\sum_{i=0}^{n}\frac{1}{i!}t^i\left[a+b(|f|_p^2+|g|_p^2)\right]^i,$$

$f, g \in E_{\mathbf{c}}$, $(x,t) \in [0,l]\times\mathbf{R}_+$. Then, for $x \in [0,l]$, $t > 0$, let

$$H_n(y,s) = G_{t-s}(x,y)V_n(y,s) \lozenge K(y,s), \quad (y,s) \in [0,l]\times[0,t).$$

In a similar way, we see that the function $\widehat{H_n(\cdot,\cdot)}(f,g)$ is continuous for any $f, g \in E_{\mathbf{c}}$. Furthermore, by a tedious computation we find that

$$\left|\widehat{H_n(y,s)}(f,g)\right| \leqslant \alpha G_{t-s}(x,y)\exp\left\{\beta(|f|_p^2+|g|_p^2)\right\},$$

$f, g \in E_{\mathbf{c}}$, $(y,s) \in [0,l]\times[0,t)$, where

$$\alpha = ck_1\exp\{a+at\}, \qquad \beta = b+bt+d.$$

Hence, the integral

$$\int_0^t\int_0^l G_{t-s}(x,y)V_n(y,s)\lozenge K(y,s)\,\mathrm{d}y\,\mathrm{d}s = \int_0^t\int_0^l H_n(y,s)\,\mathrm{d}y\,\mathrm{d}s$$

exists and belongs to $\mathcal{L}$. Define

$$V_{n+1}(x,t) \equiv V_0(x,t) + \int_0^t\int_0^l G_{t-s}(x,y)V_n(y,s)\lozenge K(y,s)\,\mathrm{d}y\,\mathrm{d}s,$$

$(x,t) \in [0,l]\times\mathbf{R}_+$. Then, we have

$$\left|\widehat{V_{n+1}(x,t)}(f,g)\right|$$
$$\leqslant \left|\widehat{V_0(x,t)}(f,g)\right|\mathrm{d}s +$$
$$+\left[a+b(|f|_p^2+|g|_p^2)\right]\int_0^t\int_0^l G_{t-s}(x,y)\left|\widehat{V_n(y,s)}(f,g)\right|\mathrm{d}y\,\mathrm{d}s$$
$$= \mathrm{I}+\mathrm{II}.$$

Since

$$\begin{aligned}
\mathrm{I} &\leqslant c\psi(x,t)\exp\{d(|f|_p^2+|g|_p^2)\},\\
\mathrm{II} &\leqslant c\exp\{d(|f|_p^2+|g|_p^2)\}\sum_{i=0}^{n}\frac{1}{i!}\left[a+b(|f|_p^2+|g|_p^2)\right]^{i+1}\times\\
&\quad\times\int_0^t\int_0^l G_{t-s}(x,y)\psi(y,s)s^i\,\mathrm{d}y\,\mathrm{d}s,\\
&\leqslant c\psi(x,t)\exp\{d(|f|_p^2+|g|_p^2)\}\sum_{i=1}^{n+1}\frac{1}{i!}t^i\left[a+b(|f|_p^2+|g|_p^2)\right]^i,
\end{aligned}$$

it follows that

$$\begin{aligned}
&\left|\widehat{V_{n+1}(x,t)}(f,g)\right|\\
&\quad\leqslant c\psi(x,t)\exp\{d(|f|_p^2+|g|_p^2)\}\sum_{i=0}^{n+1}\frac{1}{i!}t^i\left[a+b(|f|_p^2+|g|_p^2)\right]^i,
\end{aligned}$$

$f,g\in E_{\mathbf{c}}$, $(x,t)\in[0,l]\times\mathbf{R}_+$. Obviously, the function $\widehat{V_{n+1}(\cdot,\cdot)}(f,g)$ is continuous for any $f,g\in E_{\mathbf{c}}$. Therefore, by induction we come to the conclusion. □

THEOREM 4.2. *Let the noise process* $\{K(x,t)\mid(x,t)\in[0,l]\times\mathbf{R}_+\}$ *and the initial value process* $\{J(x)\mid x\in[0,l]\}$ *in Equation* (1.2) *satisfy the conditions given in Theorem* 4.1. *Then there exists a unique continuous* $\mathcal{L}$*-valued function* $V\colon[0,l]\times\mathbf{R}_+\to\mathcal{L}$ *which solves Equation* (1.2). *Moreover, the solution* V *has the following estimate*

$$\left|\widehat{V(x,t)}(f,g)\leqslant A\mathrm{e}^{at}\exp\{(bt+d)(|f|_p^2+|g|_p^2)\},\right.$$

$f,g\in E_{\mathbf{c}}$, $(x,t)\in[0,l]\times\mathbf{R}_+$, *where* $A=ck_1$.

Proof (Existence). Consider the sequence of $\mathcal{L}$-valued functions $\{V_n\}_{n\geqslant0}$ given in Theorem 4.1. We assert that for any $n\geqslant1$,

$$\begin{aligned}
&\left|\widehat{V_n(x,t)}(f,g)-\widehat{V_{n-1}(x,t)}(f,g)\right|\\
&\quad\leqslant c\psi(x,t)\exp\{d(|f|_p^2+|g|_p^2)\}t^n\left[a+b(|f|_p^2+|g|_p^2)\right]^n/n!,
\end{aligned}$$

$f,g\in E_{\mathbf{c}}$, $(x,t)\in[0,l]\times\mathbf{R}_+$. In fact, this is obviously true for $n=1$. Assume that the inequality is proven for some $n\geqslant1$. Then, we have

$$\begin{aligned}
&\left|\widehat{V_{n+1}(x,t)}(f,g)-\widehat{V_n(x,t)}(f,g)\right|\\
&\quad=\left|\int_0^t\int_0^l G_{t-s}(x,y)\left[\widehat{V_n(y,s)}(f,g)-\right.\right.\\
&\qquad\left.\left.-\widehat{V_{n-1}(y,s)}(f,g)\right]\widehat{K(y,s)}(f,g)\,\mathrm{d}y\,\mathrm{d}s\right|
\end{aligned}$$

$$\leqslant \left[a + b(|f|_p^2 + |g|_p^2)\right] \int_0^t \int_0^l G_{t-s}(x, y) \times$$
$$\times \left| \widehat{V_n(y, s)}(f, g) - \widehat{V_{n-1}(y, s)}(f, g) \right| \mathrm{d}y\, \mathrm{d}s$$
$$\leqslant c \exp\left\{d(|f|_p^2 + |g|_p^2)\right\} \times$$
$$\times \left[a + b(|f|_p^2 + |g|_p^2)\right]^{n+1} \int_0^t \int_0^l G_{t-s}(x, y) \psi(y, s) \frac{s^n}{n!} \mathrm{d}y\, \mathrm{d}s$$
$$\leqslant c\psi(x, t) \exp\left\{d(|f|_p^2 + |g|_p^2)\right\} t^{n+1} \times$$
$$\times \left[a + b(|f|_p^2 + |g|_p^2)\right]^{n+1} / (n+1)!,$$

$f, g \in E_{\mathbf{c}}$, $(x, t) \in [0, l] \times \mathbf{R}_+$. Hence, by induction, we come to the assertion.

Now, for $(x, t) \in [0, l] \times \mathbf{R}_+$, consider the sequence $\{V_n(x, t)\}_{n \geqslant 0} \subset \mathcal{L}$. For any $f, g \in E_{\mathbf{c}}$, we have

$$\left| \widehat{V_{m+n}(x, t)}(f, g) - \widehat{V_n(x, t)}(f, g) \right|$$
$$\leqslant \sum_{i=n+1}^{m+n} \left| \widehat{V_i(x, t)}(f, g) - \widehat{V_{i-1}(x, t)}(f, g) \right|$$
$$\leqslant \sum_{i=n+1}^{m+n} ck_1 \exp\left\{d(|f|_p^2 + |g|_p^2)\right\} t^i \left[a + b(|f|_p^2 + |g|_p^2)\right]^i / i!,$$

$\forall m, n \geqslant 1$, hence $\{\widehat{V_n(x, t)}(f, g)\}_{n \geqslant 0}$ converges in $\mathbf{C}$. On the other hand, we have

$$\begin{aligned} \left| \widehat{V_n(x, t)}(f, g) \right| &\leqslant c\psi(x, t) \exp\left\{d(|f|_p^2 + |g|_p^2)\right\} \times \\ &\quad \times \sum_{i=0}^{n} \frac{1}{i!} t^i \left[a + b(|f|_p^2 + |g|_p^2)\right]^i \\ &\leqslant A \mathrm{e}^{at} \exp\left\{(bt + d)(|f|_p^2 + |g|_p^2)\right\}, \end{aligned}$$

$\forall f, g \in E_{\mathbf{c}}$, $n \geqslant 1$, where $A = ck_1$. Therefore, it follows from Lemma 2.1 that the sequence $\{V_n(x, t)\}_{n \geqslant 0}$ converges in $\mathcal{L}$.

Now, put

$$V(x, t) = \lim_{n \to \infty} V_n(x, t), \quad (x, t) \in [0, l] \times \mathbf{R}_+.$$

Then $\{V(x, t) \mid (x, t) \in [0, l] \times \mathbf{R}_+\} \subset \mathcal{L}$. Furthermore, we have

$$\left| \widehat{V(x, t)}(f, g) \right| \leqslant A \mathrm{e}^{at} \exp\left\{(bt + d)(|f|_p^2 + |g|_p^2)\right\},$$

$f, g \in E_{\mathbf{c}}$, $(x, t) \in [0, l] \times \mathbf{R}_+$. Thus, for any $(x, t) \in [0, l] \times \mathbf{R}_+$, the integral

$$\int_0^t \int_0^l G_{t-s}(x, y) V(y, s) \Diamond K(y, s)\, \mathrm{d}y\, \mathrm{d}s$$

exists and belongs to $\mathcal{L}$.

For any $f, g \in E_{\mathbf{c}}$ and $(x, t) \in [0, l] \times \mathbf{R}_+$, by the dominated convergence theorem we see that

$$\int_0^t \int_0^l G_{t-s}(x, y) \widehat{V_n(y, s)}(f, g) \widehat{K(y, s)}(f, g) \, \mathrm{d}y \, \mathrm{d}s$$
$$\to \int_0^t \int_0^l G_{t-s}(x, y) \widehat{V(y, s)}(f, g) \widehat{K(y, s)}(f, g) \, \mathrm{d}y \, \mathrm{d}s.$$

On the other hand, we have

$$\begin{aligned}\widehat{V_n(x, t)}(f, g) &= \widehat{V_0(x, t)}(f, g) + \\ &\quad + \int_0^t \int_0^l G_{t-s}(x, y) \widehat{V_{n-1}(y, s)}(f, g) \widehat{K(y, s)}(f, g) \, \mathrm{d}y \, \mathrm{d}s,\end{aligned}$$

$n \geqslant 1$, $f, g \in E_{\mathbf{c}}$, $(x, t) \in [0, l] \times \mathbf{R}_+$. Therefore,

$$\begin{aligned}\widehat{V(x, t)}(f, g) &= \widehat{V_0(x, t)}(f, g) + \\ &\quad + \int_0^t \int_0^l G_{t-s}(x, y) \widehat{V(y, s)}(f, g) \widehat{K(y, s)}(f, g) \, \mathrm{d}y \, \mathrm{d}s,\end{aligned}$$

$f, g \in E_{\mathbf{c}}$, $(x, t) \in [0, l] \times \mathbf{R}_+$, which implies that $\{V(x, t) \mid (x, t) \in [0, l] \times \mathbf{R}_+\}$ is a solution to Equation (1.2).

Next we show that the solution $V\colon [0, l] \times \mathbf{R}_+ \to \mathcal{L}$ is continuous. It suffices to show that for any $T > 0$, V is continuous on $[0, l] \times [0, T]$. In fact, for any $f, g \in E_{\mathbf{c}}$, we have

$$\begin{aligned}&\left|\widehat{V_n(x, t)}(f, g) - \widehat{V_{n-1}(x, t)}(f, g)\right| \\ &\quad \leqslant c\psi(x, t) \exp\left\{d(|f|_p^2 + |g|_p^2)\right\} t^n \left[a + b(|f|_p^2 + |g|_p^2)\right]^n / n! \\ &\quad \leqslant ck_1 \exp\left\{d(|f|_p^2 + |g|_p^2)\right\} T^n \left[a + b(|f|_p^2 + |g|_p^2)\right]^n / n!,\end{aligned}$$

$(x, t) \in [0, l] \times [0, T]$, which implies that

$$\begin{aligned}&\sup\left\{\left|\widehat{V_n(x, t)}(f, g) -\right.\right. \\ &\quad \left.\left. - \widehat{V(x, t)}(f, g)\right| \mid (x, t) \in [0, l] \times [0, T]\right\} \to 0 \quad (n \to \infty).\end{aligned}$$

Hence, the function $\widehat{V(\cdot, \cdot)}(f, g)\colon [0, l] \times [0, T] \to \mathbf{C}$ is continuous. On the other hand, we have

$$\begin{aligned}\left|\widehat{V(x, t)}(f, g)\right| &\leqslant A\mathrm{e}^{at} \exp\left\{(bt + d)(|f|_p^2 + |g|_p^2)\right\} \\ &\leqslant A\mathrm{e}^{aT} \exp\left\{(bT + d)(|f|_p^2 + |g|_p^2)\right\},\end{aligned}$$

$f, g \in E_{\mathbf{c}}$, $(x, t) \in [0, l] \times [0, T]$. Therefore, $V\colon [0, l] \times [0, T] \to \mathcal{L}$ is continuous.

(Uniqueness) Let U: $[0, l] \times \mathbf{R}_+ \to \mathcal{L}$ be another continuous solution to Equation (1.2). Then, for any $f, g \in E_{\mathrm{c}}$, we have

$$\begin{aligned}
&\left|\widehat{U(x,t)}(f,g) - \widehat{V(x,t)}(f,g)\right| \\
&\quad = \left|\int_0^t \int_0^l G_{t-s}(x,y)\left[\widehat{U(y,s)}(f,g) - \widehat{V(y,s)}(f,g)\right] K(y,s)(f,g)\,\mathrm{d}y\,\mathrm{d}s\right| \\
&\quad \leqslant \left[a + b(|f|_p^2 + |g|_p^2)\right] \times \\
&\qquad \times \int_0^t \int_0^l G_{t-s}(x,y)\left|\widehat{U(y,s)}(f,g) - \widehat{V(y,s)}(f,g)\right| \mathrm{d}y\,\mathrm{d}s \\
&\quad \leqslant \left[a + b(|f|_p^2 + |g|_p^2)\right]\left\{\int_0^t \int_0^l [G_{t-s}(x,y)]^2\,\mathrm{d}y\,\mathrm{d}s\right\}^{1/2} \times \\
&\qquad \times \left\{\int_0^t \int_0^l \left|\widehat{U(y,s)}(f,g) - \widehat{V(y,s)}(f,g)\right|^2 \mathrm{d}y\,\mathrm{d}s\right\}^{1/2} \\
&\quad \leqslant \sigma^{1/2}\left[a + b(|f|_p^2 + |g|_p^2)\right] \times \\
&\qquad \times \left\{\int_0^t \int_0^l \left|\widehat{U(y,s)}(f,g) - \widehat{V(y,s)}(f,g)\right|^2 \mathrm{d}y\,\mathrm{d}s\right\}^{1/2},
\end{aligned}$$

$(x, t) \in [0, l] \times \mathbf{R}_+$, where $\sigma = k_2 \int_0^{+\infty} \mathrm{e}^{-2s} s^{-1/2}\,\mathrm{d}s < +\infty$. Thus

$$\begin{aligned}
&\int_0^l \left|\widehat{U(x,t)}(f,g) - \widehat{V(x,t)}(f,g)\right|^2 \mathrm{d}x \\
&\quad \leqslant l\sigma\left[a + b(|f|_p^2 + |g|_p^2)\right]^2 \int_0^t \int_0^l \left|\widehat{U(x,t)}(f,g) - \widehat{V(x,t)}(f,g)\right|^2 \mathrm{d}y\,\mathrm{d}s,
\end{aligned}$$

$t \in \mathbf{R}_+$.

By Gronwall inequality, it follows that

$$\int_0^l \left|\widehat{U(x,t)}(f,g) - \widehat{V(x,t)}(f,g)\right|^2 \mathrm{d}x = 0, \quad t \in \mathbf{R}_+,$$

which implies that

$$\widehat{U(x,t)}(f,g) = \widehat{V(x,t)}(f,g), \quad (x,t) \in [0,l] \times \mathbf{R}_+.$$

The arbitrariness of f and g in the equations yields that

$$U(x,t) = V(x,t), \quad (x,t) \in [0,l] \times \mathbf{R}_+,$$

i.e. $U = V$. □

5. Markov-Like Property of the Solution

In this section, we show that the solution to Equation (1.2) has a Markov-like property.

THEOREM 5.1. *Let the noise process* $\{K(x,t) \mid (x,t) \in [0,l] \times \mathbf{R}_+\}$ *and the initial value process* $\{J(x) \mid x \in [0,l]\}$ *in Equation* (1.2) *satisfy the conditions given in Theorem* 4.1. *Then the solution* V *of Equation* (1.2) *has the following property:*

$$\begin{aligned} V(x,t+r) &= \int_0^l G_t(x,y)V(y,r)\,\mathrm{d}y + \\ &\quad + \int_0^t \int_0^l G_{t-s}(x,y)V(y,s+r)\Diamond K(y,s+r)\,\mathrm{d}y\,\mathrm{d}s, \end{aligned}$$

$t > 0,\ r \geqslant 0,\ x \in [0,l]$.

Proof. For any $f, g \in E_{\mathbf{c}}$, we have

$$\begin{aligned} &\widehat{V(x,t+r)}(f,g) \\ &= \int_0^l G_{t+r}(x,y)\widehat{J(y)}(f,g)\,\mathrm{d}y + \\ &\quad + \int_0^{t+r} \int_0^l G_{t+r-s}(x,y)\widehat{V(y,s)}(f,g)\widehat{K(y,s)}(f,g)\,\mathrm{d}y\,\mathrm{d}s \\ &= \int_0^l \int_0^l G_t(x,z)G_r(z,y)\widehat{J(y)}(f,g)\,\mathrm{d}z\,\mathrm{d}y + \\ &\quad + \int_0^r \int_0^l \int_0^l G_t(x,z)G_{r-s}(z,y)\widehat{V(y,s)}(f,g)\widehat{K(y,s)}(f,g)\,\mathrm{d}z\,\mathrm{d}y\,\mathrm{d}s + \\ &\quad + \int_r^{t+r} \int_0^l G_{t+r-s}(x,y)\widehat{V(y,s)}(f,g)\widehat{K(y,s)}(f,g)\,\mathrm{d}y\,\mathrm{d}s \\ &= \int_0^l G_t(x,z)\bigg[\int_0^l G_r(z,y)\widehat{J(y)}(f,g)\,\mathrm{d}y + \\ &\quad + \int_0^r \int_0^l G_{r-s}(z,y)\widehat{V(y,s)}(f,g)\widehat{K(y,s)}(f,g)\,\mathrm{d}y\,\mathrm{d}s\bigg]\mathrm{d}z + \\ &\quad + \int_0^t \int_0^l G_{t-s}(x,y)\widehat{V(y,s+r)}(f,g)\widehat{K(y,s+r)}(f,g)\,\mathrm{d}y\,\mathrm{d}s \\ &= \int_0^l G_t(x,z)\widehat{V(z,r)}(f,g)\,\mathrm{d}z + \\ &\quad + \int_0^t \int_0^l G_{t-s}(x,y)\widehat{V(y,s+r)}(f,g)\widehat{K(y,s+r)}(f,g)\,\mathrm{d}y\,\mathrm{d}s. \end{aligned}$$

Hence the conclusion follows. □

In the following, we always assume that $p > 5/6$ is given and that the noise process $\{K(x,t) \mid (x,t) \in [0,l] \times \mathbf{R}_+\}$ in Equation (1.2) is such that

$$\begin{aligned} K(x,t) &\equiv F_1(x,t)\partial^*(x,t) + F_2(x,t)\partial^*(x,t)\partial(x,t) + \\ &\quad + F_3(x,t)\partial(x,t), \end{aligned}$$

where $\partial^*(x,t)$, $\partial^*(x,t)\partial(x,t)$ and $\partial(x,t)$ are the creation, number and annihilation operators at (x,t), respectively (cf. [2]); F_1, F_2 and F_3 are continuous and bounded complex-valued functions on $[0,l] \times \mathbf{R}_+$.

It is easy to prove that there exist constants $\rho_1, \rho_2 > 0$ such that

$$\left|\widehat{K(x,t)}f,g\right| \leqslant \rho_1 + \rho_2(|f|_p^2 + |g|_p^2),$$

$f,g \in E_\mathbf{c}$, $(x,t) \in [0,l] \times \mathbf{R}_+$. Moreover, for any $f,g \in E_\mathbf{c}$ the function $\widehat{K(\cdot,\cdot)}(f,g)\colon [0,l] \times \mathbf{R}_+ \to \mathbf{C}$ is continuous (see [2]).

THEOREM 5.2. *Let the initial value process* $\{J(x) \mid x \in [0,l]\}$ *in Equation* (1.2) *be such that* (1) *for any* $f,g \in E_\mathbf{c}$, $\widehat{J(\cdot)}(f,g)\colon [0,l] \to \mathbf{C}$ *is continuous,* (2) *there exist constants* $c,d > 0$ *such that*

$$\left|\widehat{J(x)}(f,g)\right| \leqslant c \exp\left\{d(|f|_p^2 + |g|_p^2)\right\},$$

$f,g \in E_\mathbf{c}$, $x \in [0,l]$.

Then the solution V *of Equation* (1.2) *has the following Markov-like property:*

$$\widehat{V(x,t+r)}(f,g) = \int_0^l G_t(x,y)\widehat{V(y,r)}(f,g)\,\mathrm{d}y,$$

$t > 0$, $r \geqslant 0$, $f,g \in E_\mathbf{c}^r$, *where*

$$E_\mathbf{c}^r = \left\{f \mid f \in E_\mathbf{c},\ \operatorname{supp} f \subset \mathbf{R} \times (-\infty, r)\right\}.$$

Proof. For any $f,g \in E_\mathbf{c}^r$, $0 \leqslant s \leqslant t$ and $x \in [0,l]$, we have

$$\begin{aligned}\widehat{K(x,s+r)}(f,g) &= F_1(x,s+r)g(x,s+r) + \\ &\quad + F_2(x,s+r)f(x,s+r)g(x,s+r) + \\ &\quad + F_3(x,s+r)f(x,s+r) = 0.\end{aligned}$$

By Theorem 5.1 we come to the conclusion. □

EXAMPLE 1. The quantum cable equation driven by Gaussian white noises:

$$\begin{aligned}&\frac{\partial V(x,t)}{\partial t} = \frac{\partial^2 V(x,t)}{\partial x^2} - V(x,t) + V(x,t) \Diamond (\partial^*(x,t) + \partial(x,t)),\\ &\frac{\partial V(0,t)}{\partial x} = \frac{\partial V(l,t)}{\partial x} = 0, \quad t \geqslant 0, \\ &V(x,0) = I, \quad 0 \leqslant x \leqslant l,\end{aligned} \tag{5.1}$$

where $\partial^*(x,t) + \partial(x,t)$ is known as quantum Gaussian white noise in space and time (cf. [5]). It is easy to see that Equation (5.1) satisfies the conditions given in Theorem 4.1. Hence Equation (5.1) has a unique continuous solution which has the Markov-like property.

EXAMPLE 2. The quantum cable equation driven by Poisson white noises:

$$
\begin{aligned}
\frac{\partial V(x,t)}{\partial t} &= \frac{\partial^2 V(x,t)}{\partial x^2} - V(x,t) + V(x,t) \diamondsuit [\partial^*(x,t)\partial(x,t) + \\
&\quad + \sqrt{\lambda}(\partial^*(x,t) + \partial(x,t)) + \lambda], \\
\frac{\partial V(0,t)}{\partial x} &= \frac{\partial V(l,t)}{\partial x} = 0, \quad t \geqslant 0, \\
V(x,0) &= I, \quad 0 \leqslant x \leqslant l,
\end{aligned} \tag{5.2}
$$

where $\partial^*(x,t)\partial(x,t) + \sqrt{\lambda}(\partial^*(x,t) + \partial(x,t)) + \lambda$ is known as quantum Poisson white noise in space and time (cf. [5]). Similarly, Equation (5.2) also has a unique continuous solution.

References

1. Applebaum, D.: Quantum martingale measures and stochastic partial differential equations in Fock space, *J. Math. Phys.* **39** (1998), 3019–3030.
2. Hida, T., Kuo, H. H., Potthoff, J. and Streit, L.: *White Noise – An Infinite Dimensional Calculus*, Kluwer Acad. Publ., Dordrecht, 1993.
3. Holden, H., Lindstrøm, T., Øksendal, B., Ubøe, J. and Zhang, T.: The stochastic Wick type Burgers equation, In: A. Etheridge (ed.), *Stochastic Partial Differential Equations*, Cambridge Univ. Press, 1995, pp. 141–161.
4. Holden, H., Øksendal, B., Ubøe, J. and Zhang, T.: *Stochastic Partial Differential Equations*, Birkhäuser, Basel, 1996.
5. Huang, Z. Y.: Quantum white noise, *Nagoya Math. J.* **129** (1993), 23–42.
6. Huang, Z. Y. and Luo, S. L.: Quantum white noises and free fields, *Infin. Dimens. Anal. Quantum Probab. Relat. Topics* **1** (1998), 69–82.
7. Huang, Z. Y. and Luo, S. L.: Wick calculus of generalized operators and its applications to quantum stochastic calculus, *Infin. Dimens. Anal. Quantum Probab. Relat. Top.* **1** (1998), 455–466.
8. Huang, Z. Y., Wang, C. S. and Wang, X. J.: Quantum integral equation of Volterra type with generalized operator-valued kernels, to appear in *Infin. Dimens. Anal. Quantum Probab. Relat. Top.* **3** (2000).
9. Huang, Z. Y. and Yan, J. A.: *Introduction to Infinite Dimensional Stochastic Analysis*, Science Press, Beijing/Kluwer Academic Publ., Dordrecht, 2000.
10. Kallianpur, G. and Wolpert, R.: Infinite dimensional stochastic differential equation models for spatially distributed neurons, *Appl. Math. Opt.* **12** (1984), 125–172.
11. Lindstrøm, T., Øksendal, B. and Ubøe, J.: Wick multiplication and Itô–Skorohod stochastic differential equations, In: S. Albeverio *et al.* (eds), *Ideas and Methods in Mathematical Analysis, Stochastics and Applications*, Cambridge Univ. Press, 1992, pp. 183–206.
12. Obata, N.: *White Noise Calculus and Fock Space*, Lecture Notes in Math. 1577, Springer, New York, 1994.
13. Penrose, R.: *Shadows of the Mind*, Oxford Univ. Press, 1994.
14. Walsh, J.: A stochastic model of neural response, *Adv. Appl. Probab.* **13** (1981), 231–281.
15. Walsh, J.: An introduction to stochastic partial differential equations, In: *Lecture Notes in Math.* 1180, Springer, New York, 1986, pp. 266–439.

Acta Applicandae Mathematicae **63:** 165–174, 2000.

Large Deviation Theorems for Gaussian Processes and Their Applications in Information Theory

SHUNSUKE IHARA
School of Informatics and Sciences, Nagoya University, Nagoya, 464-8601 Japan.
e-mail: ihara@math.nagoya-u.ac.jp

(Received: 29 December 1998)

Abstract. We discuss on the large deviation theorems for stationary Gaussian processes and their applications in information theory. The topics investigated here include error probability of string matching, error probabilities for random codings, and a conditional limit theorem which justifies the maximum entropy principle.

Mathematics Subject Classifications (2000): 60F10, 60G10, 94A15, 94A24.

Key words: large deviation theorem, stationary Gaussian process, random coding, channel coding theorem, maximum entropy principle.

1. Introduction

Various problems in information theory and mathematical statistics can be solved by investigating asymptotic behavior of stochastic processes. The large deviation theorems (LDT's) are useful to study such asymptotic behavior. In this paper, using LDT's, we solve some problems concerning the asymptotic behavior of discrete time stationary Gaussian processes. The topics investigated here are error probability of string matching, error probabilities for random codings, a conditional limit theorem related to the maximum entropy principle, and error probabilities in hypothesis testing.

2. Preparation

This section is devoted to introduce notations and terminologies which we use later. Let $X = \{X_n; n = 1, 2, \ldots\}$ and $Y = \{Y_n\}$ be real stationary processes with spectral density functions (SDF) $f(\lambda)$ and $g(\lambda)$, respectively. Throughout the paper, stationary processes are to be regular (or purely nondeterministic) and with mean zero. The covariance of X has the spectral representation $E[X_k X_{k+n}] = \int_{-\pi}^{\pi} e^{in\lambda} f(\lambda)\, d\lambda$. Denote by $\mathcal{S}$ the class of all SDF's. The set $\mathcal{S}$ consists of all $f \in L^1[-\pi, \pi]$ such that

$$f(\lambda) \geqslant 0, \qquad \log f(\lambda) \in L^1[-\pi, \pi] \quad \text{and} \quad f(-\lambda) = f(\lambda).$$

We denote by μ_X and μ_X^n the probability distributions of $X = \{X_n\}$ and $X_1^n \equiv (X_1, \ldots, X_n)$, respectively. For probability measures μ and ν on $\mathbf{R}^\infty$, the divergence (per unit time) $\overline{D}(\mu\|\nu)$ of μ with respect to ν is defined by $\overline{D}(\mu\|\nu) = \underline{\lim}_{n\to\infty} n^{-1} D(\mu^n\|\nu^n)$, where μ^n denotes the restriction of μ on $\mathbf{R}^n$, and

$$D(\mu^n\|\nu^n) = \int_{\mathbf{R}^n} \log \frac{\mathrm{d}\mu^n}{\mathrm{d}\nu^n}(\mathbf{x})\,\mathrm{d}\mu^n(\mathbf{x})$$

if $\mu^n \ll \nu^n$; otherwise $D(\mu^n\|\nu^n) = \infty$. The divergence is also called the relative entropy or the Kullback–Leibler information. The divergence (per unit time) of X with respect to Y is defined by $\overline{D}(X\|Y) = \overline{D}(\mu_X\|\mu_Y)$. If X and Y are Gaussian processes, it it known that

$$\overline{D}(X, Y) = \overline{D}(f\|g) \equiv \frac{1}{4\pi}\int_{-\pi}^{\pi}\left\{\frac{f(\lambda)}{g(\lambda)} - 1 - \log\frac{f(\lambda)}{g(\lambda)}\right\}\mathrm{d}\lambda.$$

The mutual information (per unit time) $\overline{I}(X, Y)$ between X and Y is defined by $\overline{I}(X, Y) = \overline{D}(\mu_{XY}\|\mu_X \times \mu_Y)$, where μ_{XY} is the distribution of $(X, Y) = \{(X_n, Y_n)\}$.

3. Large Deviation Theorem

Let $\{Z_n\}_{n=1,2,\ldots}$ be a sequence of d-dimensional random variables and consider the logarithmic generating function

$$\Lambda_n(\theta) = \log E[\exp\langle\theta, Z_n\rangle], \quad \theta \in \mathbf{R}^d,$$

where $\langle\cdot,\cdot\rangle$ denotes the inner product. We define

$$\Lambda(\theta) = \lim_{n\to\infty} \frac{1}{n}\Lambda_n(n\theta), \tag{1}$$

if the limit exists at θ. Let $\mathcal{D}$ be the set of all θ such that the limit (1) exists and $\Lambda(\cdot)$ is a function of C^1 class in a neighborhood of θ, and $\mathcal{V} = \{\nabla\Lambda(\theta); \theta \in \mathcal{D}\}$, where $\nabla\Lambda$ denotes the gradient of Λ. We fix a point $a^* = \nabla\Lambda(\theta^*) \in \mathcal{V}^\circ$ in the following, where $\theta^* \in \mathcal{D}$, $\theta^* \neq 0$ and $\mathcal{V}^\circ$ denotes the interior of $\mathcal{V}$, and consider a hyperplane

$$\Gamma = \{x \in \mathbf{R}^d; \langle\theta^*, x\rangle > \langle\theta^*, a^*\rangle\}. \tag{2}$$

A function $\psi(\theta)$ is defined by

$$\psi(\theta) = \langle\theta, \nabla\Lambda(\theta)\rangle - \Lambda(\theta), \quad \theta \in \mathcal{D}.$$

We denote by $B_d(a, r)$ an open ball in $\mathbf{R}^d$ with center a and radius r. Then we can prove the following LDT:

THEOREM 1 (cf. [6]). *Assume that $\mathcal{D}$ and $\mathcal{V}^\circ$ are nonempty.*

(a) (Upper estimate)

$$\overline{\lim_{n\to\infty}} \frac{1}{n}\log P(Z_n \in \overline{\Gamma}) \leqslant -\psi(\theta^*).$$

(b) (Lower estimate) *Let A be an open set such that $A \cap \mathcal{V}^\circ \neq \phi$. Then, for any $\theta \in \mathcal{D}$ such that $\nabla\Lambda(\theta) \in A \cap \mathcal{V}^\circ$,*

$$\varliminf_{n\to\infty} \frac{1}{n} \log P(Z_n \in A) \geqslant -\psi(\theta).$$

(c) *Let $A \subset \Gamma$ be an open set such that $A \cap B_d(a^*, \delta) \neq \phi$ for any $\delta > 0$. Then*

$$\lim_{n\to\infty} \frac{1}{n} \log P(Z_n \in A) = \lim_{n\to\infty} \frac{1}{n} \log P(Z_n \in \overline{A}) = -\psi(\theta^*).$$

The LDT can be applied to quadratic forms of Gaussian processes. Let $X = \{X_n\}$ be a stationary Gaussian process with SDF $f(\lambda)$, and consider a d-dimensional process $Z = \{Z_n \equiv (Z_{n1}, \ldots, Z_{nd})\}$ given by

$$Z_{nk} = \frac{1}{2\pi n} \sum_{p,q=1}^{n} \alpha_{p-q}(k) X_p X_q, \quad k = 1, \ldots, d, \tag{3}$$

where $\alpha_j(k)$ is a real constant. Since we can express $\alpha_j(k)$ in the form $\alpha_j(k) = \int_{-\pi}^{\pi} u_k(\lambda) e^{ij\lambda}\, d\lambda$, where $u_k(\lambda) \in L^2[-\pi, \pi]$, Z_{nk} can be written as

$$Z_{nk} = \int_{-\pi}^{\pi} u_k(\lambda) I_n(\lambda)\, d\lambda,$$

where

$$I_n(\lambda) = (2n\pi)^{-1} \left| \sum_{j=1}^{n} X_j e^{-ij\lambda} \right|^2$$

is the periodogram. We may assume that $u_k(\lambda)$ is a real valued function. For simplicity, we assume that $u_k(\lambda)$ and $f(\lambda)$ are continuous functions. For each θ, we define a SDF f_θ by

$$\frac{1}{f_\theta(\lambda)} - \frac{1}{f(\lambda)} = -4\pi \sum_{k=1}^{d} \theta_k u_k(\lambda). \tag{4}$$

For each $g \in L^2[-\pi, \pi]$ we denote $\mathbf{u}_d(g) = ((u_1, g), \ldots, (u_d, g))$, where (u, v) is the inner product of $L^2[-\pi, \pi]$.

Applying Theorem 1, we can prove the following theorem:

THEOREM 2 (cf. [1, 2]). *Assume that $u_k(\lambda)$, $k = 1, \ldots, d$, are linearly independent. For any fixed vector $\theta^* \in \Theta_d^\circ$, where $\Theta_d = \{\theta \in \mathbf{R}^d;\ f_\theta \in \mathcal{S}\}$, define Γ by* (2) *with $a^* = \mathbf{u}_d(f_{\theta^*})$.*

(a) (Upper estimate)

$$\varlimsup_{n\to\infty} \frac{1}{n} \log P(Z_n \in \overline{\Gamma}) \leqslant -\overline{D}(f_{\theta^*} \| f).$$

(b) (Lower estimate) *Let A be an open set such that $\mathbf{u}_d(f_\theta) \in A$ ($\theta \in \Theta_d$). Then*

$$\underline{\lim}_{n\to\infty} \frac{1}{n} \log P(Z_n \in A) \geqslant -\overline{D}(f_\theta \| f).$$

(c) *Let $A \subset \Gamma$ be any open set such that $A \cap B_d(a^*, \delta) \neq \phi$ for any $\delta > 0$. Then*

$$\lim_{n\to\infty} \frac{1}{n} \log P(Z_n \in A) = \lim_{n\to\infty} \frac{1}{n} \log P(Z_n \in \overline{A}) = -\overline{D}(f_{\theta^*} \| f).$$

4. String Matching and Random Coding

Let X and Y be mutually independent stationary Gaussian processes with SDF f and g, respectively. We denode by $\mathcal{U}_D$ a class of all processes $U = \{U_n\}$ satisfying

$$\overline{\lim}_{n\to\infty} E[\rho_n(X, U)^2] \leqslant D,$$

where

$$\rho_n(X, U) = \left\{ n^{-1} \sum_{k=1}^{n} |X_k - U_k|^2 \right\}^{1/2}.$$

Applying the LDT we can prove the following theorem concerning the asymptotic behavior of the probability $P(\rho_n(Y, x)^2 < D)$ of string matching with mean-squared distortion D, where $x = \{x_n\}$ is a sample path of X.

THEOREM 3 ([7]). *Assume that the SDF's $f(\lambda)$ and $g(\lambda)$ are continuous. If $0 < D < D_0 \equiv E[|X_n - Y_n|^2]$, then for μ_X-a.e. $x = \{x_n\}$*

$$\lim_{n\to\infty} \frac{1}{n} \log P(\rho_n(Y, x)^2 < D) = -R^*(D),$$

where

$$R^*(D) = \inf\left\{ \underline{\lim}_{n\to\infty} \frac{1}{n} \{I(X_1^n, U_1^n) + D(U_1^n \| Y_1^n)\}; U \in \mathcal{U}_D \right\}.$$

It is well known that the source coding theorem and the channel coding theorem can be proved by using the so-called random coding arguments. Using the law of large numbers, we first show that the average of the probability of error over random choice of codebooks is small. Then we conclude that there exists at least one good codebook. This is the random coding argument.

If we can apply the LDT to evaluate the probability of error, we may show that, for each random choice of codebook, the probability of error is small if the codelength is large enough.

For Gaussian information sources we can prove this property. Let an information source $X = \{X_n\}$ be a stationary Gaussian process with SDF f. The rate-distortion function $\overline{R}(D) = \overline{R}(D; X)$ is defined as

$$\overline{R}(D) = \inf\{\overline{I}(X, U); U \in \mathcal{U}_D\}.$$

The source coding theorem says that $\overline{R}(D)$ is the minimum coding rate to transmit the source X with distortion D and with arbitrarily small error. It is known that

$$\overline{R}(D) = \frac{1}{4\pi}\int_{-\pi}^{\pi} \log\max\left(\frac{f(\lambda)}{\theta^2}, 1\right) \mathrm{d}\lambda,$$

where $\theta > 0$ is a constant uniquely determined by $\int_{-\pi}^{\pi} \min(f(\lambda), \theta^2)\,\mathrm{d}\lambda = D$. Let $Y = \{Y_n\}$ be a stationary Gaussian processes with SDF $\widetilde{g}(\lambda) = \max(f(\lambda) - \theta^2, 0)$ and $Y^{(m)} = \{Y_n^{(m)}\}$, $m = 1, 2, \ldots$, be independent copies of Y. Using the observed data $y^{(m)} = \{y_n^{(m)}\}$ of $Y^{(m)}$, we define a coding scheme for the source X. The encoder $\varphi_n\colon \mathbf{R}^n \to \mathcal{M}_n$, where $\mathcal{M}_n = \{1, 2, \ldots, M_n\}$ and M_n is an integer, and the decoder $\psi_n\colon \mathcal{M}_n \to \mathbf{R}^n$ are given by

$$\varphi_n(x) = \begin{cases} m, & \text{if } \rho_n(y^{(j)}, x)^2 \geqslant D,\ j < m, \text{ and } \rho_n(y^{(m)}, x)^2 < D, \\ 1, & \text{if } \rho_n(y^{(j)}, x)^2 \geqslant D,\ j \in \mathcal{M}_n, \end{cases} \tag{5}$$

and

$$\psi_n(m) = (y_1^{(m)}, \ldots, y_n^{(m)}), \quad m \in \mathcal{M}_n. \tag{6}$$

Now we can prove the following coding theorem which means that the scheme given by (5) and (6) provides an optimal coding.

THEOREM 4. *For the Gaussian source X, if the coding rate* $\underline{\lim}_{n\to\infty} n^{-1}\log M_n$ *is greater than* $\overline{R}(D)$, *there exists a process* $\widetilde{X} = \{\widetilde{X}_n\}$ *such that*

$$\rho_n(X, \widetilde{X})^2 < D$$

and

$$\lim_{n\to\infty} P\big(\psi_n(\varphi_n(X_1^n)) \neq \widetilde{X}_1^n\big) = 0.$$

Proof. It suffices to prove that, for μ_X-a.e. $x = \{x_n\}$,

$$\lim_{n\to\infty} P(\exists m \leqslant M_n,\ \rho_n(Y^{(m)}, x)^2 < D) = 1.$$

Since $\widetilde{g}(\lambda) \leqslant f(\lambda)$, the process $X = \{X_n\}$ can be decomposed as $X_n = U_n + V_n$, where $U = \{U_n\}$ is a stationary Gaussian process with SDF $\widetilde{g}(\lambda)$ and $V = \{V_n\}$ is independent of U. Then we know that

$$\inf\left\{\underline{\lim}_{n\to\infty} \frac{1}{n}\{I(X_1^n, U_1^n) + D(U_1^n \| (Y^{(m)})_1^n)\}; U \in \mathcal{U}_D\right\}$$
$$= \inf\{\overline{I}(X, U); U \in \mathcal{U}_D\} = \overline{R}(D).$$

Thus, applying Theorem 3, we have

$$\lim_{n\to\infty} \frac{1}{n} \log P\big(\rho_n(Y^{(m)}, x)^2 < D\big) = -\overline{R}(D).$$

For any $\delta > 0$ there exists $n_1 = n_1(x)$ such that $M_n \geqslant \mathrm{e}^{n(\overline{R}(D)+2\delta)}$ and

$$P\big(\rho_n(Y^{(m)}, x)^2 < D\big) \geqslant \exp\{-n(\overline{R}(D) + \delta)\}, \quad n \geqslant n_1.$$

Therefore

$$\begin{aligned} P\big(\rho_n(Y^{(m)}, x)^2 &\geqslant D,\ m = 1, \ldots, M\big) \\ &= \prod_{m=1}^{M} P\big(\rho_n(Y^{(m)}, x)^2 \geqslant D\big) \\ &\leqslant \big(1 - \mathrm{e}^{-n(\overline{R}(D)+\delta)}\big)^M. \end{aligned}$$

For any $\varepsilon > 0$, $(1 - \mathrm{e}^{-n(\overline{R}(D)+\delta)})^{M_n} < \varepsilon$ for sufficiently large n. Thus there exists n_2 such that

$$P\big(\rho_n(Y^{(m)}, x)^2 \geqslant D,\ m = 1, \ldots, M_n\big) < \varepsilon,$$

so that

$$P\big(\exists m \leqslant M_n,\ \rho_n(Y^{(m)}, x)^2 < D\big) \geqslant 1 - \varepsilon, \quad n \geqslant n_2.$$

The proof of the theorem is complete. □

We now turn to discussing the coding theorem for a Gaussian channel without feedback. The model of the channel is given by

$$Y_n = X_n + Z_n, \quad n = 1, 2, \ldots, \tag{7}$$

where the noise $Z = \{Z_n\}$ is a stationary Gaussian process with SDF $g(\lambda)$ and the input signal $X = \{X_n\}$ is independent of Z. We assume that an average power constraint

$$\lim_{n\to\infty} \frac{1}{n} \sum_{k=1}^{n} E[X_k^2] \leqslant A$$

is imposed on the input signals, where A is a positive constant. It is known that the capacity (per unit time) $\overline{C}$ of the channel is given by

$$\overline{C} = \frac{1}{4\pi} \int_{-\pi}^{\pi} \log\Big[\max\Big(\frac{\alpha}{g(\lambda)}, 1\Big)\Big]\, \mathrm{d}\lambda,$$

where $\alpha > 0$ is uniquely determined by $\int_{\{\lambda\in[-\pi,\pi];\, g(\lambda)<\alpha\}} (\alpha - g(\lambda))\, \mathrm{d}\lambda = A$. If an input signal X is a stationary Gaussian process with SDF $f(\lambda) = \max(\alpha - g(\lambda), 0)$, then $\overline{I}(X, Y) = \overline{C}$, where Y is the output signal corresponding to X.

Let us define encoding and decoding schemes. For each n, a message ξ_n is defined as

$$P(\xi_n = m) = \frac{1}{M_n}, \quad m \in \mathcal{M}_n = \{1, 2, \ldots, M_n\},$$

where M_n is an integer. Let $X^{(m)} = \{X_n^{(m)}\}$, $m = 1, 2, \ldots$, be independent copies of $X = \{X_n\}$. The encoding function $\varphi_n \colon \mathcal{M}_n \to \mathbf{R}^n$ is given by

$$\varphi_n(m) = (x^{(m)})_1^n, \quad m \in \mathcal{M}_n,$$

where $x^{(m)} = \{x_n^{(m)}\}$ is the observed data of $X^{(m)}$. To define decoding scheme we put

$$A_n = \left\{ (\mathbf{x}, \mathbf{y}) \in \mathbf{R}^n ; \left| \frac{1}{n} \log \frac{p_{XY}^n(\mathbf{x}, \mathbf{y})}{p_X^n(\mathbf{x}) p_Y^n(\mathbf{y})} - \overline{C} \right| < \delta \right\},$$

where $p_X^n(\mathbf{x})$, $p_Y^n(\mathbf{y})$, $p_{XY}^n(\mathbf{x}, \mathbf{y})$ are the probability density functions of X_1^n, Y_1^n, (X_1^n, Y_1^n), respectively. The decoding function $\psi_n \colon \mathbf{R}^n \to \mathcal{M}_n \cup \{0\}$ is defined as follows. For each $y_1^n \in \mathbf{R}^n$, we define $\psi_n(y_1^n) = m$ ($m \in \mathcal{M}_n$) if and only if there exists only one $m \in \mathcal{M}_n$ such that $((x^{(m)})_1^n, y_1^n) \in A_n$; otherwise we define $\psi_n(y_1^n) = 0$.

We can prove the following theorem:

THEOREM 5. *For the Gaussian channel* (7) *define the code* (φ_n, ψ_n) *as above. If the coding rate* $\overline{\lim}_{n\to\infty} n^{-1} \log M_n$ *is less than the capacity* $\overline{C}$, *then for almost all* $x^{(m)}$, $m = 1, 2, \ldots$,

$$\lim_{n\to\infty} P(\xi_n \neq \psi_n(Y_1^n)) = 0.$$

The proof will be given elsewhere.

5. Maximum Entropy Principle

The maximum entropy (ME) method has been proposed to estimate the probability distribution. Let $X = \{X_n\}$ be a stationary process with unknown distribution P^*. Suppose that, based on observations, we know that the true distribution belongs to a class Π of distributions. Then the maximum entropy (ME) principle chooses, as the optimum estimation of P^*, the distribution Q^* that has the maximum entropy in Π:

$$\overline{h}(Q^*) \geqslant \overline{h}(Q), \quad \forall Q \in \Pi,$$

where $\overline{h}(Q)$ denotes the differential entropy (per unit time) of Q. On the other hand, the the minimum relative entropy (MRE) principle chooses the distribution $Q^* \in \Pi$ that minimizes the relative entropy in Π:

$$\overline{D}(Q^* \| P^*) \leqslant \overline{D}(Q \| P^*), \quad \forall Q \in \Pi, \tag{8}$$

as the optimum estimation. The MRE principle may be regarded as an extention of the ME principle.

Denote by $\widehat{Q}_n$ the emperical distribution of the observed data X_1^n. We consider a random variable $Z = F(X)$ with any fixed functional F. Let ν_n be the distribution of Z under the condition $\widehat{Q}_n \in \Pi$, and ν^* be the distribution of Z under the condition that the distribution of X is Q^*, where Q^* is the MRE distribution given by (8). It is known (cf. [3, 6]) that, if X is an i.i.d. or a Markov chain with finite alphabets,

$$\lim_{n\to\infty} \nu_n = \nu^*. \tag{9}$$

We may say that the property (9), called the conditional limit theorem, justifies the ME and MRE principles. Csiszár [4] conjectured that the conditional limit theorem holds for processes in a wide class of stationary ergodic processes.

We can show that the conditional limit theorem holds for Gaussian processes. Let $X = \{X_n\}$ be a stationary Gaussian process with SDF f. As an important class we consider a class Π which is described in terms of the covariance function. Since the covariance function is determined by the SDF, a class $\mathcal{G} \subset \mathcal{S}$ corresponds to Π. Given linearly independent functions $u_k(\lambda) \in L^2[-\pi, \pi]$, $k = 1, 2, \ldots, d$, and a measurable set $A \subset \mathbf{R}^d$, we consider a class

$$\mathcal{G} = \{g \in \mathcal{S};\ \mathbf{u}_d(g) \in A\}$$

and the class Π of all stationary probability distributions with a SDF belonging to $\mathcal{G}$. We may assume that $u_k(-\lambda) = \overline{u_k(\lambda)}$. Based on the observation of X_1^n, the event $\{\widehat{Q}_n \in \Pi\}$ can be simply written as

$$Z_n \in A, \tag{10}$$

where $Z = \{Z_n \equiv (Z_{n1}, \ldots, Z_{nd})\}$ is the d-dimensional process given by (3).

We assume that there exists a SDF which satisfies

$$\overline{D}(f^* \| f) = \inf\{\overline{D}(g | f);\ g \in \mathcal{G}\}.$$

Note that the distribution Q^* of the Gaussian process with SDF f^* satisfies (8). Since the periodogram $I_n(\lambda)$ is an estimator of the SDF, to prove the conditional limit theorem it suffices to show that, under the condition (10), $\int_{-\pi}^{\pi} v(\lambda) I_n(\lambda)\,\mathrm{d}\lambda$ converges to $\int_{-\pi}^{\pi} v(\lambda) f^*(\lambda)\,\mathrm{d}\lambda$ as $n \to \infty$ for each $v \in L^2[-\pi, \pi]$.

THEOREM 6. *Let $X = \{X_n\}$ be a stationary Gaussian process with SDF f. Assume that functions f and u_k are continuous, and $f(\lambda) > 0$, $\forall\lambda$. Fix a SDF $f^* = f_{\theta^*}$ given by* (4), *where $\theta^* \in \{\theta \in \mathbf{R}^d;\ f_\theta \in \mathcal{S}\}^\circ$. Let $A \subset \Gamma_d(\theta^*)$ be an open set such that $A \cap B_d(\mathbf{u}_d(f^*), \delta) \neq \phi$ for all $\delta > 0$, where $\Gamma_d(\theta)$ is a set defined by* (2) *with $a^* = \mathbf{u}_d(f_\theta)$. Then for any $v \in L^2[-\pi, \pi]$ and any $\varepsilon > 0$,*

$$\lim_{n\to\infty} P\left(\left|\int_{-\pi}^{\pi} v(\lambda) I_n(\lambda)\,\mathrm{d}\lambda - (v, f^*)\right| < \varepsilon \;\Big|\; Z_n \in A\right) = 1. \tag{11}$$

The conditional limit theorem (11) can be proved by applying Theorem 2 to $Z = \{Z_n\}$ and $(Z, Y) = \{(Z_n, Y_n)\}$.

6. Hypothesis Testing

The LDT can be applied to evaluate the asymptotic behavior of error probabilities in hypothesis testing. Let $X = \{X_n\}$ be a stationary Gaussian process with unknown SDF f. We consider a problem of testing of following alternative hypotheses H_0 and H_1:

$$H_0\colon\ f = f_0, \qquad H_1\colon\ f = f_1,$$

where f_0 and f_1 are SDF's. We are required to decide which of H_0 and H_1 is true, based on the observed sample $x_1^n \in \mathbf{R}^n$ of X_1^n. For each n, let A_n be an acceptance region. The hypothesis H_0 is accepted if $x_1^n \in A_n$, while H_1 is accepted if $x_1^n \notin A_n$. Denote by α_n the error probability of accepting H_1 when H_0 is actually true. Similarly denote by β_n the error probability of accepting H_0 when H_1 is actually true:

$$\alpha_n = P(X_1^n \notin A_n | H_0 \text{ is true}), \quad \beta_n = P(X_1^n \in A_n | H_1 \text{ is true}).$$

The problem is to analyze the asymptotic behavior of the minimum β_n under the exponential-type constraint $\lim_{n\to\infty} n^{-1} \log \alpha_n \leqslant -\alpha$, where $\alpha > 0$ is a given constant. In this case, it is well known that the Neyman–Pearson test is optimal. The probability distribution $P_{m,n}$ of X_1^n under the hypothesis H_m has a density function

$$p_{m,n}(\mathbf{x}) = \frac{1}{(2\pi)^{n/2} |\Gamma_{m,n}|^{1/2}} \exp\left\{ -\frac{1}{2} \langle \Gamma_{m,n}^{-1} \mathbf{x}, \mathbf{x} \rangle \right\}, \quad \mathbf{x} \in \mathbf{R}^d,$$

where $\Gamma_{m,n}$ is the covariance matrix of X_1^n under H_m. The Neyman–Pearson test is a test given by an acceptance region

$$A_n(t) = \left\{ \mathbf{x} \in \mathbf{R}^n;\ \frac{1}{n} \log \frac{p_{0,n}(\mathbf{x})}{p_{1,n}(\mathbf{x})} \geqslant t \right\}.$$

We assume that $f_0(\lambda)/f_1(\lambda)$ and $f_1(\lambda)/f_0(\lambda)$ are bounded. Let us define a SDF f_θ by

$$\frac{1}{f_\theta(\lambda)} = \frac{(1-\theta)}{f_0(\lambda)} + \frac{\theta}{f_1(\lambda)}.$$

There exist constants $\theta_0 < 0$ and $\theta_1 > 1$ such that $f_\theta \in \mathcal{S}$ and $f_\theta(\lambda)/f_m(\lambda)$ $(m = 0, 1)$ is bounded for each $\theta_0 < \theta < \theta_1$. Applying Theorem 2 to

$$Z_n = \frac{1}{n} \log \frac{p_{0,n}(X_1^n)}{p_{1,n}(X_1^n)} = \frac{1}{2n} \log \frac{|\Gamma_{1,n}|}{|\Gamma_{0,n}|} - \frac{1}{2n} \langle (\Gamma_{0,n}^{-1} - \Gamma_{1,n}^{-1}) X_1^n, X_1^n \rangle,$$

we can prove the following theorem:

THEOREM 7. *For any* $\alpha \in (0, \overline{D}(f_{\theta_1} \| f_0)]$ *there exist constants* t_α *and* $\theta(\alpha) > 0$ *such that*

$$\lim_{n\to\infty} \frac{1}{n} \log P_{0,n}(A_n(t_\alpha)^c) = -\alpha$$

and

$$\alpha = \overline{D}(f_{\theta(\alpha)} \| f_0).$$

If $0 < \alpha < \overline{D}(f_1 \| f_0)$ $(0 < \theta(\alpha) < 1)$, *then*

$$\lim_{n\to\infty} \frac{1}{n} \log P_{1,n}(A_n(t_\alpha)) = -\overline{D}(f_{\theta(\alpha)} \| f_1). \tag{12}$$

If $\alpha > \overline{D}(f_1 \| f_0)$ $(\theta(\alpha) > 1)$, *then*

$$\lim_{n\to\infty} \frac{1}{n} \log\{1 - P_{1,n}(A_n(t_\alpha))\} = -\overline{D}(f_{\theta(\alpha)} \| f_1). \tag{13}$$

The asymptotic behavior (12) was proved in [5], and the property (13) was proved in [8].

7. Concluding Remark

We can prove the similar results to Theorems 3–7 for i.i.d. processes with arbitrary state spaces.

References

1. Bercu, B., Gamboa, F. and Rouault, A.: Large deviations for quadratic forms of Gaussian stationary processes, *Stochastic Proc. Appl.* **71** (1997), 75–90.
2. Bryc, W. and Dembo, A.: Large deviations for quadratic functionals of Gaussian processes, *J. Theoretical Probab.* **10** (1997), 307–332.
3. Cover, T. M. and Thomas, J. A.: *Elememts of Information Theory*, Wiley, New York, 1991.
4. Csiszár, I.: Is the maximum entropy principle operationally justifiable? In: *Open Problems in Communication and Computation*, Springer, New York, 1987, pp. 36–37.
5. Dacunha-Castelle, D.: Formule de chernoff pour des rapports de vraisemblance, In: *Grandes déviations et applications statistiques* (*Astérisque* 68), Soc. Math. France (1979), pp. 25–31.
6. Dembo, A. and Zeitouni, O.: *Large Deviations Techniques and Applications*, Jones and Bartlett Pub., 1993.
7. Ihara, S. and Kubo, M.: The asymptotics of string matching probabilities for Gaussian random sequences, Preprint Ser. in Math. Sciences, Nagoya Univ., 1998.
8. Ihara, S. and Muramatsu, J.: On the converse theorem in hypothesis testing for stationary Gaussian sequences, *Studies Inform. Sci. Nagoya Univ.* **5** (1997), 19–27.
9. Yang, E.-H. and Kieffer, J.: On the performance of data compression algorithms based upon string matching, *IEEE Trans. Inform. Theory* **IT-44** (1998), 47–65.

Acta Applicandae Mathematicae **63:** 175–184, 2000.

Generalized Functions in Signal Theory

FRIEDRICH JONDRAL
Institut für Nachrichtentechnik, Universität Karlsruhe (T.H.), D-76128 Karlsruhe, Germany.
e-mail: int@etec.uni-karlsruhe.de

(Received: 8 January 1999)

Abstract. Within a self-contained signal theory, generalized functions have to be taken into account, because without them notions like impulse response or transmission function cannot be defined. Starting from the requirements that have to be taken for a function space, if it should be suitable for a signal theory, generalized functions are introduced. Moreover, the connections between such a signal theory and the theory of white noise are discussed.

Mathematics Subject Classifications (2000): 46F12, 60G15, 60G20.

Key words: generalized functions, signal analysis, bandlimited signals, white noise.

1. Introduction

As a consequence of the increasing digitalization of communication systems, properties of time functions and process models, representing signals or systems, are placed at the foreground of interest.

The transmission of a message between two places by means of electromagnetic waves may in fact be interpreted as the transfer of one exactly determined time function. But every radio transmission line (Figure 1) has to be designed for a broad variety of possible signal functions, since

(i) the receiver does not know the time development of the current signal function in advance (otherwise the transmission would be useless, since the signal function would not carry any information for the receiver in this case),
(ii) the transmission is generally disturbed by unpredictable random effects.

First of all, it seems to be very important that, within the transmission model, the set of signal functions has to be suitably chosen and that this choice guarantees the inclusion of

- the signals Fourier transforms as well as,
- the impulse responses describing the signal processing systems, and
- the appertaining transmission functions.

The desirable properties of a signal function space Φ are

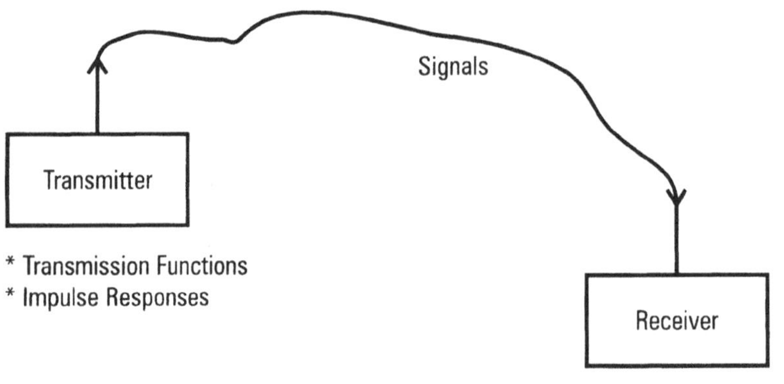

Figure 1. Transmission line.

(1) Φ should be a linear space (vector space) over the field of the real numbers $\mathbb{R}$, i.e. for all signal functions φ, $\psi \in \Phi$ and every real number a, we get

$$\varphi + \psi \in \Phi, \qquad a\varphi \in \Phi. \tag{1}$$

(2) Since we want to proceed from the time domain to the frequency domain and vice versa without any problems, the Fourier transform $\mathcal{F}$ and the inverse Fourier transform $\mathcal{F}^{-1}$ should be one-to-one mappings from Φ to Φ, i.e. if

$$\hat{\varphi}(\omega) = \mathcal{F}\{\varphi(t)\} = \int_{-\infty}^{\infty} \varphi(t)\mathrm{e}^{-j\omega t}\,\mathrm{d}t \tag{2}$$

is the Fourier transform of $\varphi(t) \in \Phi$, we should have

$$\hat{\varphi}(\omega) \in \Phi. \tag{3}$$

(3) For the investigation of linear time invariant systems, the following two properties are important:

$$\begin{aligned} &\varphi(t) \in \Phi \to \varphi(t - t_0) \in \Phi \quad \text{for any } t_0 \in \mathbb{R}, \\ &\varphi(t), \psi(t) \in \Phi \to \varphi * \psi = \int_{-\infty}^{\infty} \varphi(t - \tau)\psi(\tau)\,\mathrm{d}\tau \in \Phi. \end{aligned} \tag{4}$$

(4) The time derivative of every signal function should be a signal function again. This means that every signal function should be infinitely often differentiable.

2. Test Function Spaces

One special function space, possessing the properties (1)–(4) is the Schwartz space of rapidly decreasing functions $\mathcal{S}$ [1]. Its exact definition is:

$\mathcal{S}(\mathbb{R})$, or briefly $\mathcal{S}$, is the space of all infinitely often differentiable real valued functions of the real variable t, which converges itself as well as all its derivatives to zero for $|t| \to \infty$ more rapidly than every power of $|t|^{-1}$, i.e. for every $\varphi(t) \in \mathcal{S}$ we get

$$\lim_{|t|\to\infty} \left| t^k \frac{\mathrm{d}^\alpha}{\mathrm{d}t^\alpha} \varphi(t) \right| = 0 \quad \text{for all } k, \alpha \in \mathbb{N}. \tag{5}$$

Remarks. (i) The function space $\mathcal{S}$ is countably normed and complete.

(ii) The Fourier transform $\mathcal{F}$ and the inverse Fourier transform $\mathcal{F}^{-1}$ are topological mappings of $\mathcal{S}$ onto $\mathcal{S}$, i.e. in addition to property (2), we get that $\mathcal{F}$ and $\mathcal{F}^{-1}$ are defined on the *whole* space $\mathcal{S}$ and are *continuous* mappings of $\mathcal{S}$ onto $\mathcal{S}$.

(iii) The functions of $\mathcal{S}$ are square-integrable:

$$\int_{-\infty}^{\infty} \varphi^2(t)\,\mathrm{d}t < \infty \quad \text{for any } \varphi \in \mathcal{S}, \tag{6}$$

i.e. signals represented by functions of $\mathcal{S}$ are finite energy signals.

From the signal theory point of view, there remain some open questions opposed to a self-contained understanding. The two most important of them are

(1) There are signal functions, for example $s(t) = \cos \omega_0(t)$, that are not of finite energy (and as a consequence they are not members of $\mathcal{S}$). Such signals are not realizable in the physical sense but, from a theoretical point of view, they are nevertheless of great importance. *How to incorporate such signal functions*?

(2) *What are the means to describe the reaction of a linear time invariant system upon the input of a signal*? The use of the notion 'impulse response' is prohibited for the time being, since the Dirac impulse δ also is not found in $\mathcal{S}$.

To approach the answer to these two questions, two further test function spaces are introduced [1], namely:

$\mathcal{D}$ the space of all infinitely often differentiable real-valued functions of one real variable t that have compact support (i.e. for every function $\varphi \in \mathcal{D}$ we can find a closed interval on the real line, on the complement of which φ vanishes).

$\mathcal{E}$ the space $C^\infty(\mathbb{R})$ of all infinitely often differentiable real-valued functions of one real variable t, equipped with the topology introduced by the system

$$\|\varphi\|_p = \sup_{\alpha \leqslant p} \sup_{t \in K} \left| \frac{\mathrm{d}^\alpha}{\mathrm{d}t^\alpha} \varphi(t) \right| \tag{7}$$

of infinitely many norms and with the metric derived from this system. K varies over all compact subsets of $\mathbb{R}$, α and p are natural numbers.

For the test function spaces $\mathcal{D}$, $\mathcal{S}$ and $\mathcal{E}$ the following inclusions are valid:

$$\mathcal{D} \subset \mathcal{S} \subset \mathcal{E}. \tag{8}$$

The answer to our two questions is given in the next section.

3. Spaces of Generalized Functions

A continuous linear functional defined on a test function space Λ (i.e. a continuous linear mapping of the test function space into the set of real numbers $\mathbb{R}$) is called a *generalized function* or a *distribution*.

An *example* of a generalized function on $\mathcal{E}$ (and therefore (cf. (8)) also on $\mathcal{S}$ and $\mathcal{D}$) is the Dirac impulse δ, which satisfies

$$\delta\{\varphi(t)\} = \varphi(0) \quad \text{for any } \varphi \in \mathcal{E}. \tag{9}$$

Just like the test functions, the generalized functions defined on Λ form a vector space, the so-called dual space Λ' of the test function space Λ. So there are

$\mathcal{D}'$ the space of distributions over $\mathcal{D}$,
$\mathcal{S}'$ the space of tempered distributions,
$\mathcal{E}'$ the space of distributions with compact support.*

Corresponding to (8), we get

$$(\mathcal{D} \subset \mathcal{S} \subset \mathcal{E} \subset)\mathcal{E}' \subset \mathcal{S}' \subset \mathcal{D}'. \tag{10}$$

The application of a generalized function to a test function is written in the form

$$f\{\varphi\} = \langle f, \varphi \rangle = \int_{-\infty}^{\infty} f(t)\varphi(t)\,\mathrm{d}t. \tag{11}$$

As already mentioned, the Fourier transform $\mathcal{F}$, as well as the inverse Fourier transform $\mathcal{F}^{-1}$, are one-to-one mappings of $\mathcal{S}$ onto $\mathcal{S}$.

With the definition

$$\langle \mathcal{F}\{f\}, \varphi \rangle := \langle f, \mathcal{F}\{\varphi\} \rangle, \quad \text{for any } \varphi \in \mathcal{S}, \tag{12}$$

the Fourier transform is extended to a topological mapping of $\mathcal{S}'$ onto $\mathcal{S}'$ [1], i.e. $\mathcal{F}$ and $\mathcal{F}^{-1}$ are continuous one-to-one mappings from $\mathcal{S}'$ onto $\mathcal{S}'$.

Within the theory of bandlimited signals, Fourier transform and inverse Fourier transform on the space $\mathcal{E}'$ of distributions with compact support are of special interest. The following two propositions are very important [5]:

(1) Let $\hat{f} \in \mathcal{E}'$ be a generalized function with compact support: The inverse Fourier transform $f(t)$ of $\hat{f}(\omega)$ belongs to the class

$$\Theta_M = \left\{ a(t) \in C^{\infty}(\mathbb{R});\ \forall \alpha \in \mathbb{N} \cup \{0\} : \exists C_\alpha \in \mathbb{R} \wedge \exists m_\alpha \in \mathbb{N} : \left| \frac{\mathrm{d}^\alpha}{\mathrm{d}t^\alpha} a(t) \right| \leqslant C_\alpha (1 + |t|)^{m_\alpha} \right\} \tag{13}$$

---

* The generalized function f vanishes in the region $G \subset \mathbb{R}$, iff $\langle f, \varphi \rangle = 0$ for every $\varphi \in \Lambda$ with $\operatorname{supp} \varphi \subset G$. The set of all $x \in \mathbb{R}$, that do not have any neighborhood on which $g \in \Lambda'$ vanishes, is called support of f.

and is given by

$$\begin{aligned} f(t) &= \frac{1}{2\pi}\langle \hat{f}(\omega), \eta(\omega)\mathrm{e}^{-j\omega t}\rangle \\ &= \frac{1}{2\pi}\int_{-\infty}^{\infty} \hat{f}(\omega)\eta(\omega)\mathrm{e}^{j\omega t} \\ &= \mathcal{F}^{-1}\{\hat{f}\}. \end{aligned} \tag{14}$$

Here $\eta(\omega) \in \mathcal{D}$ is constant 1 on a neighborhood of the support of $f(\omega)$ (such a function η always exists [10]).

(2) Let $\hat{g}(\omega)$ be a tempered distribution and $\hat{f}(\omega)$ be a generalized function with compact support. Then we get

$$\begin{aligned} &\text{(i)} \quad \hat{g} * \hat{f} \in \mathcal{S}', \\ &\text{(ii)} \quad \mathcal{F}^{-1}\{\hat{g} * \hat{f}\} = \mathcal{F}^{-1}\{\hat{g}\} \cdot \mathcal{F}^{-1}\{\hat{f}\}. \end{aligned} \tag{15}$$

Remark. Obviously, it seems to be recommendable to treat all test functions and all generalized functions as signals. In the sense of distribution theory, the members of $\mathcal{S}'$ fulfil all properties of a signal function space as required in Section 1.

The property of a signal to be bandlimited can therefore be expressed as the following definition: *The signal $s(t)$ is bandlimited, if its Fourier transform $\hat{s}(\omega)$ is a generalized function with compact support.*

Remark. As a consequence of this definition, bandlimited signals belong to the class Θ_M, i.e. they are represented by functions which are infinitely often differentiable with respect to the time variable and, as well as all their derivatives, they increase no faster than a polynomial.

EXAMPLE. The signal $s(t) = \cos\omega_0 t$ is bandlimited, because (14) yields

$$\begin{aligned} &\mathcal{F}^{-1}\{\pi[\delta(\omega-\omega_0)+\delta(\omega+\omega_0)]\} \\ &\quad = \tfrac{1}{2}\langle\delta(\omega-\omega_0), \eta(\omega)\mathrm{e}^{-j\omega t}\rangle + \tfrac{1}{2}\langle\delta(\omega+\omega_0), \eta(\omega)\mathrm{e}^{-j\omega t}\rangle \\ &\quad = \tfrac{1}{2}\eta(-\omega_0)\mathrm{e}^{j\omega_0 t} + \tfrac{1}{2}\eta(\omega_0)\mathrm{e}^{-j\omega_0 t} = \cos\omega_0 t. \end{aligned}$$

Here $\eta(\omega)$ is chosen such that this function is identically 1 on a closed interval U including $[-\omega_0, \omega_0]$. Obviously, U contains the support of the generalized function $\pi[\delta(\omega-\omega_0)+\delta(\omega+\omega_0)]$ (Figure 2).

In principle, bandlimited signals cannot be timelimited, i.e. it is impossible to find a $T \in \mathbb{R}$ for a bandlimited signal $s(t)$, such that $s(t) = 0$ for all $|t| > T$. This follows from the identity theorem of analytic functions [8], a consequence of which is that the (unique) continuation of a holomorphic function that vanishes on

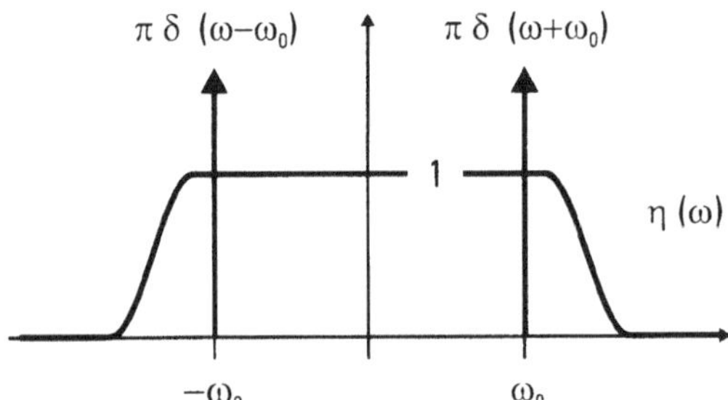

Figure 2. $\mathcal{F}\{\cos(\omega_0 t)\}$.

an interval of the real axis, is identically zero on the whole complex plane, i.e. this function also has to be identically zero on the whole real axis.

On the set of bandlimited signals the structure of a vector space may now be introduced:

The Fourier transform is a topological mapping of $\mathcal{S}'$ onto $\mathcal{S}'$. Due to the inclusion $\mathcal{E}' \subset \mathcal{S}'$ the Fourier transform is also a one-to-one mapping of $\mathcal{E}'$. Since $\mathcal{E}'$ is a linear space,

$$\mathrm{BBS} = \{s(t); \mathcal{F}\{s(t)\} = \hat{s}(\omega) \in \mathcal{E}'\} \subset \Theta_M \tag{16}$$

is also a vector space, the space of bandlimited signals.

In communications theory, an important example of a generalized function is given by the Cauchy main value $\mathrm{CH}1/t$: Let $x(t)$ be a test function, then

$$\tilde{x}(t) = \left\langle \mathrm{CH}\frac{1}{t}, x(t) \right\rangle = \frac{1}{\pi} \int_{-\infty}^{\infty} \frac{x(u)}{t-u}\, du = x(t) * \frac{1}{\pi t}, \tag{17}$$

where the integral is to be considered as its Cauchy main value. $x(t)$ may be written in the equivalent form

$$\tilde{x}(t) = \mathcal{H}\{x(t)\} = \lim_{\epsilon \to 0} \frac{1}{\pi} \int_{|t-u|>\epsilon} \frac{x(u)}{t-u}\, du \tag{18}$$

and $\tilde{x}(t)$ is called the Hilbert transform of $x(t)$.

The Fourier transform $\hat{\tilde{x}}(\omega)$ of $\tilde{x}(t)$ is

$$\hat{\tilde{x}}(\omega) = \mathcal{F}\{\tilde{x}(t)\} = (-j \operatorname{sign} \omega)\hat{x}(\omega), \tag{19}$$

where sign denotes the signum function.

Remark. Equation (19) shows the close relation between the Hilbert transform and the Fourier transform of a test function. Just like the Fourier transform (cf. (12)), the Hilbert transform may be extended to generalized functions.

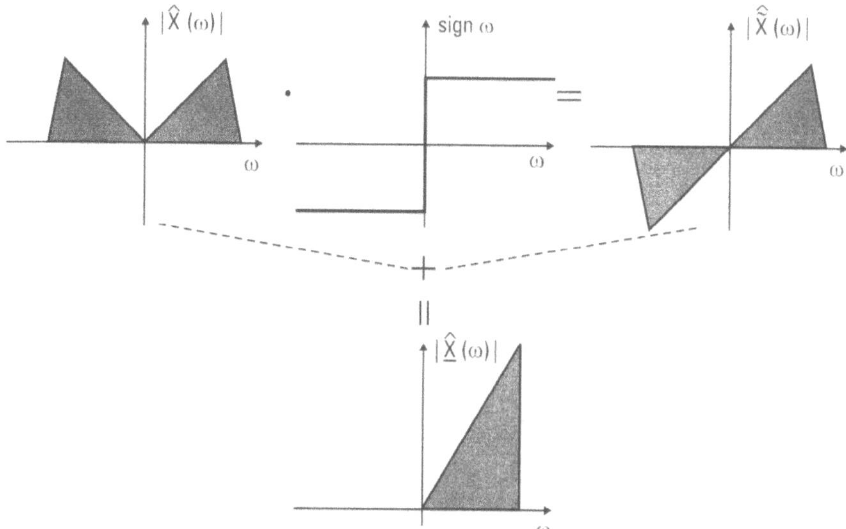

Figure 3. Relation of different spectra.

The importance of the Cauchy main value and therefore of the Hilbert transform is given by the fact that the analytic signal belonging to $x(t)$ is defined by

$$\underline{x}(t) = x(t) + j\tilde{x}(t). \tag{20}$$

Remarks. (i) From the knowledge of the real part $x(t)$ and the imaginary part $\tilde{x}(t)$ of a signal $\underline{x}(t)$, its amplitude and its phase may be easily computed:

$$|\underline{x}(t)| = \sqrt{x^2(t) + \tilde{x}^2(t)}, \qquad \arg\{\underline{x}(t)\} = \arctan\frac{\tilde{x}(t)}{x(t)}. \tag{21}$$

(ii) The Fourier transform of the analytic signal $\underline{x}(t)$ is

$$\hat{\underline{x}}(\omega) = \begin{cases} 2\hat{x}(\omega) & \text{if } \omega > 0, \\ \hat{x}(\omega) & \text{if } \omega = 0, \\ 0 & \text{if } \omega < 0. \end{cases} \tag{22}$$

The relations between the signals spectrum (2), the spectrum of its Hilbert transform (19), and the spectrum of the analytic signal (22) are sketched in Figure 3.

4. Realizations of White Noise Processes

Starting from the discussion of the preceding sections, we are now going to treat some aspects of white noise processes. For simplification it is assumed that the processes are all zero mean, stationary, ergodic, and Gaussian. Our main interest is the analytic behavior of the processes' realizations. A stochastic process is 'white'

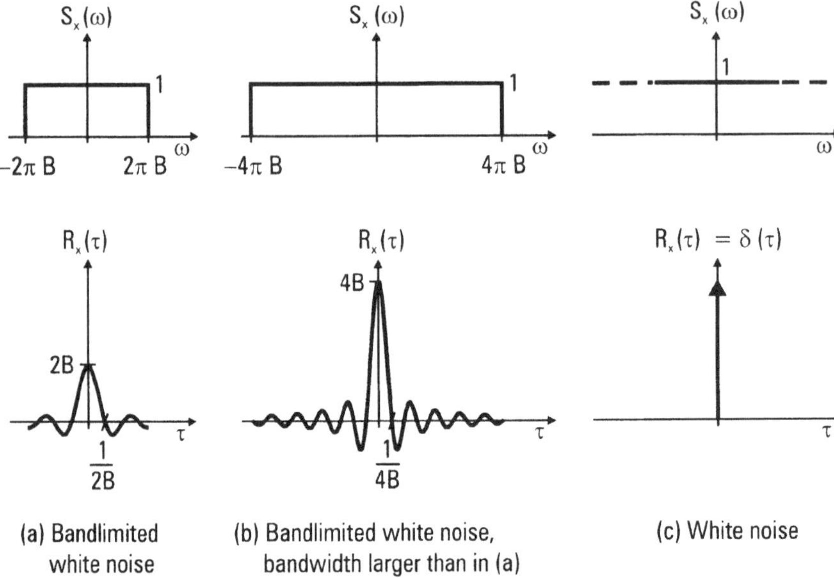

Figure 4. Spectral power densities and autocorrelations.

iff its spectral power density $S_X(\omega)$ is constant inside a specific domain of the frequency axis and vanishes on its complement. Under the assumption that the constant is 1, we get as the spectral power density of a bandlimited white noise of bandwidth B:

$$S_X(\omega) = \begin{cases} 1 & \text{if } |\omega| \leqslant 2\pi B, \\ 0 & \text{elsewhere} \end{cases} \tag{23}$$

(cf. Figure 4a). As a consequence of the Wiener theorem [2], the autocorrelation of the process $X(t)$ is the inverse Fourier transform of $S_X(\omega)$:

$$R_X(\tau) = \frac{1}{2\pi} \int_{-\infty}^{\infty} S_X(\omega) e^{j\omega\tau}\, d\omega = \frac{\sin 2\pi B\tau}{\pi\tau}. \tag{24}$$

This autocorrelation function is also shown in Figure 4a.

Remarks. (i) Two random variables of a bandlimited white noise are uncorrelated iff their appertaining time difference is exactly $\tau = (2B)^{-1}$. This means, for Gaussian processes, that these random variables are statistically independent.

(ii) For Gaussian processes stationarity and weak stationarity are equivalent.

(iii) To every (weak) stationary process $Y(t)$ with mean value μ_Y and autocorrelation function $R_Y(\tau)$, there exists a Gaussian process that has exactly the same mean and the same autocorrelation. Moreover, this so-called equivalent Gaussian process is uniquely defined.

(iv) Because of our discussion in Section 3 (see (13)), it follows, that the realizations of a bandlimited, white, stationary, ergodic noise $X(t)$ are almost surely analytic functions.

The autocorrelation function (24) shows that the random variables $X(t_i)$ and $X(t_k)$ are correlated in general. They are uncorrelated only if $|t_i - t_k| = n(2B)^{-1}$ ($n \in \mathbb{N}$). If the spectral interval over which $S_X(\omega)$ is constant 1 becomes larger, $R_X(\tau)$ becomes narrower and the time difference for the uncorrelation becomes smaller (Figure 4b). Continuing this process beyond all limits leads to $S_X(\omega) = 1$ for all ω and $R_X(\tau)$ converges towards a Dirac impulse (Figure 4c).

Since $S_X(\omega) = 1$ for all ω, the not bandlimited white noise process is of infinite power, i.e. white noise is a theoretical process not existing in nature. But an investigation of such processes is very interesting in spite of this, because

(1) Applying the theory of white noise leads to enormous simplifications for the analytic treatment of many problems.
(2) Many stochastic processes appearing in practical problems can be approximated by white noise.

The last statement is *not* to be understood in the sense that there exists a sequence of stochastic processes that converges (for example, in mean square) to a white noise. Rather, we have to remember that the Dirac impulse $\delta(t)$ is not a function. There is a similarity with white noise since it is not a stochastic process in the usual sense but a generalized stochastic process.

In a rather formal manner, white noise may be interpreted as the time derivative of a Brownian motion process $B(t)$ satisfying the following conditions:

(i) $B(0) = 0$ almost surely,
(ii) $\{B(t); t \geqslant 0\}$ is a Gaussian process,
(iii) for every t and $t + h$ with $h > 0$, we get

$$E\{B(t+h) - B(t)\} = 0, \qquad \operatorname{var}\{B(t+h) - B(t)\} = h.$$

$B(t)$ is a process with independent increments, its realizations are almost surely continuous but also almost surely nowhere differentiable. From this statement it becomes obvious that the realizations of a white noise process are highly irregular.

The exact definition of the white noise process is as follows (cf. [4]): Let $\mathfrak{B}$ be the σ-algebra generated by the cylinder sets of $\mathcal{S}'$. Because of the Bochner–Minlos theorem, it is possible to define a probability measure μ on the measurable space $(\mathcal{S}', \mathfrak{B})$ for which

$$\int_{\mathcal{S}'} \mathrm{e}^{j\langle x,\xi\rangle}\,\mathrm{d}\mu(x) = \mathrm{e}^{-\|\xi\|^2}, \quad \xi \in \mathcal{S}, \tag{25}$$

is valid. Here $\langle\cdot,\cdot\rangle$ is the canonic bilinear form connecting $\mathcal{S}'$ and $\mathcal{S}$ (cf. (11)), $\|\cdot\|$ denotes the $L^2(\mathbb{R})$-norm.

The measure space $(\mathcal{S}', \mathfrak{B}, \mu)$ *is called white noise.*

Remarks. (i) The space of tempered distributions may be interpreted as the space of realizations of the white noise process.

(ii) For $\xi \in \mathcal{S}$ fixed, $\langle x, \xi \rangle$ is a random variable defined on $(\mathcal{S}', \mathfrak{B}, \mu)$.

(iii) μ is a standard Gaussian measure. Therefore with $\|\xi\| = 1$, $\langle x, \xi \rangle$ is a standard Gaussian random variable.

(iv) If $\xi, \psi \in \mathcal{S}$ are $L^2(\mathbb{R})$-orthogonal, $\langle x, \xi \rangle$ and $\langle x, \psi \rangle$ are independent random variables on $(\mathcal{S}', \mathfrak{B}, \mu)$.

(v) The spectral power density of the white noise process is $S_X(\omega) = 1$ for all $\omega \in \mathbb{R}$.

Acknowledgement

I gratefully acknowledge the continuing support of Professor Takeyuki Hida over many years, especially during the period from 1975 to 1979 when I was with Prof. Dr.rer.nat. Ernst Henze as a PhD student at the Technical University of Braunschweig. One of the most impressive experiences in my personal and scientific development was my visit to the University of Nagoya in the winter semester 1977/78 as a guest of Professor Hida.

References

1. Constantinescu, F.: *Distributionen und ihre Anwendung in der Physik*, Teubner, Stuttgart, 1974.
2. Fischer, F. A.: *Einführung in die statistische Übertragungstheorie*, Bibliographisches Institut, Mannheim, 1969 (BI Hochschultaschenbuch 130/130a).
3. Henze, E.: *Einführung in die Maßtheorie*, 2nd edn, Bibliographisches Institut, Zürich, 1985.
4. Hida, T.: *Brownian Motion*, Springer-Verlag, Berlin, 1980.
5. Hörmander, L.: *Linear Partial Differential Operators*, 4th edn, Springer-Verlag, Berlin, 1976.
6. Jondral, F.: Some remarks about generalized functionals of complex white noise, *Nagoya Math. J.* **81** (1981), 113–122.
7. Jondral, F.: The Hilbert transform of a sampled random signal, *ntzArchiv* **4** (1982), 227–232.
8. Knopp, K.: *Funktionentheorie I – Grundlagen der allgemeinen Theorie der analytischen Funktionen*, 12th edn, Sammlung Göschen 668, de Gruyter, Berlin, 1970.
9. Kolmogorov, A. N. and Fomin, S. V.: *Reelle Funktionen und Funktionalanalysis*, VEB Deutscher Verlag der Wissenschaften, Berlin, 1975.
10. Walter, W.: *Einführung in die Theorie der Distributionen*, Bibliographisches Institut, Zürich, 1974.
11. Wong, E. and Hajek, B.: *Stochastic Processes in Engineering Systems*, Springer-Verlag, Berlin, 1985.

Acta Applicandae Mathematicae **63:** 185–201, 2000.

Ergodic Properties of Random Positive Semigroups

HIROSHI KUNITA
Graduate School of Mathematics, Kyushu University, Fukuoka 812, Japan

(Received: 16 April 1999)

Abstract. In this paper, we study the unique ergodicity of random positive semigroups and their asymptotic behavior as time tends to infinity.

Mathematics Subject Classification (2000): 37Axx.

Key words: ergodicity, random positive semigroups, asymptotic behavior.

1. Introduction

In [3] and [7], we studied infinitesimal generators of random positive semigroups with independent increments, or positive semigroups in random environments with independent increments.

In [3], random positive semigroups of diffusion type are studied, where their random infinitesimal generators are represented as second-order stochastic partial differential operators. In [7], random positive semigroups of the jump type are studied, where their random infinitesimal generators are represented as random integro-differential operators.

In this paper, we shall study the unique ergodicity of random positive semigroups and their asymptotic behavior as time tends to infinity.

In the next section, we shall introduce random positive semigroups with independent increments and represent their random infinitesimal generators following [7].

In Section 3, we shall regard the random positive semigroup $\{T_{s,t}, t \geqslant s\}$ as a stochastic process with values in the space of positive measures on $\mathbf{R}^d$ and show that the stochastic process is a Markov process.

In Section 4, we consider the case where $\{T_{s,t}\}$ is temporally homogeneous. We shall discuss its unique ergodicity in connection with the asymptotic property of the mean semigroup $\{S_{s,t}\}$ on $\mathbf{C}$, i.e., $S_{s,t}f(x) = E[T_{s,t}f(x)]$.

It will be shown that the ergodic property of the measure-valued Markov process depends on the transience, null recurrence, and positive recurrence of the mean sesmigroup $\{S_{s,t}\}$ (see Theorems 4.2 and 4.3).

In Section 5, we will study the asymptotic behavior of the temporally homogeneous random positive semigroup $T_{s,t}f$ as t tends to infinity, by applying results of Section 4. It will turn out that these properties depend again on the transience, null recurrence, and positive recurrence of the mean semigroup $\{S_{s,t}\}$ (see Theorems 5.1 and 5.2).

2. Random Positive Semigroups and their Infinitesimal Generators

Let $\mathbf{C} = C(\mathbf{R}^d)$ be the totality of real continuous functions on $\mathbf{R}^d$ such that $\lim_{x\to\infty} f(x)$ exists and equals 0.

It is a real separable Banach space with the supremum norm $\| \ \|$.

Given a positive integer m, let $\mathbf{C}^m$ be the set of all $f \in \mathbf{C}$ which are m-times continuously differentiable and their derivatives belong to $\mathbf{C}$. We set $\mathbf{C}^\infty = \bigcup_m \mathbf{C}^m$. We denote by $\mathbf{C}^m_{\mathrm{loc}}$ the set of all m-times continuously differentiable functions on $\mathbf{R}^d$.

Let $\{T_{s,t}, 0 \leqslant s \leqslant t < \infty\}$ be a family of stochastic processes with values in linear operators on $\mathbf{C}$, cadlag with respect to t ($\geqslant s$), defined on a probability space $(\Omega, \mathcal{F}, P)$.

It is called a *random positive semigroup* if (i) $f \geqslant 0$ implies $T_{s,t}f \geqslant 0$ a.s. P for any s, t, (ii) for each s $\lim_{t\to s} T_{s,t}f = f$ holds a.s. P for $f \in \mathbf{C}$, and (iii) for each $s < t < u$ $T_{s,t}T_{t,u}f = T_{s,u}f$ holds a.s. P for $f \in \mathbf{C}$.

It is called *temporally homogeneous* if (iv) the law of $T_{s,t}$ coincides with the law of $T_{s+h,t+h}$ for any $s < t$ and $h > 0$. It is called *Markovian* if (v) $T_{s,t}1 \equiv 1$ holds a.s. P for any $s < t$. Further, it is said to have *independent increments* if (vi) $T_{t_i,t_{i+1}}, i = 0, \dots, n-1$ are independent for any $0 \leqslant t_0 < t_1 < \cdots < t_n < \infty$.

In this paper, we only consider the random positive semigroup with independent increments.

Let $A(t), t \geqslant 0$ be a family of random linear maps from $\mathbf{C}^\infty$ to $\mathbf{C}_{\mathrm{loc}}$ such that $(A(t_{i+1}) - A(t_i))f, i = 0, \dots, n-1$ are independent for any $0 \leqslant t_0 < t_1 < \cdots < t_n < \infty$. It is called the *random infinitesimal generator* of $\{T_{s,t}\}$, if it satisfies

$$T_{s,t}f = f + \int_s^t T_{s,r-}A(\mathrm{d}r)f \quad \text{a.s. } P\ \forall s < t, \tag{2.1}$$

for any $f \in \mathbf{C}^\infty$, where the right-hand side is Itô's stochastic integral.

We introduce two assumptions for $\{T_{s,t}\}$. Set

$$S_{s,t}f(x) = E[T_{s,t}f(x)]. \tag{2.2}$$

Then $\{S_{s,t}\}$ is a family of deterministic positive semigroup of linear operators on $\mathbf{C}$. It is called the *mean semigroup* of $\{T_{s,t}\}$.

(A.1) For each $f \in \mathbf{C}^\infty$, the limit

$$L(t)f(x) = \lim_{h\to 0} \frac{S_{t,t+h}f(x) - f(x)}{h} \tag{2.3}$$

exists uniformly in x and it is a continuous function of (t, x).

The family of operators $\{L(t)\}$ is called the *infinitesimal generator* of $\{S_{s,t}\}$.

$(\mathrm{A}.2)_m$ For any $s < t$, $T_{s,t}$ maps $\mathbf{C}^{m+2}$ to $\mathbf{C}^m_{\mathrm{loc}}$ a.s. P. Further, for any $N > 0$ there exists a positive constant c_N such that

$$\sup_{|x|\leqslant N} E[|D^\alpha T_{s,t}f(x) - D^\alpha f(x)|^2] \leqslant c_N|t-s|\|f\|^2_{|\alpha|+2}, \quad \forall f \in \mathbf{C}^{|\alpha|+2} \tag{2.4}$$

hold for any α with $|\alpha| \leqslant m$.

THEOREM 2.1 (cf. [7]). *Let $\{T_{s,t}\}$ be a random positive semigroup with independent increments satisfying* (A.1), $(\mathrm{A}.2)_m$ *for some $m \geqslant [d/2] + 1$. Then it admits a unique random infinitesimal generator $A(t)$, which maps $\mathbf{C}^{m+2}$ into $\mathbf{C}^\gamma_{\mathrm{loc}}$, where $\gamma = m - [d/2] - 1$.*

Further, the random infinitesimal generator $A(t)$ is represented as a random integro-differential operator:

$$A(t)f(x) = \int_0^t L(s)f(x)\,\mathrm{d}s + \sum_i F_i(x,t)\frac{\partial f}{\partial x_i}(x) + G(x,t)f(x) + \\ + \int_0^t \int_{\mathbf{V}^+} \{Tf(x) - f(x)\}\tilde{N}(\mathrm{d}r\,\mathrm{d}T). \tag{2.5}$$

Here, each term of the above representation is interpreted as follows:

(i) *$L(t)$ is a second-order integro-differential operator represented by*

$$L(t)f(x) = \frac{1}{2}\sum_{i,j} a_{ij}(x,t)\frac{\partial^2 f}{\partial x_i \partial x_j}(x) + \sum_i b_i(x,t)\frac{\partial f}{\partial x_i}(x) + \\ + c(x,t)f(x) + \int_{\mathbf{R}^d} (f(y) - f(x) - \\ - \sum_i \frac{y_i - x_i}{1+|y-x|^2}\frac{\partial f}{\partial x_i}(x))n_t(x,\mathrm{d}y). \tag{2.6}$$

Here $a_{ij}(x,t)$, $b_i(x,t)$, $c(x,t)$ are continuous in (x,t).

The matrix $(a_{ij}(x,t))$ is symmetric and nonnegative definite, $n_t(x,\mathrm{d}y)$ is a Lévy measure such that $n_t(x,\{x\}) = 0$ and $\int_{\mathbf{R}^d}\phi_x(y)n_t(x,\mathrm{d}y) < \infty$ for any t, x, where ϕ_x is a function of $\mathbf{C}^2$ such that $\phi_x(x) = 0$, $\phi_x(y) > 0$ if $y \neq x$, $\liminf_{y\to\infty}\phi_x(y) > 0$ *and $\phi_x(y) = \mathrm{O}(|x-y|^2)$ near x.*

(ii) *$(F_1(x,t),\ldots,F_d(x,t),G(x,t))$ is a continuous Brownian motion with values in $C^\gamma_{\mathrm{loc}}(\mathbf{R}^d,\mathbf{R}^{d+1})$. Its mean is* 0 *and the covariance is*

$$\mathrm{Cov}(F_i(x,t),F_j(y,t)) = \int_0^t f_{ij}(x,y,s)\,\mathrm{d}s, \tag{2.7}$$

$$\mathrm{Cov}(G(x,t),G(y,t)) = \int_0^t g(x,y,s)\,\mathrm{d}s, \tag{2.8}$$

$$\mathrm{Cov}(F_i(x,t),G(y,t)) = \int_0^t h_i(x,y,s)\,\mathrm{d}s, \tag{2.9}$$

where f_{ij}, g, h_i *are continuous in* (x, y, t) *and* $\mathbf{C}^{\gamma}_{\mathrm{loc}}$*-functions of* (x, y).

Furthermore, the matrices $(a_{ij}(x, t)) - (f_{ij}(x, x, t))$ *are nonnegative definite for any* x *and* t.

(iii) $N(\mathrm{d}t\,\mathrm{d}T)$ *is a Poisson counting measure on* $\mathbf{V}^+$ *with an intensity measure of the form* $\mathrm{d}t\mu_t(\mathrm{d}T)$, *where* $\mathbf{V}^+$ *is the cone of positive linear operators* T; $\mathbf{C}^{m+2} \to \mathbf{C}^{\gamma}$.

We set $\tilde{N}(\mathrm{d}t\,\mathrm{d}T) = N(\mathrm{d}t\,\mathrm{d}T) - \mathrm{d}t\mu_t(\mathrm{d}T)$.

The measures $n_t(x, \cdot) - \int_{\mathbf{V}^+} \mu_t(\mathrm{d}T)T(x, \cdot)$ *are positive (nonnegative) for any* x, t, *where* $T(x, \mathrm{d}y)$ *is the kernel such that* $Tf(x) = \int T(x, \mathrm{d}y)f(y)$ *holds for* $f \in \mathbf{C}$.

The intensity measure satisfies

$$\int_0^t \int_{\mathbf{V}^+} \sup_{|x| \leqslant N} |D^{\alpha}Tf(x) - D^{\alpha}f(x)|^2 \mu_s(\mathrm{d}T)\,\mathrm{d}s \leqslant c_N \|f\|^2_{m+2}, \tag{2.10}$$

for any α *with* $|\alpha| \leqslant \gamma$. *Furthermore, we have the following:*

(1) $\{T_{s,t}\}$ *is Markovian if and only if* $c(x, t) = 0$, $G(x, t) = 0$ *and* $N(\mathrm{d}t\,\mathrm{d}T)$ *is a Poisson random measure concentrated on* $\mathbf{V}_1^+ = \{T \in \mathbf{V}^+; T1 = 1\}$.

(2) $\{T_{s,t}\}$ *is temporally homogeneous if and only if the integro-differential operator* $L(t)$ *does not depend on* t, *and the Brownian motion* $(F(x, t), G(x, t))$ *and the Poisson counting measure* $N(\mathrm{d}t\,\mathrm{d}T)$ *are temporally homogeneous.*

Remark. The Brownian motion $(F(x, t), G(x, t))$ is temporally homogeneous if and only if the functions f_{ij}, g, h_i of (2.7)–(2.9) do not depend on t. The Poisson counting measure $N(\mathrm{d}t\,\mathrm{d}T)$ is temporally homogeneous if and only if its intensity measure μ_t does not depend on t.

Next, we consider the converse problem.

THEOREM 2.2. *Suppose that a given random integro-differential operator* $A(t)$ *of* (2.5) *satisfies the following properties:*

(i) *There exists a bi-Lipschitz continuous function* $a_{ij}(x, y, t)$ *such that* $a_{ij}(x, y, t) - f_{ij}(x, y, t)$ *is nonnegative definite and* $a_{ij}(x, x, t) = a_{ij}(x, t)$ *holds.*

(ii) *There exists a family of measures* $\{\nu_t, t > 0\}$ *on* $C^2(\mathbf{R}^d; \mathbf{R}^d)$ *such that* $n_t(x, \mathrm{d}y) = \nu_t(\{v \in \mathbf{C}^2; v(x) \in \mathrm{d}y\})$.

(iii) *On the product probability space* $(\Omega \times W, P \times Q)$, *there exists a Poisson point process* $\bar{p}_s, s \geqslant 0$ *with values in* $C^2(\mathbf{R}^d; \mathbf{R}^d)$ *such that* $p_s f = E_Q[f(\bar{p}_s)]$ *defines a Poisson point process on* $\mathbf{V}$ *with the intensity measure* ν_t.

Then there exists a unique random positive semigroup $\{T_{s,t}\}$, *whose infinitesimal generator is* $A(t)$.

Proof. We consider an SDE on the product probability space $(\Omega \times W, P \times Q)$. Let $X^c(x, t)$ be a Brownian motion with spatial parameter x defined on the probability space (W, Q) with mean 0 and covariance $\int_0^t (a_{ij}(x, s) - f_{ij}(x, x, s))\,\mathrm{d}s$. Let $Y(\mathrm{d}t\,\mathrm{d}v)$ be a Poisson random measure on (W, Q) with the intensity measure ν_t.

Let $\varphi_{s,t}$ be the stochastic flow generated by the stochastic differential equation:

$$\begin{aligned}\varphi_{s,t}(x) &= x + \int_s^t F(\varphi_{s,r}(x), \mathrm{d}r) + \int_s^t X(\varphi_{s,r}(x), \mathrm{d}r) + \\ &+ \int_s^t \int_{C^2(\mathbf{R}^d;\mathbf{R}^d)} \{v(\varphi_{s,r-}) - \varphi_{s,r-}\}(Y(\mathrm{d}r\,\mathrm{d}v) - \nu_r(\mathrm{d}v)\,\mathrm{d}r) + \\ &+ \int_s^t \int_{C^2(\mathbf{R}^d;\mathbf{R}^d)} \{v(\varphi_{s,r-}) - \varphi_{s,r-}\}(\bar{N}(\mathrm{d}r\,\mathrm{d}v) - \bar{\nu}_r(\mathrm{d}v)\,\mathrm{d}r), \quad (2.11)\end{aligned}$$

where $\bar{N}(\mathrm{d}t\,\mathrm{d}v)$ is the Poisson random measure defined by $\bar{p}_t$. Set

$$\xi_{s,t}(x) = \exp\left\{\int_s^t G(\varphi_{s,r-}(x), \mathrm{d}r) + \int_s^t d(\varphi_{s,r-}(x), r)\,\mathrm{d}r\right\}, \tag{2.12}$$

where

$$d(x,t) = c(x,t) - \frac{1}{2}\sum_i \frac{\partial h_i}{\partial y_i}(x,y,t)\bigg|_{y=x} + g(x,x,t). \tag{2.13}$$

Define

$$T_{s,t}f(x) = E_Q[f(\varphi_{s,t}(x))\xi_{s,t}(x)]. \tag{2.14}$$

Then it is the required random positive semigroup.

The proof can be carried out similarly as in [3]. □

3. Measure-Valued Markov Processes Associated with Random Positive Semigroups

Let $\mathcal{M}_+(\mathbf{R}^d)$ be the set of all finite positive measures on $\mathbf{R}^d$ and let $\mathcal{M}_1(\mathbf{R}^d)$ be the set of all elements of $\mathcal{M}_+(\mathbf{R}^d)$ with total mass 1. Elements of $\mathcal{M}_+(\mathbf{R}^d)$ or $\mathcal{M}_1(\mathbf{R}^d)$ are denoted by ν, etc. The integral of the function f of $\mathbf{C}$ by the measure ν is denoted by $\nu(f)$, i.e.,

$$\nu(f) = \int_{\mathbf{R}^d} f(x)\nu(\mathrm{d}x). \tag{3.1}$$

For a given ν of $\mathcal{M}_+(\mathbf{R}^d)$, we set

$$T_{s,t}^{(\nu)}(f) = \int_{\mathbf{R}^d} T_{s,t}f(x)\nu(\mathrm{d}x), \tag{3.2}$$

where $f \in \mathbf{C}$. Then for any $s < t$ and for almost all ω, $T_{s,t}^{(\nu)}(\omega)$ can be regarded as a positive continuous linear functional on $\mathbf{C}$ such that $T_{s,t}^{(\nu)}(\omega)(1) < \infty$. Therefore $T_{s,t}^{(\nu)}(\omega)$ is an element of $\mathcal{M}_+(\mathbf{R}^d)$. Thus, $T_{s,t}^{(\nu)}, t \in [s,\infty)$ can be regarded as a stochastic process with values in $\mathcal{M}_+(\mathbf{R}^d)$ defined on $(\Omega, \mathcal{F}, P)$. In particular, if $\{T_{s,t}\}$ is Markovian, $T_{s,t}^{(\nu)}$ can be regarded as a stochastic process with values in $\mathcal{M}_1(\mathbf{R}^d)$ provided $\nu \in \mathcal{M}_1(\mathbf{R}^d)$.

The $\mathcal{M}_+(\mathbf{R}^d)$-valued process $T_{s,t}^{(\nu)}, t \geqslant s$ satisfies the following Itô stochastic differential equation

$$T_{s,t}^{(\nu)}(f) = \nu(f) + \int_s^t T_{s,r-}^{(\nu)}(A(\mathrm{d}r)f) \tag{3.3}$$

for $f \in \mathbf{C}^\infty$.

We shall show that the above $T_{s,t}^{(\nu)}$ is a Markov process with state space $\mathcal{M}_+(\mathbf{R}^d)$.

PROPOSITION 3.1. *The $\mathcal{M}_+(\mathbf{R}^d)$-valued process $T_{s,t}^{(\nu)}, t \in [s, \infty)$ is a Markov process with transition probability*

$$\Sigma_{s,t}(\nu, \Gamma) \equiv P(T_{s,t}^{(\nu)} \in \Gamma), \tag{3.4}$$

where Γ is a Borel subset of $\mathcal{M}_+(\mathbf{R}^d)$.

Proof. Since $T_{r,t}^{(\nu)} = T_{s,t}^{(T_{r,s}^{(\nu)})}$ holds for $r < s < t$ and since $T_{s,t}^{(\nu)}$ is independent of the σ-field $\sigma(A(u) - A(s); u \leqslant s)$, we have

$$P(T_{r,t}^{(\nu)} \in \Gamma | \sigma(A(u) - A(s); u \leqslant s)) = \Sigma_{s,t}(T_{r,s}^{(\nu)}, \Gamma). \tag{3.5}$$

This shows the Markov property of $T_{s,t}^{(\nu)}$. The proof is complete. □

Remark. Given a probability measure Ψ on $\mathcal{M}_+(\mathbf{R}^d)$, set

$$\nu(f) = \int_{\mathcal{M}_+(\mathbf{R}^d)} \Psi(\mathrm{d}\nu')\nu'(f) \quad \text{for } f \in \mathbf{C}.$$

Then ν can be regarded as an element of $\mathcal{M}_+(\mathbf{R}^d)$. It is called the *barycenter* of Ψ.

Let $S_{s,t}(x, \cdot)$ be the transition measure associated with the semigroup $\{S_{s,t}\}$, i.e.,

$$S_{s,t}f(x) = \int S_{s,t}(x, \mathrm{d}y)f(y). \tag{3.6}$$

Then the two measures $S_{s,t}(x, \cdot)$ and $\Sigma_{s,t}(\nu, \cdot)$ are related by

$$\int_{\mathbf{R}^d} S_{s,t}(\nu, \mathrm{d}y)f(y) = \int_{\mathcal{M}_+(\mathbf{R}^d)} \Sigma_{s,t}(\nu, \mathrm{d}\nu')\nu'(f), \quad f \in \mathbf{C}, \tag{3.7}$$

where $S_{s,t}(\nu, \mathrm{d}y) = \int \nu(\mathrm{d}x)S_{s,t}(x, \mathrm{d}y)$. Therefore $S_{s,t}(\nu, \cdot) \in \mathcal{M}_+(\mathbf{R}^d)$ is called the *barycenter* of the measure $\Sigma_{s,t}(\nu, \cdot)$ on $\mathcal{M}_+(\mathbf{R}^d)$.

In order to define the semigroup associated with the transition probability (3.4), we shall define a topology for the set $\mathcal{M}_+(\mathbf{R}^d)$ and introduce a space of continuous functions over $\mathcal{M}_+(\mathbf{R}^d)$.

Let $\overline{\mathbf{R}}^d \equiv \mathbf{R}^d \cup \{\infty\}$ be the one point compactification of $\mathbf{R}^d$. Let $\mathcal{M}_+(\overline{\mathbf{R}}^d)$ be the set of all finite measures on $\overline{\mathbf{R}}^d$ and let $\mathcal{M}_1(\overline{\mathbf{R}}^d)$ be the set of all probability measures on $\overline{\mathbf{R}}^d$. In our later discussion we often regard that any element μ of $\mathcal{M}_+(\mathbf{R}^d)$ belongs to $\mathcal{M}_+(\overline{\mathbf{R}}^d)$ by setting $\mu(\{\infty\}) = 0$. Similarly, $\mathcal{M}_1(\mathbf{R}^d)$ is regarded as a subset of $\mathcal{M}_1(\overline{\mathbf{R}}^d)$. The topology of $\mathcal{M}_+(\mathbf{R}^d)$ (or $\mathcal{M}_1(\mathbf{R}^d)$) is defined as the restriction of the weak star topology of $\mathcal{M}_+(\overline{\mathbf{R}}^d)$ (or $\mathcal{M}_1(\overline{\mathbf{R}}^d)$), i.e., it is given by the metric

$$d(\nu, \mu) = \sum_n \frac{1}{2^n} \frac{|\nu(f_n) - \mu(f_n)|}{1 + |\nu(f_n) - \mu(f_n)|}, \tag{3.8}$$

where $\{f_n\}$ is a dense subset of $\mathbf{C}$. Note that $\mathcal{M}_+(\overline{\mathbf{R}}^d)$ and $\mathcal{M}_1(\overline{\mathbf{R}}^d)$ are compact spaces but $\mathcal{M}_+(\mathbf{R}^d)$ and $\mathcal{M}_1(\mathbf{R}^d)$ are not compact ones.

We denote by $C(\mathcal{M}_+(\overline{\mathbf{R}}^d))$ (or $C(\mathcal{M}_1(\overline{\mathbf{R}}^d))$) the set of all continuous functions on $\mathcal{M}_+(\overline{\mathbf{R}}^d)$ (or $\mathcal{M}_1(\overline{\mathbf{R}}^d)$). We denote by $C(\mathcal{M}_+(\mathbf{R}^d))$ (or $C(\mathcal{M}_1(\mathbf{R}^d))$) the restriction of functions of $C(\mathcal{M}_+(\overline{\mathbf{R}}^d))$ (or $C(\mathcal{M}_1(\overline{\mathbf{R}}^d))$) to the subspace $\mathcal{M}_+(\mathbf{R}^d)$ (or $\mathcal{M}_1(\mathbf{R}^d)$).

For $F \in C(\mathcal{M}_+(\mathbf{R}^d))$, we set

$$\Sigma_{s,t} F(\nu) = \int_{\mathcal{M}_+(\mathbf{R}^d)} \Sigma_{s,t}(\nu, \mathrm{d}\nu') F(\nu'). \tag{3.9}$$

The following proposition is easily verified:

PROPOSITION 3.2. *$\Sigma_{s,t}$ maps $C(\mathcal{M}_+(\mathbf{R}^d))$ into itself. In particular, if $\{T_{s,t}\}$ is Markovian, $\Sigma_{s,t}$ maps $C(\mathcal{M}_1(\mathbf{R}^d))$ into itself. Further, the family of linear operators $\{\Sigma_{s,t}, 0 < s < t < \infty\}$ has the semigroup property.*

4. Unique Ergodicity of Temporally Homogeneous Measure-Valued Processes

In this and the next section, we assume that the random positive semigroup $\{T_{s,t}\}$ is temporally homogeneous.

Then both transition measures $S_{s,t}(x, \cdot)$ (defined by (3.7)) and $\Sigma_{s,t}(\nu, \cdot)$ (defined by (3.9)) depend on $t - s$ only. We denote them by $S_{t-s}(x, \cdot)$ and $\Sigma_{t-s}(\nu, \cdot)$, respectively. Similarly, the operators $S_{s,t}$ and $\Sigma_{s,t}$ are denoted by S_{t-s} and Σ_{t-s}, respectively. They have the semigroup properties $S_s S_t = S_{s+t}$ and $\Sigma_s \Sigma_t = \Sigma_{s+t}$ for any $s, t > 0$.

We will further assume the following condition throughout this and the next sections:

(A.3) The mean semigroup $\{S_t\}$ is Markovian, i.e., $S_t 1 = 1$ for any t.

Further, there exists a strictly positive continuous function $s_t(x, y)$ such that $S_t(x, dy) = s_t(x, y)\, dy$ holds for any t and x.

We shall study the unique ergodic property of the semigroups $\{S_t\}$ and $\{\sum_t\}$.

We shall first consider the semigroup $\{S_t\}$. Associated with this semigroup, there exists a Markov process. We shall construct it on a functional space. Let $\mathbf{D} = D([0, \infty); \mathbf{R}^d)$ be the set of all cadlag maps from $[0, \infty)$ into $\mathbf{R}^d$. For $w \in \mathbf{D}$, set $x_t(w) = w(t)$ as usual. Let $P_x, x \in \mathbf{R}^d$ be a family of Markovian probability measures on $\mathbf{D}$ such that $P_x(x_0 = x) = 1$ and

$$E_x[f(x_t)|\sigma(x_u; u \leqslant s)] = S_{t-s}f(x_s) \quad \text{a.s. } P_x \tag{4.1}$$

holds for all $s < t$ and $f \in \mathbf{C}$. The family of pairs $(x_t, P_x; x \in \mathbf{R}^d)$ is called a *Markov process associated with the semigroup* $\{S_t\}$.

We shall classify the Markov process (x_t, P_x) or its semigroup $\{S_t\}$ into recurrent and transient ones. It is called *recurrent in the sense of Harris* if

$$\int_0^\infty \chi_E(x_s)\, ds = \infty \quad \text{a.s. } P_x \tag{4.2}$$

holds for all $x \in \mathbf{R}^d$ whenever the Lebesgue measure of the Borel set E is positive. It is called *transient* if

$$\sup_{x \in \mathbf{R}^d} E_x\left[\int_0^\infty \chi_K(x_s)\, ds\right] < \infty \tag{4.3}$$

holds for any compact subset K of $\mathbf{R}^d$.

It is known that the Markov process is either recurrent in the sense of Harris or transient.

Furthermore, if it is recurrent in the sense of Harris, it has a unique (up to constant multiple) invariant Radon measure Λ. It is mutually absolutely continuous with respect to the Lebesgue measure. The process is called *positive recurrent* if it has an invariant probability measure and is called *null recurrent* if it has an infinite invariant measure.

The unique ergodic property of the semigroup $\{S_t\}$ is summarized as follows:

PROPOSITION 4.1. (1) *If the semigroup* $\{S_t\}$ *is transient or null recurrent, then*

$$\exists \lim_{t \uparrow \infty} S_t f(x) = f(\infty), \quad \forall x \in \mathbf{R}^d \tag{4.4}$$

holds for all $f \in \mathbf{C}$.

(2) *If the semigroup* $\{S_t\}$ *is positive recurrent with the invariant probability* Λ, *then*

$$\exists \lim_{t \uparrow \infty} S_t f(x) = \int_{\mathbf{R}^d} f(y) \Lambda(dy), \quad \forall x \in \mathbf{R}^d \tag{4.5}$$

holds for all $f \in \mathbf{C}$.

Proof can be found, e.g., in [6], Theorem 1.3.10.

We will next discuss the unique ergodic property of the semigroup $\{\Sigma_t\}$. Its ergodic property is closely related to the ergodic property of the corresponding mean semigroup $\{S_t\}$.

Two cases will be discussed separately. The first is the case where $T_{s,t}$ is Markovian, i.e., $T_{s,t}1 = 1$ a.s. P for any $s < t$.

THEOREM 4.2. *Assume that $\{T_{s,t}\}$ is Markovian.*

(1) *If the mean semigroup $\{S_t\}$ is transient or null recurrent, then*

$$\lim_{t\uparrow\infty} \Sigma_t F(\nu) = F(\infty), \quad \forall \nu \in \mathcal{M}_1(\mathbf{R}^d) \tag{4.6}$$

holds for any F of $C(\mathcal{M}_1(\mathbf{R}^d))$, where $F(\infty) = \lim_{\nu\to\delta_\infty} F(\nu)$. ($\delta_\infty$ is the unit measure concentrated at the point ∞.)

(2) *If the mean semigroup $\{S_t\}$ is positive recurrent with the invariant probability Λ, then $\{\Sigma_t\}$ has a unique invariant probability Ψ with barycenter Λ. Further,*

$$\exists \lim_{t\uparrow\infty} \Sigma_t F(\nu) = \int_{\mathcal{M}_1(\mathbf{R}^d)} F(\nu')\Psi(d\nu'), \quad \forall \nu \in \mathcal{M}_1(\mathbf{R}^d) \tag{4.7}$$

holds for any F of $C(\mathcal{M}_1(\mathbf{R}^d))$.

Proof. Suppose first that $\{S_t\}$ is transient or null recurrent. Then $S_t f(x)$ converges to $f(\infty)$ as $t \uparrow \infty$ by Proposition 4.1. Since $E[|T_{0,t}^{(\nu)}(f)|] \leqslant \int S_t(|f|)(y)\nu(dy)$ holds, $T_{0,t}^{(\nu)}(f)$ converges to 0 in $L^1(\Omega, \mathcal{F}, P)$ if $f(\infty) = 0$. This shows that $T_{0,t}^{(\nu)}$ converges to δ_∞ in $L^1(\Omega, \mathcal{F}, P)$ with respect to the weak star topology of $\mathcal{M}_1(\overline{\mathbf{R}}^d)$. Therefore, if $\tilde{F}$ is a continuous function on $\mathcal{M}_1(\overline{\mathbf{R}}^d)$ such that $\tilde{F} = F$ holds on $\mathcal{M}_1(\mathbf{R}^d)$, we have $\lim_{t\uparrow\infty} E[F(T_{0,t}^{(\nu)})] = \tilde{F}(\delta_\infty)$. Therefore (1) is proved.

We will next prove (2).

Let $\varphi_{s,t}(x)$ be the stochastic flow constructed in the proof of Theorem 2.2.

We can define a stationary stochastic process $\bar{x}_t, t \in (-\infty, \infty)$ with values in $\mathbf{R}^d$ on the product probability space such that $P \times Q(\bar{x}_t \in A) = \Lambda(A)$ holds for any t, $\sigma(\bar{x}_u; u \leqslant s)$ is independent of the flow $\varphi_{s,t}$ and satisfies $\bar{x}_t = \varphi_{s,t}(\bar{x}_s)$. It satisfies the same SDE (2.11). Define an $\mathcal{M}_1(\mathbf{R}^d)$-valued stochastic process $\bar{T}_t, t \in (-\infty, \infty)$ on $(\Omega, \mathcal{F}, P)$ by the partial expectation (conditional expectation) by Q:

$$\bar{T}_t(f) = E_Q[f(\bar{x}_t)]. \tag{4.8}$$

Then it is stationary Markov process with the transition probability $\Sigma_t(\nu, A)$. Consequently, $\Psi(\Gamma) \equiv P(\bar{T}_t \in \Gamma)$ is an invariant probability of Σ_t.

Since $E[\bar{T}_t(f)] = E_{P\times Q}[f(\bar{x}_t)] = \Lambda(f)$, Λ is the barycenter of Ψ.

Further, we can show, similarly as in [5], Theorem 2.6, that (4.7) holds for any convex function of $C(\mathcal{M}_1(\mathbf{R}^d))$. Now since $\mathcal{M}_1(\overline{\mathbf{R}}^d)$ is a compact space, any elements $\tilde{F}$ of $C(\mathcal{M}_1(\overline{\mathbf{R}}^d))$ can be approximated uniformly by a sequence of functions $\tilde{G}_n - \tilde{H}_n, n = 1, 2, \ldots$, where $\tilde{G}_n$, $\tilde{H}_n$ are continuous convex functions on $\mathcal{M}_1(\overline{\mathbf{R}}^d)$. (Stone–Weierstrass theorem.) Therefore (4.7) holds for any function of $C(\mathcal{M}_1(\mathbf{R}^d))$. The proof is complete. □

We shall next consider the case where $\{T_{s,t}\}$ is non-Markovian. We will see that the semigroup $\{\Sigma_t\}$ always looks null recurrent or transient.

THEOREM 4.3. *Assume that* $\{T_{s,t}\}$ *is not Markovian. Then*

$$\exists \lim_{t\uparrow\infty} \Sigma_t F(\nu) = 0, \quad \forall \nu \in \mathcal{M}_+(\mathbf{R}^d) \tag{4.9}$$

holds for all $F \in C(\mathcal{M}_+(\mathbf{R}^d))$ *such that* $\lim_{\nu\to c\delta_\infty} F(\nu) = 0$ *for any* $c \geqslant 0$.

It is enough to prove that $T_{0,t}^{(\nu)}(f)$ converges to 0 in probability as $t \uparrow \infty$ for any f of $\mathbf{C}$ with $f(\infty) = 0$.

If the mean semigroup $\{S_t\}$ is transient or null recurrent, $T_{0,t}^{(\nu)}(f)$ converges to 0 in $L^1(\Omega, \mathcal{F}, P)$, since $E[T_{0,t}^{(\nu)}(f)] = \int S_t f(x)\nu(\mathrm{d}x)$ and $S_t f(x)$ converges to 0 as $t \uparrow \infty$ by Proposition 4.1.

However, if $\{S_t\}$ is positive recurrent, the argument is not so simple as the above, because $E[T_{0,t}^{(\nu)}(f)]$ does not converge to 0 if $\Lambda(f) \neq 0$, where Λ is the invariant probability of $\{S_t\}$.

In the remainder of this section we shall prove that $T_{0,t}^{(\nu)}(f)$ converges to 0 a.s. P for any ν, though it does not converge to 0 in $L^1(\Omega, \mathcal{F}, P)$ if $\Lambda(f) \neq 0$.

For this purpose we will consider the normalized process $\tau_{s,t}^{(\nu)}$ taking values in $\mathcal{M}_1(\mathbf{R}^d)$ defined by $\tau_{s,t}^{(\nu)}(f) = T_{s,t}^{(\nu)}(f)/T_{s,t}^{(\nu)}(1)$. We will first show that it is again a Markov process.

LEMMA 4.4. *The stochastic process* $\tau_{s,t}^{(\nu)}, t \geqslant s$ *is a temporally homogeneous Markov process with values in* $\mathcal{M}_1(\mathbf{R}^d)$.

Proof. We shall prove the lemma by obtaining a stochastic differential equation governing $\tau_{s,t}^{(\nu)}$.

For simplicity, we only consider the case $s = 0$. We set $\sigma_t = T_{0,t}^{(\nu)}1$, $\tau_t = \tau_{0,t}^{(\nu)}$ and $T_t = T_{0,t}^{(\nu)}$. Then, $\tau_t = \sigma_t^{-1}T_t$. We first consider σ_t.

Since $L(t)1 = 0$, σ_t satisfies the SDE

$$\begin{aligned}\sigma_t &= 1 + \int_0^t T_{r-}(G(\mathrm{d}r)1) + \int_0^t \int_{\mathbf{V}+} T_{r-}(T1 - 1)\tilde{N}(\mathrm{d}r\,\mathrm{d}T) \\ &= 1 + \int_0^t \sigma_{r-}\tau_{r-}(G(\mathrm{d}r)1) + \int_0^t \int_{\mathbf{V}+} \sigma_{r-}\tau_{r-}(T1 - 1)\tilde{N}(\mathrm{d}r\,\mathrm{d}T)\end{aligned} \tag{4.10}$$

in view of (2.1). Therefore, it is a martingale with mean 1. Set

$$X(t) = \int_0^t \tau_{r-}(G(\mathrm{d}r)1) + \int_0^t \int_{\mathbf{V}+} \tau_{r-}(T1 - 1)\tilde{N}(\mathrm{d}r\,\mathrm{d}T). \tag{4.11}$$

It is a martingale with mean 0 and Equation (4.10) is written as

$$\sigma_t = 1 + \int_0^t \sigma_{r-}\, \mathrm{d}X(r). \tag{4.12}$$

Therefore by Dolean's formula, σ_t is represented as

$$\sigma_t = \exp\big(X(t) - \tfrac{1}{2}\langle X^c\rangle(t)\big) \prod_{s\leqslant t}(1 + \Delta X(s)) \exp -\Delta X(s), \tag{4.13}$$

where $X^c(t)$ is the continuous martingale part of $X(t)$, $\langle X^c\rangle(t)$ is its quadratic variation and $\Delta X(s) = X(s) - X(s-)$.

It holds that

$$\langle X^c\rangle(t) = \int_0^t \tau_{r-} \otimes \tau_{r-}(g)\, \mathrm{d}r, \tag{4.14}$$

where $\nu \otimes \nu$ denotes the product measure of ν and g is the function of (2.8).

We shall consider the inverse σ_t^{-1}. Note that the compensated sum is evaluated as

$$\tilde{\sum_{s\leqslant t}}\{(1 + \Delta X(s))^{-1} - 1\} = \int_0^t \int_{\mathbf{V}^+} ((\tau_{r-}T1)^{-1} - 1)\tilde{N}(\mathrm{d}r\, \mathrm{d}T).$$

Then, σ_t^{-1} satisfies the equation

$$\sigma_t^{-1} = 1 + \int_0^t \sigma_{r-}^{-1}\, \mathrm{d}Y(r), \tag{4.15}$$

where

$$\begin{aligned} Y(t) \;=\; & -\int_0^t \tau_{r-}(G(\mathrm{d}r)1) + \int_0^t \tau_{r-} \otimes \tau_{r-}(g)\, \mathrm{d}r + \\ & + \int_0^t \int_{\mathbf{V}^+} (\tau_{r-}(T1)^{-1} - \tau_{r-}(T1))\, \mathrm{d}r\,\mu(\mathrm{d}T) + \\ & + \int_0^t \int_{\mathbf{V}^+} (\tau_{r-}(T1)^{-1} - \tau_{r-}(T1))\tilde{N}(\mathrm{d}r\, \mathrm{d}T) \end{aligned} \tag{4.16}$$

and $\mathrm{d}r\,\mu(\mathrm{d}T)$ is the intensity measure of $N(\mathrm{d}r\, \mathrm{d}T)$.

We shall now obtain the SDE governing τ_t. By Itô's formula (or integration by parts formula), we have

$$\begin{aligned} \tau_t(f) \;=\; & 1 + \int_0^t \sigma_{r-}^{-1}\, \mathrm{d}T_{r-}(f) + \int_0^t \mathrm{d}\sigma_{r-}^{-1} T_{r-}(f) + [T_t(f), \sigma_t^{-1}] \\ =\; & 1 + \int_0^t \tau_{r-}(A(\mathrm{d}r)f) + \int_0^t \tau_{r-}(f)\, \mathrm{d}Y(r) + \\ & + \left[\int_0^t T_{r-}(A(\mathrm{d}r)f), \int_0^t \sigma_{r-}^{-1}\, \mathrm{d}Y(r)\right]. \end{aligned} \tag{4.17}$$

For the computation of the last term, we remark that the joint quadratic variation of $A(t)f$ and $Y(t)$ is computed as

$$[A(t)f, Y(t)] = -\int_0^t \tau_{r-}(gf)\,\mathrm{d}r - \int_0^t \tau_{r-}\left(\sum_i h_i \frac{\partial f}{\partial x_i}\right)\mathrm{d}r + \\ + \int_0^t \int_{\mathbf{V}+} (\tau_{r-}(T1)^{-1} - 1)(Tf - f)N(\mathrm{d}r\,\mathrm{d}T).$$

Then the last term of (4.17) is written as

$$-\int_0^t \tau_{r-} \otimes \tau_{r-}(gf)\,\mathrm{d}r - \int_0^t \tau_{r-} \otimes \tau_{r-}\left(\sum_i h_i \frac{\partial f}{\partial x_i}\right)\mathrm{d}r + \\ + \int_0^t \int_{\mathbf{V}+} (\tau_{r-}(T1)^{-1} - 1)\tau_{r-}(Tf - f)N(\mathrm{d}r\,\mathrm{d}T).$$

Consequently, τ_t satisfies a SDE based on a Brownian motion $(F^1, \ldots, F^d, G)$ and a Poisson random measure $N(\mathrm{d}t\,\mathrm{d}T)$. It has a unique solution. Therefore, it is a Markov process. The proof is complete. □

Now let $\theta_h, h > 0$ be shift operators acting on the sample space Ω satisfying

$$F(t)(\theta_h\omega) - F(s)(\theta_h\omega) = F(t+h)(\omega) - F(s+h)(\omega),$$
$$G(t)(\theta_h\omega) - G(s)(\theta_h\omega) = G(t+h)(\omega) - G(s+h)(\omega),$$
$$N((s,t], \mathrm{d}T)(\theta_h\omega) = N((s+h, t+h], \mathrm{d}T)(\omega), \quad \text{a.s. } P.$$

Then σ_t satisfies the multiplicative property, $\sigma_{t+s}(\omega) = \sigma_t(\omega)\sigma_s(\theta_t\omega)$ a.s. P. We shall transform the measure P by the multiplicative functional σ_t. Define a probability measure $\tilde{P}$ by

$$\tilde{P}(A) = \int_A \sigma_t\,\mathrm{d}P, \quad A \in \mathcal{F}_t, \tag{4.18}$$

where $\mathcal{F}_t = \sigma(F(s), G(s), N((0,s] \times E); s \leqslant t)$.

LEMMA 4.5. *The process* $(\tau_{0,t}^{(\nu)}, \tilde{P})$ *is also a Markov process with values in* $\mathcal{M}_1(\mathbf{R}^d)$. *Let* $\{\tilde{\Sigma}_t\}$ *be its transition probabilities. Then its barycenter coincides with the mean semigroup* S_t, *i.e.,*

$$\int \tilde{\Sigma}_t(\nu, \mathrm{d}\tilde{\nu})\tilde{\nu}(f) = \int S_t(\nu, \mathrm{d}y) f(y), \tag{4.19}$$

where $S_t(\nu, \mathrm{d}y) = \int \nu(\mathrm{d}x) S_t(x, \mathrm{d}y)$.

Proof. The Markov property of $(\tau_{0,t}^{(\nu)}, \tilde{P})$ follows from the fact that σ_t is a multiplicative functional.

Further, it satisfies

$$\tilde{E}[\tau_{0,t}^{(\nu)}(f)] = E[T_{0,t}^{(\nu)}(f)] = \nu(S_t f).$$

Then the barycenter of $\{\tilde{\Sigma}\}$ coincides with $\{S_t\}$. The proof is complete. □

LEMMA 4.6. *σ_t^{-1} is a $\tilde{P}$-martingale. Further, $\lim_{t\to\infty}\sigma_t^{-1} = 0$ a.s. $\tilde{P}$.*

Proof. Let $s < t$ and F be a bounded $\mathcal{F}_t$-measurable function. Then we have $\tilde{E}[F|\mathcal{F}_s] = E[\sigma_t F|\mathcal{F}_s]\sigma_s^{-1}$. Setting $F = \sigma_t^{-1}$, we have $\tilde{E}[\sigma_t^{-1}|\mathcal{F}_s] = \sigma_s^{-1}$, proving that σ_t^{-1} is a positive $\tilde{P}$-martingale. Therefore $\sigma_\infty^{-1} = \lim_{t\to\infty}\sigma_t^{-1}$ exists a.s. $\tilde{P}$.

Since σ_t^{-1} is governed by a linear SDE driven by $Y(t)$, the limit σ_∞^{-1} is 0 on the set

$$A = \{\langle Y^c\rangle(\infty) = \infty\} \cup \{\langle Y^d\rangle(\infty) = \infty\}, \tag{4.20}$$

where

$$\langle Y^c\rangle(t) = \int_0^t \tau_{r-} \otimes \tau_{r-}(g)\,\mathrm{d}r,$$

$$\langle Y^d\rangle(t) = \int_0^t \int_{\mathbf{V}^+} (\tau_{r-}T1 - 1)^2\,\mathrm{d}r\mu(\mathrm{d}T).$$

We shall prove $\tilde{P}(A) = 1$. Since $T_{s,t}^{(\nu)}$ is non-Markovian, we have

$$G(x,t) \neq 0 \quad \text{or} \quad \int_0^t \int_{\mathbf{V}^+} (T1(x) - 1)N(\mathrm{d}r\,\mathrm{d}T) \neq 0$$

for some x with positive probability.

Suppose $G(x,t) \neq 0$ for some x with positive probability x. Then $g(x,x) > 0$ for some x. We will prove $\langle Y^c\rangle(\infty) = \infty$ a.s. $\tilde{P}$.

We can prove, similarly as in the proof of Theorem 4.2, that $\{\tilde{\Sigma}_t\}$ has a unique invariant probability $\tilde{\Psi}$ and that its barycenter coincides with the invariant probability Λ of the semigroup $\{S_t\}$.

By the ergodic theorem of the Markov process $(\tau_t, \tilde{P})$,

$$\lim_{T\to-\infty}\frac{1}{T}\int_0^T \tau_{r-}\otimes\tau_{r-}(g)\,\mathrm{d}r = \int_{\mathcal{M}_1(\mathbf{R}^d)} \nu\otimes\nu(g)\tilde{\Psi}(\mathrm{d}\nu) \quad \text{a.s. } \tilde{P}. \tag{4.21}$$

The above is positive a.s. $\tilde{P}$.

Indeed, since g is nonnegative definite and the support of the measure Λ is the whole space, $\Lambda\otimes\Lambda(g) > 0$. Set $F(\nu) \equiv \nu\otimes\nu$. It is a convex function on $\mathcal{M}_1(\mathbf{R}^d)$. Since Λ is the barycenter of $\tilde{\Psi}$, we have $\tilde{\Psi}(F) \geqslant F(\Lambda)$ by Jensen's inequality. Therefore $\int \nu\otimes\nu(g)\tilde{\Psi}(\mathrm{d}\nu) \geqslant \Lambda\otimes\Lambda(g) > 0$ holds. We have thus proved $\langle Y^c\rangle(\infty) = \infty$ a.s. $\tilde{P}$.

Suppose next that $\iint (T1(x) - 1)N(\mathrm{d}r\,\mathrm{d}T) \neq 0$ with positive probability for some x. Then $\int_{\mathbf{V}^+}(T1(x) - 1)^2\mu(\mathrm{d}T) > 0$ holds for some x.

We shall prove $\langle Y^d\rangle(\infty) = \infty$ holds a.s. $\tilde{P}$. The ergodic theorem proves that the equality

$$\lim_{T\to\infty}\frac{1}{T}\int_0^T\int_{\mathbf{V}^+}(\tau_{r-}(T1)^{-1} - 1)^2\mu(\mathrm{d}T)\,\mathrm{d}r$$

$$= \int_{\mathcal{M}(\mathbf{R}^d)}\int_{\mathbf{V}^+}(\nu(T1)^{-1} - 1)^2\mu(\mathrm{d}T)\tilde{\Psi}(\mathrm{d}\nu) \tag{4.22}$$

holds a.s. $\tilde{P}$ and the left-hand side is positive.

We have thus proved $\langle Y^d \rangle(\infty) = \infty$ a.s. $\tilde{P}$. Therefore we have $\tilde{P}(A) = 1$. The proof is complete. □

Proof of Theorem 4.3. It is enough to prove $\sigma_\infty = \lim_{t\to\infty} \sigma_t = 0$ a.s. P. Let $P = P_a + P_s$ be the Lebesgue decomposition of the measure P with respect to $\tilde{P}$, where P_a is the absolute continuous part of P wuth respect to $\tilde{P}$. Then it holds that $P_a = \sigma_\infty^{-1} \tilde{P}$. Therefore we have $P_a \equiv 0$ by Lemma 4.6.

This proves that P and $\tilde{P}$ are mutually singular. Then the absolute continuous part $\tilde{P}_a$ of the measure $\tilde{P}$ with respect to P is again identically 0. Since it holds $\tilde{P}_a = \sigma_\infty P$, we obtain $\sigma_\infty = 0$ a.s. P. The proof is complete. □

5. Asymptotic Behavior of Temporally Homogeneous Random Positive Semigroups

Let $\{T_{s,t}\}$ be a temporally homogeneous random positive semigroup. In this section we shall discuss the asymptotic behavior of $T_{s,t} f(x)$ as t tends to infinity. We first consider the case where $\{T_{s,t}\}$ is Markovian. Let $\{S_t\}$ be the mean semigroup of $\{T_{s,t}\}$. We will see in the next theorem, that $\lim_{t\uparrow\infty} T_{s,t} f(x)$ exists in $L^p(\Omega, \mathcal{F}, P)$ for any $p \geqslant 1$ and is equal to the limit of $S_t f(x)$ if the mean semigroup $\{S_t\}$ is transient or null recurrent. However, if the mean semigroup $\{S_t\}$ is positive recurrent, $\lim_{t\uparrow\infty} T_{s,t} f(x)$ does not exist in general, but the limit of the time average $1/T \int_0^T T_{s,t} f(x)\,\mathrm{d}t$ (as $T \to \infty$) exists a.s. P and is equal to the limit of $S_t f(x)$. Indeed the existence of the limit of $T_{s,t} f(x)$ is ensured only for a special initial function.

THEOREM 5.1. *Assume that $\{T_{s,t}\}$ is Markovian.*

(1) *If the mean semigroup $\{S_t\}$ is transient or null recurrent, for any $f \in \mathbf{C}$, $T_{s,t} f(x)$ converges to $f(\infty)$ in $L^p(\Omega, \mathcal{F}, P)$ as $t \to \infty$ for any $p \geqslant 1$ for all x.*

(2) *If the mean semigroup $\{S_t\}$ is positive recurrent with the invariant probability Λ, then for any $f \in \mathbf{C}$*

$$\exists \lim_{T\to\infty} \frac{1}{T} \int_s^T T_{s,t} f(x)\,\mathrm{d}t = \Lambda(f) \quad a.s.\ P \tag{5.1}$$

and

$$\exists \lim_{t\to\infty} E[|T_{s,t} f(x) - \Lambda(f)|^p] = \int_{\mathcal{M}_1(\mathbf{R}^d)} |\nu(f) - \Lambda(f)|^p \Psi(\mathrm{d}\nu) \tag{5.2}$$

holds for all x, s, where Ψ is the invariant probability of the semigroup $\{\Sigma_t\}$ defined by (3.4).

(3) *Further, suppose that the initial function $f \in \mathbf{C}$ is twice differentiable with bounded derivatives. Then the above limit* (5.2) *is* 0 *if and only if the function f satisfies*

$$\sum_{ij} f_{ij}(x,x)\frac{\partial f}{\partial x_i}(x)\frac{\partial f}{\partial x_j}(x) = 0 \quad a.e.\ \Lambda, \tag{5.3}$$

$$\int_{\mathbf{V}^+} (Tf(x) - f(x))^2 \mu(dT) = 0 \quad a.e.\ \Lambda, \tag{5.4}$$

where (f_{ij}) is defined by (2.7).

Proof. Suppose first that the mean semigroup $\{S_t\}$ is transient or null recurrent. For a given $f \in \mathbf{C}$, set $F(\nu) = |\nu(f)|^p$. It is a function belonging to $C(\mathcal{M}_1(\mathbf{R}^d))$ and satisfies $F(\infty) = \lim_{\nu\to\delta_\infty} F(\nu) = |f(\infty)|^p$. Then $\sum_t F(\delta_x)$ converges to $|f(\infty)|^p$ as $t \uparrow \infty$ for any x. Since $E[|T_{s,t}f(x)|^p] = \Sigma_{t-s}F(\delta_x)$ holds, we get $\lim_{t\uparrow\infty} E[|T_{s,t}f(x)|^p] = 0$ if $f(\infty) = 0$. Now if $f(\infty) \neq 0$, consider $T_{s,t}\tilde{f}(x)$, where $\tilde{f}(x) = f(x) - f(\infty)$. Then it converges to 0 in $L^p(\Omega, \mathcal{F}, P)$. Therefore the assertion (1) follows.

Suppose next that $\{S_t\}$ is positive recurrent with the invariant measure Λ.

Let $\varphi_{s,t}(x)$ be the stochastic flow constructed in the proof of Theorem 2.2 and let (x_t, P_x) be the Markov process with the semigroup $\{S_t\}$. Observe the relation

$$\frac{1}{T}\int_s^T T_{s,t}f(x)\,dt = E_Q\left[\frac{1}{T}\int_s^T f(\varphi_{s,t}(x))\,dt\right].$$

Note that the law of $1/T\int_s^T f(\varphi_{s,t}(x))\,dt$ with respect to $P \times Q$ coincides with the law of $1/T\int_s^T f(x_{t-s})\,dt$ with respect to P_x. Then by the ergodic theorem,

$$\exists \lim_{T\to\infty} \frac{1}{T}\int_s^T f(\varphi_{s,t}(x))\,dt = \Lambda(f)$$

holds a.s. $P \times Q$ for any x. See Kunita [6], Theorem 1.3.12. Therefore we obtain

$$\lim_{T\to\infty} E_Q\left[\frac{1}{T}\int_s^T f(\varphi_{s,t}(x))\,dt\right] = \Lambda(f) \quad \text{a.s. } P$$

for any x, proving (5.1). On the other hand, we have

$$E[|T_{s,t}f(x) - \Lambda(f)|^p] = \Sigma_t\tilde{F}(\delta_x), \tag{5.5}$$

where $\tilde{F}(\nu) = |\nu(f) - \Lambda(f)|^p$. The right-hand side converges to

$$\int \tilde{F}(d\nu)\Psi(d\nu) = \int |\nu(f) - \Lambda(f)|^p\Psi(d\nu),$$

since $\tilde{F}$ belongs to $C(\mathcal{M}_1(\mathbf{R}^d))$. See Theorem 4.2. Therefore (5.2) is established.

It remains to prove (3). Let $\bar{T}_t, t \in (-\infty, \infty)$ be the stationary $\mathcal{M}_1(\mathbf{R}^d)$-valued process with the law Ψ defined by (4.8). Then it satisfies

$$E[|\bar{T}_t(f) - \Lambda(f)|^p] = \int_{\mathcal{M}_1(\mathbf{R}^d)} |\nu(f) - \Lambda(f)|^p \Psi(\mathrm{d}\nu). \tag{5.6}$$

Further, it satisfies a SDE

$$\bar{T}_t(f) = \bar{T}_s(f) + \int_s^t \bar{T}_r(A(\mathrm{d}r)f)\,\mathrm{d}r, \tag{5.7}$$

where

$$\begin{aligned} A(t)f(x) &= tLf(x) + \sum_i F^i(x,t)\frac{\partial f}{\partial x_i}(x) + \\ &\quad + \int_0^t \int_{\mathbf{V}+} (Tf(x) - f(x))\tilde{N}(\mathrm{d}r\,\mathrm{d}T). \end{aligned}$$

Now suppose that f satisfies (5.3) and (5.4). Since

$$\left\langle \sum_i F^i(x,t)\frac{\partial f}{\partial x_i}(x) \right\rangle = t\sum_{ij} f_{ij}(x,x)\frac{\partial f}{\partial x_i}(x)\frac{\partial f}{\partial x_j}(x) = 0 \quad \text{a.e. } \Lambda,$$

$$\begin{aligned} &\left\langle \int_0^t \int_{\mathbf{V}+} (Tf(x) - f(x))\tilde{N}(\mathrm{d}r\,\mathrm{d}T) \right\rangle \\ &\quad = t\int_{\mathbf{V}+} (Tf(x) - f(x))^2 \mu(\mathrm{d}T) = 0 \quad \text{a.e. } \Lambda \end{aligned}$$

hold by (5.3) and (5.4), we have $A(t)f = tLf$ a.e. Λ. Therefore, we have

$$\bar{T}_t(f) = \bar{T}_s(f) + \int_s^t \bar{T}_r(Lf)\,\mathrm{d}r.$$

Since $\bar{T}_t f$ is stationary, it should be equal to $\Lambda(f)$ a.s. P for any t.

This proves

$$\int_{\mathcal{M}_1(\mathbf{R}^d)} |\nu(f) - \Lambda(f)|^p \Psi(\mathrm{d}\nu) = 0 \tag{5.8}$$

by (5.6). Therefore $T_{s,t}f(x)$ converges to $\Lambda(f)$ in $L^p(\Omega, \mathcal{F}, P)$ as $t \to \infty$ in view of (5.2).

Conversely, suppose that $T_{s,t}f(x)$ converges to $\Lambda(f)$ in $L^p(\Omega, \mathcal{F}, P)$ as $t \to \infty$. Then (5.8) holds. Hence, $\bar{T}_t(f) = \Lambda(f)$ holds for any t. Since $\bar{T}_t$ is the stationary solution of (5.7), $A(t)f = tLf$. Then we have (5.3) and (5.4). The proof is complete. □

We next consider the non-Markovian case. We will see in the next theorem that the solution converges to 0 a.s., regardless of the transience, the null recurrence or the positive recurrence of the semigroup $\{S_t\}$.

THEOREM 5.2. *Assume* $G \neq 0$ *or* $\int (Tf - f)\tilde{N}(\mathrm{d}r\,\mathrm{d}T) \neq 0$. *Then* $\lim_{t\uparrow\infty} T_{s,t}f(x)$ *exists in probability for any* x. *The limit is* 0 *a.s. if either one of the following conditions is satisfied:*

(a) *The mean semigroup* $\{S_t\}$ *is transient or null recurrent and* $f(\infty) = 0$.
(b) *The mean semigroup* $\{S_t\}$ *is positive recurrent.*

Proof. We have shown in the proof of Theorem 4.3 that (a) or (b) implies that $\lim_{t\uparrow\infty} T_{s,t}f(x)$ exists and equals 0.

What is remained to show is that the existence of $\lim_{t\uparrow\infty} T_{s,t}f(x)$ in the case where $\{S_t\}$ is transient or null recurrent and $f(\infty) \neq 0$.

We consider the case $s = 0$ only. Let us consider the $\mathcal{M}_1(\mathbf{R}^d)$-valued Markov process $(\tau_{0,t}^{(\nu)}, \tilde{P})$ of Lemma 4.5. Since (4.19) holds, we can prove, similarly as in the proof of Theorem 5.1, that $\tau_{0,t}^{(\nu)}(f)$ converges to $f(\infty)$ in $L^1(\Omega, \mathcal{F}, \tilde{P})$. This implies $\tau_{0,t}^{(\nu)}(\omega)(f)$ converges to $f(\infty)$ in P-probability if $\omega \in \{\sigma_\infty > 0\}$. Since $T_{0,t}^{(\nu)}(f) = \tau_{0,t}^{(\nu)}(f)\sigma_t$, it converges to $f(\infty)\sigma_\infty$ on the set $\{\sigma_\infty > 0\}$. On the other hand, $T_{0,t}^{(\nu)}(f)$ converges to 0 on the set $\{\sigma_\infty = 0\}$. The proof is complete. □

References

1. Hashiminsky, R. Z.: Ergodic properties of recurrent diffusion processes and stabilization of the solution of the Cauchy problem for parabolic equations, *Theoret. Probab. Appl.* **5** (1960), 179–196.
2. Jacod, J.: *Calcul stochastique et problemes de martingales*, Lecture Notes in Math. 714, Springer, New York, 1979.
3. Kifer, Y. and Kunita, H.: Random positive semigroups and their infinitesimal generators, In: Davies, Truman and Elworthy (eds), *Stochastic Analysis and Applications*, Proc. Fifth Gregynog Conference 1995, World Scientific, Singapore, 1995.
4. Kunita, H.: Asymptotic behavior of the nonlinear filtering errors of Markov processes, *J. Multivariate Anal.* **1** (1971), 365–393.
5. Kunita, H.: Ergodic properties of nonlinear filtering process, In: K. S. Alexander and J. C. Watkins (eds), *Spatial Stochastic Processes*, Birkhäuser, Basel, 1991.
6. Kunita, H.: *Stochastic Flows and Stochastic Differential Equations*, Cambridge Univ. Press, 1990.
7. Kunita, H.: Infinitesimal generators of random positive semigroups, *Taiwanese J. Math.*, to appear.

Acta Applicandae Mathematicae **63:** 203–218, 2000.

Wiener–Itô Theorem in Terms of Wick Tensors *

H.-H. KUO[1], Y.-J. LEE[2] and C.-Y. SHIH[2]
[1] *Department of Mathematics, Louisiana State University, Baton Rouge, LA 70803, U.S.A.*
[2] *Department of Mathematics, Cheng Kung University, Tainan 701, Taiwan*

(Received: 8 March 1999)

Abstract. Wick tensors are used to describe homogeneous chaos and to define multiple Wiener integrals. The Wiener–Itô decomposition is expressed by the formula

$$\varphi(x) = \sum_{n=0}^{\infty} \frac{1}{n!} \int_B [D^n(\mu\varphi)(0)]^{\sim}(x+iy, \dots, x+iy)\, \mathrm{d}\mu(y).$$

We use this formula to interpret Hida's original idea of generalized multiple Wiener integrals as generalized functions acting on the space of test functions.

Mathematics Subject Classifications (2000): Primary 60H40, 28C20; Secondary 60H05, 46F25.

Key words: white noise analysis, Wiener measure, multiple Wiener integral, homogeneous chaos, Hermite polynomials, polarization identity.

1. Introduction

Let C be the space of all real-valued continuous functions x on $[0, 1]$ vanishing at 0. The space C is a Banach space with the supremum norm. The Wiener measure on C is the unique probability measure w defined on the Borel field of C such that for any $0 < t_1 < t_2 < \cdots < t_n$ and any $U \in \mathcal{B}(\mathbb{R}^n)$,

$$\begin{aligned} w&\big\{x \in C \,|\, \big(x(t_1), x(t_2), \dots, x(t_n)\big) \in U\big\} \\ &= \frac{1}{\sqrt{(2\pi)^n t_1(t_2-t_1)\cdots(t_n-t_{n-1})}} \times \\ &\quad \times \int_U \exp\left[-\frac{1}{2}\left(\frac{u_1^2}{t_1} + \frac{(u_2-u_1)^2}{t_2-t_1} + \cdots + \frac{(u_n-u_{n-1})^2}{t_n-t_{n-1}}\right)\right] \mathrm{d}u_1 \mathrm{d}u_2 \dots \mathrm{d}u_n. \end{aligned}$$

The probability space (C, w) is known as the classical Wiener space [7].

Let $L^2(w)$ denote the Hilbert space of real-valued square w-integrable functions on C. The well-known Wiener–Itô theorem states that each function φ in $L^2(w)$ can be expressed as $\varphi = \sum_{n=0}^{\infty} I_n(f_n)$ and the $L^2(w)$-norm $\|\varphi\|$ of φ is given by $\|\varphi\|^2 = \sum_{n=0}^{\infty} n! |\widetilde{f}_n|^2$, where I_n is the multiple Wiener integral of order n, $f_n \in L^2(\mathbb{R}^n)$, $\widetilde{f}_n$ is the symmetrization of f_n, and $|\cdot|$ is the $L^2(\mathbb{R}^n)$-norm.

* Research supported by the National Science Council of Taiwan.

In the work of Wiener [15] each $\varphi \in L^2(w)$ is decomposed into an orthogonal sum of homogeneous chaos, i.e., $\varphi = \sum_{n=0}^{\infty} \varphi_n$, where φ_n is the nth homogeneous chao and φ_n's are orthogonal. In [4], Itô defined multiple Wiener integrals and showed that they are the homogeneous chaos, i.e., $\varphi_n = I_n(f_n)$ and $\|\varphi_n\| = \sqrt{n!}|\widetilde{f}_n|$.

Traditionally, the homogeneous chaos have been defined in terms of orthogonal projections and the multiple Wiener integrals through special elementary functions. See the original papers by Wiener [15] and Itô [4] or the book by Kallianpur [5]. In this article we will use the Wick tensor to express homogeneous chaos and to define multiple Wiener integrals. We remark that the concept of Wick tensor stems from the work of Wiener and is related to the $\mathcal{L}^2$-lifting introduced by Kallianpur in [6]. The Wick tensor is extensively used in white noise analysis in the books by Hida *et al.* [3], Kuo [9], and Obata [14]. The definition of multiple Wiener integrals in terms of Wick tensor is elementary and transparent.

For the discussion of homogeneous chaos in Section 2 we will consider the general case by taking an abstract Wiener space [7]. In Section 3 we introduce the Wick tensor and use it to describe the homogeneous chaos and the Wiener–Itô theorem. In Section 4 we will study the Wiener–Itô expansion from the Wick tensor viewpoint. Multiple Wiener integrals in terms of Wick tensors will be discussed in Section 5.

The application to white noise analysis will be given in Section 6.

2. Homogeneous Chaos

Let (H, B) be an abstract Wiener space, i.e., H is a real separable Hilbert space with norm $|\cdot|$ and B is the completion of H with respect to a measurable norm on H. For the detail, see [7]. Let μ be standard Gaussian measure on B, i.e., for any $h \in H$,

$$\int_B e^{i\langle x,h\rangle}\, d\mu(x) = e^{-\frac{1}{2}|h|^2}.$$

Here for any h, $\langle\cdot, h\rangle$ is a random variable on B. Let $L^2(\mu)$ be the real Hilbert space of square μ-integrable functions with norm denoted by $\|\cdot\|$.

Let $F_0 = \mathbb{R}$ and for $n \geqslant 1$, let F_n be the $L^2(\mu)$-closure of the linear space spanned by 1 and the random variables $\langle\cdot, h_1\rangle \dots \langle\cdot, h_k\rangle$ for any $k \leqslant n$ and $h_j \in H$. Then clearly $\{F_n | n \geqslant 0\}$ is an increasing sequence of closed subspace of $L^2(\mu)$, i.e.,

$$F_0 \subset F_1 \subset \cdots \subset F_n \subset \cdots \subset L^2(\mu).$$

Let $G_0 = \mathbb{R}$ and for $n \geqslant 1$, let G_n be the orthogonal complement of F_{n-1} in F_n, i.e.,

$$F_n = F_{n-1} + G_n \quad \text{(orthogonal direct sum)}.$$

DEFINITION 2.1. The elements in G_n are called *homogeneous chaos* of degree n.

A cylindrical polynomial on B is a function of the form

$$f(x) = p\big(\langle x, h_1\rangle, \ldots, \langle x, h_n\rangle\big),$$

where p is a polynomial on $\mathbb{R}^n$ and $h_1, \ldots, h_n \in H$. The following theorem is a consequence of the fact that the linear span of all cylindrical polynomials on B is dense in $L^2(\mu)$.

THEOREM 2.2 (Wiener–Itô decomposition theorem). *$L^2(\mu)$ is the orthogonal direct sum of G_n, $n \geqslant 0$, $L^2(\mu) = \sum_{n=0}^{\infty} G_n$.*

The Hermite polynomial $:x^n:_{\sigma^2}$ of degree n with parameter $\sigma^2 > 0$ is defined by

$$:x^n:_{\sigma^2} = (-\sigma^2)^n e^{x^2/2\sigma^2} D_x^n e^{-x^2/2\sigma^2}. \tag{2.1}$$

We have the following generating function of Hermite polynomials

$$\exp\left[tx - \frac{1}{2}\sigma^2 t^2\right] = \sum_{n=0}^{\infty} \frac{t^n}{n!} :x^n:_{\sigma^2}. \tag{2.2}$$

Let $h \in H$ and $h \neq 0$. In Equation (2.2) put $x = \langle \cdot, h\rangle$ and $\sigma^2 = |h|^2$ to get

$$\exp\left[t\langle \cdot, h\rangle - \frac{1}{2}|h|^2 t^2\right] = \sum_{n=0}^{\infty} \frac{t^n}{n!} :\langle \cdot, h\rangle^n:_{|h|^2}. \tag{2.3}$$

From this equation we can easily check that $:\langle \cdot, h\rangle^n:_{|h|^2} \in G_n$ for all $n \geqslant 0$. Thus for any nonzero vector $h \in H$, Equation (2.3) is the Wiener–Itô decomposition of the random variable $\exp[t\langle \cdot, h\rangle - |h|^2 t^2/2]$.

3. Wick Tensors

In this section we will use Wick tensor to describe the homogeneous chaos. It is well known that the Hermite polynomial defined in Equation (2.1) can be expressed by

$$:x^n:_{\sigma^2} = \sum_{k=0}^{[n/2]} \binom{n}{2k} (2k-1)!!(-\sigma^2)^k x^{n-2k}, \tag{3.1}$$

where $(-1)!! = 1$ by convention. This equality is the motivation for the definition of Wick tensor.

DEFINITION 3.1. For $x \in B$, the Wick tensor $:x^{\otimes n}:$ is defined by

$$:x^{\otimes n}: = \sum_{k=0}^{[n/2]} \binom{n}{2k} (2k-1)!!(-1)^k \tau^{\otimes k} \widehat{\otimes} x^{\otimes(n-2k)},$$

where τ is the trace operator $\langle \tau, h^{\otimes 2} \rangle = |h|^2, \ h \in H$.

Remark. When x is a tempered distribution, the Wick tensor $:x^{\otimes n}:$ is also a tempered distribution. However, for a vector x in an abstract Wiener space B, we interpret $:x^{\otimes n}:$ as follows. For any $h \in H$,

$$\langle :x^{\otimes n}:, h^{\otimes n} \rangle = \sum_{k=0}^{[n/2]} \binom{n}{2k} (2k-1)!!(-1)^k |h|^{2k} \langle x, h \rangle^{n-2k}. \tag{3.2}$$

Recall that $\langle \cdot, h \rangle$ is a random variable defined on the probability space (B, μ). Thus $\langle :\cdot^{\otimes n}:, h^{\otimes n} \rangle$ is also a random variable.

LEMMA 3.2. *Let $h_1, \ldots, h_n \in H$. Then $\langle :x^{\otimes n}:, h_1 \otimes \cdots \otimes h_n \rangle$ is defined μ-a.e. for $x \in B$. Moreover, we have*

$$\langle :\cdot^{\otimes n}:, h_1 \otimes \cdots \otimes h_n \rangle = \langle :\cdot^{\otimes n}:, h_1 \widehat{\otimes} \cdots \widehat{\otimes} h_n \rangle, \quad \mu\text{-a.e.}$$

Proof. The assertions follow from the definition of $:x^{\otimes n}:$ and the same interpretation for Equation (3.2). □

LEMMA 3.3. *For any orthogonal vectors $h_1, \ldots, h_k \in H$ and nonnegative integers n_j's such that $n_1 + \cdots + n_k = n$, we have*

$$\langle :x^{\otimes n}:, h_1^{\otimes n_1} \widehat{\otimes} \cdots \widehat{\otimes} h_k^{\otimes n_k} \rangle = :\langle x, h_1 \rangle^{n_1}:_{|h_1|^2} \cdots :\langle x, h_k \rangle^{n_k}:_{|h_k|^2}, \quad \mu\text{-a.e.}$$

Proof. First we prove a special case, i.e., for any $h \neq 0$ in H, we have

$$\langle :x^{\otimes n}:, h^{\otimes n} \rangle = :\langle x, h \rangle^n:_{|h|^2}, \quad \mu\text{-a.e. } x \in B. \tag{3.3}$$

To prove this equality, just use Equations (3.1) and (3.2). For the general case, we may assume that the vectors h_j are nonzero. Apply Equation (2.3) to each j and then take the product to get

$$\prod_{j=1}^{k} \mathrm{e}^{t_j \langle x, h_j \rangle - \frac{1}{2}|h_j|^2 t_j^2} = \prod_{j=1}^{k} \sum_{n_j=0}^{\infty} \frac{t_j^{n_j}}{n_j!} :\langle x, h_j \rangle^{n_j}:_{|h_j|^2}. \tag{3.4}$$

Since the vectors h_j's are orthogonal, we can easily check that

$$\begin{aligned} \prod_{j=1}^{k} \mathrm{e}^{t_j \langle x, h_j \rangle - \frac{1}{2}|h_j|^2 t_j^2} &= \mathrm{e}^{\langle x, t_1 h_1 + \cdots + t_k h_k \rangle - \frac{1}{2}|t_1 h_1 + \cdots + t_k h_k|^2} \\ &= \sum_{n=0}^{\infty} \frac{1}{n!} \big\langle :x^{\otimes n}:, (t_1 h_1 + \cdots + t_k h_k)^{\otimes n} \big\rangle. \end{aligned} \tag{3.5}$$

By comparing the coefficients of $t_1^{n_1} \cdots t_k^{n_k}$ in the right-hand sides of Equations (3.4) and (3.5), we get the conclusion of the lemma. □

THEOREM 3.4. *For any $h \in H$ and $n \geqslant 0$, we have $\langle :\cdot^{\otimes n}:, h^{\otimes n}\rangle \in G_n$ and*

$$\exp\left[t\langle x, h\rangle - \frac{1}{2}|h|^2 t^2\right] = \sum_{n=0}^{\infty} \frac{t^n}{n!}\langle :x^{\otimes n}:, h^{\otimes n}\rangle, \quad \mu\text{-a.e. } x \in B. \tag{3.6}$$

Proof. The conclusion is trivial if $h = 0$. So let $h \neq 0$. Recall that $:\langle\cdot, h\rangle^n:_{|h|^2} \in G_n$. Hence, the conclusion follows from Equations (3.3) and (2.3). □

LEMMA 3.5. *For any $u, v \in H$, we have*

$$\int_B \langle :x^{\otimes n}:, u^{\otimes n}\rangle \langle :x^{\otimes n}:, v^{\otimes n}\rangle \, d\mu(x) = n!(u, v)^n, \tag{3.7}$$

where $(\cdot, \cdot)$ is the inner product on H.

Proof. Apply Equation (3.6) to get

$$e^{t\langle x,u\rangle - \frac{1}{2}|u|t^2} = \sum_{n=0}^{\infty} \frac{t^n}{n!}\langle :x^{\otimes n}:, u^{\otimes n}\rangle,$$

$$e^{s\langle x,v\rangle - \frac{1}{2}|v|s^2} = \sum_{m=0}^{\infty} \frac{s^m}{m!}\langle :x^{\otimes m}:, v^{\otimes m}\rangle.$$

Multiply these two equations together to get

$$e^{t\langle x,u\rangle + s\langle x,v\rangle - \frac{1}{2}|u|t^2 - \frac{1}{2}|v|s^2} = \sum_{n,m=0}^{\infty} \frac{t^n s^m}{n!m!}\langle :x^{\otimes n}:, u^{\otimes n}\rangle \langle :x^{\otimes m}:, v^{\otimes m}\rangle. \tag{3.8}$$

But it is easy to check that

$$\int_B e^{t\langle x,u\rangle + s\langle x,v\rangle - \frac{1}{2}|u|t^2 - \frac{1}{2}|v|s^2} \, d\mu(x) = e^{ts\langle u,v\rangle}. \tag{3.9}$$

Thus, we can integrate both sides of Equation (3.8) and then compare the coefficients of $t^n s^n$ to get Equation (3.7). □

LEMMA 3.6. *Let $h_1, \ldots, h_n \in H$. Then $\langle :\cdot^{\otimes n}:, h_1 \otimes \cdots \otimes h_n\rangle \in G_n$. Moreover, for any $u_j, v_j \in H$, $j = 1, \ldots, n$, we have*

$$\int_B \langle :x^{\otimes n}:, u_1 \otimes \cdots \otimes u_n\rangle \langle :x^{\otimes n}:, v_1 \otimes \cdots \otimes v_n\rangle \, d\mu(x) = n!(u_1 \widehat{\otimes} \cdots \widehat{\otimes} u_n, v_1 \widehat{\otimes} \cdots \widehat{\otimes} v_n), \tag{3.10}$$

where $(\cdot, \cdot)$ is the inner product on $H^{\otimes n}$ induced by the inner product on H.

Proof. By the polarization identity on page 358 in [9]

$$\begin{aligned}&\langle :x^{\otimes n}:, h_1 \widehat{\otimes} \cdots \widehat{\otimes} h_n\rangle \\ &\quad = \frac{1}{2^n n!} \sum \epsilon_1 \cdots \epsilon_n \big\langle :x^{\otimes n}:, (\epsilon_1 h_1 + \cdots + \epsilon_n h_n)^{\otimes n}\big\rangle,\end{aligned} \tag{3.11}$$

where the summation is over all $\epsilon_1 = \pm 1, \ldots, \epsilon_n = \pm 1$. Therefore, by Theorem 3.4 we have $\langle :\cdot^{\otimes n}:, h_1 \widehat{\otimes} \cdots \widehat{\otimes} h_n\rangle \in G_n$. But then by Lemma 3.2 we also have $\langle :\cdot^{\otimes n}:, h_1 \otimes \cdots \otimes h_n\rangle \in G_n$.

Now we can apply Equation (3.11) and Lemma 3.5 to derive that

$$\begin{aligned}&\int_B \langle :x^{\otimes n}:, u_1 \widehat{\otimes} \cdots \widehat{\otimes} u_n\rangle \langle :x^{\otimes n}:, v_1 \widehat{\otimes} \cdots \widehat{\otimes} v_n\rangle \, d\mu(x) \\ &\quad = \frac{1}{(2^n n!)^2} \sum_{\epsilon} \sum_{\delta} \epsilon_1 \cdots \epsilon_n \delta_1 \cdots \delta_n n! \times \\ &\quad\quad \times (\epsilon_1 u_1 + \cdots + \epsilon_n u_n, \delta_1 v_1 + \cdots + \delta_n v_n)^n \\ &\quad = n! \frac{1}{(2^n n!)^2} \sum_{\epsilon} \sum_{\delta} \epsilon_1 \cdots \epsilon_n \delta_1 \cdots \delta_n \bigg(\sum_{i,j} \epsilon_i \delta_j (u_i, v_j) \bigg)^n .\end{aligned} \tag{3.12}$$

But by the polarization identity, we have

$$\begin{aligned}&\sum_{\delta} \delta_1 \cdots \delta_n \bigg(\sum_{i,j} \epsilon_i \delta_j (u_i, v_j) \bigg)^n \\ &\quad = 2^n n! \bigg(\sum_i \epsilon_i u_i, v_1 \bigg) \cdots \bigg(\sum_i \epsilon_i u_i, v_1 \bigg).\end{aligned} \tag{3.13}$$

Apply again the polarization identity to get

$$\begin{aligned}&\sum_{\epsilon} \epsilon_1 \cdots \epsilon_n \bigg(\sum_i \epsilon_i u_i, v_1 \bigg) \cdots \bigg(\sum_i \epsilon_i u_i, v_1 \bigg) \\ &\quad = 2^n n! (u_1 \otimes \cdots \otimes u_n, v_1 \widehat{\otimes} \cdots \widehat{\otimes} v_n).\end{aligned} \tag{3.14}$$

Therefore, from Equations (3.12), (3.13), and (3.14),

$$\begin{aligned}&\int_B \langle :x^{\otimes n}:, u_1 \widehat{\otimes} \cdots \widehat{\otimes} u_n\rangle \langle :x^{\otimes n}:, v_1 \widehat{\otimes} \cdots \widehat{\otimes} v_n\rangle \, d\mu(x) \\ &\quad = n!(u_1 \otimes \cdots \otimes u_n, v_1 \widehat{\otimes} \cdots \widehat{\otimes} v_n) = n!(u_1 \widehat{\otimes} \cdots \widehat{\otimes} u_n, v_1 \widehat{\otimes} \cdots \widehat{\otimes} v_n).\end{aligned}$$

This equality and Lemma 3.2 imply the equality in Equation (3.10). □

THEOREM 3.7. *For any $u, v \in H^{\otimes n}$, we have $\langle :\cdot^{\otimes n}:, u\rangle \in G_n$ and*

$$\int_B \langle :x^{\otimes n}:, u\rangle \langle :x^{\otimes n}:, v\rangle \, d\mu(x) = n!(\widetilde{u}, \widetilde{v}), \tag{3.15}$$

where $\widetilde{w}$ denotes the symmetrization of $w \in H^{\otimes n}$.

Proof. Let $\{e_1, e_2, \ldots\}$ be an orthonormal basis for H. Then the set

$$\left\{e_1^{\otimes n_1} \widehat{\otimes} e_2^{\otimes n_2} \widehat{\otimes} \cdots \mid n_j \geqslant 0, n_1 + n_2 + \cdots = n\right\}$$

is a complete orthogonal systems for $H^{\widehat{\otimes} n}$. We can use this fact and Lemma 3.6 to show that $\langle :\cdot^{\otimes n}:, u\rangle \in G_n$ for any $u \in H^{\otimes n}$ and Equation (3.15) holds for any $u, v \in H^{\widehat{\otimes} n}$. But by Lemma 3.2,

$$\langle :\cdot^{\otimes n}, w\rangle = \langle :\cdot^{\otimes n}, \widetilde{w}\rangle, \qquad \mu\text{-a.e.}$$

for any $w \in H^{\otimes n}$. Hence, Equation (3.15) holds for any $u, v \in H^{\otimes n}$. □

THEOREM 3.8. *Suppose $\varphi \in G_n$. Then there exists $u \in H^{\widehat{\otimes} n}$ such that*

$$\varphi(x) = \langle :x^{\otimes n}:, u\rangle, \qquad \mu\text{-a.e.}$$

Proof. Let $\{e_1, e_2, \ldots\}$ be an orthonormal basis for H. For any nonnegative integers n_j's such that $n_1 + n_2 + \cdots = n$, let

$$\psi_{n_1, n_2, \ldots} = :\langle\cdot, e_1\rangle^{n_1}:_1:\langle\cdot, e_2\rangle^{n_2}:_1 \cdots.$$

Recall that $:\langle\cdot, h\rangle^n:_{|h|^2} \in G_n$ for any nonzero vector h in H. Hence $\psi_{n_1, n_2, \ldots} \in G_n$. And then it follows from the definition of G_n that the following set

$$\{\psi_{n_1, n_2, \ldots} \mid n_j \geqslant 0, n_1 + n_2 + \cdots = n\}$$

is a complete orthogonal system for G_n. Thus if $\varphi \in G_n$, then

$$\varphi = \sum \alpha_{n_1, n_2, \ldots} \psi_{n_1, n_2, \ldots},$$

where the summation is over all nonnegative integers n_j such that $n_1 + n_2 + \cdots = n$. But by Lemma 3.3

$$\psi_{n_1, n_2, \ldots} = \left\langle :\cdot^{\otimes n}:, e_1^{\otimes n_1} \widehat{\otimes} e_2^{\otimes n_2} \widehat{\otimes} \cdots \right\rangle.$$

Hence, $\varphi = \langle :\cdot^{\otimes n}:, u\rangle$ with $u = \sum \alpha_{n_1, n_2, \ldots} e_1^{\otimes n_1} \widehat{\otimes} e_2^{\otimes n_2} \widehat{\otimes} \cdots \in H^{\widehat{\otimes} n}$. □

THEOREM 3.9 (Wiener–Itô theorem). *Let $\varphi \in L^2(\mu)$. Then there exist $u_n \in H^{\widehat{\otimes} n}$ for all $n \geqslant 0$ such that*

$$\varphi(x) = \sum_{n=0}^{\infty} \langle :x^{\otimes n}:, u_n\rangle, \qquad \mu\text{-a.e. } x \in B,$$

and the $L^2(\mu)$-norm of φ is given by

$$\|\varphi\|^2 = \sum_{n=0}^{\infty} n! |u_n|^2.$$

Proof. Just apply Theorems 2.2, 3.7, and 3.8. □

At the end of this section we give orthonormal bases for G_n and $L^2(\mu)$. Let $\psi_{n_1,n_2,\dots}$ be defined as in the proof of Theorem 3.8, i.e.,

$$\psi_{n_1,n_2,\dots} = :\langle\cdot, e_1\rangle^{n_1}:_1:\langle\cdot, e_2\rangle^{n_2}:_1 \cdots = \left\langle :\cdot^{\otimes n}:, e_1^{\otimes n_1} \widehat{\otimes} e_2^{\otimes n_2} \widehat{\otimes} \cdots \right\rangle.$$

It is easy to check that

$$\int_B \psi_{n_1,n_2,\dots}(x)^2 \, \mathrm{d}\mu(x) = n_1!n_2!\cdots.$$

Therefore, the set

$$\left\{ \frac{1}{\sqrt{n_1!n_2!\cdots}} \psi_{n_1,n_2,\dots} \mid n_j \geqslant 0, n_1 + n_2 + \cdots = n \right\}$$

is an orthonormal basis for G_n and the set

$$\left\{ \frac{1}{\sqrt{n_1!n_2!\cdots}} \psi_{n_1,n_2,\dots} \mid n_j \geqslant 0, n_1 + n_2 + \cdots = n, n \geqslant 0 \right\}$$

is an orthonormal basis for $L^2(\mu)$.

4. Wiener–Itô Expansion

In this section we will discuss the Wiener–Itô expansion of a function φ in $L^2(\mu)$. In particular, we will use Wick tensors to give a simple proof of a formula due to Lee [11].

First recall that (H, B) is a fixed abstract Wiener space. An n-linear map $T\colon H \times \cdots \times H \to \mathbb{R}$ is said to be of *Hilbert–Schmidt type* (see [7, 8]) if there exists an orthonormal basis $\{e_n \mid n \geqslant 1\}$ for H such that

$$\|T\|_2^2 = \sum_{i_1,\dots,i_n} T(e_{i_1}, \dots, e_{i_n})^2 < \infty.$$

Let $\mathcal{L}^n_{(2)}(H)$ be the set of all Hilbert–Schmidt type n-linear maps on H. It is a Hilbert space with the inner product

$$\langle\langle S, T\rangle\rangle = \sum_{i_1,\dots,i_n} S(e_{i_1}, \dots, e_{i_n}) T(e_{i_1}, \dots, e_{i_n}).$$

THEOREM 4.1. *Let $\varphi \in L^2(\mu)$. Then the n-linear map*

$$T(h_1, \dots, h_n) = \int_B \varphi(x) \langle :x^{\otimes n}:, h_1 \otimes \cdots \otimes h_n\rangle \, \mathrm{d}\mu(x), \qquad h_1, \dots, h_n \in H,$$

is of Hilbert–Schmidt type and

$$\|T\|_2 \leqslant \sqrt{n!}\|\varphi\|.$$

Proof. By Theorem 3.9 φ can be represented as

$$\varphi = \sum_{k=0}^{\infty} \langle :\cdot^{\otimes k}:, u_k \rangle,$$

where $u_k \in H^{\widehat{\otimes} k}$. By Lemma 3.6, $\langle :\cdot^{\otimes n}:, h_1 \otimes \cdots \otimes h_n \rangle \in G_n$. Hence by the orthogonality of the G_n's,

$$T(h_1, \ldots, h_n) = \int_B \langle :x^{\otimes n}:, u_n \rangle \langle :x^{\otimes n}:, h_1 \otimes \cdots \otimes h_n \rangle \, \mathrm{d}\mu(x).$$

Therefore, by Theorem 3.7,

$$T(h_1, \ldots, h_n) = n!(u_n, h_1 \widehat{\otimes} \cdots \widehat{\otimes} h_n).$$

Let $\{e_n \mid n \geqslant 1\}$ be an orthonormal basis for H. Then by the last equality and Theorem 3.7,

$$\begin{aligned}
\sum_{i_1,\ldots,i_n} T(e_{i_1}, \ldots, e_{i_n})^2 &= (n!)^2 \sum_{i_1,\ldots,i_n} (u_n, e_{i_1} \widehat{\otimes} \cdots \widehat{\otimes} e_{i_n})^2 \\
&= (n!)^2 |u_n|^2 \\
&= n! \| \langle :\cdot^{\otimes n}:, u_n \rangle \|^2 \\
&\leqslant n! \|\varphi\|^2.
\end{aligned}$$

This shows that T is of Hilbert–Schmidt type and $\|T\|_2 \leqslant \sqrt{n!}\|\varphi\|$. □

THEOREM 4.2. *Let $\varphi \in L^2(\mu)$. Then the function defined by*

$$(\mu\varphi)(h) = \int_B \varphi(x + h) \, \mathrm{d}\mu(x), \quad h \in H,$$

is infinitely Fréchet differentiable on H with nth *derivative at* 0 *given by*

$$D^n(\mu\varphi)(0)(h_1, \ldots, h_n) = \int_B \varphi(x) \langle :x^{\otimes n}:, h_1 \otimes \cdots \otimes h_n \rangle \, \mathrm{d}\mu(x). \tag{4.1}$$

Moreover, $D^n(\mu\varphi)(0) \in \mathcal{L}^n_{(2)}(H)$ for all $n \geqslant 1$ and

$$\|D^n(\mu\varphi)(0)\|_2 \leqslant \sqrt{n!}\|\varphi\|.$$

Proof. Let μ_h be the translation of μ by $h \in H$, i.e., $\mu_h(\cdot) = \mu(\cdot - h)$. Then by the translation formula (see Kuo [7]) we have

$$\frac{\mathrm{d}\mu_h}{\mathrm{d}\mu}(x) = \exp\left[\langle x, h \rangle - \frac{1}{2}|h|^2\right].$$

Hence the function $\mu\varphi$ on H can be rewritten as

$$(\mu\varphi)(h) = \int_B \varphi(x) \exp\left[\langle x, h \rangle - \frac{1}{2}|h|^2\right] \mathrm{d}\mu(x).$$

Then by using Equation (3.6) we get

$$(\mu\varphi)(h) = \sum_{n=0}^{\infty} \frac{1}{n!} \int_B \varphi(x) \langle :x^{\otimes n}:, h^{\otimes n} \rangle \, d\mu(x). \tag{4.2}$$

Obviously, this is the Taylor expansion of the function $\mu\varphi$ on H. Hence $\mu\varphi$ is infinitely Fréchet differentiable and its nth derivative $D^n(\mu\varphi)(0)$ at 0 is given by Equation (4.1). Then by Theorem 4.1 $D^n(\mu\varphi)(0)$ is of Hilbert–Schmidt type and $\|D^n(\mu\varphi)(0)\|_2 \leqslant \sqrt{n!}\|\varphi\|$. □

LEMMA 4.3. *For any $u \in H^{\otimes n}$, we have*

$$\int_B \langle (x+iy)^{\otimes n}, u \rangle \, d\mu(y) = \langle :x^{\otimes n}:, u \rangle. \tag{4.3}$$

Proof. By the polarization identity it suffices to prove Equation (4.3) for the special case $u = h^{\otimes n}$, $h \in H$. We may assume that $|h| = 1$. By the second Fourier–Gauss transform formula on page 355 of Ref. [9]

$$\begin{aligned} \int_B \langle (x+iy)^{\otimes n}, h^{\otimes n} \rangle \, d\mu(y) &= \int_B \big(\langle x, h\rangle + i\langle y, h\rangle\big)^n d\mu(y) \\ &= :\langle x, h\rangle^n:_1 \\ &= \langle :x^{\otimes n}:, h^{\otimes n} \rangle. \end{aligned}$$

This proves the lemma. □

Now, let the Wiener–Itô expansion of $\varphi \in L^2(\mu)$ be given as in Theorem 3.9 by

$$\varphi(x) = \sum_{k=0}^{\infty} \langle :x^{\otimes k}:, u_k \rangle, \quad u_k \in H^{\widehat{\otimes} k}.$$

Then it follows from Equation (4.1) that

$$D^n(\mu\varphi)(0)(h_1, \ldots, h_n) = \int_B \langle :x^{\otimes n}:, u_n \rangle \langle :x^{\otimes n}:, h_1 \otimes \cdots \otimes h_n \rangle \, d\mu(x).$$

Hence by Theorem 3.7, we get

$$D^n(\mu\varphi)(0)(h_1, \ldots, h_n) = n!(h_1 \otimes \cdots \otimes h_n, u_n).$$

Observe from this equation that the n-linear map $D^n(\mu\varphi)(0)$ on H extends to a random variable $[D^n(\mu\varphi)(0)]^\sim$ on B^n defined by

$$[D^n(\mu\varphi)(0)]^\sim(x_1, \ldots, x_n) = n!\langle x_1 \otimes \cdots \otimes x_n, u_n \rangle.$$

In particular, when $x_1 = \cdots = x_n = x$, we have the following random variable defined on B

$$[D^n(\mu\varphi)(0)]^\sim(x, \ldots, x) = n!\langle x^{\otimes n}, u_n \rangle. \tag{4.4}$$

The next theorem has been proved by Lee [11]. We will use Wick tensors to give a simple proof.

THEOREM 4.4. *Let $\varphi \in L^2(\mu)$. Then the Wiener–Itô expansion of φ is given by*

$$\varphi(x) = \sum_{n=0}^{\infty} \frac{1}{n!} \int_B [D^n(\mu\varphi)(0)]^{\sim}(x + iy, \ldots, x + iy)\, d\mu(y). \tag{4.5}$$

Proof. Suppose the Wiener–Itô expansion of φ is given by

$$\varphi(x) = \sum_{n=0}^{\infty} \langle :x^{\otimes n}:, u_n \rangle. \tag{4.6}$$

It follows from Equation (4.4) and Lemma 4.3 that

$$\int_B [D^n(\mu\varphi)(0)]^{\sim}(x + iy, \ldots, x + iy)\, d\mu(y) = n! \langle :x^{\otimes n}:, u_n \rangle. \tag{4.7}$$

Obviously, Equations (4.6)–(4.7) imply Equation (4.5). □

5. Multiple Wiener Integrals

We now take the abstract Wiener space B in Sections 2–4 to be the classical Wiener space C in Section 1. The corresponding Hilbert space for H is the space C' consisting of all absolutely continuous functions g on $[0, 1]$ with $g' \in L^2([0, 1])$. The inner product in C' is given by $(g, h) = \int_0^1 g'(t)h'(t)\, dt$ (see Kuo [7]).

Define $B(t, x) = \langle x, 1_{[0,t)} \rangle$, $t \in [0, 1]$, $x \in C$. Then $B(t)$ is a Brownian motion. In this section we will show that multiple Wiener integrals with respect to $B(t)$ can be expressed in terms of Wick tensors. Recall that Itô [4] defined a multiple Wiener integral of order n

$$I_n(f) = \int_{[0,1]^n} f(t_1, \ldots, t_n)\, dB(t_1) \cdots dB(t_n) \tag{5.1}$$

for any $f \in L^2([0, 1]^n)$. This multiple Wiener integral $I_n(f)$ is first defined for special elementary functions and then for $f \in L^2([0, 1]^n)$ by approximation. It has the following properties:

(P-1) $I_n(f) = I_n(\widetilde{f})$, where $\widetilde{f}$ is the symmetrization of f.

(P-2) $E|I_n(f)|^2 = n!|\widetilde{f}|^2$.

(P-3) $EI_m(f)I_n(g) = 0$ if $m \neq n$.

(P-4) For any orthogonal functions $f_1, \ldots, f_k$ in $L^2([0, 1])$ and nonnegative integers n_j's with $n_1 + \cdots + n_k = n$, we have

$$I_n\big(f_1^{\otimes n_1} \otimes \cdots \otimes f_k^{\otimes n_k}\big) = H_{n_1}\big(|f_1|^2; \textstyle\int_0^1 f_1(t)\, dB(t)\big) \cdots H_{n_k}\big(|f_k|^2; \int_0^1 f_k(t)\, dB(t)\big), \tag{5.2}$$

where $H_n(\sigma^2; x)$ is the Hermite polynomial of degree n with parameter σ^2, i.e., $H_n(\sigma^2; x) = :x^n:_{\sigma^2}$ (see Equation (2.1)).

(P-5) For any symmetric function $f \in L^2([0,1]^n)$,

$$I_n(f) = n! \int_0^1 \int_0^{t_n} \cdots \int_0^{t_2} f(t_1, t_2, \ldots, t_n)\, \mathrm{d}B(t_1) \cdots \mathrm{d}B(t_n),$$

where the right-hand side is an iterated Itô integral.

(P-6) For any $\varphi \in L^2(w)$ (w: the Wiener measure on C) there exists a sequence $\{f_n\}$ of symmetric functions, $f_n \in L^2([0,1]^n)$ such that

$$\varphi = \sum_{n=0}^{\infty} I_n(f_n).$$

In fact, the multiple Wiener integral I_n of order n is the unique linear map from $L^2([0,1]^n)$ into $L^2(w)$ such that properties (P-1) and (P-4) above are satisfied. It is related to the Wick tensor of order n by the next theorem.

THEOREM 5.1. *For any* $f \in L^2([0,1]^n)$, *we have*

$$I_n(f) = \langle :\cdot^{\otimes n}:, f \rangle, \quad w\text{-a.e.}$$

Proof. First note that $\langle \cdot, f \rangle = \int_0^1 f(t)\, \mathrm{d}B(t)$. Define a linear map J_n from $L^2([0,1]^n)$ into $L^2(w)$ by $J_n(f) = \langle :\cdot^{\otimes n}:, f \rangle$. Then Lemmas 3.2 and 3.3 imply that the linear map J_n has the same properties (P-1) and (P-4) as I_n. Hence, by the remark preceding this theorem, we have $I_n = J_n$ and the conclusion follows. □

By the above theorem we see that properties (P-1), (P-2), and (P-4) follow respectively from Lemma 3.2, Theorem 3.7, and Lemma 3.3. Property (P-3) follows from Theorem 3.7 and the orthogonality of the G_n's. Property (P-6) is the Wiener–Itô decomposition in Theorem 3.9. Below we give a simple proof of property (P-5) by using Wick tensors.

THEOREM 5.2. *For any symmetric function* $f \in L^2([0,1]^n)$,

$$I_n(f) = n! \int_0^1 \int_0^{t_n} \cdots \int_0^{t_2} f(t_1, t_2, \ldots, t_n)\, \mathrm{d}B(t_1) \cdots \mathrm{d}B(t_n), \tag{5.3}$$

where the right-hand side is an iterated Itô integral.

Proof. By the polarization identity on page 358 in [9] it suffices to prove Equation (5.3) for the case $f = g^{\otimes n}$, $g \in L^2([0,1])$. We show this assertion by induction. So let $g \in L^2([0,1])$ and assume that Equation (5.3) holds for $n-1$ and any f of the form $f = h^{\otimes(n-1)}$, $h \in L^2([0,1])$. By taking $h = g \cdot 1_{[0,t)}$ we get

$$I_{n-1}\big((g \cdot 1_{[0,t)})^{\otimes(n-1)}\big) = (n-1)! \int_0^t \int_0^{t_{n-1}} \cdots \int_0^{t_2} g(t_1) g(t_2) \cdots g(t_{n-1})\, \mathrm{d}B(t_1) \cdots \mathrm{d}B(t_{n-1}). \tag{5.4}$$

Now, let $\theta(t) = \int_0^t g(s)^2\,\mathrm{d}s$ and $X(t) = \int_0^t g(s)\,\mathrm{d}B(s)$. It is well known that the Hermite polynomial $H_n(t; x)$ satisfies the following identities

$$\frac{\partial}{\partial t} H_n(t; x) = -\frac{1}{2}\frac{\partial^2}{\partial x^2} H_n(t; x), \qquad \frac{\partial}{\partial x} H_n(t; x) = n H_{n-1}(t; x).$$

With these two identities, we can apply Itô's formula to the function $H_n(\theta(t); X(t))$ to get

$$H_n(\theta(t); X(t)) = \int_0^t n g(s) H_{n-1}(\theta(s); X(s))\,\mathrm{d}B(s). \tag{5.5}$$

Therefore, by Equations (5.4)–(5.5) and property (P-4),

$$\begin{aligned}
&\int_0^1 \int_0^{t_n} \cdots \int_0^{t_2} g(t_1)g(t_2)\cdots g(t_n)\,\mathrm{d}B(t_1)\cdots \mathrm{d}B(t_n) \\
&= \int_0^1 g(s)\frac{1}{(n-1)!} I_{n-1}\big((g \cdot 1_{[0,s)})^{\otimes(n-1)}\big)\,\mathrm{d}B(s) \\
&= \frac{1}{(n-1)!}\int_0^1 g(s) H_{n-1}(\theta(s); X(s))\,\mathrm{d}B(s) \\
&= \frac{1}{n!} H_n(\theta(1); X(1)) \\
&= \frac{1}{n!} I_n(g^{\otimes n}).
\end{aligned}$$

Hence, we have proved Equation (5.3) for n and any f of the form $f = g^{\otimes n}$, $g \in L^2([0, 1])$. This completes the proof by induction. □

The Wick tensor as defined in Section 3 satisfies the following recursion formula:

$$:x^{\otimes 1}: = x, \quad :x^{\otimes(n+1)}: = x \mathbin{\widehat{\otimes}} :x^{\otimes n}: - n :x^{\otimes(n-1)}: \mathbin{\widehat{\otimes}} \tau, \quad n \geqslant 1, \tag{5.6}$$

where $:x^{\otimes 0}: = 1$ by convention. This formula provides an intrinsic proof of the next theorem in [4].

THEOREM 5.3. *For any $f \in L^2([0, 1])$ and $g \in L^2([0, 1]^n)$, we have*

$$I_{n+1}(f \otimes g) = I_1(f)I_n(g) - \sum_{k=1}^{n} I_{n-1}(f \otimes_k g), \tag{5.7}$$

where $f \otimes_k g$ is the function on $[0, 1]^{n-1}$ defined by

$$f \otimes_k g(t_1, \ldots, t_{n-1}) = \int_0^1 f(t)g(t_1, \ldots, t_{k-1}, t, t_k, \ldots, t_{n-1})\,\mathrm{d}t.$$

Proof. By Theorem 5.1 we have

$$I_{n+1}(f \otimes g)(x) = \langle :x^{\otimes(n+1)}:, f \otimes g \rangle. \tag{5.8}$$

Note that

$$\begin{aligned} & f \mathbin{\widehat{\otimes}} g(t_1, t_2, \ldots, t_{n+1}) \\ & \quad = \frac{1}{n+1}\big(f(t_1)\widetilde{g}(t_2, t_3, \ldots, t_{n+1}) + \\ & \qquad + f(t_2)\widetilde{g}(t_1, t_3, \ldots, t_{n+1}) + \cdots + f(t_{n+1})\widetilde{g}(t_1, t_2, \ldots, t_n)\big), \end{aligned} \tag{5.9}$$

where $\widetilde{g}$ is the symmetrization of g. By using Equation (5.9) we can easily check that

$$\begin{aligned} \langle x \mathbin{\widehat{\otimes}} :x^{\otimes n}:, f \otimes g \rangle &= \langle x \otimes :x^{\otimes n}:, f \mathbin{\widehat{\otimes}} g \rangle \\ &= \langle x, f \rangle \langle :x^{\otimes n}:, g \rangle \\ &= I_1(f)(x) I_n(g)(x). \end{aligned} \tag{5.10}$$

On the other hand, we again use Equation (5.9) to get

$$\begin{aligned} \langle :x^{\otimes(n-1)}: \mathbin{\widehat{\otimes}} \tau, f \otimes g \rangle &= \langle :x^{\otimes(n-1)}: \otimes \tau, f \mathbin{\widehat{\otimes}} g \rangle \\ &= \left\langle :x^{\otimes(n-1)}:, \frac{1}{n} \sum_{k=1}^{n} f \otimes_k g \right\rangle \\ &= \frac{1}{n} \sum_{k=1}^{n} I_{n-1}(f \otimes_k g)(x). \end{aligned} \tag{5.11}$$

Obviously, Equation (5.7) follows from Equations (5.8), (5.10), and (5.11). □

6. Application to White Noise Analysis

In this section we will use some the results in Section 4 to interpret Hida's original definition of generalized Brownian functionals in his 1975 monograph [1]. This idea was first given by Lee in [12]. Consider the simplest example, i.e., the white noise $\dot{B}(t)$. It is defined in [1] as the limit

$$\begin{aligned} \dot{B}(t) &= \lim_{\Delta \to 0} \frac{B(t+\Delta) - B(t)}{\Delta} \\ &= \lim_{\Delta \to 0} \frac{1}{\Delta} \langle \cdot, 1_{[t,t+\Delta]} \rangle \\ &= \langle \cdot, \delta_t \rangle, \end{aligned}$$

where $\langle \cdot, \delta_t \rangle$ is a generalized Wiener integral as defined in [2]. Then for a test function given by $\varphi = \sum_{n=0}^{\infty} \langle :\cdot^{\otimes n}:, u_n \rangle$, we have

$$\langle\langle \dot{B}(t), \varphi \rangle\rangle = \langle \delta_t, u_1 \rangle,$$

where $\langle\langle\cdot,\cdot\rangle\rangle$ is the pairing between generalized functions and test functions. Thus by Equation (4.4) we would get

$$\langle\langle \dot{B}(t), \varphi\rangle\rangle = [D(\mu\varphi)(0)]^{\sim}(\delta_t).$$

Observe that in this equation we do not need to use the Wiener–Itô decomposition of the test function φ. What we need is the differentiability of $\mu\varphi$ in order to define $\dot{B}(t)$ as the generalized function such that

$$\langle\langle \dot{B}(t), \varphi\rangle\rangle = [D(\mu\varphi)(0)](\delta_t) \tag{6.1}$$

for all test functions φ. Here the extension is unnecessary when we require φ to be sufficiently smooth.

Similarly, consider the renormalization $:\dot{B}(t)^2:$ which is defined in [1, 2] as the generalized multiple Wiener integral

$$:\dot{B}(t)^2: = \langle :\cdot^{\otimes 2}:, \delta_t^{\otimes 2}\rangle.$$

Then for a test function $\varphi = \sum_{n=0}^{\infty}\langle :\cdot^{\otimes n}:, u_n\rangle$, we have

$$\begin{aligned}\langle\langle :\dot{B}(t)^2:, \varphi\rangle\rangle &= 2\langle \delta_t^{\otimes 2}, u_2\rangle \\ &= [D^2(\mu\varphi)(0)](\delta_t, \delta_t).\end{aligned} \tag{6.2}$$

Thus we can use Equation (6.2) to define the generalized function $:\dot{B}(t)^2:$ as acting on test functions without using the Wiener–Itô decomposition of φ. In general, we can define the renormalization $:\dot{B}(t)^n:$ as the generalized function such that for any test function φ

$$\langle\langle :\dot{B}(t)^n:, \varphi\rangle\rangle = [D^n(\mu\varphi)(0)](\delta_t, \dots, \delta_t). \tag{6.3}$$

With this idea we can actually define generalized multiple Wiener integral $I_n(u)$ of a Schwartz distribution u on $\mathbb{R}^n$ [2] as the generalized function such that for any test function φ

$$\langle\langle I_n(u), \varphi\rangle\rangle = [D^n(\mu\varphi)(0)](u). \tag{6.4}$$

In view of Equations (6.1)–(6.4) we need to impose the differentiability condition (in fact, analyticity, see below) on test functions. Moreover, we need to impose the growth condition. For example, take the above Schwartz distribution u and consider the renormalization $\Phi = :\mathrm{e}^{\langle\cdot,u\rangle}:$ defined by

$$\Phi = :\mathrm{e}^{\langle\cdot,u\rangle}: = \sum_{n=0}^{\infty}\frac{1}{n!}\langle :\cdot^{\otimes n}:, u^{\otimes n}\rangle = \sum_{n=0}^{\infty}\frac{1}{n!}I_n(u^{\otimes n}). \tag{6.5}$$

By using Equations (6.4) and (6.5) we would get the following equation for any test function φ,

$$\langle\langle \Phi, \varphi\rangle\rangle = \sum_{n=0}^{\infty}\frac{1}{n!}[D^n(\mu\varphi)(0)](u^{\otimes n}). \tag{6.6}$$

Now, observe that informally we have the series expansion of $\mu\varphi$

$$(\mu\varphi)(u) = \sum_{n=0}^{\infty} \frac{1}{n!}[D^n(\mu\varphi)(0)](u^{\otimes n}). \tag{6.7}$$

Therefore, from Equation (6.6) we would get

$$\langle\langle \Phi, \varphi \rangle\rangle = (\mu\varphi)(u) = \int_{\mathcal{S}'(\mathbb{R})} \varphi(u+x)\, \mathrm{d}\mu(x), \tag{6.8}$$

where $\mathcal{S}'(\mathbb{R})$ is the space of Schwartz distributions.

In order to ensure the existence of the integral in Equation (6.8) we need to impose some condition on the test function φ. It turns out that an appropriate condition is the analyticity which implies the validity of series expansion of φ (and hence of $\mu\varphi$) as in Equation (6.7).

From the above discussion we see that it is possible to develop a theory of generalized functions on the space $\mathcal{S}'(\mathbb{R})$ without using the Wiener–Itô expansion. Namely, we can define a space $(\mathcal{E})$ of test functions satisfying the analyticity and growth conditions. A topology can be introduced on $(\mathcal{E})$ and then we can define the space of generalized functions on $\mathcal{S}'(\mathbb{R})$ as the dual space $(\mathcal{E})^*$. For details, see the papers [12, 13] and the books [3, 9].

References

1. Hida, T.: *Analysis of Brownian Functionals*, Carleton Mathematical Lecture Notes 13, 1975.
2. Hida, T.: Generalized multiple Wiener integrals, *Proc. Japan Acad.* **54A** (1978), 55–58.
3. Hida, T., Kuo, H.-H., Potthoff, J. and Streit, L.: *White Noise: An Infinite Dimensional Calculus*, Kluwer Acad. Publ., 1993.
4. Itô, K.: Multiple Wiener integral, *J. Math. Soc. Japan* **3** (1951), 157–169.
5. Kallianpur, G.: *Stochastic Filtering Theory*, Springer-Verlag, New York, 1980.
6. Kallianpur, G.: Traces, natural extensions and Feynman distributions, In: K. Itô and T. Hida (eds), *Gaussian Random Fields, the Third Nagoya Lévy Seminar*, World Scientific, Singapore, 1991, pp. 14–27.
7. Kuo, H.-H.: *Gaussian Measures in Banach Spaces*, Lecture Notes in Math. 463, Springer-Verlag, 1975.
8. Kuo, H.-H.: Integration in Banach spaces, In: H. Elton Lacey (ed.), *Notes in Banach Spaces*, University of Texas Press, Austin, 1980, pp. 1–38.
9. Kuo, H.-H.: *White Noise Distribution Theory*, CRC Press, Boca Raton, 1996.
10. Lee, Y.-J.: Unitary operators on the space of L^2-functions over abstract Wiener spaces, *Soochow J. Math.* **13** (1987), 165–174.
11. Lee, Y.-J.: On the convergence of Wiener–Ito decomposition, *Bull. Inst. Math., Acad. Sinica* **17** (1989), 305–312.
12. Lee, Y.-J.: Generalized functions on infinite dimensional spaces and its application to white noise calculus, *J. Funct. Anal.* **82** (1989), 429–464.
13. Lee, Y.-J.: Analytic version of test functionals, Fourier transform and a characterization of measures in white noise calculus, *J. Funct. Anal.* **100** (1991), 359–380.
14. Obata, N.: *White Noise Calculus and Fock Space*, Lecture Notes in Math. 1577, Springer, New York, 1994.
15. Wiener, N.: The homogeneous chaos, *Amer. J. Math.* **60** (1938), 897–936.

Acta Applicandae Mathematicae **63:** 219–231, 2000.

Donsker's Delta Function of Lévy Process

YUH-JIA LEE and HSIN-HUNG SHIH
Department of Mathematics, National Cheng Kung University, Tainan, Taiwan 70101, R.O.C.

(Received: 31 March 1999)

Abstract. Let $X = \{X(t) : t \in \mathbb{R}\}$ be a Lévy process and β a non-decreasing, right continuous, bounded function with $\beta(-\infty) = 0$ ($((1+u^2)/u^2)\mathrm{d}\beta(u)$ is the Lévy measure). In this paper we define the Donsker delta function $\delta(X(t)-a)$, $t > 0$ and $a \in \mathbb{R}$, as a generalized Lévy functional under the condition that $\beta(0)-\beta(0-) > 0$. This leads us to define $F(X(t))$ for any tempered distribution F, and as an application, we derive an Itô formula for $F(X(t))$ when β has jumps at 0 and 1.

Mathematics Subject Classifications (2000): primary: 60H40, 46F25; secondary: 60B05.

Key words: white noise, Lévy process, Donsker's delta function.

1. Introduction

Let δ be the Dirac delta function and $\{B(t)\}$ a Brownian motion in $\mathbb{R}$. It is well known that Donsker's delta function $\delta(B(t)-a)$, $t > 0$ and $a \in \mathbb{R}$, may be realized as a generalized functional in (Gaussian) white noise analysis (see, e.g., [2, 9] and the papers cited there). In definition, the Donsker's delta function may be defined by the following formula

$$\delta(B(t) - a) = \frac{1}{2\pi} \int_{-\infty}^{+\infty} \mathrm{e}^{ir(B(t)-a)}\, \mathrm{d}r. \tag{1.1}$$

In this paper we are interested in defining Donsker's delta function for a non-Gaussian Lévy process. Let β be a nondecreasing, right continuous, bounded function with $\beta(-\infty) = 0$ and $\beta(0) - \beta(0-) > 0$. Let $X = \{X(t) : t \in \mathbb{R}\}$ be a Lévy process in one dimension with $X(0) = 0$ with the Lévy measure being given by $((1+u^2)/u^2)\,\mathrm{d}\beta(u)$ (see (2.1)). We shall always assume that $\{X(t)\}$ is right continuous with left limit. In this paper, we adapt the Lévy white noise calculus scheme to define and study Donsker's delta function of the Lévy process. We shall show that Donsker's delta function $\delta(X(t) - a)$ of X may be realized as a generalized Lévy functional.

To start with, in Section 2 we give a brief introduction to the analysis of Lévy functionals and then, in Section 3, we introduce the spaces of test and generalized Lévy functionals. For details, we refer the reader to Lee and Shih [12, 13]. Donsker's delta function $\delta(X(t) - a)$ of X is defined as a generalized Lévy functional in Section 4. When $\{B(t)\}$ is replaced by $\{X(t)\}$, we show that the identity (1.1) also makes sense as a Bochner integral and, as a consequence, we define

$F(X(t))$ as a generalized Lévy functional for a tempered distribution F. To demonstrate an application, in Section 5 we derive an Itô formula of $F(X(t))$ for a simple Lévy process X by which we mean that the corresponding function β has the form $p1_{[0,\infty)} + q1_{[1,\infty)}$, $p, q > 0$.

2. Preliminaries

Let $X = \{X(t) : t \in \mathbb{R}\}$ be a real-valued Lévy process with $X(0) = 0$, that is, X is an additive process, continuous in probability and the characteristic function of the increment $X(t) - X(s)$ $(t > s)$ is given by $\exp\{(t-s)f_X(r)\}$, where

$$f_X(r) = i\mu r + \int_{-\infty}^{+\infty} \left(\mathrm{e}^{iru} - 1 - \frac{iru}{1+u^2}\right)\frac{1+u^2}{u^2}\,\mathrm{d}\beta(u). \tag{2.1}$$

Here the function β is a nondecreasing, right continuous, bounded function with $\beta(-\infty) = 0$; and μ is a real constant. Let $\mathcal{S}$ be the Schwartz space of rapidly decreasing functions and $\mathcal{S}'$ the space of tempered distributions on $\mathbb{R}$. Under the condition $\int_{-\infty}^{+\infty} |u|\,\mathrm{d}\beta(u) < +\infty$, $\mathcal{S}'$ carries a probability measure Λ on $(\mathcal{S}', \mathcal{B})$ (see [12]) such that

$$\int_{\mathcal{S}'} \mathrm{e}^{i(x,\eta)}\Lambda(\mathrm{d}x) = \exp\left\{\int_{-\infty}^{+\infty} f_X(\eta(t))\,\mathrm{d}t\right\}, \quad \eta \in \mathcal{S}, \tag{2.2}$$

where $(\cdot,\cdot)$ is the $\mathcal{S}'$-$\mathcal{S}$ pairing; $\mathcal{B}$ the Borel field of $\mathcal{S}'$. Throughout this paper, we assume that the function β satisfies the moment condition $\int_{-\infty}^{+\infty} |u|^n\,\mathrm{d}\beta(u) < \infty$ for any $n \in \mathbb{N}$. Then $(\mathcal{S}', \mathcal{B}, \Lambda)$ serves as the underlying probability space. In addition, we shall assume that $(\mathcal{S}', \mathcal{B}, \Lambda)$ is complete; otherwise, we replace it by its completion.

For each $\eta \in \mathcal{S}$, the mean $\mathrm{E}[(\cdot,\eta)]$ and the variance $\mathrm{Var}[(\cdot,\eta)]$ of the random variable $(\cdot,\eta)$ are given by

$$\mathrm{E}[(\cdot,\eta)] = \kappa_1 \int_{-\infty}^{+\infty} \eta(t)\,\mathrm{d}t \quad \text{and} \quad \mathrm{Var}[(\cdot,\eta)] = \kappa_2 \int_{-\infty}^{+\infty} \eta(t)^2\,\mathrm{d}t, \tag{2.3}$$

where

$$\kappa_1 = \mu + \int_{-\infty}^{+\infty} u\,\mathrm{d}\beta(u) \quad \text{and} \quad \kappa_2 = \int_{-\infty}^{+\infty} (1+u^2)\,\mathrm{d}\beta(u).$$

Let $\rho \in L^1 \cap L^2(\mathbb{R}, \mathrm{d}t)$ and let $\{\eta_n\}$ be a sequence in $\mathcal{S}$ so that $\eta_n \to \rho$ in $L^1 \cap L^2(\mathbb{R}, \mathrm{d}t)$ with respect to the norm $|\cdot|_{L^1(\mathbb{R},\mathrm{d}t)} + |\cdot|_{L^2(\mathbb{R},\mathrm{d}t)}$. Then, by (2.3), $\{(\cdot,\eta_n)\}$ forms a Cauchy sequence in $L^2(\mathcal{S}', \Lambda)$. Define $\langle\cdot,\rho\rangle$ by the L^2-limit of $\{(\cdot,\eta_n)\}$. Then $\langle\cdot,\rho\rangle$ is defined almost everywhere as a random variable with mean and variance as those given in (2.3) with η being replaced by ρ. When ρ is the indicator $1_{(s,t]}$ of $(s,t]$, the characteristic function of $\langle\cdot,\rho\rangle$ is exactly $\exp\{(t-s)f_X(1)\}$. Thus we may represent the Lévy process X on $(\mathcal{S}', \mathcal{B}, \Lambda)$ by

$$X(t;x) = \begin{cases} \langle x, 1_{[0,t]}\rangle, & \text{if } t \geqslant 0; \\ -\langle x, 1_{[t,0]}\rangle, & \text{if } t < 0,\ x \in \mathcal{S}'. \end{cases}$$

Note that, formally, $\dot{X}(t;x) = x(t)$ for $x \in \mathcal{S}'$ which are independent and form a generalized coordinate system. Thus, members of $L^2(\mathcal{S}', \Lambda)$ are also referred as Lévy white noise functionals.

For any real locally convex space V, V_c will denote the complexification of V. For $\eta = \eta_1 + i\eta_2 \in L^1_c \cap L^2_c(\mathbb{R}, \mathrm{d}t)$ with $\eta_1, \eta_2 \in L^1 \cap L^2(\mathbb{R}, \mathrm{d}t)$, we define $\langle \cdot, \eta \rangle$ as $\langle \cdot, \eta_1 \rangle + i\langle \cdot, \eta_2 \rangle$. Then $\langle x, \eta \rangle$ is defined for a.e. $x \in \mathcal{S}'$ such that its mean and variance satisfy the equalities (2.3).

2.1. THE LÉVY–ITÔ INTEGRALS

For notational convenience, we denote the measure associated with β also by β. Also, we denote $\mathbb{R} \setminus \{0\}$ (resp. $\mathbb{R}^2 \setminus \{(t, 0) : t \in \mathbb{R}\}$) by $\mathbb{R}_*$ (resp. $\mathbb{R}^2_*$); and let $\mathcal{B}_b(\mathbb{R}^2_*)$ be the class of all bounded Borel sets $E \subseteq \mathbb{R}^2_*$, away from the t-axis.

For $E \in \mathcal{B}_b(\mathbb{R}^2_*)$, let $N(E; \cdot)$ be a random variable on $(\mathcal{S}', \mathcal{B})$ defined by

$$N(E; x) = \left|\{(t, u) \in E : X(t; x) - X(t-; x) = u\}\right|.$$

$N(E; \cdot)$ is Poisson distributed with intensity measure ν, where ν is the measure $\mathrm{d}t \otimes (1 + u^2/u^2)\,\mathrm{d}\beta(u)$ on $\mathcal{B}(\mathbb{R}^2_*)$ and the family $\{N(E; \cdot) - \nu(E) : E \in \mathcal{B}_b(\mathbb{R}^2_*)\}$ of random measures are independent with zero mean. For simplicity, we shall denote '$N(E; \cdot) - \nu(E)$' by '$N_0(E; \cdot)$' for $E \in \mathcal{B}_b(\mathbb{R}^2_*)$. Let $B = \{B(t) : t \in \mathbb{R}\}$ be a process defined by

$$\sigma B(t) := X(t) - \kappa_1 t - \int_{\mathbb{R}^2_*} u 1_{[0\wedge t, 0\vee t]}(s)\,\mathrm{d}N_0(s, u), \tag{2.4}$$

where $\sigma^2 = \beta(0) - \beta(0-)$, and the integral on the right-hand side of (2.4) is a stochastic integral with respect to N_0. Then B is a one-dimensional Wiener process which is independent of the system $\{N(E) : E \in \mathcal{B}_b(\mathbb{R}^2_*)\}$ (see [4]).

Let λ be a positive measure on $\mathcal{B}(\mathbb{R}^2)$ given by $\mathrm{d}\lambda = \mathrm{d}t \otimes (1 + u^2)\,\mathrm{d}\beta(u)$. Define a $L^2(\mathcal{S}', \Lambda)$-valued function M on $\{E \in \mathcal{B}(\mathbb{R}^2) : \lambda(E) < +\infty\}$ by

$$M(E) = \sigma \int_{-\infty}^{+\infty} 1_E(t, 0)\,\mathrm{d}B(t) + \int_{\mathbb{R}^2_*} u 1_E(t, u)\,\mathrm{d}N_0(t, u). \tag{2.5}$$

Then the family of random measures $\{M(E; \cdot) : E \in \mathcal{B}(\mathbb{R}^2)\}$ is also independent with zero mean. According to Itô [7], for $\phi_n \in L^2_c(\mathbb{R}^2, \lambda)^{\widehat{\otimes} n}$ ($\widehat{\otimes}$ denotes the symmetric tensor), the nth order Lévy–Itô integral $I_n(\phi_n)$ of ϕ_n is defined as

$$I_n(\phi_n) = \int_{\mathbb{R}^2} \cdots \int_{\mathbb{R}^2} \phi_n(s_1, \ldots, s_n)\,\mathrm{d}M(s_1) \ldots \mathrm{d}M(s_n).$$

Then every $\varphi \in L^2(\mathcal{S}', \Lambda)$ enjoys the following orthogonal decomposition

$$\varphi = \mathrm{E}[\varphi] + \sum_{n=1}^{\infty} \oplus I_n(\phi_n),$$

known as the Lévy–Itô decomposition. In notation, we write $\varphi \sim (\phi_n)$. Moreover, we have the following isometry:

$$\mathrm{E}[|\varphi|^2] = |\mathrm{E}[\varphi]|^2 + \sum_{n=1}^{\infty} n!|\phi_n|^2_{L^2_c(\mathbb{R}^2,\lambda)^{\widehat{\otimes}n}}. \tag{2.6}$$

2.2. THE T-TRANSFORM AND S-TRANSFORM

As in the cases of Brownian functionals or Poisson functionals (see [6, 7]), the S-transform and T-transform for a function $\varphi \in L^2(\mathcal{S}', \Lambda)$ also play an important role in the study of generalized Lévy functionals. Their definitions are given as follows:

DEFINITION 2.1 [12]. Let

$$\mathcal{M} = \{g \in L^2(\mathbb{R}^2, \lambda) : g^* \in L^1_c \cap L^\infty_c(\mathbb{R}^2_*, \nu)\} \quad (g^*(t,u) := ug(t,u))$$

and let $\varphi \in L^2(\mathcal{S}', \Lambda)$.

(i) The T-transform $T\varphi(\eta)$ of φ for $\eta \in L^1 \cap L^2(\mathbb{R}, \mathrm{d}t)$ is defined by

$$T\varphi(\eta) = \int_{\mathcal{S}'} \varphi(x) \mathrm{e}^{i\langle x,\eta\rangle} \Lambda(\mathrm{d}x).$$

(ii) The S-transform $S\varphi$ of φ from $L^2_c(\mathbb{R}^2, \lambda)$ into $\mathbb{C}$ is defined by

$$S\varphi(g) = \int_{\mathcal{S}'} \varphi(x)\, \mathrm{Exp}(g)(x) \Lambda(\mathrm{d}x).$$

Here $\mathrm{Exp}(g) = \gamma_g \cdot \vartheta_g$, if $g \in \mathcal{M}$, where γ_g and ϑ_g are given by

$$\gamma_g(y) = \exp\left\{-\int_{\mathbb{R}^2_*} g^*(x)\,\mathrm{d}\nu(x)\right\} \prod_{t \in J_X(y)} [1 + g^*(t, X(t;y) - X(t-;y))],$$

in which $J_X(y) = \{t \in \mathbb{R} : X(t;y) - X(t-;y) \neq 0\}$, and

$$\vartheta_g(y) = \exp\left\{-\frac{\sigma^2}{2}\int_{-\infty}^{+\infty} g(t,0)^2\,\mathrm{d}t + \sigma \int_{-\infty}^{+\infty} g(t,0)\,\mathrm{d}B(t;y)\right\}$$

for $y \in \mathcal{S}'$; otherwise, $\mathrm{Exp}(g)$ is defined as a L^2-limit of $\{\gamma_{g_m} \cdot \vartheta_{g_m}\}$ for an approximating sequence $\{g_m\} \subset \mathcal{M}$ with $g_m \to g$ in $L^2(\mathbb{R}^2, \lambda)$.

In white noise analysis, S-transform and T-transform are related by the formula:

$$S\varphi(\eta) = \mathrm{E}[\mathrm{e}^{i\langle\cdot,\eta\rangle}]T\varphi(-i\eta), \quad \eta \in L^1 \cap L^2(\mathbb{R}, \mathrm{d}t).$$

For Lévy processes including both Brownian and Poisson parts, a similar relation has been proved in Lee and Shih [12] as follows:

THEOREM 2.2 [12]. *Let* $\varphi \in L^2(\mathcal{S}', \Lambda)$ *and* $\eta \in L^1 \cap L^2(\mathbb{R}, \lambda)$. *Then*

$$T\varphi(\eta) = \mathrm{E}[\mathrm{e}^{i\langle \cdot, \eta \rangle}] \cdot S\varphi(\phi_\eta),$$

where $\phi_\eta\colon \mathbb{R}^2 \to \mathbb{C}$ *is defined by* $\phi_\eta(t, u) = (\mathrm{e}^{iu\eta(t)} - 1)/u$ *if* $u \neq 0$; *if* $u = 0$, *we define* $\phi_\eta(t, u) = i\eta(t)$.

3. Test and Generalized Lévy Functionals

Let A denote the self-adjoint extension in $L^2(\mathbb{R}, \mathrm{d}t)$ of $A\rho(t) = -\rho''(t) + (1 + t^2)\rho(t)$, $\rho \in \mathcal{S}$. Let $\{e_n : n \in \mathbb{N}_0\}$ the complete orthonormal basis (CONS) of $L^2(\mathbb{R}, \mathrm{d}t)$ consisting of eigenfunctions of A with corresponding eigenvalues $\{2n + 2 : n \in \mathbb{N}_0\}$. For any $p \in \mathbb{R}$ and $\eta \in L^2(\mathbb{R}, \mathrm{d}t)$, define $|\eta|_p$ as $|A^p\eta|_{L^2(\mathbb{R},\mathrm{d}t)}$. Let $\mathcal{S}_p$ be the completion of the class $\{\eta \in L^2(\mathbb{R}, \mathrm{d}t) : |\eta|_p < +\infty\}$ with respect to $|\cdot|_p$-norm. Then $\mathcal{S}_p$ forms a real separable Hilbert space and we have the dense inclusion:

$$\mathcal{S} = \operatorname*{pr\,lim}_{r\to\infty} \mathcal{S}_r \subset \mathcal{S}_p \subset \mathcal{S}_q \subset L^2(\mathbb{R}, \mathrm{d}t) \subset \mathcal{S}_{-q} \subset \mathcal{S}_{-p} \subset \mathcal{S}' = \operatorname*{ind\,lim}_{r\to\infty} \mathcal{S}_{-r},$$

where $p > q \geqslant 0$. ('pr lim' and 'ind lim' denote 'projective limit' and 'inductive limit', respectively.)

Next, we construct a Gel'fand triple on $L^2(\mathbb{R}, \beta_0)$ as follows, where $\mathrm{d}\beta_0 = (1+u^2)\,\mathrm{d}\beta(u)$. The Lusin theorem (see [15]) assures that $\{1, u, u^2, \ldots\}$ (not necessarily infinite) is total in $L^2(\mathbb{R}, \beta_0)$. Applying the Gram–Schmidt orthogonalization process to $\{1, u, u^2, \ldots\}$, one obtains a CONS $\{\zeta_0, \zeta_1, \ldots\}$ of $L^2(\mathbb{R}, \beta_0)$. Thus, $\{e_n \otimes \zeta_m : n, m \in \mathbb{N}_0\}$ forms a CONS of $L^2(\mathbb{R}^2, \mathrm{d}\lambda)$. Let A_β be a linear operator densely defined on $L^2(\mathbb{R}, \beta_0)$ such that $A_\beta\zeta_n = (2n+2)\zeta_n$, $n = 0, 1, \ldots$. For each $p \geqslant 0$, define $|\psi|_{p,\beta}$ as $|A_\beta^p\psi|_{L^2(\mathbb{R},\beta_0)}$, and let

$$\mathcal{E}_p = \{\psi \in L^2(\mathbb{R}, \beta_0) : |\psi|_{p,\beta} < +\infty\}.$$

Then $\mathcal{E}_p$ is a real separable Hilbert space, and let $\mathcal{E} = \operatorname{pr\,lim}_{p\to\infty} \mathcal{E}_p$. Then $\mathcal{E}$ is a nuclear space and $\mathcal{E} \subset L^2(\mathbb{R}, \beta_0) \subset \mathcal{E}' = \operatorname{ind\,lim}_{p\to\infty} \mathcal{E}'_p$ forms a Gel'fand triple. Moreover, we have

$$\mathcal{E} \subset \mathcal{E}_p \subset \mathcal{E}_q \subset L^2(\mathbb{R}, \mathrm{d}t) \subset \mathcal{E}'_q \subset \mathcal{E}'_p \subset \mathcal{E}',$$

where $p > q \geqslant 0$ and the norm on $\mathcal{E}'_p$'s are given by $|\psi|_{-p,\beta} = |A_\beta^{-p}\psi|_{L^2(\mathbb{R},\beta_0)}$.

Now, for $p \in \mathbb{R}$, let $\mathcal{N}_p$ be the Hilbert space tensor product $\mathcal{S}_p \otimes \mathcal{E}_p$ with norm $|\cdot|_{p;p,\beta}$ defined by $|e_n \otimes \zeta_m|_{p;p,\beta} = |e_n|_p \cdot |\zeta_m|_{p,\beta}$; and let $\mathcal{N} = \mathcal{S} \otimes \mathcal{E}$, $\mathcal{N}' = \mathcal{S}' \otimes \mathcal{E}'$. Then $\mathcal{N}$ is a nuclear space induced by the family $\{(\mathcal{S}_p \otimes \mathcal{E}_p, |\cdot|_{p;p,\beta}) : p \geqslant 0\}$, and again we have a Gel'fand triple $\mathcal{N} \subset L^2(\mathbb{R}^2, \lambda) \subset \mathcal{N}'$ such that the inclusions $\mathcal{N} \subset \mathcal{N}_p \subset L^2(\mathbb{R}^2, \lambda) \subset \mathcal{N}_{-p} \subset \mathcal{N}'$, $p \geqslant 0$, are all continuous.

To construct the spaces of test and generalized functions, we note that for $\varphi \in L^2(\mathcal{S}', \Lambda)$, $S\varphi$ is an analytic function on $L_c^2(\mathbb{R}^2, \lambda)$ with the nth Fréchet derivative $D^n S\varphi(g)$ being of n-linear Hilbert–Schmidt type on $L_c^2(\mathbb{R}^2, \lambda)$. Further,

$$\|\varphi\|_{L^2(\mathcal{S}',\Lambda)}^2 = \sum_{n=0}^{\infty} \frac{1}{n!} |D^n S\varphi(0)|^2_{\mathcal{L}^n_{(2)}[L^2(\mathbb{R}^2,\lambda)]},$$

where $\mathcal{L}^n_{(2)}[H]$ is the space of all n-linear Hilbert–Schmidt operator on a Hilbert space H. Let Γ_p be the second quantization of $A^p \otimes A_\beta^p$, $p \in \mathbb{R}$; and define $\|\varphi\|_p = \|\Gamma_p(\varphi)\|_{L^2(\mathcal{S}',\Lambda)}$. Let $\mathcal{L}_p$ be the completion of the class $\{\varphi \in L^2(\mathcal{S}', \Lambda) : \|\varphi\|_p < +\infty\}$ with respect to $\|\cdot\|_p$-norm. Then $\mathcal{L}_p$, $p \in \mathbb{R}$, forms a separable Hilbert space and let $\mathcal{L} = \operatorname{pr\,lim}_{p\to\infty} \mathcal{L}_p$. Then $\mathcal{L}$ is a nuclear space. Moreover, we have the continuous inclusion:

$$\mathcal{L} \subset \mathcal{L}_p \subset \mathcal{L}_q \subset L^2(\mathcal{S}', \Lambda) \subset \mathcal{L}_{-q} \subset \mathcal{L}_{-p} \subset \mathcal{L}' = \operatorname*{ind\,lim}_{r\to\infty} \mathcal{L}_{-r},$$

where $p > q \geqslant 0$. $\mathcal{L}$ will serves as *the space of test functions* in our investigation and members of the dual space $\mathcal{L}'$ of $\mathcal{L}$ are called *generalized Lévy white noise functions*. In the sequel, the bilinear pairing of $\mathcal{L}'$ and $\mathcal{L}$ will be denoted by $\langle\langle\cdot,\cdot\rangle\rangle$.

PROPOSITION 3.1 ([13]). (i) *For $p > 0$ and $F \in \mathcal{L}_{-p}$, there exists F_n's $\in \mathcal{N}_{-p,c}^{\widehat{\otimes} n}$ such that*

$$\langle\langle F, \varphi\rangle\rangle = \sum_{n=0}^{\infty} n!(F_n, \phi_n) \quad \text{and} \quad \|F\|_{-p}^2 = \sum_{n=0}^{\infty} n!|F_n|^2_{\mathcal{N}_{-p,c}^{\widehat{\otimes} n}},$$

where $\varphi \sim (\phi_n) \in \mathcal{L}_p$ and $(\cdot,\cdot)$ is the $\mathcal{N}_{-p,c}^{\widehat{\otimes} n}$-$\mathcal{N}_{p,c}^{\widehat{\otimes} n}$ pairing. We write it as $F \sim (F_n)$.

(ii) *For $g \in \mathcal{N}_p$ with $p \in \mathbb{R}$, let $\mathcal{E}_M(g) \sim (1/n! g^{\otimes n})$. Then $\mathcal{E}_M(g) \in \mathcal{L}_p$ and $\|\mathcal{E}_M(g)\|_p = \mathrm{e}^{\frac{1}{2}|g|^2_{p;p,\beta}}$.*

We note that, when $p > 0$, $\mathcal{E}_M(g) = \operatorname{Exp}(g)$ for $g \in \mathcal{N}_{p,c}$. According to Proposition 3.1(ii), we define the S-transform SF of $F \in \mathcal{L}_{-p}$, $p \in \mathbb{R}$, as a complex-valued function on $\mathcal{N}_{p,c}$ by $SF(g) = \langle\langle F, \mathcal{E}_M(g)\rangle\rangle$. Then SF is an analytic function on $\mathcal{N}_{p,c}$ satisfying the growth condition:

$$|SF(g)| \leqslant \|F\|_{-p}\, \mathrm{e}^{\frac{1}{2}|g|^2_{p;p,\beta}}, \quad g \in \mathcal{N}_{p,c},\ p \in \mathbb{R}. \tag{3.1}$$

Moreover, we have the following characterization theorem.

THEOREM 3.2 ([13]). *Suppose that a complex-valued function G is analytic on $\mathcal{N}_{p,c}$ and satisfies the following growth condition:*

$$\exists p \in \mathbb{R}, \exists c > 0 \ni |G(\xi)| \leqslant c \cdot \mathrm{e}^{\frac{1}{2}|\xi|^2_{-p;-p,\beta}}, \quad \xi \in \mathcal{N}_c.$$

Then there exists a unique $F \in \mathcal{L}_{p-1}$ such that $SF = G$.

4. Donsker's Delta Function of Lévy Processes

In this section, we always assume that $\sigma^2 = \beta(0) - \beta(0-) > 0$. Let $f \in \mathcal{S}$. Then the inversion formula of Fourier transform implies that

$$f(X(t) - a) = \frac{1}{\sqrt{2\pi}} \int_{-\infty}^{+\infty} \hat{f}(r)\, \mathrm{e}^{ir(X(t)-a)}\, \mathrm{d}r,$$

where $\hat{f}(r) = 1/\sqrt{2\pi} \int_{-\infty}^{+\infty} f(s)\, \mathrm{e}^{-irs}\, \mathrm{d}s$. Clearly, $f(X(t) - a) \in L^1(\mathcal{S}', \Lambda)$. Let $t > 0$ be fixed and regard $f(X(t))$ as a generalized function in $\mathcal{L}'$. Then, for $\varphi \in \mathcal{L}$,

$$\begin{aligned}
\langle\langle f(X(t) - a), \varphi\rangle\rangle &= \int_{\mathcal{S}'} f(X(t; x) - a)\varphi(x)\Lambda(\mathrm{d}x) \\
&= \frac{1}{\sqrt{2\pi}} \int_{-\infty}^{+\infty} \int_{\mathcal{S}'} \hat{f}(r)\varphi(x)\, \mathrm{e}^{ir(X(t;x)-a)}\Lambda(\mathrm{d}x)\, \mathrm{d}r \\
&= \frac{1}{\sqrt{2\pi}} \int_{-\infty}^{+\infty} \hat{f}(r)\, \mathrm{e}^{-ira} T\varphi(r 1_{[0,t]})\, \mathrm{d}r \\
&= \int_{-\infty}^{+\infty} \hat{f}(r)\, \mathrm{e}^{-ira} G_{t,\varphi}(r)\, \mathrm{d}r,
\end{aligned}$$

where $G_{t,\varphi}(r) = 1/\sqrt{2\pi}\, T\varphi(r 1_{[0,t]})$. Choose a sequence $\{\delta_n\} \subset \mathcal{S}$ such that $\delta_n \to \delta$ in $\mathcal{S}'$. Then $\widehat{\delta_n} \to \hat{\delta}$ in $\mathcal{S}'$. We shall show that Donsker's delta function $\delta(X(t)-a)$, $t > 0$ and $a \in \mathbb{R}$ may be defined by

$$\langle\langle \delta(X(t) - a), \varphi\rangle\rangle := \lim_{n\to\infty} \int_{-\infty}^{+\infty} \widehat{\delta_n}(r)\, \mathrm{e}^{-ira} G_{t,\varphi}(r)\, \mathrm{d}r. \tag{4.1}$$

We have to show that (i) the limit in (4.1) does exists, (ii) such a limit is independent of the choice of the sequence $\{\delta_n\}$; and (iii) $\delta(X(t) - a) \in \mathcal{L}'$. The proof will be accomplished in Theorem 4.3.

By Theorem 2.2 we have

$$G_{t,\varphi}(r) = \frac{\mathrm{E}[\mathrm{e}^{irX(t)}]}{\sqrt{2\pi}} \times S\varphi(\phi_t(r)),$$

where $\phi_t\colon \mathbb{R} \to L^2(\mathbb{R}^2, \lambda)$ is given by

$$\phi_t(r)(s, u) = \frac{\mathrm{e}^{iru 1_{[0,t]}(s)} - 1}{u},$$

if $u \neq 0$; otherwise, $\phi_t(r) = ir 1_{[0,t]}(s)$.

LEMMA 4.1. *Let $\varphi \in \mathcal{L}$ and $t > 0$ be fixed. Then $S\varphi(\phi_t(\cdot))$ is infinitely differentiable on $\mathbb{R}$. Moreover, for any $n \in \mathbb{N}$, there exists a constant $c(n) > 0$, depending only on n, such that for any $p \geqslant 0$ and $r \in \mathbb{R}$,*

$$\left| \frac{\mathrm{d}^n}{\mathrm{d}r^n} S\varphi(\phi_t(r)) \right| \leqslant c(n) \|\varphi\|_p\, \mathrm{e}^{2^{-(4p+1)} c(n)(r^2+1)t}. \tag{4.2}$$

Proof. Let $p \geqslant 0$ be fixed. It is easy to see that ϕ_t is an infinitely differentiable $\mathcal{N}_{-p,c}$-valued function on $\mathbb{R}$. Indeed, for any $r \in \mathbb{R}$ and $m \in \mathbb{N}$,

$$\begin{aligned} \left|\frac{\mathrm{d}^m}{\mathrm{d}r^m}\phi_t(r)\right|_{-p;-p,\beta} &\leqslant \frac{1}{4^p}\left|\frac{\mathrm{d}^m}{\mathrm{d}r^m}\phi_t(r)\right|_{L^2(\mathbb{R}^2,\lambda)} \\ &\leqslant \frac{\sqrt{t}}{4^p}\left\{\int_{-\infty}^{+\infty} u^{2m-2}(1+u^2)\,\mathrm{d}\beta(u)\right\}^{1/2} \end{aligned} \tag{4.3}$$

and

$$|\phi_t(r)|_{-p;-p,\beta} \leqslant \frac{1}{4^p}|\phi_t(r)|_{L^2(\mathbb{R}^2,\lambda)} \leqslant \frac{|r|\sqrt{\kappa_2 t}}{4^p}, \tag{4.4}$$

where κ_2 is defined in (2.3). Observe that the nth derivative of $S\varphi(\phi_t(r))$ with respect to r can be expressed as a linear combination of the terms of the form

$$D^k S\varphi(\phi_t(r))(\phi_t^{(l_1)}(r), \ldots, \phi_t^{(l_k)}(r)),$$

where $1 \leqslant k, l_1, \ldots, l_k \leqslant n$ and D denotes the Fréchet derivative of $S\varphi(\cdot)$. By the Cauchy integral formula, we have for $g, g_1, \ldots, g_n \in \mathcal{N}_{-p,c}$ and $m \in \mathbb{N}$,

$$\begin{aligned} &|D^m S\varphi(g)(g_1, \ldots, g_m)| \\ &\quad \leqslant \|\varphi\|_p \exp\left\{\frac{(m+1)^2}{2}\left(|g|^2_{-p;-p,\beta} + \sum_{i=1}^m |g_i|^2_{-p;-p,\beta}\right)\right\}. \end{aligned} \tag{4.5}$$

Apply the estimations (4.3) and (4.4) to (4.5), we obtain the inequality (4.2). □

PROPOSITION 4.2. *For an arbitrary $\varphi \in \mathcal{L}$ and $t > 0$, $G_{t,\varphi} \in \mathcal{S}_c$. Moreover, for any $q > 0$ and $0 < a < b < +\infty$, there exists a positive real number p and a constant $c > 0$, depending only on q, a, b, such that $\sup_{s\in[a,b]} |G_{s,\varphi}|_q \leqslant c\|\varphi\|_p$.*

Proof. We note that

$$G_{t,\varphi}(r) = f_1(r) f_2(r) f_3(r) S\varphi(\phi_t(r)),$$

where

$$f_1(r) = \frac{\mathrm{e}^{irt\kappa_1}}{\sqrt{2\pi}}, \qquad f_2(r) = \mathrm{e}^{-\frac{1}{2}\sigma^2 r^2 t},$$

and

$$f_3(r) = \exp\left\{t\int_{|u|>0} (\mathrm{e}^{iru} - 1 - iru)\frac{1+u^2}{u^2}\,\mathrm{d}\beta(u)\right\}$$

for $r \in \mathbb{R}$. Then

$$\begin{aligned} &\frac{\mathrm{d}^n}{\mathrm{d}r^n} G_{t,\varphi}(r) \\ &\quad = \sum_{\substack{\alpha_1+\cdots+\alpha_4=n \\ \alpha_1,\ldots,\alpha_4\in\mathbb{N}\cup\{0\}}} \frac{n!}{\alpha_1!\cdots\alpha_4!} f_1^{(\alpha_1)}(r) f_2^{(\alpha_2)}(r) f_3^{(\alpha_3)}(r)\left(\frac{\mathrm{d}^{\alpha_4}}{\mathrm{d}r^{\alpha_4}} S\varphi(\phi_t(r))\right). \end{aligned}$$

It is easy to see that $f_3^{(\alpha_3)}$ is a bounded function and $\lim_{r\to\infty} f_2^{(\alpha_2)}(r)\mathrm{e}^{\frac{1}{4}\sigma^2 r^2 t} = 0$. Then, by applying (3.1) and Lemma 4.1, there exists a constant $d(n) > 0$, depending only on n, a, b, such that

$$\left|\frac{\mathrm{d}^n}{\mathrm{d}r^n} G_{t,\varphi}(r)\right| \leqslant \mathrm{d}(n)\|\varphi\|_p\, \mathrm{e}^{(2^{-(4p+1)}\mathrm{d}(n)-\frac{\sigma^2}{4})r^2 t}, \quad \forall p \geqslant 0. \tag{4.6}$$

Choose a positive real number p such that $2^{-(4p+1)}\mathrm{d}(n) - (\sigma^2/4) < 0$. Then we have, $\forall n, m \in \mathbb{N}$,

$$\lim_{r\to\infty} (1+r^2)^m \frac{\mathrm{d}^n}{\mathrm{d}r^n} G_{t,\varphi}(r) = 0.$$

This proves that $G_{t,\varphi} \in \mathcal{S}_c$. Clearly, for any $q > 0$, there exist $n_q \in \mathbb{N}$ and $c_q > 0$ so that for every $p \geqslant 0$

$$\begin{aligned}
&\sup_{s\in[a,b]} |G_{s,\varphi}|_q \\
&\quad\leqslant c_q \sup_{s\in[a,b]} \sup_{|\alpha|\leqslant n_q} \sup_{r\in\mathbb{R}} (1+r^2)^{n_q} \left|\frac{\mathrm{d}^\alpha}{\mathrm{d}r^\alpha} G_{s,\varphi}(r)\right| \\
&\quad\leqslant \|\varphi\|_p c_q \tilde{d} \sup_{s\in[a,b]} \sup_{r\in\mathbb{R}} (1+r^2)^{n_q}\, \mathrm{e}^{(2^{-(4p+1)}\tilde{d}-\frac{\sigma^2}{4})r^2 s},
\end{aligned}$$

where $\tilde{d} = \max_{|\alpha|\leqslant n_q} \mathrm{d}(\alpha)$.

Pick p so large that $2^{-(4p+1)}\tilde{d} - (\sigma^2/4) < 0$ and let

$$c = c_q \tilde{d} \sup_{r\in\mathbb{R}} (1+r^2)^{n_q}\, \mathrm{e}^{(2^{-(4p+1)}\tilde{d}-\frac{\sigma^2}{4})r^2 a}.$$

Then $\sup_{s\in[a,b]} |G_{s,\varphi}|_q \leqslant c\|\varphi\|_p$ for any $\varphi \in \mathcal{L}$. This completes the proof. □

As an application of Theorem 4.2, we represent $\delta(X(t) - a)$ in terms of a Bochner integral and, as a consequence, we define $F(X(t) - a)$ for any tempered distribution F.

THEOREM 4.3. *Let $t > 0$ and $a \in \mathbb{R}$ be fixed. Then*

$$\delta(X(t) - a) = \frac{1}{2\pi} \int_{-\infty}^{+\infty} \mathrm{e}^{ir(X(t)-a)}\, \mathrm{d}r, \quad \text{in } \mathcal{N}_{-p,c}, \tag{4.7}$$

for any $p > 0$ which is sufficiently large so that $\sigma^2 - 2^{-4p}\kappa_2 > 0$, where the integral in (4.7) *exists in the sense of Bochner integral. Moreover, for any $\varphi \in \mathcal{L}$,*

$$\langle\!\langle \delta(X(t) - a), \varphi \rangle\!\rangle = \frac{1}{2\pi} \int_{-\infty}^{+\infty} \mathrm{e}^{-ira} G_{t,\varphi}(r)\, \mathrm{d}r. \tag{4.8}$$

Proof. Let $\rho(r) \equiv \mathrm{e}^{-ira} G_{t,\varphi}(r)$ for $r \in \mathbb{R}$. By Theorem 4.2, $G_{t,\varphi} \in \mathcal{S}_c$ so that $\rho \in \mathcal{S}_c$. Then, for any sequence $\{\delta_n\} \subseteq \mathcal{S}'$ such that $\delta_n \to \delta$ in $\mathcal{S}'$, we have $\widehat{\delta_n} \to \hat{\delta}$ in $\mathcal{S}'$ and

$$\lim_{n\to\infty} \int_{-\infty}^{+\infty} \widehat{\delta_n}(r)\,\mathrm{e}^{-ira} G_{t,\varphi}(r)\,\mathrm{d}r = \lim_{n\to\infty} (\widehat{\delta_n}, \rho) = \left(\lim_{n\to\infty} \widehat{\delta_n}, \rho\right) = (\hat{\delta}, \rho),$$

where $(\cdot, \cdot)$ is the $\mathcal{S}_c'$-$\mathcal{S}_c$ pairing. Since $\hat{\delta} = 1/\sqrt{2\pi}$ in $\mathcal{S}'$, we obtain the formula (4.8). Next, we note that for any $r \in \mathbb{R}$,

$$\langle\!\langle \mathrm{e}^{ir(X(t)-a)}, \varphi \rangle\!\rangle = \mathrm{e}^{-ira} G_{t,\varphi}(r) = \mathrm{E}[\mathrm{e}^{ir(X(t)-a)}] \times S\varphi(\phi_t(r)),$$

by Theorem 2.2. Then from (3.1) and (4.4), it follows that

$$|\langle\!\langle \mathrm{e}^{ir(X(t)-a)}, \varphi \rangle\!\rangle| \leqslant \|\varphi\|_p\, \mathrm{e}^{-\frac{1}{2}(\sigma^2 - 2^{-4p}\kappa_2) r^2 t},$$

where $\sigma^2 - 2^{-4p}\kappa_2 > 0$ by the choice of p. It follows that

$$\|\mathrm{e}^{ir(X(t)-a)}\|_{-p} \leqslant \mathrm{e}^{-\frac{1}{2}(\sigma^2 - 2^{-4p}\kappa_2) r^2 t}.$$

Observe that the term on the right-hand side of the above inequality is Lebesgue integrable on $\mathbb{R}$ as a function of r and the formula (4.7) follows. Clearly the formula (4.8) follows from (4.7). □

For $s \in \mathbb{R}$ and $\varphi \in \mathcal{S}$, let

$$g_{a,\varphi}(s) = \langle\!\langle \delta(X(t) - a - s), \varphi \rangle\!\rangle.$$

Then it follows from (4.8) that $g_{a,\varphi}(s) = 1/\sqrt{2\pi}\,\widehat{G_{t,\varphi}}(s + a)$ and then, by Proposition 4.2, $g_{a,\varphi} \in \mathcal{S}$. This leads to the following definition:

DEFINITION 4.4. Let $t > 0$ and $a \in \mathbb{R}$. For $F \in \mathcal{S}'$, $F(X(t) - a)$ is defined as a generalized Lévy functional by

$$\langle\!\langle F(X(t) - a), \varphi \rangle\!\rangle = (F_{[s]}, \langle\!\langle \delta(X(t) - a - s), \varphi \rangle\!\rangle),$$

where $F_{[s]}$ means that F acts on the test functions in the variable s.

5. Generalized Itô Formula for a Simple Lévy Process

In this section, we assume that $\{X(t)\}$ is a simple Lévy process by which we mean that the measure β_0 is of the form $\sigma^2\delta_0 + b_1\delta_1$, where $\sigma^2, b_1 > 0$ and δ_i is the Dirac measure concentrated on the point i, $i = 0, 1$. In this case, a function $g \in L^2(\mathbb{R}, \beta_0)$ if and only if $g = \alpha_0 1_{\{0\}} + \alpha_1 1_{\{1\}}$ β_0 a.e. for some $\alpha_0, \alpha_1 \in \mathbb{C}$. In this section, we shall derive an Itô formula for the process $\{F(X(t))\}$ with $F \in \mathcal{S}'$.

Let $\{\zeta_0, \zeta_1\}$ is a CONS of $L^2(\mathbb{R}, \beta_0)$. For any $(t, u) \in \mathbb{R}^2$, let $\delta_{(t,u)}$ be the evaluating map on $L^2(\mathbb{R}^2, \lambda)$, i.e.,

$$(\delta_{(t,u)}, g) = \sum_{n,m} (g, e_n \otimes \zeta_m) e_n(t) \otimes \zeta_m(u),$$

where $(\cdot, \cdot)$ is the $\mathcal{N}_c'$-$\mathcal{N}_c$ pairing. Then $\delta_{(t,u)} \in \mathcal{N}_{-1,c}$ for all $(t, u) \in \mathbb{R}^2$. Moreover, we have

LEMMA 5.1. *For any $u \in \mathbb{R}$ and $g \in \mathcal{N}_c$, there exists a positive real number c_u such that, for all $\alpha_i, t_i \in \mathbb{R}$, $i = 1, 2$,*

$$|\alpha_1 \delta_{(t_1,u)} - \alpha_2 \delta_{(t_2,u)}|_{-1;-1,\beta} \leqslant c_u |\alpha_1 \delta_{t_1} - \alpha_2 \delta_{t_2}|_{-1}.$$

Let $\phi_t \colon \mathbb{R} \to L^2(\mathbb{R}^2, \lambda)$ be defined as in Section 4. Then, for any $g \in \mathcal{N}_c$ and $r \in \mathbb{R}$,

$$\begin{aligned} &\left|\left(\frac{\phi_{t+\epsilon}(r) - \phi_t(r)}{\epsilon} - ir\sigma^2 \delta_{(t,0)} - 2b_1(\mathrm{e}^{ir} - 1)\delta_{(t,1)}, g\right)\right| \\ &\quad\leqslant \frac{|r|(\sigma^2 + 2b_1))}{\epsilon} \int_t^{t+\epsilon} (|\delta_{(s,0)} - \delta_{(t,0)}, g)| + |(\delta_{(s,1)} - \delta_{(t,1)}, g)|)\, \mathrm{d}s \\ &\quad\leqslant \frac{c'|g|_1|r|}{\epsilon} \int_t^{t+\epsilon} |\delta_s - \delta_t|_{-1}\, \mathrm{d}s, \end{aligned} \tag{5.1}$$

by Lemma 5.1, where $(\cdot, \cdot)$ is the $\mathcal{N}_c'$-$\mathcal{N}_c$ pairing and c' is a constant. From (5.1) it follows that the function $t \mapsto \phi_t(r)$ is differentiable with respect to t in $|\cdot|_{-1;-1,\beta}$-norm; moreover,

$$\frac{\mathrm{d}}{\mathrm{d}t}\phi_t(r) = ir\sigma^2 \delta_{(t,0)} + 2b_1(\mathrm{e}^{ir} - 1)\delta_{(t,1)}. \tag{5.2}$$

DEFINITION 5.2. For $(t, u) \in \mathbb{R}^2$, a linear operator $\partial_{(t,u)}$ from $\mathcal{L}$ to $\mathcal{L}'$ is defined by

$$\partial_{(t,u)}\varphi := S^{-1}(DS\varphi(\cdot)\delta_{(t,u)}), \quad \varphi \in \mathcal{L}.$$

Note that $\partial_{(t,u)}$ is continuous from $\mathcal{L}$ into itself. Moreover, we note that for $t_1, t_2 \in \mathbb{R}$, $q > 0$, and $\varphi \in \mathcal{L}$,

$$\|\partial_{(t_1,u)}\varphi - \partial_{(t_2,u)}\varphi\|_q \leqslant |\delta_{(t_1,u)} - \delta_{(t_2,u)}|_{-q;-q,\beta} \|\varphi\|_{q+1}. \tag{5.3}$$

For more details about $\partial_{(t,u)}$, we refer the reader to [13].

It immediately follows from (5.2) that we have the following lemma:

LEMMA 5.3. *Let $\varphi \in \mathcal{L}$ and $G_{t,\varphi}(r)$, $r \in \mathbb{R}, t > 0$, be defined as in* (4.1). *Then*

$$\begin{aligned}\frac{\mathrm{d}}{\mathrm{d}t} G_{t,\varphi}(r) &= \left\{ ir\kappa_1 - \frac{\sigma^2 r^2}{2} + 2b_1(\mathrm{e}^{ir} - 1 - ir) \right\} G_{t,\varphi}(r) + \\ &\quad + ir\sigma^2 G_{t,\partial_{(t,0)}\varphi}(r) + 2b_1(\mathrm{e}^{ir} - 1) G_{t,\partial_{(t,1)}\varphi}(r),\end{aligned}$$

where κ_1 is defined in (2.3).

Let $\varphi \in \mathcal{L}$ and $u \in \mathbb{R}$ be fixed. Consider two $\mathcal{S}_c$-valued mappings ρ_1, ρ_2 on $\mathbb{R}$ by $\rho_1(t) = G_{t,\varphi}$ and $\rho_2(t) = G_{t,\partial_{(t,u)}\varphi}$. Then, by applying (5.3), we have

LEMMA 5.4 [14]. *ρ_1 and ρ_2 are both $\mathcal{S}_c$-valued continuous mappings on $\mathbb{R}$.*

Let $F \in \mathcal{S}'$ and $\varphi \in \mathcal{L}$. For $0 < a < b < +\infty$ and $t, t + \epsilon \in (a, b)$,

$$\left(\frac{F(X(t+\epsilon)) - F(X(t))}{\epsilon}, \varphi \right) = \int_{-\infty}^{+\infty} \widehat{F}(r) \cdot \frac{G_{t+\epsilon,\varphi}(r) - G_{t,\varphi}(r)}{\epsilon} \, \mathrm{d}r, \quad (5.4)$$

where $(\cdot, \cdot)$ is the $\mathcal{L}'$-$\mathcal{L}$ pairing and the integral is understood to be the $\mathcal{S}_c'$-$\mathcal{S}_c$ pairing. By Lemma 5.3, the term on the right-hand side of (5.4) becomes

$$\begin{aligned}&\int_{-\infty}^{+\infty} \widehat{F}(r) \frac{1}{\epsilon} \int_t^{t+\epsilon} \frac{\mathrm{d}}{\mathrm{d}s} G_{s,\varphi}(r) \, \mathrm{d}s \, \mathrm{d}r \\ &= \int_{-\infty}^{+\infty} \left[\kappa_1 \widehat{F'}(r) + \frac{\sigma^2}{2} \widehat{F''}(r) \right] \frac{1}{\epsilon} \int_t^{t+\epsilon} G_{s,\varphi}(r) \, \mathrm{d}s \, \mathrm{d}r + \\ &\quad + 2b_1 \int_{-\infty}^{+\infty} (\widehat{\tau_1 F}(r) - 1 - \widehat{F'}(r)) \frac{1}{\epsilon} \int_t^{t+\epsilon} G_{s,\varphi}(r) \, \mathrm{d}s \, \mathrm{d}r + \\ &\quad + \sigma^2 \int_{-\infty}^{+\infty} \widehat{F'}(r) \frac{1}{\epsilon} \int_t^{t+\epsilon} G_{s,\partial_{(s,0)}\varphi}(r) \, \mathrm{d}s \, \mathrm{d}r + \\ &\quad + 2b_1 \int_{-\infty}^{+\infty} (\widehat{\tau_1 F}(r) - 1) \frac{1}{\epsilon} \int_t^{t+\epsilon} G_{s,\partial_{(s,1)}\varphi}(r) \, \mathrm{d}s \, \mathrm{d}r,\end{aligned}$$

where $\tau_1 F(\cdot) = F(\cdot + 1)$, and F', F'' are the distributional derivatives of F. Letting $\epsilon \to 0$ and applying Lemma 5.4, we obtain the following theorem:

THEOREM 5.5 (Itô formula). *Let $F \in \mathcal{S}'$ and X be a simple Lévy process, i.e., $\beta_0 = \sigma^2 \delta_0 + b_1 \delta_1$, $\sigma^2, b_1 > 0$. Then*

$$\begin{aligned}\frac{\mathrm{d}}{\mathrm{d}t} F(X(t)) &= \kappa_1 F'(X(t)) + \frac{\sigma^2}{2} F''(X(t)) + \sigma^2 \partial_{(t,0)}^{\dagger} F'(X(t)) + \\ &\quad + 2b_1((\tau_1 F)(X(t)) - F(X(t)) - F'(X(t))) + \\ &\quad + 2b_1 \partial_{(t,1)}^{\dagger}((\tau_1 F)(X(t)) - F(X(t))), \qquad (5.5)\end{aligned}$$

where $\partial_{(t,u)}^{\dagger}$ is the adjoint operator of $\partial_{(t,u)}$ for $u = 0, 1$.

Remark 5.6. The Itô formula can also be derived. For an arbitrary β satisfying $\sigma^2 > 0$. Since the computation is more involved, we shall prove it in our forthcoming paper [14].

References

1. Doob, J. L.: *Stochastic Processes*, Wiley, New York, 1953.
2. Hida, T., Kuo, H.-H., Potthoff, J. and Streit, L.: *White Noise: An Infinite Dimensional Calculus*, Kluwer Acad. Publ., Dordrecht, 1993.
3. Ikeda, N. and Watanabe, S.: *Stochastic Differential Equations and Diffusion Processes*, North-Holland, Amsterdam, 1981.
4. Itô, K.: On stochastic processes, *Japan. J. Math.* **18** (1942), 261–301.
5. Itô, K.: Spectral type of shift transformations of differential process with stationary increments, *Trans. Amer. Math. Soc.* **81** (1956), 253–263.
6. Ito, Y.: Generalized Poisson functionals, *Probab. Theory Related Fields* **77** (1988), 1–28.
7. Ito, Y. and Kubo, I.: Calculus on Gaussian and Poisson white noises, *Nagoya Math. J.* **111** (1988), 41–84.
8. Kubo, I. and Takenaka, S.: Calculus on Gaussian white noises II, *Proc. Japan Acad. Ser. A* **56**(8) (1980), 411–416.
9. Kuo, H.-H.: Donsker's delta function as a generalized brownian functional and its application, In: *Lecture Notes in Control and Inform. Sci.* 49, Springer, New York, 1983, pp. 167–178.
10. Kuo, H.-H.: *White Noise Distribution Theory*, CRC Press, Boca Raton, 1996.
11. Lee, Y.-J.: Generalized functions on infinite dimensional spaces and its application to white noise calculus, *J. Funct. Anal.* **82** (1989), 429–464.
12. Lee, Y.-J. and Shih, H.-H.: The Segal–Bargmann transform for Lévy functionals, *J. Funct. Anal.* **168** (1999), 46–83.
13. Lee, Y.-J. and Shih, H.-H.: Analysis of generalized Lévy functionals, Preprint.
14. Lee, Y.-J. and Shih, H.-H.: Itô formula for generalized Lévy functionals, In: T. Hida and K. Saitô (eds), *Quantum Information* II, World Scientific, 2000, pp. 87–105.
15. Rudin, W.: *Real and Complex Analysis*, 3rd edn, McGraw-Hill, New York, 1987.

Acta Applicandae Mathematicae **63:** 233–243, 2000.

Approximation of Hunt Processes by Multivariate Poisson Processes

ZHI-MING MA[1,2], MICHAEL RÖCKNER[2] and WEI SUN[1]
[1]*Institute of Applied Mathematics, the Chinese Academy of Sciences, Beijing 100080, P.R. China*
[2]*Fakultät für Mathematik, Universität Bielefeld, Postfach 100131, 33501 Bielefeld, Germany*

(Received: 10 February 1999)

Abstract. We prove that arbitrary Hunt processes on a general state space can be approximated by multivariate Poisson processes starting from each point of the state space. The key point is that no additional regularity assumption on the state space and on the underlying transition semigroup is used.

Mathematics Subject Classifications (2000): Primary: 60J40, Secondary: 60J10, 31C15, 60J45.

Key words: Hunt process, multivariate Poisson process, weak convergence.

1. Introduction

In [MRZ98], it was proved that any Hunt process associated with a Dirichlet form on a general state space can be approximated by Markov chains (more precisely, by multivariate Poisson processes, see (2.1) below) whose construction is based on the Yoshida approximation of the generator and tightness arguments. Among other things, the above-mentioned approximation scheme gave a new proof for the existence of Hunt processes for a large class of Dirichlet forms (i.e. strictly quasi-regular ones). But, as mentioned in the introduction of [MRZ98], the price for the new existence theorem is that one gets only the approximation of the path space measures P_x for quasi-every x in the state space.

This paper is a continuation of the discussion initiated in [MRZ98]. In fact, we shall prove that *any* Hunt process (associated to a Dirichlet form or not) can be approximated by multivariate Poisson processes and this approximation works for all P_x, i.e. *each* point x of the state space (see Theorem 2.1 below). This generalizes the approximations result in [MRE98] considerably. However, we do not recover the existence result of [MRE98] mentioned above.

We should mention that under some additional regularity assumptions on the state space (e.g., if it is a locally compact separable metric space) and on the underlying transition semigroup (e.g., if it is a Feller semigroup), the Yoshida approximation by Markov chains to the underlying process is well known (cf. the beautiful exposition in [EK86]). However, because of the involved tightness argument, it is difficult (at least to our knowledge) to relax the regularity assumptions.

The strategy we use in this paper is to change the topology of the state space, and then to implement the tightness argument in the new topological space. Note that although we change the topology during the proof, the limit process is always a Hunt process w.r.t. the original topology, as it was pointed out in [MRZ98, Theorem 4.4]. In this paper we shall eventually prove, moreover, that also the weak convergence takes place w.r.t. the original topology (see Section 5 below) which completes the results in [MRZ98].

2. Statement of the Main Result

Throughout this paper we assume that E is a metrizable Lusin space [DMe75, III, 16]. Let d be a compatible metric on E. We adjoin an extra point Δ (the cemetery) to E and write E_Δ for $E \cup \{\Delta\}$. If E is locally compact, then Δ can be considered either as an isolated point of E_Δ, or as a point 'at infinity' of E_Δ with the topology of the one point compactification. We select one of the above two topologies and fix it. If E is not locally compact then we simply consider Δ as an isolated point of E_Δ. Let $(\Omega, \mathcal{F}, (Z_t)_{t\geqslant 0}, (P_x)_{x\in E_\Delta})$ be a Hunt process with state space E (see [MR92, IV, 1.13] for the definition of Hunt processes on nonlocally compact spaces). The transition semigroup and resolvent of kernels related to $(Z_t)_{t\geqslant 0}$ are defined in a standard way, i.e.

$$P_t f(x) := P_t(x, f) := E_x\big[f(Z_t)\big], \quad \forall x \in E_\Delta,\ t \geqslant 0,\ f \in B_b(E_\Delta),$$

$$R_\alpha f(x) := R_\alpha(x, f) := E_x\left[\int_0^\infty \mathrm{e}^{-\alpha t} f(Z_t)\,\mathrm{d}t\right]$$

$$= \int_0^\infty \mathrm{e}^{-\alpha t} P_t f(x)\,\mathrm{d}t, \quad \forall \alpha > 0,\ x \in E_\Delta,\ f \in B_b(E_\Delta).$$

Here and henceforth, $B_b(E_\Delta)$ denotes all the bounded Borel functions on E_Δ. For fixed $n \in \mathbb{N}$, let $\{Y^n(k),\ k = 0, 1, \ldots\}$ be a Markov chain in E_Δ with some initial distribution ν and transition function nR_n, and let $(\Pi_t^n)_{t\geqslant 0}$ be a Poisson process with parameter n, i.e.

$$P(\Pi_t^n = k) = \mathrm{e}^{-nt}\frac{(nt)^k}{k!}.$$

Assume that $(\Pi_t^n)_{t\geqslant 0}$ is independent of $\{Y^n(k),\ k = 0, 1, \ldots\}$ and define

$$Z_t^n = Y^n(\Pi_t^n), \quad \forall t \geqslant 0. \tag{2.1}$$

Then $(Z_t^n)_{t\geqslant 0}$ is a strong Markov process in E_Δ. Following [J74] we call $(Z_t^n)_{t\geqslant 0}$ multivariate Poisson process with intensity n and state space E_Δ. Define

$$P_t^n f := \mathrm{e}^{-nt} \sum_{k=0}^{\infty} \frac{(nt)^k}{k!}(nR_n)^k f, \quad \forall f \in B_b(E_\Delta). \tag{2.2}$$

It is well known that $(P_t^n)_{t\geqslant 0}$ is the transition semigroup of $(Z_t^n)_{t\geqslant 0}$, i.e., for all $f \in B_b(E_\Delta)$, $t, s \geqslant 0$, we have

$$E\big[f(Z_{t+s}^n) \mid \sigma(Z_{s'}^n,\ s' \leqslant s)\big] = (P_t^n f)(Z_s^n)$$

(see [EK86, IV, 2]). Note that $(P_t^n)_{t\geqslant 0}$ is a strongly continuous contraction semigroup on the Banach space $(B_b(E_\Delta),\ \|\cdot\|_{E_\Delta})$ ($\|f\|_{E_\Delta} := \sup_{x\in E_\Delta} |f(x)|$), and the corresponding generator is given by

$$L^n u(x) = n(nR_n u(x) - u(x)) = n \int_{E_\Delta} (u(y) - u(x)) nR_n(x, \mathrm{d}y) \tag{2.3}$$

for all $u \in B_b(E_\Delta)$.

Let $D_{E_\Delta}[0, \infty)$ be the space of all cadlag functions from $[0, \infty)$ to E_Δ, equipped with the Skorokhod topology related to d (see [EK86, III]). Let P_x (resp. P_x^n) be the law of $(Z_t)_{t\geqslant 0}$ (resp. $(Z_t^n)_{t\geqslant 0}$) on $D_{E_\Delta}[0, \infty)$ with initial distribution δ_x for $x \in E_\Delta$.

Now we can state our main result.

THEOREM 2.1 (cf. Theorem 5.3 below). *For any $x \in E_\Delta$, P_x^n weakly converges to P_x on $D_{E_\Delta}[0, \infty)$ (with respect to the Skorokhod topology related to d), when $n \to \infty$.*

3. Construction of a Compact Metric Space $(\overline{E}, \rho)$

Without loss of generality we may assume that $\Omega = \Omega_{E_\Delta} := D_{E_\Delta}[0, \infty)$ and $(Z_t)_{t\geqslant 0} = (X_t)_{t\geqslant 0}$ be the coordinate process on Ω_{E_Δ}. As usual, we set $X_\infty := \Delta$. Let $(\mathcal{F}_t^n)_{t\geqslant 0}$, $(\mathcal{F}_t)_{t\geqslant 0}$ be the natural right continuous filtration of $(X_t)_{t\geqslant 0}$ completed w.r.t. $(P_x^n)_{x\in E_\Delta}$, $(P_x)_{x\in E_\Delta}$, respectively. By a routine argument following [Ge75], one can check that

$$M_n := (\Omega_{E_\Delta},\ (X_t)_{t\geqslant 0},\ (\mathcal{F}_t^n)_{t\geqslant 0},\ (P_x^n)_{x\in E_\Delta})$$

and

$$M := (\Omega_{E_\Delta},\ (X_t)_{t\geqslant 0},\ (\mathcal{F}_t)_{t\geqslant 0},\ (P_x)_{x\in E_\Delta})$$

are Hunt processes. For a Borel subset $B \in \mathcal{B}(E_\Delta)$, we define

$$\begin{aligned}\sigma_B &:= \inf\{t \geqslant 0 \mid X_t \in B\},\\ \tau_B &:= \inf\{t \geqslant 0 \mid X_t \in B \text{ or } X_{t-} \in B\}.\end{aligned}$$

(Here and henceforth we make the convention that $X_{0-} = X_0$.) Note that if B is open, then σ_B coincides with τ_B.

From now on we fix $x_0 \in E_\Delta$. Following LeJan [L83], we define

$$\mathrm{Cap}_{x_0}(B) := E_{x_0}[\mathrm{e}^{-\tau_B}], \quad B \in \mathcal{B}(E_\Delta).$$

It has been shown in [L83] that Cap_{x_0} is a Choquet capacity on E_Δ. Moreover, for any $f \in \mathcal{B}_b(E_\Delta)$, $R_\alpha f$ is Cap_{x_0}-quasi-continuous, i.e., there exists an increasing sequence of closed set $(F_k)_{k\in\mathbb{N}}$ such that $\mathrm{Cap}_{x_0}(E_\Delta\backslash F_k) \downarrow 0$ and $R_\alpha f$ is continuous on F_k for each $k \in N$.

Remark. In [L83] it was assumed that E is a locally compact metrizable space. But the argument in [L83] goes through for a metrizable Lusin space E. See Section 5 in Chapter IV of [MR92].

Let Q_+, Q_+^* denote the nonnegative, respectively, strictly positive rational numbers. Let J_0 be a countable family of nonnegative bounded continuous functions on E_Δ which separates the points of E_Δ and contains the constant function 1. Let J_1 be the smallest family of excessive functions containing $\{R_\alpha u \mid u \in J_0, \alpha \in Q_+^*\}$ and having the properties that if $u, v \in J_1, \alpha, c_1, c_2 \in Q_+^*$, then $R_\alpha u$, $c_1u + c_2v$, $u \wedge v$ are all in J_1. Clearly, J_1 separates the points of E_Δ and contains the constant function 1.

LEMMA 3.1. *There exists an increasing sequence of compact subsets $\{E_k\}_{k\in\mathbb{N}}$ of E_Δ such that*

(i) $E_{x_0}[\mathrm{e}^{-\sigma_{E_\Delta\backslash F_k}}] \to 0$ *as* $k \to \infty$*;*
(ii) $u|_{E_k}$ *is continuous for each* $u \in J_1$ *and* $k \in \mathbb{N}$.

Proof. By [L83, Lemma 2] (see also the argument of [MR92, IV, Section 5]) each $u \in J_1$ is Cap_{x_0}-quasi-continuous. By [FOT94, Lemma 7.1.1], J_1 is a countable family. Therefore, it follows from, e.g., [MR92, III, 3.3] that there exists a common increasing sequence $(F_k)_{k\in\mathbb{N}}$ of closed subsets of E_Δ such that $\mathrm{Cap}_{x_0}(E_\Delta\backslash F_k) \to 0$ and $u|_{E_k}$ is continuous for all $u \in J_1$ and $k \in \mathbb{N}$. On the other hand, by [MR92, Theorem IV, 1.15], there exists an increasing sequence of compact subsets $(F_k')_{k\in\mathbb{N}}$ of E_Δ such that $P_{x_0}[\lim_{k\to\infty}\sigma_{E_\Delta\backslash F_k'} < \infty] = 0$. Let $E_k = F_k \cap F_k'$. Then $(E_k)_{k\in\mathbb{N}}$ satisfies assertions (i) and (ii). □

Note that by Lemma 3.1(i) we may assume that $x_0 \in E_k$ for all $k \in \mathbb{N}$. Since J_1 is countable. We may write $J_1 = \{u_j \mid j \in \mathbb{N}\}$. Set $Y := \bigcup_{k\geqslant 1} E_k$, where E_k is as specified in the above lemma. We define for $x, y \in Y$

$$\rho(x, y) = \sum_{j=1}^{\infty} \frac{1}{2^j}\big(|u_j(x) - u_j(y)| \wedge 1\big). \tag{3.1}$$

Let $\overline{E}$ be the completion of Y with respect to ρ. Then $\overline{E}$ is a compact metric space. Let $J_1 - J_1$ denote all the functions which are expressed by a difference of two functions in J_1. For $u \in J_1 - J_1$, we denote by $\overline{u}$ the ρ-continuous extension of $u|_Y$ to $\overline{E}$. Let J be the collection of all such $\overline{u}$. Then J is a lincar lattice and contains 1. Hence J is ρ-uniformly dense in $C_b(\overline{E})$ (:= all the bounded ρ-continuous functions on $\overline{E}$).

Remark. The metric ρ induces a topology on E_Δ which is called the Ray-topology and has been intensively discussed in the literature (see [Ge75, S88]).

Note that the ρ-topology and the d-topology coincide on E_k for each k, hence by Lemma 3.1 $(Z_t)_{t\geqslant 0}$ with initial distribution δ_{x_0} can be regarded as a cadlag process with state space $\overline{E}$. Let $(Z_t^n)_{t\geqslant 0}$ be defined by (2.1). By making use of the following lemma, one can show that for each $n \in \mathbb{N}$, $(Z_t^n)_{t\geqslant 0}$ with initial distribution δ_{x_0} can also be regarded as a cadlag process with state space $\overline{E}$.

LEMMA 3.2 (cf. [MRZ98, Lemma 3.8 and 3.9]). *For $n \geqslant 2$, $U \subset E_\Delta$ open, and $x \in E_\Delta$, we have*

$$E_x^n[\mathrm{e}^{-2\sigma_U}] \leqslant E_x[\mathrm{e}^{-\sigma_U}]. \tag{3.2}$$

Hence, for each $n \in \mathbb{N}$,

$$E_{x_0}^n[\mathrm{e}^{-\sigma_{E_\Delta\setminus E_k}}] \to 0, \quad \text{as } k \to \infty. \tag{3.3}$$

Proof. For any $x \in E_\Delta$ we denote $E_x[\mathrm{e}^{-\sigma_U}]$ by $\mathrm{e}(x)$. Since $(n-1)R_n\mathrm{e}(x) \leqslant \mathrm{e}(x)$ for each $n \geqslant 2$, it follows that $((n-1)R_n)^k\mathrm{e}(x) \leqslant \mathrm{e}(x)$ for all $k \in \mathbb{N}$. Hence

$$\begin{aligned} P_t^n e(x) &= \mathrm{e}^{-nt}\sum_{k=0}^{\infty}\frac{(nt)^k}{k!}(nR_n)^k\mathrm{e}(x) \\ &= \mathrm{e}^{-nt}\sum_{k=0}^{\infty}\frac{(nt)^k}{k!}\left(\frac{n}{n-1}\right)^k((n-1)R_n)^k\mathrm{e}(x) \\ &\leqslant \mathrm{e}^{-nt}\sum_{k=0}^{\infty}\frac{\left(\frac{n^2}{n-1}t\right)^k}{k!}\mathrm{e}(x) = \mathrm{e}^{(1+\frac{1}{n-1})t}\mathrm{e}(x). \end{aligned}$$

This gives

$$\mathrm{e}^{-2t}P_t^n\mathrm{e}(x) \leqslant \mathrm{e}(x), \quad x \in E_\Delta.$$

But $\lim_{t\to 0}\mathrm{e}^{-2t}P_t^n f(x) = f(x)$ holds for all $x \in E_\Delta$ and $f \in \mathcal{B}_b(E_\Delta)$. Hence $\mathrm{e}(x)$ is 2-excessive related to (P_t^n) and, consequently, we have

$$E_x^n\left[\mathrm{e}^{-2\sigma_U}\mathrm{e}(X_{\sigma_U})\right] \leqslant \mathrm{e}(x). \tag{3.4}$$

Since obviously for any $x \in E_\Delta$,

$$X_{\sigma_U} \in U \quad P_x^n\text{-a.s. on } \{\sigma_U < \infty\},$$

we have that for any $x \in E_\Delta$,

$$\mathrm{e}^{-2\sigma_U} \leqslant \mathrm{e}^{-2\sigma_U}e(X_{\sigma_U}), \quad P_x^n\text{-a.s.}$$

Therefore, (3.2) follows from (3.4). (3.3) is a direct consequence of Lemma 3.1(i) and (3.2). □

Let $\widetilde{P}_{x_0}$ (resp. $\widetilde{P}^n_{x_0}$) be the law of (Z_t) (resp. (Z^n_t)) on $D_{\overline{E}}[0,\infty)$ which is equipped with the Skorokhod topology related to ρ. The corresponding expectation will be denoted by $\widetilde{E}_{x_0}[\cdot]$ (resp. $\widetilde{E}^n_{x_0}[\cdot]$). The coordinate process on $D_{\overline{E}}[0,\infty)$ will be denoted again by $(X_t)_{t\geqslant 0}$.

LEMMA 3.2. *Let $f \in C_b(\overline{E})$. Then for any $T > 0$ and $\varepsilon > 0$, there exists a function $g \in J_1 - J_1$ such that*

$$\widetilde{E}_{x_0}\Big[\sup_{t\leqslant T}|\overline{R_1 g} - f|(X_t)\Big] \leqslant \varepsilon \tag{3.5}$$

and

$$\widetilde{E}^n_{x_0}\Big[\sup_{t\leqslant T}|\overline{R_1 g} - f|(X_t)\Big] \leqslant \varepsilon \quad \textit{for all } n \in \mathbb{N}. \tag{3.6}$$

Proof. Since J is uniformly dense in $C(\overline{E})$. We can take a function $u \in J_1 - J_1$ such that $\|\overline{u} - f|_{\overline{E}} \leqslant \varepsilon/3$. By Lemma 3.1 we can find $k_0 \in \mathbb{N}$ such that

$$\widetilde{E}_{x_0}[\mathrm{e}^{-\sigma_{\overline{E}\setminus E_{k_0}}}] = E_{x_0}[\mathrm{e}^{-\sigma_{E_\Delta\setminus E_{k_0}}}] \leqslant \frac{\varepsilon}{6}\mathrm{e}^{-2(T+1)}\big(\|\overline{u}\|_{\overline{E}} + 1\big)^{-1}.$$

By (3.2) for all $n \in \mathbb{N}$, we then have

$$\widetilde{E}^n_{x_0}[\mathrm{e}^{-2\sigma_{\overline{E}\setminus E_{k_0}}}] = E^n_{x_0}[\mathrm{e}^{-2\sigma_{E_\Delta\setminus E_{k_0}}}] \leqslant \frac{\varepsilon}{6}\mathrm{e}^{-2(T+1)}\big(\|\overline{u}\|_{\overline{E}} + 1\big)^{-1}.$$

The above two inequalities imply that

$$\widetilde{P}_{x_0}[\sigma_{\overline{E}\setminus E_{k_0}} \leqslant T+1] \leqslant \frac{1}{6}\frac{\varepsilon}{\|\overline{u}\|_{\overline{E}}+1}\mathrm{e}^{-(T+1)} \leqslant \frac{1}{6}\frac{\varepsilon}{\|\overline{u}\|_{\overline{E}}+1},$$

$$\widetilde{P}^n_{x_0}[\sigma_{\overline{E}\setminus E_{k_0}} \leqslant T+1] \leqslant \frac{1}{6}\frac{\varepsilon}{\|\overline{u}\|_{\overline{E}}+1}.$$

Since u is the difference of two excessive functions and E_{k_0} is compact, there exists $\alpha \in Q^*_+$ such that $\|\alpha R_\alpha u - u\|_{E_{k_0}} \leqslant \varepsilon/3$. Then

$$\widetilde{E}_{x_0}\Big[\sup_{t\leqslant T}|\overline{\alpha R_\alpha u} - f|(X_t)\Big] \leqslant \frac{\varepsilon}{3} + \widetilde{E}_{x_0}\Big[\sup_{t\leqslant T}|\overline{\alpha R_\alpha u} - \overline{u}|(X_t)\Big] \leqslant \varepsilon.$$

Similarly, for $n \in \mathbb{N}$,

$$\widetilde{E}^n_{x_0}\Big[\sup_{t\leqslant T}|\overline{\alpha R_\alpha u} - f|(X_t)\Big] \leqslant \varepsilon.$$

The proof is completed by employing the resolvent equation to rewrite $\alpha R_\alpha u$ as $R_1 g$ for some $g \in J_1 - J_1$. □

4. Convergence of the Finite-Dimensional Distributions

LEMMA 4.1. *Let $f = R_1 u$ for some $u \in \mathcal{B}_b(E_\Delta)$. Then for any $T > 0$,*

$$\sup_{t \leqslant T} \sup_{x \in E_\Delta} |P_t^n f(x) - P_t f(x)| \longrightarrow 0, \quad n \to \infty. \tag{4.1}$$

In particular, if $u \in J_1 - J_1$, then $(t, x) \to P_t f(x)$ is uniformly continuous on $[0, T] \times Y$ where $Y := \bigcup_{k \geqslant 1} E_k$ is equipped with ρ-topology.

Proof. We assume first that $f = R_\alpha R_1 u$ for some $\alpha \in Q_+^*$ and $u \in \mathcal{B}_b(E_\Delta)$. In this case we have (cf. the proof of [MRZ98, Theorem 4.3])

$$\sup_{t \leqslant T} \sup_{x \in E_\Delta} \left|P_t^{n_1} f(x) - P_t^{n_2} f(x)\right| \leqslant T\left(\frac{1}{n_1} + \frac{1}{n_2}\right) \|w\|_{E_\Delta} \tag{4.2}$$

with $w := R_1(\alpha R_\alpha u - u) - (\alpha R_\alpha u - u) \in \mathcal{B}_b(E_\Delta)$.

Indeed by (2.2) we have for $n_1, n_2 \in \mathbb{N}$

$$\begin{aligned} P_t^{n_1} f - P_t^{n_2} f &= \int_0^t \frac{\mathrm{d}}{\mathrm{d}s} (P_s^{n_1} P_{t-s}^{n_2} f) \,\mathrm{d}s \\ &= \int_0^t P_s^{n_1} P_{t-s}^{n_2} (L^{n_1} - L^{n_2}) f \,\mathrm{d}s, \end{aligned}$$

consequently

$$\sup_{t \leqslant T} \sup_{x \in E_\Delta} \left|P_t^{n_1} f(x) - P_t^{n_2} f(x)\right| \leqslant T \sup_{x \in E_\Delta} \left|L^{n_1} f(x) - L^{n_2} f(x)\right|. \tag{4.3}$$

Using the explicit formula (2.3) one can check that

$$(L^{n_1} - L^{n_2}) f = R_{n_1} w - R_{n_2} w$$

with w specified as above. Consequently (4.2) follows from (4.3).

Taking into account the right continuity of $t \to P_t^n f(x)$, we conclude from (4.2) that for each $x \in E_\Delta$, there exists a right continuous function, say $P_t^* f(x)$, $t \in [0, \infty)$, such that for any $T > 0$

$$\sup_{t \leqslant T} \sup_{x \in E_\Delta} |P_t^n f(x) - P_t^* f(x)| \longrightarrow 0, \quad n \to \infty. \tag{4.4}$$

On the other hand, if we define for $\beta > 0$, $n \in \mathbb{N}$

$$R_\beta^n f(x) = E_x^n \left[\int_0^\infty \mathrm{e}^{-\beta t} f(X_t) \,\mathrm{d}t \right], \quad \forall x \in E_\Delta.$$

Then one can check that (cf. [MRZ98, Lemma 4.1])

$$R_\beta^n f = \left(\frac{n}{\beta + n}\right)^2 R_{\frac{\beta n}{\beta + n}} f + \frac{1}{\beta + n} f.$$

Letting $n \to \infty$ we see that for all $\beta > 0$

$$\int_0^\infty e^{-\beta t} P_t^n f(x)\,dt = R_\beta^n f(x) \to R_\beta f(x) = \int_0^\infty e^{-\beta t} P_t f(x)\,dt.$$

By the uniqueness of the Laplace transform, we conclude that $P_t^* f(x) = P_t f(x)$ for all $t \in [0, \infty)$. Hence (4.1) is verified for $f = R_\alpha R_1 u$, which extends to the general situation immediately since by the resolvent equation $\lim_{\alpha\to\infty} \|(\alpha - 1) R_\alpha R_1 f - R_1 f\|_{E_\Delta} = 0$ for all $f \in \mathcal{B}_b(E_\Delta)$. The last assertion of the lemma follows from (4.1) and the fact that $(t, x) \to P_t^n f(x)$ is uniformly continuous on $[0, T] \times Y$ for each $n \in \mathbb{N}$ and $f \in J_1 - J_1$. □

PROPOSITION 4.2. *For any* $m \in \mathbb{N}$, $f \in C_b(\overline{E}^m)$, $0 = t_1 < t_2 < \cdots < t_m$,

$$\widetilde{E}_{x0}^n[f(X_{t_1}, X_{t_2}, \ldots, X_{t_m})] \to \widetilde{E}_{x0}[f(X_{t_1}, X_{t_2}, \ldots, X_{t_m})], \quad n \to \infty. \tag{4.5}$$

Proof. By virtue of Stone–Weirestrass theorem, we need only to show that for any $m \in \mathbb{N}$, $f_1, f_2, \ldots, f_m \in C_b(\overline{E})$, $0 = t_1 < t_2 < \cdots < t_m$,

$$\begin{aligned}&\widetilde{E}_{x0}^n[f_1(X_{t_1}) f_2(X_{t_2}) \ldots f_m(X_{t_m})]\\ &\quad\to \widetilde{E}_{x0}[f_1(X_{t_1}) f_2(X_{t_2}) \ldots X_{t_m})], \quad n \to \infty.\end{aligned} \tag{4.6}$$

Note that (4.6) holds for $m = 1$ by (4.1). Suppose (4.6) holds for $m = k$, we are going to show that then it holds also for $m = k + 1$. To this end let $f_1, f_2, \ldots, f_k \in C_b(\overline{E})$ and $f_{k+1} = \overline{R_1 u}$ for some $u \in J_1 - J_1$. Let φ be the ρ-continuous extension of $(P_{t_{k+1}-t_k} f_{k+1})|_Y$. Then $f_k\varphi \in C_b(\overline{E})$. By the induction assumption and the Markov property of $(Z_t)_{t\geqslant 0}$, we get

$$\widetilde{E}_{x0}^n\big[f_1(X_{t_1}) \ldots (f_k\varphi)(X_{t_k})\big] \to \widetilde{E}_{x0}\big[f_1(X_{t_1}) \ldots f_k(X_{t_k}) f_{k+1}(X_{t_{k+1}})\big]. \tag{4.7}$$

But if we denote the ρ-continuous extension of $(P_{t_{k+1}-t_k}^n f_{k+1})|_Y$ by φ_n, then by the Markov property of $(Z_t^n)_{t\geqslant 0}$ we have

$$\widetilde{E}_{x0}^n\big[f_1(X_{t_1}) \ldots (f_k\varphi_n)(X_{t_k})\big] = \widetilde{E}_{x0}^n\big[f_1(X_{t_1}) \ldots f_k(X_{t_k}) f_{k+1}(X_{t_{k+1}})\big]. \tag{4.8}$$

By the above lemma we have $\|\varphi_n - \varphi\|_{\overline{E}} \to 0$ as $n \to \infty$. Therefore, (4.8) implies

$$\begin{aligned}&\big|\widetilde{E}_{x0}^n\big[f_1(X_{t_1}) \ldots f_{k+1}(X_{t_{k+1}})\big] - \widetilde{E}_{x0}^n\big[f_1(X_{t_1}) \ldots (f_k\varphi)(X_{t_k})\big]\big|\\ &\quad\leqslant \|f_1\|_{\overline{E}} \|f_2\|_{\overline{E}} \cdots \|_k\|_{\overline{E}} \|\varphi_n - \varphi\|_{\overline{E}} \to 0\end{aligned}$$

as $n \to \infty$, which together with (4.7) inplies that

$$\widetilde{E}_{x0}^n\big[f_1(X_{t_1}) \ldots f_{k+1}(X_{t_{k+1}})\big] \to \widetilde{E}_{x0}\big[f_1(Xt_1) \ldots f_{k+1}(X_{t_{k+1}})\big]$$

for $f_1, f_2, \ldots, f_k \in C_b(\overline{E})$ and $f_{k+1} = \overline{R_1 u}$ with $u \in J_1 - J_1$. The proof is completed by employing Lemma 3.3. □

5. Weak Convergence of $\{P_{x_0}^n\}_{n\in\mathbb{N}}$

Let $D_R[0,\infty)$ be the space of all real-valued cadlag functions on $[0,\infty)$.

LEMMA 5.1. *For any $f \in C_b(\overline{E})$, the laws of $\{f \circ Z^n\}_{n\in\mathbb{N}}$ with initial distribution δ_{x_0} form a tight family on $D_R[0,\infty)$.*

Proof. Let $f \in C_b(\overline{E})$. Then by Lemma 3.3 for every $\varepsilon > 0$ and $T > 0$ there exists $g \in J_1 - J_1$ such that if we set $\xi^n(t) = \overline{R_1 g}(Z_t^n)$, then

$$E_{x_0}^n\Big[\sup_{t\leqslant T} |\xi^n(t) - f(Z_t^n)|\Big] \leqslant \varepsilon.$$

By the strong Markov property $\xi^n(t)$ is expressed by

$$\xi^n(t) - \xi^n(0) = M_t^n + \int_0^t L^n(R_1 g)(Z_s^n)\,\mathrm{d}s$$

where $(M_t^n)_{t\geqslant 0}$ is a $P_{x_0}^n$-martingale. By (2.3) we have $L^n(R_1 g) = nR_n(R_1 g - g)$. Therefore, by the contraction property of nR_n, we have

$$\sup_{n\in\mathbb{N}} E_{x_0}^n\left[\int_0^T \sup_{t\leqslant T}\big|L^n(R_1 g)(Z_t^n)\big|\mathrm{d}s\right] \leqslant T\|R_1 g - g\|_{E_\Delta} < \infty.$$

Thus the lemma is proved by applying [EK86, III, Theorem 9.4]. □

PROPOSITION 5.2. *As $n \to \infty$, $\widetilde{P}_{x_0}^n$ converges weakly to $\widetilde{P}_{x_0}$ on $D_{\overline{E}}[0,\infty)$ with respect to the Skorokhod topology related to ρ.*

Proof. By virtue of the above lemma, we see that by [EK86, III, Theorem 9.1] $\{\widetilde{P}_{x_0}^n\}_{n\in\mathbb{N}}$ is relatively compact on $D_{\overline{E}}[0,\infty)$. But by Proposition 4.2 and [EK86, III, Theorem 7.8] any weak limit of a subsequence of $\{\widetilde{P}_{x_0}^n\}_{n\in\mathbb{N}}$ must have the same finite-dimensional distributions as $\widetilde{P}_{x_0}$ and, hence, coincides with $\widetilde{P}_{x_0}$. □

THEOREM 5.3. *As $n \to \infty$, $P_{x_0}^n$ converges weakly to P_{x_0} on $D_{E_\Delta}[0,\infty)$ with respect to the Skorokhod topology related to d.*

Proof. Let $\varepsilon > 0$ be arbitrary. By Lemma 3.1(i) and Lemma 3.2 we can inductively take a subsequence $\{k_l\}_{l\geqslant 1} \subset \mathbb{N}$ such that

$$\inf_n\big\{\widetilde{P}_{x_0}^n(B_l)\big\} \wedge \widetilde{P}_{x_0}(B_l) > 1 - \frac{\varepsilon}{2^l},$$

where

$$B_l := \left\{w \in D_{\overline{E}}[0,\infty) \mid w(t) \in E_{k_l} \text{ for all } t \in \left[0, -\log\frac{\varepsilon}{2^l}\right)\right\}.$$

Set $B = \bigcap_{l\geqslant 1} B_l$. Then B is a closed subset of $D_{\overline{E}}[0,\infty)$ (since each B_l is so by [EK86, III, Proposition 5.2 and III, Lemma 5.1]). Since the ρ-topology and the d-topology coincide on each E_k, by [EK86, III, Proposition 6.5] one can check that B is also a closed subset of $D_{E_\Delta}[0,\infty)$ and the two topologics coincide on B. Moreover, $\widetilde{P}_{x_0}$ (resp. $\widetilde{P}^n_{x_0}$) coincides with P_{x_0} (resp. $P^n_{x_0}$) on B. Let $f \in C_b(D_{E_\Delta}[0,\infty))$. Then $f|_B$ is continuous on $B \subset D_{\overline{E}}[0,\infty)$ w.r.t. the trace topology of $D_{\overline{E}}[0,\infty)$. By Tietz's extension theorem f can be continuously extended to $D_{\overline{E}}[0,\infty)$, without enlarging the uniform norm $\|f\|_\infty := \sup_{x\in D_{E_\Delta}[0,\infty)} |f(x)|$. Let $\overline{f}$ denote this extension. Thus

$$
\begin{aligned}
&\overline{\lim}_{n\to\infty}\left|\int_{D_{E_\Delta}[0,\infty)} f\,\mathrm{d}P^n_{x_0} - \int_{D_{E_\Delta}[0,\infty)} \mathrm{d}P_{x_0}\right| \\
&\leqslant \overline{\lim}_{n\to\infty}\Bigg[\left|\int_B f\,\mathrm{d}P^n_{x_0} - \int_B f\,\mathrm{d}P_{x_0}\right| + \\
&\quad + \left|\int_{D_{E_\Delta}[0,\infty)\setminus B} f\,\mathrm{d}P^n_x\right| + \left|\int f_{D_{E_\Delta}[0,\infty)\setminus B} f\,\mathrm{d}P_{x_0}\right|\Bigg] \\
&\leqslant \overline{\lim}_{n\to\infty}\Bigg[\left|\int_B f\,\mathrm{d}P^n_{x_0} - \int_B f\,\mathrm{d}P_{x_0}\right| + 2\varepsilon\|f\|_\infty \\
&\leqslant \overline{\lim}_{n\to\infty}\Bigg[\left|\int_{D_{\overline{E}}[0,\infty)} \widetilde{f}\,\mathrm{d}\widetilde{P}^n_{x_0} - \int_{D_{\overline{E}}[0,\infty)} \widetilde{f}\,d\,\widetilde{P}_{x_0}\right| + \left|\int_{D_{\overline{E}}[0,\infty)\setminus B} \widetilde{f}\,\mathrm{d}\widetilde{P}^n_{x_0}\right| + \\
&\quad + \left|\int_{D_{\overline{E}}[0,\infty)\setminus B} \widetilde{f}\,\mathrm{d}\widetilde{P}_{x_0}\right|\Bigg] + 2\varepsilon\|f\|_\infty \\
&\leqslant \overline{\lim}_{n\to\infty}\left|\int_{D_{\overline{E}}[0,\infty)} \widetilde{f}\,\mathrm{d}\widetilde{P}^n_{x_0} - \int_{D_{\overline{E}}[0,\infty)} \widetilde{f}\,\mathrm{d}\widetilde{P}_{x_0}\right| + 4\varepsilon\|f\|_\infty = 4\varepsilon\|f\|_\infty.
\end{aligned}
$$

Since ε was arbitrary, we conclude that

$$\lim_{n\to\infty}\int_{D_{\overline{E}}[0,\infty)} f\,\mathrm{d}P^n_{x_0} = \int_{D_{E_\Delta}[0,\infty)} f\,\mathrm{d}P_{x_0}.$$

Since $f \in C_b(D_{E_\Delta}[0,\infty))$ was arbitrary, this completes the proof. □

Acknowledgement

Financial support of SFB 343 Bielefeld, the Chinese National Natural Science Foundation, the Mathematical Center of State Education Commission, the Tian-Yuan Mathematics Foundation and the Morningside Center of Mathematics is gratefully acknowledged.

References

[AMR93] Albeverio, S., Ma, S. Z. M. and Röckner, M.: A remark on the support of Cadlag processes, In: *Stochastic Processes*, Springer, New York, 1993, pp. 1–5.

[DMe75] Dellacherie, C. and Meyer, P. A.: *Probabilité et potentiel*, Chapters II, V, 1975.

[EK86] Ethier, S. N. and Kurtz, T. G.: *Markov Processes: Characterization and Convergence*, Wiley, New York, 1986.

[FOT94] Fukushima, M., Oshima, Y. and Takeda, M.: *Dirichlet Forms and Symmetric Markov Processes*, de Gruyter, Berlin, 1994.

[Ge75] Getoor, R.: *Markov Processes: Ray Processes and Right Processes*, Lecture Notes in Math. 440, Springer, New York, 1975.

[J74] Jacod, J.: Multivariate point processes: predictable projection. Radon–Nikodym derivatives, representation of martingales, *Z. Wahrscheinlichkeitstheorie und Verw. Gebiete* **31** (1974/1975), 235–253.

[L83] LeJan, Y.: Quasi-continuous functions and Hunt processes, *J. Math. Soc. Japan* **35**(1) (1983), 37–42.

[MR92] Ma, Z. M. and Röckner, M.: *Introduction to the Theory of (Non-symmetric) Dirichlet Forms*, Springer, New York, 1992.

[MRZ98] Ma, Z. M., Röckner, M. and Zhang, T. S.: Approximation of arbitrary Dirichlet processes by Markov chains, *Ann. Inst. H. Poincaré* **34** (1998), 1–22.

[S88] Sharpe, M. J.: *General Theory of Markov Processes*, Academic Press, San Diego, 1988.

[SZ96] Stroock, D. W. and Zheng, W.: Markov chain approximations to symmetric diffusions, *Ann. Inst. H. Poincaré* **33** (1997), 619–649.

Note Added in Proof. After this paper was finished, both P. Fitzsimmons and T. Kurtz kindly pointed out to us that the main results of this paper can be obtained more easily by using the paths rather than the distribution of the limiting process. Nevertheless, we think that our different method of proof has its own interest, in particular, in regard to constructing the limiting process [MRZ98].

Acta Applicandae Mathematicae **63:** 245–252, 2000.

Bayes Formula for Optimal Filter with n-ple Markov Gaussian Errors

PRANAB K. MANDAL[1] and V. MANDREKAR[2]
[1]*EURANDOM, PO Box 513 5600 MB, Eindhoven, The Netherlands.*
e-mail: mandal@eurandom.tue.nl
[2]*Michigan State University, East Lansing, MI 48824, U.S.A.*

(Received: 22 December 1999)

Abstract. We consider the nonlinear filtering problem where the observation noise process is n-ple Markov Gaussian. A Kallianpur–Striebel type Bayes formula for the optimal filter is obtained.

Mathematics Subject Classifications (2000): 60G35, 60G15, 62M20, 93E11.

Key words: nonlinear filtering, Bayes formula, n-ple Markov Gaussian process.

Introduction

Professor Hida has been interested in nonlinear problems and the study of n-ple Markov processes. The purpose of this work is to obtain a Bayes formula for an optimal filter in the case where the noise process is n-ple Markov in the sense of Levy and Hida ([6, 3]). We apply here the general result obtained by us [7]. The main effort here is to compute a tractable form of Reproducing Kernel Hilbert Space (RKHS) using the Goursat representation of n-ple Markov processes obtained by one of the authors [7] under active encouragement of Professor Hida. As a consequence of our work, we are able to obtain a proper form of the Bayes formula given by Kunita [5] for the noise process being a solution of stochastic differential equation of order n. It is a great pleasure to dedicate this work to Professor T. Hida who has been a friend and mentor for one of the authors (Mandrekar) for more than 30 years.

1. Preliminaries and Notation

The general filtering problem can be described as follows. The signal or system process $\{X_t, 0 \leqslant t \leqslant T\}$ is unobservable. Information about $\{X_t\}$ is obtained by observing another process Y which is a function of X corrupted by noise, i.e.,

$$Y_t = \beta(t, X) + N_t, \quad 0 \leqslant t \leqslant T, \tag{1}$$

where β_t is measurable with respect to $\mathcal{F}_t^X$, the σ-field generated by the 'past' of the signal, i.e. $\sigma\{X_u, 0 \leqslant u \leqslant t\}$ (augmented by the inclusion of zero probability sets) and N_t is some noise process. The observation σ-field $\mathcal{F}_t^Y = \sigma\{Y_u, 0 \leqslant u \leqslant t\}$. The aim of filtering theory is to get an estimate of X_t based on $\mathcal{F}_t^Y$. This is given by the conditional distribution of X_t given $\mathcal{F}_t^Y$ or, equivalently, by the conditional expectations $E(f(X_t)|\mathcal{F}_t^Y)$ for a large enough class of functions f. Kallianpur and Striebel [4] provided an explicit Bayes formula for the conditional expectation in case N_t is a Brownian Motion W_t. In [7], this formula was extended to general N_t, a Gaussian process, with a restriction on $\beta(t, X)$. In order to explain it, we need to introduce the idea of RKHS.

Given a Gaussian process $\{N_t, 0 \leqslant t \leqslant T\}$, w.l.o.g. assume that $EN_t = 0$ and let $R_N(s,t) := EN(t)N(s)$ be its covariance function. It uniquely determines the distribution of N. Once the Gaussian process becomes clear within context, we will drop the subscript N. Given a symmetric, nonnegative definite function R (covariance), we can associate with it a unique Hilbert space of functions, called the RKHS $H(R)$ of R, satisfying

$$\begin{aligned} &\text{(a) } R(\cdot, t) \in H(R) \quad \text{for } 0 \leqslant t \leqslant T, \\ &\text{(b) For all } f \in H(R),\ (f, R(\cdot, t)) = f(t), \quad 0 \leqslant t \leqslant T. \end{aligned} \tag{2}$$

Here $(\cdot, \cdot)$ denotes the inner product in $H(R)$. Thus, there is a one-one correspondence between Gaussian processes and RKHS's generated by covariances ([1]). In fact, the simplest way to describe $H(R_N)$ is to consider the functions $f(t) = EN(t)U$, where $U \in \overline{\text{sp}}\{N(t), 0 \leqslant t \leqslant T\}$ and $(f_1, f_2) = EU_1U_2$. We denote the unique U by $\langle N, f\rangle$. Under the assumption that $\beta(\cdot, X) \in H(R_N)$ a.s., the following extension of the Kallianpur–Striebel formula was given in [7]. Below, $\langle N, f\rangle_t$ and $||\cdot||_t$ denote the $\langle N, f\rangle$, as defined above, and the norm, respectively, corresponding to the RKHS of $R_N|_{[0,t]\times[0,t]}$.

THEOREM. *Suppose that the observation process is given by* (1) *with noise N, a Gaussian process with covariance R and $\beta(\cdot, X(w)) \in H(R)$ a.s. w. Then for an $\mathcal{F}_T^X$ measurable and integrable function $g(T, X)$,*

$$E(g(T,X)|\mathcal{F}_t^Y) = \frac{\int g(T,x)\mathrm{e}^{\langle Y,\beta(\cdot,x)\rangle_t - \frac{1}{2}||\beta(\cdot,x)||_t^2}\,\mathrm{d}P_X(x)}{\int \mathrm{e}^{\langle Y,\beta(\cdot,x)\rangle_t - \frac{1}{2}||\beta(\cdot,x)||_t^2}\,\mathrm{d}P_X(x)},$$

where

$$\langle Y, \beta(\cdot, x)\rangle_t = \langle N, \beta(\cdot, x)\rangle_t + ||\beta(\cdot, x)||_t^2.$$

As stated in the introduction, we want to compute explicitly the above formula in the case of n-ple Markov processes, which involves computing the RKHS of such processes. For this, we need the form of $L^2(M)$, the space of vector-valued functions, square integrable with respect to matrix-valued measure M ([10]). Let

M be a nonnegative $n \times n$ matrix-valued measure M on a measurable space $(S, \mathcal{S})$. For two n-dimensional (row) vectors, ψ_1 and ψ_2 we define the integral

$$\int \psi_1 \, \mathrm{d}M \psi_2^* = \int (M' \psi_2, \psi_1) \, \mathrm{d}\mu$$

for any nonnegative σ-finite measure μ which dominates the measures $m_{i,j}$ $(i, j = 1, 2, \ldots, n)$, the entries of M. Here $M' = ((\mathrm{d}m_{ij}/\mathrm{d}\mu))$. It can be easily seen that the definition above does not depend on μ.

In particular, $\operatorname{tr} M = \sum_{i=1}^{n} m_{ii}$ is a dominating measure. We denote by

$$L^2(M) = \left\{ \psi : S \to \mathcal{R}^n \text{ (row), such that } \int \psi \, \mathrm{d}M \psi^* < \infty \right\}.$$

Then $L^2(M)$ is a Hilbert space under the inner product

$$(\psi_1, \psi_2)_M = \int \psi_1 \, \mathrm{d}M \psi_2^*, \quad \text{for } \psi_1, \psi_2 \in L^2(M).$$

Let $\{\underline{Z}(t), 0 \leqslant t \leqslant T\}$ be an n-dimensional (row) Gaussian martingale. Then

$$E(\underline{Z}(t) - \underline{Z}(s))^*(\underline{Z}(t) - \underline{Z}(s)) = F_z(t) - F_z(s), \tag{3}$$

where F is a nonnegative matrix-valued 'increasing' function (i.e., $F_z(t) - F_z(s)$ is a nonnegative definite matrix if $s \leqslant t$). Let M_z be the matrix-valued measure associated with F_z. Then it is easy to see that we can define

$$\int (\psi(u), \mathrm{d}\underline{Z}(u)) = \sum_{i=1}^{n} \int \psi_i(u) \, \mathrm{d}Z_i(u)$$

for all $\psi \in L^2(M_z)$ and $\overline{\mathrm{sp}}\{Z_i(u), i = 1, 2, \ldots, n, u \leqslant t\}$ equals

$$\left\{ \int_0^t (\psi(u), \mathrm{d}\underline{Z}(u)), \ \psi \in L^2(M_z) \right\}.$$

2. n-ple Markov Gaussian Processes

Following Hida [3] and Levy [6], we define a Gaussian process N to be n-ple Markov if, for each s, the set $\{E(N(t) \mid \mathcal{F}_s^N), t \geqslant s\}$ in $L^2(\Omega, \mathcal{F}_T^N, P)$ has exactly n linearly independent elements. It was shown in [8] (see also [9]) that such processes have Goursat representation, i.e.,

$$N(t) = \sum_{i=1}^{n} \varphi_i(t) Z_i(t), \tag{4}$$

where $\underline{Z} = (Z_1, \ldots, Z_n)$ is a nonsingular vector-valued martingale and $\det((\varphi_i(t_j))) \neq 0$ for any $0 \leqslant t_1 < t_2 < \cdots < t_n \leqslant T$. Here by nonsingular we mean $F_z(t)$ of (3) is a nonsingular $n \times n$ matrix for each t. In particular, N has representation (4) with $H(N:t) = H(\underline{Z}:t)$, where

$$\begin{aligned} H(N:t) &:= \overline{\text{sp}}\{N(s), s \leqslant t\}, \quad \text{and} \\ H(\underline{Z}:t) &:= \overline{\text{sp}}\{Z_i(s), s \leqslant t, 1 \leqslant i \leqslant n\}. \end{aligned} \tag{5}$$

We now derive the RKHS of N of the form (4).

LEMMA 1. *Let* $N(t) = \sum_{i=1}^{n} \varphi_i(t) Z_i(t)$, *where* $\underline{Z}(t)$ *is an n-dimensional Gaussian martingale. Then*

$$H(R_N) = \left\{ f : f(t) = \int_0^t \varphi(t)\,\mathrm{d}M_z(u)\psi^*(u),\ \psi \in L^2(M_z) \right\}$$

with

$$(f_1, f_2)_{H(R_N)} = \int_0^T \psi_1(u)\,\mathrm{d}M_z(u)\psi_2^*(u).$$

Proof.

$$\begin{aligned} R_N(\cdot, t) &= EN(\cdot)N(t) \\ &= \sum_{i=1}^{n}\sum_{j=1}^{n} \varphi_i(\cdot)\varphi_j(t) m_{ij}(t, \cdot). \end{aligned}$$

Define

$$\psi_t(u) = (\varphi_1(t)1_{[0,t]}(u), \varphi_2(t)1_{[0,t]}(u), \ldots, \varphi_n(t)1_{[0,t]}(u)).$$

Then

$$\psi_t \in L^2(M_z) \quad \text{and} \quad R_N(\cdot, t) = \int_0^{\cdot} \varphi(\cdot)\,\mathrm{d}M_z(u)\psi_t^*(u),$$

i.e., $R_N(\cdot, t) \in H(R_N)$. Also, if $f(t) = \int_0^t \varphi(t)\,\mathrm{d}M_z(u)\psi^*(u)$, then

$$\begin{aligned} (f, R_N(\cdot, t)) &= \int_0^T \psi_t(u)\,\mathrm{d}M_z(u)\psi^*(u) \\ &= \int_0^t \varphi(t)\,\mathrm{d}M_z(u)\psi^*(u) = f(t). \end{aligned}$$

Thus $H(R_N)$ is the RKHS of N. □

Note that

$$\langle N, f\rangle_t = \int_0^t (\psi(u), \mathrm{d}Z(u)).$$

Suppose $N(t)$ is a solution of the stochastic differential equation of order n given by

$$L_t N(t) = \dot{W}(t), \quad \text{with } \dot{W}(t) \text{ `white noise'} \tag{6}$$

and where

$$L(t) = a_n(t)\left(\frac{\mathrm{d}}{\mathrm{d}t}\right)^n + a_{n-1}\left(\frac{\mathrm{d}}{\mathrm{d}t}\right)^{n-1} + \cdots + a_0(t) \quad (\text{with } a_n(t) > 0). \tag{7}$$

More precisely, $N(t)$ is an $(n-1)$ times differentiable Gaussian process and satisfies

$$a_n(t)\,\mathrm{d}N^{(n-1)}(t) + \big(a_{n-1}(t)N^{(n-1)}(t) + \cdots + a_0(t)N(t)\big)\,\mathrm{d}t = \mathrm{d}W(t).$$

Then ([2]),

$$N(t) = \int_0^t \sum_{i=1}^n \varphi_i(t)g_i(u)\,\mathrm{d}W(u),$$

where $\{g_i(u)\}_{i=1}^n$ are solutions of the adjoint homogeneous equation corresponding to $L_t h = 0$ and $\{\varphi_i(t)\}_{i=1}^n$ are determined so that $F(t,u) := \sum_{i=1}^n \varphi_i(t)g_i(u)$ satisfies the following conditions:

$$\left.\frac{\partial^k F(t,u)}{\partial u^k}\right|_{u=t} = \begin{cases} 0, & k = 0, 1, \ldots, (n-2), \\ (-1)^{(n-1)}/a_n(t), & k = (n-1). \end{cases}$$

Hence, with $Z_i(t) = \int_0^t g_i(u)\,\mathrm{d}W(u)$,

$$N(t) = \sum_{i=1}^n \varphi_i(t)Z_i(t)$$

is of the form as in Lemma 1. Here $m_{ij}(B) = \int_B g_i(u)g_j(u)\,\mathrm{d}\lambda(u)$ where λ is the Lebesgue measure. The space $L^2(M_z)$ consists of the functions ψ so that

$$\int_0^T \left(\sum_{i=1}^n \psi_i(u)g_i(u)\right)^2 \mathrm{d}\lambda(u) < \infty$$

and $f \in H(R_N)$ iff

$$\begin{aligned} f(t) &= \int_0^t \left(\sum_{i=1}^n \varphi_i(t)g_i(u)\right)\left(\sum_{j=1}^n \psi_j(u)g_j(u)\right)\mathrm{d}\lambda(u) \\ &= \int_0^t F(t,u)g^*(u)\,\mathrm{d}u, \end{aligned}$$

for some $g^* \in L^2(\lambda)$, if we assume $H(N:t) = H(W:t)$. This is the case for Levy–Hida n-ple Markov processes. In this case, as in [7], we get

$$\langle N, f\rangle_t = \int_0^t g^*(u)\,\mathrm{d}W(u).$$

This also is the case if $N(t)$ is a solution of (6) as $g \in L^2(\lambda)$ and $g \perp F(t,\cdot)1_{[0,t]}(\cdot)$ for all t, implies

$$\int_0^t F(t,u)g(u)\,\mathrm{d}u = 0 \quad \forall t.$$

Then using the fact that $F(t,u)$ is the Riemann function ([2]), we get

$$g(t) = L_t \int_0^{(\cdot)} F(\cdot,u)g(u)\,\mathrm{d}u = 0.$$

3. Bayes Formula for n-ple Markov Noise Processes

We now use our Theorem to obtain the Bayes formula in the case where the observation process Y is of the form (1) with $N(t) = \sum_{i=1}^n \varphi_i(t)Z_i(t)$ with $\underline{Z}(t)$ an n-dimensional Gaussian martingale.

PROPOSITION 1. *Let X be a system process and Y be given by* (1) *with N as above and*

$$\beta(t,X) = \int_0^t \varphi(t)\,\mathrm{d}M_z(u)\psi^*(u,X), \quad \textit{with } \psi(\cdot,X) \in L^2(M_z) \text{ a.e.}$$

Then, for an $\mathcal{F}_T^X$-measurable and integrable function $g(T,X)$, we have, with $\hat{Y}(t) = \underline{Z}(t) + \int_0^t \psi(u)\,\mathrm{d}M_z(u)$,

$$E(g(T,X) \mid \mathcal{F}_t^Y) = \frac{\int g(T,x)\mathrm{e}^{\int_0^t (\psi(u,x),\,\mathrm{d}\hat{Y}(u)) - \frac{1}{2}\int_0^t \psi(u,x)\,\mathrm{d}M_z(u)\psi^*(u,x)}\,\mathrm{d}P_X(x)}{\int \mathrm{e}^{\int_0^t (\psi(u,x),\,\mathrm{d}\hat{Y}(u)) - \frac{1}{2}\int_0^t \psi(u,x)\,\mathrm{d}M_z(u)\psi^*(u,X)}\,\mathrm{d}P_X(x)}.$$

In the case where $Z_i(t) = \int_0^t g_i(u)\,\mathrm{d}W(u)$, we get a more appealing one-dimensional form because in this case

$$\beta(t,X) = \int_0^t F(t,u)\phi^*(u,X)\,\mathrm{d}\lambda(u),$$

where

$$\phi^*(u,X) = \sum_{i=1}^n g_i(u)\psi_i(u,X)$$

and

$$E(g(T,X) \mid \mathcal{F}_t^Y) = \frac{\int g(T,x) e^{\int_0^t \phi^*(u,x)\,d\hat{Y}(u) - \frac{1}{2}\int_0^t \phi^*(u,x)^2\,d\lambda(u)}\,dP_X(x)}{\int e^{\int_0^t \phi^*(u,x)\,d\hat{Y}(u) - \frac{1}{2}\int_0^t \phi^*(u,x)^2\,d\lambda(u)}\,dP_X(x)},$$

where $\hat{Y}(t) = \int_0^t \phi^*(u,X)\,du + W(t)$.

In particular, when the noise process (N_t) has derivative n_t satisfying the stochastic differential equation (6) and the observation process (y_t) is defined by $y_t = h(X_t) + n_t$, where $h(\cdot) \equiv h(X_\cdot)$ is n times continuously differentiable, we get the following (corrected) formula due to Kunita [5]:

$$E(g(X_t) \mid \mathcal{F}_t^Y) = \frac{\int g(x_t) e^{\int_0^t L_s h(x_\cdot)\,d\hat{Y}_s - \frac{1}{2}\int_0^t (L_s h(x_\cdot))^2\,ds}\,dP_X(x)}{\int e^{\int L_s h(x_\cdot)\,d\hat{Y}_s - \frac{1}{2}\int_0^t (L_s h(x_\cdot))^2\,ds}\,dP_X(x)},$$

where

$$\begin{aligned}\hat{Y}(t) &= \int_0^t L_s h(X_\cdot)\,ds + W(t)\\ &= \int_0^t a_n(u)\,dy^{(n-1)}(u) + \sum_{i=0}^{n-1} a_i(u) y^{(i)}(u)\,du.\end{aligned}$$

This follows immediately once we note that here

$$\beta(t,X)h(X_t) = \int_0^t F(t,u) L_u h(X_\cdot)\,du.$$

References

1. Chatterji, S. D. and Mandrekar, V.: Equivalence and singularity of Gaussian measures and applications, In: A. T. Barucha-Ried (ed.), *Probabilistic Analysis and Related Topics 1*, Academic Press, New York, 1978, pp. 169–197.
2. Dolph, C. L. and Woodbury, M. A.: On the relation between Green's functions and covariances of certain stochastic processes and its applications to unbiased linear prediction, *Trans. Amer. Math. Soc.* **72** (1952), 519–550.
3. Hida, T.: Canonical representations of Gaussian processes and their applications, *Mem. Coll. Sci. Kyoto A* **33** (1960), 109–155.
4. Kallianpur, G.: *Stochastic Filtering Theory*, Springer, New York, 1980.
5. Kunita, H.: Representation and stability of nonlinear filters associated with Gaussian noises, In: Stamatis Combanis *et al.* (eds), *Stochastic Processes* (*Kallianpur Fest.*), Springer, New York, 1993, pp. 201–210.
6. Levy, P.: Fonctions linearment markovian d'ordre n, *Math. Japonica* **4** (1957), 113–121.
7. Mandal, P. K. and Mandrekar, V.: Bayes formula for Gaussian noise processes and its applications, Preprint (1998), to appear in *SIAM J. Control. Optim.*

8. Mandrekar, V.: On the multiple Markov property of Levy–Hida for Gaussian processes, *Nagoya Math. J.* **54** (1974), 69–78.
9. Pitt, L.: Hida–Cramer multiplicity theory for multiple Markov processes and Goursat representations, *Nagoya Math. J.* **57** (1975), 99–128.
10. Rozanov, Yu.: Spectral theory of multidimensional stationary random processes with discrete time, *Select. Transl. Math. Stat. Probab. I*, 1961, pp. 127–132.

Acta Applicandae Mathematicae **63:** 253–273, 2000.

One Loop Approximation of the Chern–Simons Integral

Dedicated to Professor Takeyuki Hida on his 70th birthday

ITARU MITOMA
Department of Mathematics, Saga University, Saga, 840, Japan

(Received: 8 January 1999)

Abstract. It is proven that the one loop approximation of the Wilson line integral in a perturbative $\mathfrak{SU}(2)$ Chern–Simons theory is localized around the critical point in the large level.

Mathematics Subject Classifications (2000): 60H07, 81T13, 81T45.

Key words: abstract Wiener space, stationary phase, Chern–Simons integral, linking number.

1. Introduction

From Witten [12], various papers have been published concerning the relation between the Chern–Simons integral and the topological invariants of the 3-manifold, among them the linking number [3, 4, 7, 8]. According to Frölich and King [7], Albeverio and his colleague studied the integral for the commutative case using the Fresnel integral [1] and the noncommutative case by white noise distribution [2].

Very recently, Malliavin amd Taniguchi [10] obtained the complete formula for the change of variables on the abstract Wiener space. Based on the work of Bar-Natan and Witten [4], we will consider the one-loop approximation of the Chern–Simons integral, combining the idea of Itô [9] for the Feynman integral and the Malliavin–Taniguchi formula and we examine the stationary phase method (infinite-dimensional) for the integral.

Let M be a compact oriented 3-manifold, G a compact Lie group, and $\mathfrak{g}$ the corresponding Lie algebra with an Ad-invariant inner product $(\,,\,)_{\mathfrak{g}}$. Denote by $\mathcal{A}$ the collection of all G-connections ($\mathfrak{g}$-valued 1-forms on M). Let E_j, $j = 1, 2, \ldots, d$, be the orthonormal basis of $\mathfrak{g}$. Let $A = \sum_{j=1}^{d} A^j \cdot E_j$ and $B = \sum_{j=1}^{d} B^j \cdot E_j$. We have $(A, B)_{\mathcal{A}} = \sum_{i=1}^{d} A^i \wedge *B^i$, where $*$ denotes the Hodge $*$-operation. Then the Chern–Simons integral is equal to

$$\int_{\mathcal{A}/\sim} \phi(A) \exp\left(\frac{ik}{4\pi}\int_M (A, \mathrm{d}A)_{\mathcal{A}} + \tfrac{1}{3}(A, [A, A])_{\mathcal{A}}\right) \mathcal{D}(A),$$

where $[A, B] = A \wedge B + B \wedge A$.

Among the various integrands $\phi(A)$, the Wilson line

$$W_\gamma(A) = \mathrm{Tr}_R P e^{i\int_\gamma A}$$

has been the most interested in them [3, 4, 7, 8]. For the case where $\phi(A) = \prod_{j=1}^{n} W_{\gamma_j}(A)$, the relation between the value of the integral and the linking number of loops γ_j, $j = 1, 2, \dots, n$, has been conjectured and discussed in [12].

According to the adjoint representation, $\mathfrak{g}$ in the above framework is reduced to the framework of a skew symmetric matrix algebra. Therefore, for simplicity, throughout of this paper we assume that the $\mathfrak{g}$-valued 1-form is a skew symmetric matrix algebra valued 1-form.

2. Definition of the Chern–Simons Integral

We carry out the method of super fields in line with Bar-Natan and Witten [4]. Let $\mathcal{A}$ be the collection of skew symmetric matrix algebra-valued 1-forms. The Chern–Simons Lagrangian

$$L(A) = -\frac{ik}{4\pi}\,\mathrm{Tr}\int_M A \wedge \mathrm{d}A + \tfrac{2}{3} A \wedge A \wedge A.$$

Let A_0 be the critical point of $L(A)$ such that

$$\mathrm{d}A_0 + A_0 \wedge A_0 = 0. \tag{2.1}$$

Following [4], we introduce some external fields and a super charge operator δ. For the later use, we write the necessary relations following the method of superfields. Let c be a fermionic 0-form, $\hat{c}$ a fermionic 3-form, φ becomes

$$\delta B = -D_B c, \qquad \delta\hat{c} = i\varphi,$$

and δ commutes $(D^0)^*$, where D_A is the covariant exterior derivative such that

$$D_A H = D^0 H + [A, H] \quad \text{and} \quad D^0 = D_{A_0} = \mathrm{d} + [A_0, \cdot].$$

First we perturb the Lagrangian around A_0. Let B be a $\mathfrak{g}$-valued 1-form. Then by Stokes' theorem and (2.1), we have

$$\begin{aligned} &L(A_0 + B) \\ &\quad = L(A_0) - \frac{ik}{4\pi}\,\mathrm{Tr}\int_M B \wedge D^0 B + \frac{2}{3} B \wedge B \wedge B. \end{aligned}$$

We fix the Lorenz gauge such that $(D^0)^* B = 0$. Define

$$V(B) = \frac{k}{2\pi}\int_M \mathrm{Tr}\,\hat{c} * D^0 * B,$$

where $*$ is the Hodge $*$-operation. Then

$$L(A_0 + B) = L(A_0 + B) - \delta V(B). \tag{2.2}$$

Noticing that $(D^0)^*[B,c]=0$ and the relation in the method of superfields, we get

$$\begin{aligned}\delta V(B) &= \frac{k}{2\pi}\int_M \operatorname{Tr}\delta\hat{c} * D^0 * B + \hat{c} * D^0 * \delta B\\ &= \frac{k}{2\pi}\int_M \operatorname{Tr} i\varphi * D^0 * B + \hat{c} * D^0 * D^0 c,\end{aligned}$$

which gives us

$$\begin{aligned}&L(A_0+B)-\delta V(B)\\ &\quad = L(A_0) - \frac{ik}{4\pi}\int_M \operatorname{Tr}(B\wedge D^0 B + \tfrac{2}{3}B\wedge B\wedge B +\\ &\qquad + 2\varphi * D^0 * B - 2i\hat{c} * D^0 * D^0 c).\end{aligned}$$

Let Ω^p be the space of $\mathfrak{g}$-valued p-forms and $(\omega,\eta)_p=\int_M -\operatorname{Tr}\omega\wedge *\eta$. Let $J\phi=-\phi$ if ϕ is a zero or a 3-form and $J\phi=\phi$ if ϕ is a 1 or a 2-form. We introduce the inner products $(\cdot,\cdot)_+$ and $(\cdot,\cdot)_-$ on $\Omega_+=\Omega^1\oplus\Omega^3$ and $\Omega_-=\Omega^0\oplus\Omega^2$ induced from the above inner products, respectively.

Set

$$x=(B,\phi),\qquad Q=\frac{1}{4\pi}(*D^0+D^0*)J\quad\text{and}\quad \Delta = *D^0 * D^0.$$

Let

$$c'=\sqrt{k/2\pi}\,c\quad\text{and}\quad \hat{c}'=*\sqrt{k/2\pi}\,\hat{c}.$$

Then

$$\begin{aligned}&L(A_0+B)-\delta V(B)\\ &\quad = L(A_0)+ik(x,Qx)_+ -\\ &\qquad - (\hat{c}',\Delta c')_0 - \frac{ik}{4\pi}\int_M \operatorname{Tr}\tfrac{2}{3}B\wedge B\wedge B.\end{aligned}$$

Thus we can consider the Chern–Simons integral

$$\int_{\mathcal{A}/\sim} F(A)\mathrm{e}^{L(A)}\mathcal{D}(A)$$

as

$$\begin{aligned}&\mathrm{e}^{L(A_0)}\int_{(\mathcal{A}:(D^0)^*A=0)/\sim} F(A_0+A)\times\\ &\quad\times\exp\left[ik((A,\phi),Q(A,\phi))_+ - (\hat{c}',\Delta c')_0 - \frac{ik}{4\pi}\int_M \operatorname{Tr}\tfrac{2}{3}A\wedge A\wedge A\right]\times\\ &\quad\times\mathcal{D}(A)\mathcal{D}(\phi)\mathcal{D}(\hat{c}')\mathcal{D}(c').\end{aligned}$$

Replacing the Fermi integral

$$\int \mathrm{e}^{-(\hat{c}',\Delta c')_0} \mathfrak{D}(\hat{c}')\mathfrak{D}(c')$$

by the constant θ, as done in [3, 4], we arrive at the significant form of the Chern–Simons integral

$$\theta\, \mathrm{e}^{L(A_0)} \int_H F(A_0 + x_A) \exp\Bigg[ik((x, Qx)_+ - \\ - \frac{ik}{4\pi} \int_M \mathrm{Tr} \tfrac{2}{3} x_A \wedge x_A \wedge x_A \Bigg] \mathfrak{D}(x), \tag{2.3}$$

where $H = L^2(\Omega_+)$ and $x_A =$ the projection of x on $L^2(\Omega^1)$.

From now on, using (2.3) we consider the normalized one-loop approximation of the Chern–Simons integral such that

$$\frac{\int_H F(A_0 + x_A) \mathrm{e}^{ik(x,Qx)_+} \mathfrak{D}(x)}{\int_H \mathrm{e}^{ik(x,Qx)_+} \mathfrak{D}(x)}. \tag{2.4}$$

We proceed by giving the precise meaning of (2.4). First, following Itô [9], we begin by considering the m-dimensional form of (2.4). Notice that

$$A = \sum_{t=1}^{d} A^t \cdot E_t, \qquad (A^t, B^t)_{\check{1}} = \int_M A^t \wedge *B^t$$

(in the sequel, we denote $(\cdot,\cdot)_{\check{p}}$ by the inner product of $\widehat{\Omega}_p$ of real-valued p-forms) and

$$(x, Qx)_+ = \sum_{i=1}^{\infty} \mu_i \left[\sum_{t=1}^{d} \left(x^t, (e_i^{t,A}, e_i^{t,\phi})\right)_{H^t} \right]^2,$$

where

$$\mu_i, \left(e_i^A = \sum_{t=1}^{d} e_i^{t,A} \cdot E_t, \qquad e_i^\phi = \sum_{t=1}^{d} e_i^{t,\phi} \cdot E_t \right), \quad i = 1, 2, \ldots$$

are the eigenvalues of Q and the corresponding CONs (Complete Orthonormal Systems) of H and $H^t = L^2(\widehat{\Omega}_1 \oplus \widehat{\Omega}_3)$ with the inner product $(\cdot,\cdot)_{H^t}$ induced from the inner products $(\cdot,\cdot)_{\check{1}}$ and $(\cdot,\cdot)_{\check{3}}$. Then we have the m-dimensional form of (2.4):

$$\frac{\int_{H_m} F(A_0 + x_A) \exp[ik(x, Qx)_+] \frac{\mu_m(\mathrm{d}x)}{(\sqrt{2\pi})^m}}{\int_{H_m} \exp[ik(x, Qx)_+] \frac{\mu_m(\mathrm{d}x)}{(\sqrt{2\pi})^m}}, \tag{2.5}$$

where H_m is the m-dimensional approximation of H and μ_m is the m-dimensional Lebesgue measure.

Let S be a positive definite (trace class) operator and $\lambda_j > 0$ the eigenvalue. Then we have the above (2.5) is equal to

$$\lim_{n\to\infty} \frac{1}{Z_{m,n}} \int_{H_m} F(A_0 + x) \exp\left[ik(x, Qx)_+ - \frac{(S^{-1}x, x)_+}{2n}\right] \frac{\mu_m(\mathrm{d}x)}{(\sqrt{2\pi})^m}, \tag{2.6}$$

where

$$Z_{m,n} = \int_{H_m} \exp\left[ik(x, Qx)_+ - \frac{(S^{-1}x, x)_+}{2n}\right] \frac{\mu_m(\mathrm{d}x)}{(\sqrt{2\pi})^m}.$$

Since

$$\begin{aligned} Z_{m,n} &= \prod_{j=1}^{m} \sqrt{1 + n\lambda_j} \times \\ &\times \int_{H_m} \exp\left[ik(x, Qx)_+ - \frac{(S^{-1}x, x)_+}{2n}\right] \frac{\mu_m(\mathrm{d}x)}{(\sqrt{2\pi})^m \sqrt{\det(nS)}}, \end{aligned}$$

so that, letting

$$Z_{m,n} = \prod_{j=1}^{m} \sqrt{1 + n\lambda_j} \times \widehat{Z}_{m,n},$$

we have that (2.6) is equal to

$$\begin{aligned} &\lim_{n\to\infty} \frac{1}{\widehat{Z}_{m,n}} \int_{H_m} F(A_0 + x) \times \\ &\times \exp\left[ik(x, Qx)_+ - \frac{(S^{-1}x, x)_+}{2n}\right] \frac{\mu_m(\mathrm{d}x)}{(\sqrt{2\pi})^m \sqrt{\det(nS)}}. \end{aligned} \tag{2.7}$$

Setting $x = \sqrt{nS}y$, we get

$$\begin{aligned} (2.7) &= \lim_{n\to\infty} \frac{1}{\widehat{Z}_{m,n}} \int_{H_m} F(A_0 + \sqrt{nS}y_A) \exp\left[ik(\sqrt{nS}y_A, Q\sqrt{nS}y)_+\right] \times \\ &\quad \times \frac{1}{(\sqrt{2\pi})^m} \exp\left[-\frac{(y, y)_+}{2}\right] \mu_m(\mathrm{d}y). \end{aligned} \tag{2.8}$$

However, the canonical Gaussian measure cannot be defined on a Hilbert space H, so that we precisely define the extension of (2.8) by making use of the abstract Wiener space setting.

Let $\widehat{\Omega}_p$ be the set of all real valued p-forms, $L^2(\widehat{\Omega}_p)$ the Hilbert space with the inner product $(\cdot, \cdot)_{\check{p}}$ and $H^t = L^2(\widehat{\Omega}_1 \oplus \widehat{\Omega}_3)$. Let (B^t, H^t, μ_t) be the abstract

Wiener space such that a separable Hilbert space H^t is continuously embedded into B^t equipped with the norm $\|\cdot\|_{B^t}$ and μ_t is a Gaussian measure satisfying

$$\int_{B^t} \mathrm{e}^{i\langle x^t,\xi\rangle}\,\mathrm{d}\mu_t(x^t) = \mathrm{e}^{-\frac{\|\xi\|^2_{H^t}}{2}}.$$

Let $\widehat{H} = H^1 \times H^2 \times \cdots \times H^d$ be a separable Hilbert space equipped with the norm

$$\|x\|^2_H = \sum_{t=1}^{d} \|x^t\|^2_{H^t}, \qquad B = B^1 \times B^2 \times \cdots \times B^d \quad \text{and} \quad \mu = \bigotimes_{t=1}^{d} \mu_t.$$

We denote $(B^1)^* \times (B^2)^* \times \cdots \times (B^d)^*$ by B^* and for $x \in B, \xi \in B^*$, $\sum_{t=1}^{d} \langle x^t, \xi^t\rangle$ again by $\langle x, \xi\rangle$. Then we have

$$\int_B \mathrm{e}^{i\langle x,\xi\rangle}\,\mathrm{d}\mu(x) = \mathrm{e}^{-\frac{\|\xi\|^2_{\widehat{H}}}{2}}.$$

Since Q is uniformly elliptic in the Hilbert space $L^2(\Omega_+)$ and Q^{-1} is compact on the Hilbert space, then Q has countable eigenvectors forming a complete orthonormal system of the Hilbert space, so that we can construct a positive definite and self-adjoint operator S in the Hilbert space such that $\sqrt{S}^* Q\sqrt{S}$ becomes a trace class operator and $\sum_{i=1}^{\infty} \sqrt{\lambda_i} < +\infty$, where $\{\lambda_i,\ i = 1, 2, \ldots\}$ are the eigenvalues of S ([11, 13]).

Now we will give a Wiener space realization of (2.8). We denote $L^2(\widehat{\Omega}_1)$ by H^t_A and $L^2(\widehat{\Omega}_3)$ by H^t_ϕ. Since $B^t = B^t_A \otimes B^t_\phi$, the projection of $x \in B^t$ on B^t_A is denoted by x^t_A. Then the Chern–Simons integral perturbed around a critical point A_0 can be defined by

$$\lim_{n\to\infty} \frac{1}{\widehat{Z}_n} \int_B F\left(A_0 + \sum_{t=1}^{d} \sqrt{nS}x^t_A \cdot E_t\right) \mathrm{e}^{ikCS(\sqrt{nS}x)}\mu(\mathrm{d}x), \tag{2.9}$$

where

$$\widehat{Z}_n = \int_B \mathrm{e}^{ikCS(\sqrt{nS}x)}\mu(\mathrm{d}x),$$

$$\sum_{t=1}^{d} \sqrt{nS}x^t_A \cdot E_t = \sum_{i=1}^{\infty} \sqrt{n\lambda_i}\left[\sum_{t=1}^{d} \langle x^t, (e_i^{t,A}, e_i^{t,\phi})\rangle\right] e_i^A$$

and

$$CS(x) = \sum_{i=1}^{\infty} \mu_i \left[\sum_{t=1}^{d} \langle x^t, (e_i^{t,A}, e_i^{t,\phi})\rangle\right]^2$$

is similarly defined in [10]. Here $\mu_i, (e_i^A, e_i^\phi),\ i = 1, 2, \ldots$ are the eigenvalues of Q and the corresponding CONs of $H = L^2(\Omega_+)$.

By making use of the so-called Fresnel integral formula

$$\frac{1}{\sqrt{2\pi}}\int_{-\infty}^{+\infty}\exp\left[-\frac{zx^2}{2}\right]\mathrm{d}x=\frac{1}{\sqrt{z}}$$

(z is a complex number), separately, we get

$$\widehat{Z}_n=\frac{1}{\sqrt{\det(1-2\sqrt{-1}k\sqrt{nS}Q\sqrt{nS})}},$$

so that we have one-loop approximation of the normalized Chern–Simons integral equal to

$$\lim_{n\to\infty}\sqrt{\det(1-2\sqrt{-1}k\sqrt{nS}Q\sqrt{nS})}\times \times\int_B F\left(A_0+\sum_{t=1}^{d}\sqrt{nS}x_A^t\cdot E_t\right)\mathrm{e}^{ikCS(\sqrt{nS}x)}\mu(\mathrm{d}x). \tag{2.10}$$

Let $W_\gamma^\epsilon(x)$ be the Wiener space version of the ϵ-reguralization of the Wilson line $\mathrm{Tr}_R P\mathrm{e}^{i\int_\gamma A}$ (the trace of a finite representation of the holonomy of the connection A along the loop γ), precisely defined later according to [1]. Let $\gamma_j,\ j=1,2,\ldots,m$ be the closed oriented loops and $F(x)=\prod_{j=1}^m W_{\gamma_j}^\epsilon(x)$. Then $F(x)$ is sufficiently integrable which will be proved later, so that we have

THEOREM. *Let* $F(x)=\prod_{j=1}^m W_{\gamma_j}^\epsilon(x)$. *Then*

$$\lim_{n\to\infty}\sqrt{\det(1-2\sqrt{-1}k\sqrt{nS}Q\sqrt{nS})}\times \times\int_B F\left(A_0+\sum_{t=1}^{d}\sqrt{nS}x_A^t\cdot E_t\right)\mathrm{e}^{ikCS(\sqrt{nS}x)}\mu(\mathrm{d}x) \sim F(A_0)\quad \text{as } k\to\infty.$$

3. Wilson Line

Let A be a $\mathfrak{g}$-valued smooth 1-form and $\gamma(t)\colon t\in[0,1]\to M$ a smooth closed curve in M. We assume that A is a complex $\hat{d}\times\hat{d}$-matrix-valued 1-form such that $A=\sum_{j=1}^{2\hat{d}^2}A_j\cdot\widehat{E}_j$, where $\widehat{E}_j$ is a complex unit matrix.

Let $\phi_\epsilon(x-\gamma(s))$ be the Friedrichs mollifier such that

$$\sup_{0\leqslant s\leqslant 1}\left|\int_M A_j^\alpha(x)\phi_\epsilon(x-\gamma(s))\mathrm{dvol}-A_j^\alpha(\gamma(s))\right|\to 0,$$

where $A_j^\alpha(x)$ is the component such that $A_j(x) = \sum_{\alpha=1}^3 A_j^\alpha(x)\,\mathrm{d}x_\alpha$ locally.

For sufficiently small ϵ,

$$\left| \int_s^t \int_M A_j^\alpha(x)\phi_\epsilon(x-\gamma(\tau))\mathrm{dvol}\dot{\gamma}_\alpha(\tau)\,\mathrm{d}\tau \right| \leqslant c_1 \|A_j^\alpha\|_{L^2}|t-s|$$

(in the sequel we denote by $c_p(\cdot)$ the constant depending on $(\cdot)$) and

$$\mathrm{d}\int_s^t \phi_\epsilon(x-\gamma(\tau))\,\mathrm{d}x_{\alpha+1(\mathrm{mod}3)} \wedge \mathrm{d}x_{\alpha+2(\mathrm{mod}3)}\dot{\gamma}_\alpha(\tau)\,\mathrm{d}\tau = (\phi_\epsilon(x-\gamma(t)) - \phi_\epsilon(x-\gamma(s)))\,\mathrm{d}x_1 \wedge \mathrm{d}x_2 \wedge \mathrm{d}x_3,$$

so that there exists a compact 2-form $C_t^{j,\epsilon}(\gamma)$ so that

$$\lim_{\epsilon\to 0} \sup_{0\leqslant s\leqslant 1} \left| \int_{\gamma[0,s]} A_j - (A_j, *C_t^{j,\epsilon}(\gamma))_{\check{1}} \right| = 0$$

and

$$\mathrm{d}C_1^{j,\epsilon}(\gamma) = 0. \tag{3.1}$$

Since

$$|(A_j, *C_t^j(\gamma)) - *C_s^j(\gamma))_{\check{1}}| \leqslant c_2(j)\|A_j\|_{L^2}|t-s|,$$

where $\|A_j\|_{L^2}^2 = (A_j, A_j)_{\check{1}}$, we get

$$\| *C_t^j(\gamma) - *C_s^j(\gamma)\|_{L^2} \leqslant c_2|t-s|. \tag{3.2}$$

From now on we write $c_p(\cdot)$ as c_p and $*C_t^{j,\epsilon}(\gamma)$ as $*C_t^j(\gamma)$ whenever there is no confusion.

Now we proceed to extend the holonomy to the stochastic holonomy.

Let $u_t(A) = P\mathrm{e}^{i\int_{\gamma[0,t]}A}$, where P denotes the product integral defined in [6]. Denote

$$\widetilde{A}(t) = \sum_{j=1}^{2d^2} (A_j, *C_t^j(\gamma))_{\check{1}}\widehat{E}_j.$$

Then we have

$$\mathrm{d}u_t(A) = \mathrm{i}\,\mathrm{d}\widetilde{A}(t)u_t(A)$$

and, by Chen's iterated integrals (Theorem 4.3 in §1.4 of [6], p. 31),

$$u_1(A) = I + \sum_{k=1}^\infty i^k \int_0^1\int_0^{t_1}\cdots\int_0^{t_{k-1}} \mathrm{d}\widetilde{A}(t_1)\mathrm{d}\widetilde{A}(t_2)\ldots\mathrm{d}\widetilde{A}(t_k). \tag{3.3}$$

Throughout this section we use the expression

$$\sum_{u=1}^{d} x_A^u \cdot E_u = \sum_{j=1}^{2\hat{d}^2} x_j \cdot \widehat{E}_j,$$

so that

$$x_j = \sum_{u=1}^{d} x_A^u (E_u)_j,$$

where $(E_u)_j$ is the j-unit entry of E_u. Denote

$$\begin{aligned} \bar{x}(t) &= \sum_{j=1}^{2\hat{d}^2} \langle x_j, *C_t^j(\gamma)\rangle \widehat{E}_j \\ &= \sum_{j=1}^{2\hat{d}^2} \sum_{u=1}^{d} \langle x_A^u, *C_t^j(\gamma)\rangle (E_u)_j \widehat{E}_j. \end{aligned}$$

Set

$$\bar{x}_j(t) = \sum_{u=1}^{d} \langle x_A^u, *C_t^j(\gamma)\rangle (E_u)_j.$$

Since

$$\langle x_A^u, *C_t^j(\gamma)\rangle = \langle x^u, (*C_t^j(\gamma), 0)\rangle,$$

then $\bar{x}_j(t)$ becomes a Gaussian random variable such that

$$E[(\bar{x}_j(t))^2] = \sum_{u=1}^{d} \| *C_t^j(\gamma)\|^2 (E_u)_j^2.$$

According to (3.3), we can define the stochastic holonomy

$$\begin{aligned} W(\gamma, x) &= W(x) = W(x_A) \\ &= I + \sum_{k=1}^{\infty} i^k \int_0^1 \int_0^{t_1} \cdots \int_0^{t_{k-1}} \mathrm{d}\bar{x}(t_1)\, \mathrm{d}\bar{x}(t_2) \ldots \mathrm{d}\bar{x}(t_k). \end{aligned} \tag{3.4}$$

We call this $W(x)$ the ϵ-reguralized stochastic holonomy.

Since $\bar{x}(t) = \sum_{j=1}^{2\hat{d}^2} \bar{x}_j(t) \widehat{E}_j$,

$$\begin{aligned} &\int_0^1 \int_0^{t_1} \cdots \int_0^{t_{k-1}} \mathrm{d}\bar{x}(t_1)\, \mathrm{d}\bar{x}(t_2) \ldots \mathrm{d}\bar{x}(t_k) \\ &= \sum_{j_1, j_2, \ldots, j_k = 1}^{2\hat{d}^2} \int_0^1 \int_0^{t_1} \cdots \int_0^{t_k} \mathrm{d}\bar{x}_{j_1}(t_1)\, \mathrm{d}\bar{x}_{j_2}(t_2) \ldots \mathrm{d}\bar{x}_{j_k}(t_k) \times \\ &\quad \times \widehat{E}_{j_1} \widehat{E}_{j_2} \cdots \widehat{E}_{j_k}. \end{aligned}$$

Hence, to guarantee the well-definedness, we estimate

$$E\left[\left(\sum_{j_1,j_2,\ldots,j_k=1}^{2\hat{d}^2}\left|\int_0^1\int_0^{t_1}\cdots\int_0^{t_k}\mathrm{d}\bar{x}_{j_1}(t_1)\,\mathrm{d}\bar{x}_{j_2}(t_2)\ldots\mathrm{d}\bar{x}_{j_k}(t_k)\right|\right)^2\right]$$

$$\leqslant\left(\sum_{j_1,j_2,\ldots,j_k=1}^{2d^2}E\left[\left|\int_0^1\int_0^{t_1}\cdots\int_0^{t_k}\mathrm{d}\bar{x}_{j_1}(t_1)\,\mathrm{d}\bar{x}_{j_2}(t_2)\ldots\mathrm{d}\bar{x}_{j_k}(t_k)\right|^2\right]^{\frac{1}{2}}\right)^2.$$

The following lemma is well known.

LEMMA. *Let X_i, $i=1,2,\ldots,2n$ be a mean-zero Gaussian system. Then*

$$E[X_1X_2\cdots X_{2n}]=\sum_{\text{comb}}E[X_{i_1},X_{j_1}]E[X_{i_2}X_{j_2}]\cdots E[X_{i_n}X_{j_n}],$$

where $\sum_{\text{comb}}$ is the summation taken over all pairs of $\{1,2,\ldots,2n\}$ such that $i_1<i_2<\cdots<i_n$, $i_k\leqslant j_k$ for $k=1,2,\ldots,n$.

From the above lemma and the notational convention where $[ts]$ means $[ts]=t$ or $[ts]=s$,

$$E\left[\left|\int_0^1\int_0^{t_1}\cdots\int_0^{t_k}\mathrm{d}\bar{x}_{j_1}(t_1)\,\mathrm{d}\bar{x}_{j_2}(t_2)\ldots\mathrm{d}\bar{x}_{j_k}(t_k)\right|^2\right]$$

$$=E\left[\int_0^1\int_0^{t_1}\cdots\int_0^{t_k}\mathrm{d}\bar{x}_{j_1}(t_1)\,\mathrm{d}\bar{x}_{j_2}(t_2)\ldots\mathrm{d}\bar{x}_{j_k}(t_k)\times\right.$$
$$\left.\times\int_0^1\int_0^{s_1}\cdots\int_0^{s_k}\mathrm{d}\bar{x}_{j_1}(s_1)\,\mathrm{d}\bar{x}_{j_2}(s_2)\ldots\mathrm{d}\bar{x}_{j_k}(s_k)\right]$$

$$=\int_0^1\int_0^{t_1}\cdots\int_0^{t_k}\int_0^1\int_0^{s_1}\cdots\int_0^{s_k}E\big[\mathrm{d}\bar{x}_{j_1}(t_1)\,\mathrm{d}\bar{x}_{j_2}(t_2)\ldots\mathrm{d}\bar{x}_{j_k}(t_k)$$
$$\mathrm{d}\bar{x}_{j_1}(s_1)\,\mathrm{d}\bar{x}_{j_2}(s_2)\ldots\mathrm{d}\bar{x}_{j_k}(s_k)\big]$$

$$=\int_0^1\int_0^{t_1}\cdots\int_0^{t_k}\int_0^1\int_0^{s_1}\cdots\int_0^{s_k}\sum_{\text{comb}}E\big[\mathrm{d}\bar{x}_{j_{u_1}}([ts]_{u_1})\,\mathrm{d}\bar{x}_{j_{v_1}}([ts]_{v_1})\big]\times$$
$$\times E\big[\mathrm{d}\bar{x}_{j_{u_2}}([ts]_{u_2})\,\mathrm{d}\bar{x}_{j_{v_2}}([ts]_{v_2})\big]\ldots E\big[\bar{x}_{j_{u_k}}([ts]_{u_k})\,\mathrm{d}\bar{x}_{j_{v_k}}([ts]_{v_k})\big]$$

$$\leqslant\int_0^1\int_0^{t_1}\cdots\int_0^{t_k}\int_0^1\int_0^{s_1}\cdots\int_0^{s_k}\sum_{\text{comb}}E\big[(\mathrm{d}\bar{x}_{j_{u_1}}([ts]_{u_1}))^2\big]^{1/2}\times$$
$$\times E\big[(\mathrm{d}\bar{x}_{j_{v_1}}([ts]_{v_1}))^2\big]^{1/2}E\big[\mathrm{d}\bar{x}_{j_{u_2}}([ts]_{u_2}))^2\big]^{1/2}E\big[(\mathrm{d}\bar{x}_{j_{v_2}}([ts]_{v_2}))^2\big]^{1/2}\ldots$$
$$\ldots E\big[(\bar{x}_{j_{u_k}}([ts]_{u_k}))^2\big]^{1/2}E\big[(\mathrm{d}\bar{x}_{j_{v_k}}([ts]_{v_k}))^2\big]^{1/2}$$

$$\leqslant\frac{(2k)!(c_2c_3)^{2k}}{2^kk!}\left(\int_0^1\int\cdots\int_0^{t_k}\mathrm{d}t_1\,\mathrm{d}t_2\ldots\mathrm{d}t_k\right)^2$$

$$
\begin{aligned}
&\leqslant \frac{(2k)!(c_2c_3)^{2k}}{2^k k!(k!)^2} \\
&\leqslant (c_2c_3)^{2k} \frac{(1 \cdot 2 \ldots (2k-2) \cdot (2k-1) \cdot 2k) 2^k}{k!((2 \cdot 1) \cdot (2 \cdot 2) \ldots 2(k-1) \cdot 2k)^2} \\
&\leqslant \frac{(2(c_2c_3)^2)^k}{k!},
\end{aligned}
$$

where

$$
c_3 = \sqrt{d} \max_{1 \leqslant u \leqslant d,\ 1 \leqslant j \leqslant 2\hat{d}^2} |(E_u)_j|.
$$

Hence we arrive at

$$
\begin{aligned}
&\left(\sum_{j_1, j_2, \ldots, j_k = 1}^{2\hat{d}^2} E\left[\left| \int_0^1 \int \cdots \int_0^{t_k} \mathrm{d}\bar{x}_{j_1}(t_1)\, \mathrm{d}\bar{x}_{j_2}(t_2) \ldots \mathrm{d}\bar{x}_{j_k}(t_k) \right|^2 \right]^{1/2} \right)^2 \\
&\quad \leqslant \frac{\hat{d}^{4k} 2^{3k} ((c_2c_3)^2)^k}{k!}.
\end{aligned} \tag{3.5}
$$

Therefore we obtain

$$
\sum_{k=0}^{\infty} \frac{(\hat{d}^2 2\sqrt{2} c_2 c_3)^k}{\sqrt{k!}} < +\infty,
$$

which implies the well definedness.

Now we proceed to check the conditions of the formula of change of the variables in [10]. Let $\{h_j,\ j = 1, 2, \ldots\}$ be a complete orthonormal system of the Hilbert space H.

Setting

$$
\begin{aligned}
W_j(x) &= W_j(x_A) \\
&= i^j \int_0^1 \int_0^{t_1} \cdots \int_0^{t_{j-1}} \mathrm{d}\bar{x}(t_1)\, \mathrm{d}\bar{x}(t_2) \ldots \mathrm{d}\bar{x}(t_j),
\end{aligned}
$$

we can exchange the directional differential and summation by the similar estimation to the above and we have

$$
\begin{aligned}
&E\left[\sum_{h_1, h_2, \ldots, h_k} \left| D^k W(x)(h_1, h_2, \ldots, h_k) \right|^2 \right] \\
&\quad \leqslant \sum_{h_1, h_2, \ldots, h_k} \sum_{i,j=k}^{\infty} E\big[|D^k W_i(x)(h_1, h_2, \ldots, h_k)| |D^k W_j(x)(h_1, h_2, \ldots, h_k)| \big] \\
&\quad \leqslant \sum_{h_1, h_2, \ldots, h_k} \sum_{i,j=k}^{\infty} E\big[|D^k W_i(x)(h_1, h_2, \ldots, h_k)|^2 \big]^{1/2} \times
\end{aligned}
$$

$$\times E\big[|D^k W_j(x)(h_1, h_2, \ldots, h_k)|^2\big]^{1/2}$$

$$\leqslant \sum_{i,j=k}^{\infty} \sum_{h_2,\ldots,h_k} \Big(\sum_{h_1} E\big[|D^k W_i(x)(h_1, h_2, \ldots, h_k)|^2\big]\Big)^{1/2} \times$$

$$\times \Big(\sum_{h_1} E\big[|D^k W_j(x)(h_1, h_2, \ldots, h_k)|^2\big]\Big)^{1/2}.$$

Similarly we recursively use the Schwarz inequality,

$$\leqslant \sum_{i,j=k}^{\infty} \Big(\sum_{h_1,h_2,\ldots,h_k} E\big[|D^k W_i(x)(h_1, h_2, \ldots, h_k)|^2\big]\Big)^{1/2} \times$$

$$\times \Big(\sum_{h_1,h_2,\ldots,h_k} E\big[|D^k W_j(x)(h_1, h_2, \ldots, h_k)|^2\big]\Big)^{1/2}$$

$$= \Big(\sum_{i=k}^{\infty} \Big(\sum_{h_1,h_2,\ldots,h_k} E\big[|D^k W_i(x)(h_1, h_2, \ldots, h_k)|^2\big]\Big)^{1/2}\Big)^2.$$

Since

$$|D^k W_i(x)(h_1, h_2, \ldots, h_k)|^2$$

$$= \Big| \sum_{j_1,j_2,\ldots,j_i=1}^{2\hat{d}^2} \sum_{\text{comb} l_1,l_2,\ldots,l_k} \Big(\int_0^1 \mathrm{d}\bar{x}_{j_1}(t_1) \ldots \int_0^{t_{l_1-1}} \mathrm{d}\bar{h}_{1_{l_1}}(t_{l_1}) \ldots$$

$$\ldots \int_0^{t_{l_k-1}} \mathrm{d}\bar{h}_{k_{l_k}}(t_{l_k}) \ldots \int_0^{t_i} \mathrm{d}\bar{x}_{j_i}(t_i)\Big)\Big|^2$$

so that we get

$$E\big[|D^k W_i(x)(h_1, h_2, \ldots, h_k)|^2\big]$$

$$\leqslant E\Big[\Big(\sum_{j_1,j_2,\ldots,j_i=1}^{2\hat{d}^2} \sum_{\text{comb } l_1,l_2,\ldots,l_k} \Big|\int_0^1 \mathrm{d}\bar{x}_{j_1}(t_1) \ldots \int_0^{t_{l_1-1}} \mathrm{d}\bar{h}_{1_{l_1}}(t_{l_1}) \ldots$$

$$\ldots \int_0^{t_{l_k-1}} \mathrm{d}\bar{h}_{k_{l_k}}(t_{l_k}) \ldots \int_0^{t_i} \mathrm{d}\bar{x}_{j_i}(t_i)\Big|\Big)^2\Big]$$

$$\leqslant \Big(\sum_{j_1,j_2,\ldots,j_i=1}^{2\hat{d}^2} \sum_{\text{comb } l_1,l_2,\ldots,l_k} E\Big[\Big|\int_0^1 \mathrm{d}\bar{x}_{j_1}(t_1) \ldots \int_0^{t_{l_1-1}} \mathrm{d}\bar{h}_{1_{l_1}}(t_{l_1}) \ldots$$

$$\ldots \int_0^{t_{l_k-1}} \mathrm{d}\bar{h}_{k_{l_k}}(t_{l_k}) \ldots \int_0^{t_i} \mathrm{d}\bar{x}_{j_i}(t_i)\Big|^2\Big]^{1/2}\Big)^2.$$

Notice the following

$$E\Big[\Big|\int_0^1 \mathrm{d}\tilde{x}_{j_1}(t_1)\dots\int_0^{t_{l_1-1}} \mathrm{d}\bar{h}_{1_{l_1}}(t_{l_1})\dots$$
$$\dots\int_0^{t_{l_k-1}} \mathrm{d}\bar{h}_{k_{l_k}}(t_{l_k})\dots\int_0^{t_i} \mathrm{d}\bar{x}_{j_i}(t_i)\Big|^2\Big]$$
$$= E\Big[\int_0^1 \mathrm{d}\bar{x}_{j_1}(t_1)\dots\int_0^{t_{l_1-1}} \mathrm{d}\bar{h}_{1_{l_1}}(t_{l_1})\dots\int_0^{t_{l_k-1}} \mathrm{d}\bar{h}_{k_{l_k}}(t_{l_k})\dots\int_0^{t_i} \mathrm{d}\bar{x}_{j_i}(t_i)\times$$
$$\times\int_0^1 \mathrm{d}\bar{x}_{j_1}(s_1)\dots\int_0^{s_{l_1-1}} \mathrm{d}\bar{h}_{1_{l_1}}(s_{l_1})\dots\int_0^{s_{l_k-1}} \mathrm{d}\bar{h}_{k_{l_k}}(s_{l_k})\dots\int_0^{s_i} \mathrm{d}\bar{x}_{j_i}(s_i)\Big]$$
$$= \int_0^1\dots\int_0^{t_{l_1-1}}\dots\int_0^{t_{l_k-1}}\dots\int_0^{t_i}\int_0^1\dots\int_0^{s_{l_1-1}}\dots\int_0^{s_{l_k-1}}\dots$$
$$\dots\int_0^{s_i} E\big[\mathrm{d}\bar{x}_{j_1}(t_1)\,\mathrm{d}\bar{h}_{1_{l_1}}(t_{l_1})\dots\mathrm{d}\bar{h}_{k_{l_k}}(t_{l_k})\dots\mathrm{d}\bar{x}_{j_i}(t_i)\dots$$
$$\dots\mathrm{d}\bar{x}_{j_1}(s_1)\dots\mathrm{d}\bar{h}_{1_{l_1}}(s_{l_1})\dots\mathrm{d}\bar{h}_{k_{l_k}}(s_{l_k})\dots\mathrm{d}\bar{x}_{j_i}(s_i)\big]$$
$$= \int_0^1\dots\int_0^{t_{l_1-1}}\dots\int_0^{t_{l_k-1}}\dots\int_0^{t_i}\int_0^1\dots\int_0^{s_{l_1-1}}\dots\int_0^{s_{l_k-1}}\dots$$
$$\dots\int_0^{s_i} \mathrm{d}\bar{h}_{1_{l_1}}(t_{l_1})\,\mathrm{d}\bar{h}_{1_{l_1}}(s_{l_1})\dots\mathrm{d}\bar{h}_{k_{l_k}}(t_{l_k})\,\mathrm{d}\bar{h}_{k_{l_k}}(s_{l_k})$$
$$E\big[\mathrm{d}\bar{x}_{j_i}(t_i)\,\mathrm{d}\bar{x}_{j_1}(s_1)\dots\mathrm{d}\bar{x}_{j_i}(t_i)\,\mathrm{d}\bar{x}_{j_i}(s_i)\big]$$
$$= \int_0^1\dots\int_0^{t_{l_1-1}}\dots\int_0^{t_{l_k-1}}\dots\int_0^{t_i}\int_0^1\dots\int_0^{s_{l_1-1}}\dots\int_0^{s_{l_k-1}}\dots$$
$$\dots\int_0^{s_i} \mathrm{d}\bar{h}_{1_{l_1}}(t_{l_1})\,\mathrm{d}\bar{h}_{1_{l_1}}(s_{l_1})\dots\mathrm{d}\bar{h}_{k_{l_k}}(t_{l_k})\,\mathrm{d}\bar{h}_{k_{l_k}}(s_{l_k})$$
$$\sum_{\text{comb }(i-k)} E\big[\mathrm{d}\bar{x}_{j_{u_1}}([ts]_{u_1})\,\mathrm{d}\bar{x}_{j_{v_1}}([ts]_{v_1})\big]E\big[\mathrm{d}\bar{x}_{j_{u_2}}([ts]_{u_2})\,\mathrm{d}\bar{x}_{j_{v_2}}([ts]_{v_2})\big]\dots$$
$$\dots E\big[\mathrm{d}\bar{x}_{j_{u(i-k)}}([ts]_{u(i-k)})\,\mathrm{d}\bar{x}_{j_{v(i-k)}}([ts]_{v(i-k)})\big],$$

then we have

$$\sum_{h_1,h_2,\dots,h_k} E\big[|D^k W_i(x)(h_1,h_2,\dots,h_k)|^2\big]$$
$$\leqslant \sum_{h_1,h_2,\dots,h_k}\Bigg(\sum_{j_1,j_2,\dots,j_i=1}^{2\hat{d}^2}\sum_{\text{comb }l_1,l_2,\dots,l_k}\Big\{\int_0^1\dots\int_0^{t_{l_1-1}}\dots\int_0^{t_{l_k-1}}\dots$$
$$\dots\int_0^{t_i}\int_0^1\int_0^{s_{l_1-1}}\dots\int_0^{s_{l_k-1}}\dots\int_0^{s_i}\big|\mathrm{d}\bar{h}_{1_{l_1}}(t_{l_1})\,\mathrm{d}\bar{h}_{1_{l_1}}(s_{l_1})\big|\dots$$

$$\ldots \left|\mathrm{d}\bar{h}_{k_{l_k}}(t_{l_k})\,\mathrm{d}\bar{h}_{k_{l_k}}(s_{l_k})\right| \sum_{\mathrm{comb}\,(i-k)} \left|E\left[\mathrm{d}\bar{x}_{j_{u_1}}([ts]_{u_1})\,\mathrm{d}\bar{x}_{j_{v_1}}([ts]_{v_1})\right]\ldots\right.$$

$$\left.\left.\left.\ldots E\left[\mathrm{d}\bar{x}_{j_{u_{(i-k)}}}([ts]_{u_{(i-k)}})\,\mathrm{d}\bar{x}_{j_{v_{(i-k)}}}([ts]_{v_{(i-k)}})\right]\right|\right\}^{1/2}\right)^2$$

$$\leqslant (c_2\sqrt{d}c_3)^{2k}\left(\sum_{j_1,j_2,\ldots,j_i=1}^{2\hat{d}^2} \frac{i!}{(i-k)!}\left\{\int_0^1\cdots\int_0^{t_{l_1-1}}\cdots\int_0^{t_{l_k-1}}\cdots\right.\right.$$

$$\ldots\int_0^{t_i}\int_0^1\int_0^{s_{l_1-1}}\cdots\int_0^{s_{l_k-1}}\cdots\int_0^{s_i}\mathrm{d}(t_{l_1})\,\mathrm{d}(s_{l_1})\ldots\mathrm{d}(t_{l_k})\,\mathrm{d}(s_{l_k})$$

$$\sum_{\mathrm{comb}\,(i-k)} E\left[\left(\mathrm{d}\bar{x}_{j_{u_1}}([ts]_{u_1})\right)^2\right]^{1/2} E\left[\left(\mathrm{d}\bar{x}_{j_{v_1}}([ts]_{v_1})\right)^2\right]^{1/2}$$

$$E\left[\left(\mathrm{d}\bar{x}_{j_{u_2}}([ts]_{u_2})\right)^2\right]^{1/2} E\left[\left(\mathrm{d}\bar{x}_{j_{v_2}}([ts]_{v_2})\right)^2\right]^{1/2}\ldots$$

$$\left.\left.\ldots E\left[\left(\mathrm{d}\bar{x}_{j_{u_{(i-k)}}}([ts]_{u_{(i-k)}})\right)^2\right]^{1/2} E\left[\left(\mathrm{d}\bar{x}_{j_{v_{(i-k)}}}([ts]_{v_{(i-k)}})\right)^2\right]^{1/2}\right\}^{1/2}\right)^2$$

$$\leqslant \left(c_2\sqrt{d}c_3\right)^{2k}\left(\sum_{j_1,j_2,\ldots,j_i=1}^{2\hat{d}^2} \frac{i!}{(i-k)!}\left\{\int_0^1\cdots\int_0^{t_{l_1-1}}\cdots\int_0^{t_{l_k-1}}\cdots\right.\right.$$

$$\ldots\int_0^{t_i}\int_0^1\cdots\int_0^{s_{l_1-1}}\cdots\int_0^{s_{l_k-1}}\cdots\int_0^{s_i}\mathrm{d}(t_{l_1})\,\mathrm{d}(s_{l_1})\ldots\mathrm{d}(t_{l_k})\,\mathrm{d}(s_{l_k})\times$$

$$\times\frac{2(i-k)!}{2^{(i-k)}(i-k)!}(c_2c_3)^{2(i-k)}\,\mathrm{d}([ts]_{u_1})\,\mathrm{d}([ts]_{v_1})\,\mathrm{d}([ts]_{u_2})\,\mathrm{d}([ts]_{v_2})\ldots$$

$$\left.\left.\ldots\mathrm{d}([ts]_{u_{(i-k)}})\,\mathrm{d}([ts]_{v_{(i-k)}})\right\}^{1/2}\right)^2$$

$$\leqslant (\sqrt{d}c_2c_3)^{2i}\left\{(2\hat{d}^2)^i\frac{i!}{(i-k)!}\right\}^2\frac{(2(i-k))!}{2^{(i-k)}(i-k)!(i!)^2}$$

$$\leqslant (\sqrt{d}c_2c_32\hat{d}^2)^{2i}\frac{(2(i-k))!}{2^{(i-k)}((i-k)!)^3}$$

$$\leqslant (\sqrt{d}c_2c_32\hat{d}^2)^{2i}\frac{2^{(i-k)}}{(i-k)!}.$$

Since

$$\left(\sum_{i=k}^{\infty}\frac{(4\sqrt{d}c_2c_3\hat{d}^2)^i}{\sqrt{(i-k)!}}\right)^2$$

$$=\left(\sum_{i=0}^{\infty}\frac{(4\sqrt{d}c_2c_3\hat{d}^2)^{i+k}}{\sqrt{(i)!}}\right)^2$$

$$= (4\sqrt{d}c_2c_3\hat{d}^2)^{2k}\left(\sum_{i=0}^{\infty}\frac{(4\sqrt{d}c_2c_3\hat{d}^2)^i}{\sqrt{(i)!}}\right)^2 < +\infty,$$

therefore the conditions (in the sense of L^2 and L^1) in [10] are verified.

4. Stationary Phase

Let A_0 be the flat connection stated in Section 1. First we recall

$$\begin{aligned} W(\gamma, x) &= W(x) = W(x_A) \\ &= I + \sum_{k=1}^{\infty} i^k \int_0^1 \int_0^{t_1} \cdots \int_0^{t_{k-1}} \mathrm{d}\bar{x}(t_1)\,\mathrm{d}\bar{x}(t_2)\ldots\mathrm{d}\bar{x}(t_k) \end{aligned} \tag{3.4}$$

and the one loop approximation of the normalized Chern–Simons integral

$$\begin{aligned} &= \lim_{n\to\infty}\sqrt{\det(1-2\sqrt{-1}k\sqrt{nS}Q\sqrt{nS})}\int_B F\left(A_0 + \sum_{t=1}^{d}\sqrt{nS}x_A^t\cdot E_t\right)\times \\ &\quad \times \mathrm{e}^{ikCS(\sqrt{nS}x)}\mu(\mathrm{d}x). \end{aligned} \tag{2.10}$$

Let γ_j, $j = 1, 2, \ldots, m$ be the closed oriented loops and $W(\gamma_j, x) = W_j(x)$ the stochastic holonomy discussed in Section 3. Set

$$W^{\epsilon}_{\gamma_j}(x) = \operatorname{Tr} W_j(x) \quad \text{and} \quad F(x) = \prod_{j=1}^{m} W^{\epsilon}_{\gamma_j}(x).$$

Then the following formula (4.1) is known from [10]. But in this case, the integrand is concrete so that we will present a rather easy proof.

Setting

$$W^J(x) = I + \sum_{k=1}^{J} i^k \int_0^1 \int_0^{t_1} \cdots \int_0^{t_{k-1}} \mathrm{d}\bar{x}(t_1)\,\mathrm{d}\bar{x}(t_2)\ldots\mathrm{d}\bar{x}(t_k),$$

$$F_J(x) = \prod_{j=1}^{m} \operatorname{Tr} W_j^J(x)$$

and

$$\begin{aligned} &G_{N,J}(x) \\ &= F_J\left(A_0 + \sum_{i=1}^{N}\sqrt{n\lambda_i}\left[\sum_{t=1}^{d}\langle x^t, (e_i^{t,A}, e_i^{t,\phi})\rangle\right]e_i^A\right)\mathrm{e}^{ikCS_N(x)}, \end{aligned}$$

where

$$CS_N(x) = \sum_{i=1}^{N}\mu_i\left[\sum_{t=1}^{d}\langle x^t, (e_i^A, e_i^{\phi})\rangle\right]^2,$$

then we have that the right-hand side of the above (2.9) is equal to

$$\lim_{n\to\infty}\sqrt{\det(1-2\sqrt{-1}k\sqrt{nS}Q\sqrt{nS})}\lim_{N,J\to\infty}\int_B G_{N,J}(x)\mu(\mathrm{d}x).$$

Now we can apply the Fresnel integral formula separately for the integrand $G_{N,J}(x)$ and, after that, letting $N, J\to\infty$, we arrive at

$$\begin{aligned}
&\lim_{n\to\infty}\sqrt{\det(1-2\sqrt{-1}k\sqrt{nS}Q\sqrt{nS})}\times\\
&\quad\times\int_B F\left(A_0+\sum_{t=1}^{d}\sqrt{nS}x_A^t\cdot E_t\right)\mathrm{e}^{ikCS(\sqrt{nS}x)}\mu(\mathrm{d}x)\\
&=\lim_{n\to\infty}\int_B F\left(A_0+\sum_{t=1}^{d}R_{n,k}x_A^t\cdot E_t\right)\mu(\mathrm{d}x),
\end{aligned}\tag{4.1}$$

where

$$R_{n,k}=(1-2\sqrt{-1}k\sqrt{nS}Q\sqrt{nS})^{-1/2}\sqrt{nS}.$$

The remainder is to prove

$$\begin{aligned}
&\lim_{n\to\infty}\int_B F\left(A_0+\sum_{t=1}^{d}R_{n,k}x_A^t\cdot E_t\right)\mu(\mathrm{d}x)\\
&\sim F(A_0)\quad\text{as } k\to\infty.
\end{aligned}$$

Since

$$\begin{aligned}
&E\big[|\mathrm{Tr}\,W_1(x)\,\mathrm{Tr}\,W_2(x)\dots\mathrm{Tr}\,W_m(x)-\mathrm{Tr}\,W_1(A_0)\,\mathrm{Tr}\,W_2(A_0)\dots\mathrm{Tr}\,W_n(A_0)|^2\big]\\
&\quad\leqslant 2\big(E\big[|\mathrm{Tr}\,W_1(x)-\mathrm{Tr}\,W_1(A_0)|^4\big]^{1/2}E\big[|\mathrm{Tr}\,W_2(x)\dots\mathrm{Tr}\,W_m(x)|^4\big]^{1/2}+\\
&\qquad+|\mathrm{Tr}\,W_1(A_0)|^2E\big[|\mathrm{Tr}\,W_2(x)\dots\mathrm{Tr}\,W_m(x)-\mathrm{Tr}\,W_2(A_0)\dots\mathrm{Tr}\,W_m(A_0)|^2\big]
\end{aligned}$$

and $\mathrm{Tr}\,W_j(x)$ is integrable as will be shown below, so that repeating this, we reduce the above estimate to

$$E\big[|\mathrm{Tr}\,W(x)-\mathrm{Tr}\,W(A_0)|^{2p}\big].$$

In a manner similar to that in (3.5), we have

$$\begin{aligned}
E\big[|\mathrm{Tr}\,W_j(x)|^{2p}\big]\leqslant E\Bigg[\Bigg(&\sum_{k=0}^{\infty}\sum_{j_1,j_2,\dots,j_k=1}^{2\hat{d}^2}\Bigg|\int_0^1\int_0^{t_1}\cdots\\
&\cdots\int_0^{t_{k-1}}\mathrm{d}\bar{x}_{j_1}(t_1)\,\mathrm{d}\bar{x}_{j_2}(t_2)\dots\mathrm{d}\bar{x}_{j_k}(t_k)\Bigg|\Bigg)^{2p}\Bigg]
\end{aligned}$$

$$\leqslant \left(\sum_{k=0}^{\infty} E\left[\left(\sum_{j_1,j_2,\ldots,j_k=1}^{2d^2}\left|\int_0^1\int_0^{t_1}\cdots\right.\right.\right.\right.$$

$$\left.\left.\left.\left.\cdots\int_0^{t_{k-1}} \mathrm{d}\bar{x}_{j_1}(t_1)\,\mathrm{d}\bar{x}_{j_2}(t_2)\ldots\mathrm{d}\bar{x}_{j_k}(t_k)\right|\right)^{2p}\right]^{\frac{1}{2p}}\right)^{2p}$$

and since $(pk)! \leqslant (pk!)^p$,

$$E\left[\left(\sum_{j_1,j_2,\ldots,j_k=1}^{2\hat{d}^2}\left|\int_0^1\int_0^{t_1}\cdots\int_0^{t_{k-1}} \mathrm{d}\bar{x}_{j_1}(t_1)\,\mathrm{d}\bar{x}_{j_2}(t_2)\ldots\mathrm{d}\bar{x}_{j_k}(t_k)\right|\right)^{2p}\right]$$

$$\leqslant \left(2\hat{d}^2\right)^{2pk}\frac{(2pk)!(c_2c_3)^{2pk}}{2^{pk}(pk)!(k!)^{2p}}$$

$$\leqslant \left(2\hat{d}^2c_2c_3\right)^{2pk}\left(\frac{p}{k!}\right)^p 2^{pk},$$

which implies

$$E\left[|\mathrm{Tr}\,W_j(x)|^{2p}\right] \leqslant \left(\sum_{k=0}^{\infty} 2^{\frac{3}{2}k}\left(\hat{d}^2c_2c_3\right)^k\left(\frac{p}{k!}\right)^{1/2}\right)^{2p} < +\infty.$$

Since

$$|\mathrm{Tr}\,W_j(x) - \mathrm{Tr}\,W_j(A_0)| = |\mathrm{Tr}(W_j(x) - W_j(A_0))|,$$

we have

$$E\left[|\mathrm{Tr}\,W_j(A_0 + R_{n,k}(x)) - \mathrm{Tr}\,W_j(A_0)|^{2p}\right]$$

$$\leqslant \left(\sum_{m=0}^{\infty} E\left[\left(\sum_{j_1,j_2,\ldots,j_m=1}^{2\hat{d}^2}\left|\int_0^1\int_0^{t_1}\cdots\right.\right.\right.\right.$$

$$\cdots\int_0^{t_{m-1}} \mathrm{d}\overline{A_0 + R_{n,k}(x)}_{j_1}(t_1)\,\mathrm{d}\overline{A_0 + R_{n,k}(x)}_{j_2}(t_2)\ldots$$

$$\ldots\mathrm{d}\overline{A_0 + R_{n,k}(x)}_{j_m}(t_m) -$$

$$\left.\left.\left.\left.-\int_0^1\int_0^{t_1}\cdots\int_0^{t_{m-1}} \mathrm{d}\overline{A_0}_{j_1}(t_1)\,\mathrm{d}\overline{A_0}_{j_2}(t_2)\ldots\mathrm{d}\overline{A_0}_{j_m}(t_m)\right|\right)^{2p}\right]^{\frac{1}{2p}}\right)^{2p}.$$

Since

$$E\left[|\overline{R_{n,k}(x)}_j(t) - \overline{R_{n,k}(x)}_j(s)|^2\right]$$

$$= E\left[\left|\sum_{u=1}^{d}\langle R_{n,k}x_A^u, *C_t^j(\gamma) - *C_t^j(\gamma)\rangle(E_u)_j\right|^2\right]$$

$$= E\left[\left|\sum_{i=1}^{\infty}\frac{\sqrt{n\lambda_i}}{\sqrt{1-2\sqrt{-1}kn\lambda_i\mu_i}}\left[\sum_{t=1}^{d}\langle x^t, (e_i^{t,A}, e_i^{t,\phi})\rangle\right]\times\right.\right.$$

$$\left.\left.\times\left(\left(\sum_{u=1}^{d}(*C_t^j(\gamma)-*C_t^j(\gamma))(E_u)_j\cdot E_u, 0\right), (e_i^A, e_i^\phi)\right)_+\right|^2\right]$$

$$= E\left[\left(\sum_{i=1}^{\infty}\frac{\sqrt{n\lambda_i}\cos\frac{\alpha_i}{2}}{\sqrt{\sqrt{1+4k^2n^2\lambda_i^2\mu_i^2}}}\left[\sum_{t=1}^{d}\langle x^t, (e_i^{t,A}, e_i^{t,\phi})\rangle\right]\times\right.\right.$$

$$\left.\left.\times\left(\left(\sum_{u=1}^{d}(*C_t^j(\gamma)-*C_t^j(\gamma))(E_u)_j\cdot E_u, 0\right), (e_i^A, e_i^\phi)\right)_+\right)^2\right]+$$

$$+E\left[\left(\sum_{i=1}^{\infty}\frac{\sqrt{n\lambda_i}\sin\frac{\alpha_i}{2}}{\sqrt{\sqrt{1+4k^2n^2\lambda_i^2\mu_i^2}}}\left[\sum_{t=1}^{d}\langle x^t, (e_i^{t,A}, e_i^{t,\phi})\rangle\right]\times\right.\right.$$

$$\left.\left.\times\left(\left(\sum_{u=1}^{d}(*C_t^j(\gamma)-*C_t^j(\gamma))(E_u)_j\cdot E_u, 0\right), (e_i^A, e_i^\phi)\right)_+\right)^2\right]$$

$$\left(\cos\alpha_i = \frac{1}{\sqrt{1+4k^2n^2\lambda_i^2\mu_i^2}}\right)$$

$$\leqslant \frac{c_4}{2k}\left\|\sum_{u=1}^{d}(*C_t^j(\gamma)-*C_t^j(\gamma))(E_u)_j\cdot E_u\right\|_{L^2(\Omega^1)}^2$$

$$= \frac{c_4}{2k}\sum_{u=1}^{d}\|*C_t^j(\gamma)-*C_t^j(\gamma)\|_{L^2}^2(E_u)_j^2$$

$$\leqslant (c_2c_3)^2\frac{c_4}{2k}|t-s|^2 \quad (c_4>1),$$

we have

$$E\left[\left|\int_0^1\int_0^{t_1}\cdots\int_0^{t_{m-1}} \mathrm{d}\overline{A_0+R_{n,k}(x)}_{j_1}(t_1)\,\mathrm{d}\overline{A_0+R_{n,k}(x)}_{j_2}(t_2)\ldots\right.\right.$$

$$\ldots\mathrm{d}\overline{A_0+R_{n,k}(x)}_{j_m}(t_m)-$$

$$\left.\left.-\int_0^1\int_0^{t_1}\cdots\int_0^{t_{k-1}}\mathrm{d}\overline{A_0}_{j_1}(t_1)\,\mathrm{d}\overline{A_0}_{j_2}(t_2)\ldots\mathrm{d}\overline{A_0}_{j_m}(t_k)\right|^{2p}\right]$$

$$\leqslant E\left[\left(\left|\int_0^1\int_0^{t_1}\cdots\int_0^{t_{m-1}}\mathrm{d}\overline{R_{n,k}(x)}_{j_1}(t_1)\,\mathrm{d}\overline{R_{n,k}(x)}_{j_2}(t_2)\ldots\right.\right.\right.$$

$$\dots \mathrm{d}\overline{R_{n,k}(x)}_{j_m}(t_m)\bigg| + \bigg|\int_0^1\int_0^{t_1}\dots\int_0^{t_{m-1}} \mathrm{d}\overline{A_0}_{j_1}(t_1)\,\mathrm{d}\overline{R_{n,k}(x)}_{j_2}(t_2)\dots$$

$$\dots \mathrm{d}\overline{R_{n,k}(x)}_{j_m}(t_m)\bigg| + \dots$$

$$+\dots+\bigg|\int_0^1\int_0^{t_1}\dots\int_0^{t_{m-1}} \mathrm{d}\overline{A_0}_{j_1}(t_1)\,\mathrm{d}\overline{A_0}_{j_2}(t_2)\dots\mathrm{d}\overline{R_{n,k}(x)}_{j_m}(t_m)\bigg|\bigg)^{2p}\bigg]$$

$$\leqslant (2^m-1)^{2p-1}\bigg(E\bigg[\bigg|\int_0^1\int_0^{t_1}\dots\int_0^{t_{m-1}} \mathrm{d}\overline{R_{n,k}(x)}_{j_1}(t_1)\,\mathrm{d}\overline{R_{n,k}(x)}_{j_2}(t_2)\dots$$

$$\dots\mathrm{d}\overline{R_{n,k}(x)}_{j_m}(t_m)\bigg|^{2p}\bigg]$$

$$+E\bigg[\bigg|\int_0^1\int_0^{t_1}\dots\int_0^{t_{m-1}} \mathrm{d}\overline{A_0}_{j_1}(t_1)\,\mathrm{d}\overline{R_{n,k}(x)}_{j_2}(t_2)\dots\mathrm{d}\overline{R_{n,k}(x)}_{j_m}(t_m)\bigg|^{2p}\bigg]$$

$$+\dots+E\bigg[\bigg|\int_0^1\int_0^{t_1}\dots\int_0^{t_{m-1}} \mathrm{d}\overline{A_0}_{j_1}(t_1)\,\mathrm{d}\overline{A_0}_{j_2}(t_2)\dots$$

$$\dots\mathrm{d}\overline{R_{n,k}(x)}_{j_m}(t_m)\bigg|^{2p}\bigg]\bigg)$$

$$\leqslant 2^{2pm}(c_2\max\{c_3,1\})^{2pm}\times$$

$$\times\left(\sum_{r=1}^{m} mC_r\|A_0\|_{L^2}^{2p(m-r)}\left(\frac{\sqrt{c_4}}{\sqrt{2k}}\right)^{2pr}\frac{(2pr)!}{2^{pr}(pr)!(m!)^{2p}}\right)$$

$$\leqslant 2^{2pm}(c_2\max\{c_3,1\}\sqrt{c_4}\max\{\|A_0\|_{L^2},1\})^{2pm}\times$$

$$\times\left(\sum_{r=1}^{m}\left(\frac{1}{\sqrt{2k}}\right)^{2pr}\frac{m(m-1)\dots(m-r+1)}{r!}\times\frac{2^{pr}(r!p)^p}{(m!)^{2p}}\right)$$

$$\leqslant 2^{3pm+m+1}\big(c_2\max\{c_3,1\}\sqrt{c_4}\max\{\|A_0\|_{L^2},1\}\big)^{2pm}\times$$

$$\times\left(\frac{1}{\sqrt{2k}}\right)^{2p}\left(\frac{p}{m!}\right)^{p}\cdot(k>2),$$

where

$$\|A_0\|_{L^2} = \max_{1\leqslant j\leqslant \hat{d}\times\hat{d}}\|(A_0)_j\|_{L^2}.$$

Since

$$\sum_{m=1}^{\infty}\frac{1}{\sqrt{2k}}\sqrt{p}\frac{(c_2\max\{c_3,1\}\sqrt{c_4}2^3\hat{d}^2\max\{\|A_0\|_{L^0},1\})^m}{\sqrt{m!}} < +\infty,$$

from the Lebesgue dominated convergence theorem, we have the desired asymptotics.

5. Abelian Case

By making use of (2.9), we calculate the value of the Wilson line integral for the case where $G = \mathrm{U}(1)$. In this case, the dimension d of $\mathfrak{g} = 1$, so that we omit the parameter t and we choose $i = \sqrt{-1}$ as the basis. Let $\gamma_j,\ j = 1, 2, \ldots, m$ be the closed oriented loops. Then by (3.2) and (3.4), $W^{\epsilon}_{\gamma_j}(x) = \mathrm{e}^{-\langle x, *C_1(\gamma_j)\rangle}$ and

$$F(x) = \prod_{j=1}^{m} W^{\epsilon}_{\gamma_j}(x) = \mathrm{e}^{-\sum_{j=1}^{m}\langle x, *C_1(\gamma_j)\rangle}.$$

By applying the commutative case of formula (4.1), we have the commutative normalized Chern–Simons integral

$$= \lim_{n\to\infty}\int_B F(A_0 + R_{n,k}x_A)\mu(\mathrm{d}x) \tag{5.1}$$

$$= \lim_{n\to\infty}\int_B \exp\left[-\left\langle A_0 + R_{n,k}x_A, \sum_{j=1}^{m} *C_1(\gamma_j)\right\rangle\right]\mu(\mathrm{d}x)$$

$$= \exp\left[-\left\langle A_0, \sum_{j=1}^{m} *C_1(\gamma_j)\right\rangle\right] \times$$

$$\times \lim_{n\to\infty}\int_B \exp\left[-\left\langle R_{n,k}, x_A, \sum_{j=1}^{m} *C_1(\gamma_j)\right\rangle\right]\mu(\mathrm{d}x)$$

$$= \exp\left[-\left\langle A_0, \sum_{j=1}^{m} *C_1(\gamma_j)\right\rangle\right] \times$$

$$\times \lim_{n\to\infty}\int_B \exp\left[-\left\langle x, R_{n,k}(\sum_{j=1}^{m} *C_1(\gamma_j), 0)\right\rangle\right]\mu(\mathrm{d}x)$$

$$= \exp\left[-\left\langle A_0, \sum_{j=1}^{m} *C_1(\gamma_j)\right\rangle\right] \times$$

$$\times \lim_{n\to\infty}\exp\left[-\frac{1}{2}\left\| R_{n,k}\left(\sum_{j=1}^{m} *C_1(\gamma_j), 0\right)\right\|_H^2\right]$$

$$= \exp\left[-\left\langle A_0, \sum_{j=1}^{m} *C_1(\gamma_j)\right\rangle\right] \times$$

$$\times \lim_{n\to\infty}\exp\left[-\frac{1}{2}\sum_{i=1}^{\infty}\left(\left(\sum_{j=1}^{m} *C_1(\gamma_j), 0\right), (e_i^A, e_i^{\phi})\right)_H^2 \frac{n\lambda_i}{1 - 2ikn\lambda_i\mu_i}\right]$$

$$= \exp\left[-\left\langle A_0, \sum_{j=1}^{m} *C_1(\gamma_j)\right\rangle\right] \times$$

$$\times \exp\left[-\frac{1}{2}\left(\left(\sum_{j=1}^{m} *C_1(\gamma_j), 0\right), \frac{-1}{2ik} Q^{-1}\left(\sum_{j=1}^{m} *C_1(\gamma_j), 0\right)\right)_H\right]$$

$$= \exp\left[-\left\langle A_0, \sum_{j=1}^{m} *C_1(\gamma_j)\right\rangle\right] \times$$

$$\times \exp\left[-\frac{1}{2}\left(\sum_{j=1}^{m} *C_1(\gamma_j), \frac{-1}{2ik}\ \text{1-form part of } Q^{-1}\left(\sum_{j=1}^{m} *C_1(\gamma_j), 0\right)_{H_A}\right)\right].$$

Setting ω_j = 1-form part of $Q^{-1}(*C_1(\gamma_j), 0)$, as in [1], then by using $dC_1(\gamma_j) = 0$, we have $d\omega_j = C_1(\gamma_j)$, which implies

$$(*C_1(\gamma_i),\ \text{1-form part of } Q^{-1} * C_1(\gamma_j))_{H_A}$$

must be the linking number of loops γ_i and γ_j, if we choose ϵ sufficiently small so that the ϵ-tubes of γ_j are nonintersected (see [5]).

References

1. Albeverio, S. and Schafer, J.: Abelian Chern–Simons theory and link invariants, *J. Math. Phys.* **36** (1995), 2157–2169.
2. Albeverio, S. and Sengupta, A.: The Chern–Simons functional integral as an infinite dimensional distribution, *Proc. 2nd World Congress of Nonlinear Analysis*, to appear.
3. Axelrod, S. and Singer, I. M.: Chern–Simons perturbation theory, *Proc. XXth. DGM Conf.*, World Scientific, Singapore, 1992, pp. 1–36.
4. Bar-Natan, D. and Witten, E.: Perturbation expansion of Chern–Simons theory with noncompact gauge group, *Comm. Math. Phys.* **141** (1991), 423–440.
5. Bott, R. and Tu, L. W.: *Differential Forms in Algebraic Topology*, Springer, New York, 1986.
6. Dollard, J. D. and Friedman, C. N.: *Product Integration*, Encyclopedia Math. Appl. 10, Addison-Wesley, Massachusetts, 1979.
7. Frölich, J. and King, C.: The Chern–Simons theory and knot polynomials, *Comm. Math. Phys.* **126** (1989), 167–199.
8. Guadagnini, E., Martellini, M. and Mintchev, M.: Wilson lines in Chern–Simons theory and link invariants, *Nuclear Phys. B* **330** (1990), 575–607.
9. Itô, K.: Generalized uniform complex measures in the Hilbertian metric space with their application to the Feynman integral, *Proc. Fifth Berkeley Sympos. Math. Stat. Probab. II*, 1965, pp. 145–161.
10. Malliavin, P. and Taniguchi, S.: Analytic functions, Cauchy formula and stationary phase on a real abstract Wiener space, *J. Funct. Anal.* **143** (1997), 470–528.
11. Reed, M. and Simon, B.: *Method of Modern Mathematical Physics*, Vol. 1, Academic Press, San Diego, 1980.
12. Witten, E.: Quantum field theory and Jone's polynomial, *Comm. Math. Phys.* **121** (1989), 3351–3399.
13. Yosida, K.: *Functional Analysis*, Springer, New York, 1971.

Acta Applicandae Mathematicae **63:** 275–282, 2000.

Quantum Mechanics and Brownian Motions ⋆

MIKIO NAMIKI
Department of Physics, Waseda University, Tokyo 169-8555, Japan

(Received: 15 January 1999)

Abstract. From the point of view that the present formulation of quantum mechanics is very close to the theory of Brownian motions, we search for possible origins of the quantum fluctuation within the framework of new quantization schemes, such as *stochastic* and/or *microcanonical* quantizations, for increasing additional *fictitious* time other than the ordinary one. On the same basis we also show that a D-dimensional quantum system is equivalent to a (D + 1)-dimensional classical system.

Mathematics Subject Classifications (2000): 81P20, 81S20.

Key words: canonical and path-integral quantizations, stochastic and microcanonical quantizations, future dynamics.

1. Introduction

We know that we can mathematically formulate quatum mechanics in a similar way to the theory of Brownian motions (with *imaginary* diffusion constant or *imaginary* time), even though the former is given for *probability amplitude* while the latter for *probability* itself. (Remember that we prefer to use *imaginary* ordinary time instead of *real* ordinary time in these attempts.) Actually, some authors tried to formulate quantum mechanics by means of the theory of Brownian motions [1], but the so-called NO-GO theorem given by J. von Neumann prevented kind of earlier attempts from developing further [2]. However, D. Bohm broke through the barrier of the NO-GO theorem and formulated quantum mechanics as a theory of Brownian motions [4]. Undoubtedly, he wanted to eliminate quantum mechanics itself, leaving only classical dynamics as a fundamental principle. E. Nelson did not explicitly mention this kind of idea, but gave his ideas of *stochastic* quantization [4], by showing that quantum mechanics was equivalent to the theory of Brownian motions. (Remember that they keep a single ordinary time *real*, contrary to the above earlier attempts.) For this reason, it might be better to start our discussion on the background of the Planck constant from the Bohm–Nelson theory point of view **.

⋆ This article is dedicated to Prof. Takeyuki Hida on the occasion of his 70th Birthday. He is famous for his mathematics on *stochastic* processes, and for his collaboration with physicists. I heartily hope that he will continue his remarkable and fruitful support to them.

** As is well known, the present formalism of physics contains Planck constant $\hbar$, light speed c, and gravitaion constant G as given universal constants, but never asks about their origins. Besides,

Let us begin our discussion with the Bohm–Nelson theory of D-dimensional quantum system, but we are immediately led to the conclusion that we can hardly ascribe all quantum-corrrelations to classical fluctuating forces of the Bohm–Nelson type. This is not physical, even though it is accepted mathematically. In this paper, therefore, we search for possible origins of the quantum fluctuation within the framework of new quantization schemes such as *stochastic* and *microcanonical* quantizations, to give quantum mechanics in the infinite limit of an additionally introduced *fictitious* time, different from the ordinary one, assuming that $G = 0$. We also see that our procedure leads us to another view that 'a D-dimensional quantum system is equivalent to a (D + 1)-dimensional classical system'.

2. Bohm–Nelson Theory

As is well known, the standard *canonical* quantization method introduced an *operator* theory in physics, in which the operator nature of observables is specified by canonical commutation relations. While it is true that the quantum fluctuation is derived from the operator commutation relations, we do not find any kind of reasoning that $\hbar$ is rooted in a deeper physics in this formulation. For this reason we want to search for physical origins of the quantum fluctuation in another world with larger dimension, within the framework of the new quantization schemes such as *stochastic* and *microcanonical* quantizations. On the other hand, we know that the operator calculus in conventional quantum mechanics can be replaced by c-number calculus within the framework of the path-integral method. Needless to say, the c-number path-integral calculus is equivalent to the conventional canonical quantum mechanics formulated by means of operator calculus. The path-integral method is really a sort of c-number quantization, but never asks a deeper origin of the quantum fluctuation. In order to avoid the mathematically ill-posed nature of the original path-integral method, we are sometimes recommended to use a Euclidean measure $\exp(-S_E[q]/\hbar)$ (S_E being often called the Euclidean action) introduced by means of transformation $x_0 \to -ix_0$ (x_0 standing for the real time). Remember that the final goal of the above stochastic and/or microcanonical quantizations is to produce quantum mechanics of this type. We will need to have the Planck constant $\hbar$ as a given measure of *quantum fluctuation*, but need not delve into its deeper origins.

As far as we know, Bohm's theory of quantum mechanics is a nice example of making the quantum fluctuation [4]. For the sake of simplicity, let us discuss Schrödinger's equation in the one-dimensional nonrelativistic case. Bohm reformulated the Schrödinger equation in terms of R and S defined by

$$\psi = \sqrt{R} \exp \frac{iS}{\hbar}, \quad R = |\psi|^2, \quad \mathbf{p} = \nabla S, \tag{1}$$

we also have no theory to give the elementary charge e (of an electron) in terms of other fundamental parameters. These facts must be unsatisfactory from the fundamental point of view [3].

as a 'Newtonian equation', i.e.

$$\frac{d\mathbf{p}}{dx_0} = -\nabla(V + V_Q) \tag{2}$$

with the quantum mechanical potential

$$V_Q \equiv -\frac{\hbar^2}{2m}\frac{\nabla^2|\psi|}{|\psi|}. \tag{3}$$

Bohm planned to eliminate quantum mechanics within this fomulation: A quantum mechanical particle (originally obeying the classical Newtonian equation of motion) moves in unknown *ether* which makes random fluctuating forces, $\mathbf{f}_Q = -\nabla V_Q$, [4].

This line of thought, if turned inversely, is considered to be a sort of quantization method. Actually, Nelson formulated his way of stochastic quantization in an analogous way to this idea [4].

According to Bohm–Nelson's idea, the quantum fluctuation in the D-dimensional system is rooted in a classical random process in D-dimensional *real* spacetime world, where $\hbar$ is introduced as a given measure. As is well known, every quantum property must come from the classical fluctuating forces. In order to reproduce quantum mechanics fully, therefore, we have to ascribe all strange properties, such as the *nonlocal long-distant correlations*, to the *classical* random forces, for example, $\mathbf{f}_Q = -\nabla V_Q$ in the Bohm's theory. *Mathematically*, it may be accepted but *physically* not.

In order to escape from this kind of difficulty, we know at least that we can reach a satisfactory theory by introducing *stochastic* or *microcanonical* quantization method, which quantizes D-dimensional quantum systems through *classical* motions in the (D + 1)-dimensional spacetime world [5, 7]. To do this, we prefer to introduce an *additional* dependence of dynamical quantities on a *fictitious* time (say, t) but not only on the *ordinary* one (say, x^0), and then to discuss (D + 1)-dimensional classical processes. In both cases, we used to assume an additional *fictitious*-time dependence of every dynamical quantity. They are another background to producing the quantum fluctuation.

3. Stochastic and Microcanonical Quantizations

Let us briefly describe some basic ideas of these new quantization schemes, such as *stochastic* and *microcanonical* quantizations, by introducing the above additional dependences of dynamical quantities on the *fictitious* time t.

(1) The former (i.e. SQ) is based on the idea that a D-dimensional quantum system is equivalent to a hypothetical (D + 1) dimensional *classical* stochastic process [5]. We assume that Planck constant $\hbar$ is a measure of magnitude of the fluctuation force to generate the quantum fluctuation. This idea was not quite new,

because Suzuki invented a new calculation method for quantum spin systems by making use of Trotter's formula [6].

(2) If the SQ is acceptable, we can suppose that there should be a sort of irreversible process based on classical deterministic dynamics in a (D + 1)-dimensional spacetime world. Actually, we could succeed in formulating the microcanonical quantization method (MCQ) under the idea that a D-dimensional quantum system is equivalent to a (D + 1)-dimensional *classical deterministic* one [7]. Note that the Planck constant is defined with a statistical average of (D + 1)-dimensional kinetic energy over fictitious-time initial values. In this case, one might expect that the *quantum fluctuation* would come from (i) a sort of *chaotic* behaviors due to nonlinear interactions in the system and (ii) a statistical average over *fictitious*-time initial values. One may be inclined to accept the first idea because we would expect that a sort of chaotic behavior of the surrounding classical systems would provoke the quantum fluctuation. But if so, we cannot quantize any linear system (for example, the harmonic oscillator) which does not have nonlinear interactions. We have to say that this is surely unsatisfactory as quantum mechanics. Recently, we have shown that the average over the initial values is essentially important (in the harmonic oscillator case) [8]. We can accept the quantization rule of type (i), if and only if chaotic behaviors due to nonlinear interactions dominated over the statistical fluctuation coming from the initial distribution in the fictitious time. In general, we need to use the quantization rule of (ii).

On the other hand, we can expect to find chaotic behavior in classical dynamical systems with a large amount of freedom. This is true but we have to mention an interesting possibility that an appropriate classical system with a few degrees of freedom can provoke dephasing (or irreversibility) to produce the wave-function collapse in quantum measurements, [9]. The problem is what kind of interactions we have. For this reason, the idea that we cannot quantize any linear system, for example, the harmonic oscillator system, with no chaotic behavior may be hasty. Anyway, we have not to expect any *direct* connection between the quantum fluctuation and such a chaotic behavior in any quantization scheme.

3.1. SQ IN A (D+1)-DIMENSIONAL SPACETIME WORLD

The SQ is fomulated in terms of the following basic Langevin equation and the statistical properties [5]:

$$\frac{\partial q(x,t)}{\partial t} = -\left.\frac{\delta S}{\delta q}\right|_{q=q(x,t)} + \eta(x,t), \tag{4}$$

$$\langle \eta(x,t) \rangle = 0, \quad \langle \eta(x,t)\eta(x',t') \rangle = 2\hbar\delta(t-t')\delta(x-x'). \tag{5}$$

It is true that the SQ is completed by (4) and (5), but in order to discuss a possible relation of the SQ to the conventional quantum mechanics, it is more convenient to replace the Langevin method with the Fokker–Planck method, from which we obtain

$$P[q,t] \stackrel{t\to\infty}{\longrightarrow} N\exp(-S[q]/\hbar). \tag{6}$$

The last equation tells us that the SQ is equivalent to quantum mechanics of the Feynman type, i.e. it means a quantization method [5]. Also remember that $\hbar$ is a given universal constant.

In the case of a one-dimensional harmonic oscillator [8], we describe the SQ in terms of $q_n(t)$, a Fourier component of $q(x,t)$ on $\{u_n(x) = (1/\sqrt{X})\exp(k_n x)\}$ with $k_n = 2\pi n/X$ for period X. Thus, we are led to the basic Langevin equation and the statistical property

$$\frac{\mathrm{d}q_n(t)}{\mathrm{d}t} = -m\Omega_n^2 q_n + \eta_n(t), \qquad \Omega_n^2 = \omega^2 + k_n^2, \tag{7}$$

$$\langle \eta_n(t)\rangle = 0, \qquad \langle \eta_n(t)\eta_{n'}^*(t')\rangle = 2\hbar\delta_{nn'}\delta(t-t'), \tag{8}$$

whose solution is given by

$$q_n(t) = q_n^{(0)}\mathrm{e}^{-m\Omega^2 t} + \int_0^t \mathrm{d}t'\mathrm{e}^{-m\Omega_n^2(t-t')}\eta_n(t'). \tag{9}$$

Equation (9) tells us that $q_n(t)$ should have a definite fluctuation, irrespective of fictitious-time initial values, for sufficiently large ts, so that we can safely neglect these fictitious-time initial values. Note that we can obtain the continuous limit of the discretized Fourier representation through

$$\frac{1}{X}\sum_n \frac{1}{\Omega_n^2}G(k_n^2) \to \frac{1}{2\pi}\int_{-\infty}^{\infty}\frac{G(k^2)}{\omega^2+k^2}\mathrm{d}k. \tag{10}$$

Apart from the fictitious initial values, we obtain

$$(\Delta q)_t^2 \equiv \langle q_t^2\rangle = \frac{\hbar}{2m\omega}I(t), \tag{11}$$

$$(\Delta p)_t^2 = \left\langle\left(m\frac{\mathrm{d}q}{\mathrm{d}x_0}\right)^2\right\rangle = \hbar m\delta(\epsilon) - \frac{\hbar m\omega}{2}I(t), \tag{12}$$

with a small but finite ϵ and

$$I(t) \equiv \frac{2}{\sqrt{\pi}}\int_0^{\sqrt{2m\omega^2 t}} \mathrm{e}^{-z^2}\mathrm{d}z. \tag{13}$$

We easily understand that the first term of (12) vanishes for finite ϵ, and the negative value of $(\Delta p)_t^2$ comes from the use of Euclidean measure. Here we should notice that the neglection of $\delta(\epsilon)$ is also obtained by replacing k^2/Ω_n^2 with $-\omega^2/\Omega_n$. An additional ansatz that the k integral in (10) can be evaluated only from the pole, $k = i\omega$, and written as

$$\frac{1}{X}\sum_n \frac{1}{\Omega_n^2} G(k_n^2) \longrightarrow \frac{1}{2\pi}\int_{-\infty}^{\infty} \frac{G(k^2)}{\omega^2 + k^2}\mathrm{d}k = \frac{G(k = i\omega)}{2\omega} \tag{14}$$

in the continuous limit [8]. Remember that the same replacement rule can also work in the MCQ.

Finally, we obtain the following formula for the uncertainty product:

$$\Delta(t) \equiv (\Delta q)_t^2 |(\Delta p)_t^2| = \frac{\hbar^2}{4}\frac{2}{\sqrt{\pi}}\int_0^{\sqrt{2m\omega^2 t}} \mathrm{e}^{-z^2}\mathrm{d}z, \tag{15}$$

which implies that, for increasing t, $\Delta(t)$ starts from an unlikely value, say zero, and reaches the quantum-mechanical one $\hbar^2/4$ [8]. This is a possible model to generate the quantum fluctuation by means of fictitious-time evolution.

3.2. MCQ IN A (D+1)-DIMENSIONAL SPACETIME WORLD

The MCQ is fomulated in terms of the following basic (D + 1)-dimensional equation of motion [7]:

$$\mathcal{H} \equiv K[\pi(t)] + S[q(t)], \qquad K \equiv \sum_n \frac{1}{2m}\pi_n^*(t)\pi_n(t), \tag{16}$$

$$\frac{\mathrm{d}q_n(t)}{\mathrm{d}t} = \{q_n, \mathcal{H}\} = \frac{\pi_n(t)}{m}, \qquad \frac{\mathrm{d}\pi_n(t)}{\mathrm{d}t} = \{\pi_n, \mathcal{H}\} = -\frac{\partial S[q]}{\partial q_n}. \tag{17}$$

Note that $\mathcal{H}$, $\mathcal{K}$ and $S[q]$ stand for *effective* Hamiltonian, kinetic energy and potential, respectively, of the (D + 1)-dimensional deterministic motions. They are to be supplemented by the quantization rule

$$\hbar = \lim_{t\to\infty} 2\langle K[\pi(t)]\rangle. \tag{18}$$

$\langle\cdots\rangle$ stands for the statistical average over fictitious-time initial values, $q_n^{(0)}$ and $p_n^{(0)}$.

In the case of a one-dimensional harmonic oscillator, we have the *effective* 'Hamiltonian' and 'potential' as follows:

$$H_{\mathrm{MCQ}} = K[\pi(t)] + S_{\mathrm{E}}, \tag{19}$$

$$K[\pi(t)] = \sum_n \frac{1}{2m}\pi_n^*(t)\pi_n(t), \qquad S = \sum_n q_n^*(t)\frac{1}{2}m\Omega_n^2 q_n(t), \tag{20}$$

for the (D + 1)-dimensional classical motion. We have only to solve this equation and then take the fictitious-time initial distribution into account. For details, see [8].

We can see even in this case that the uncertainty product starts from an unlikely value, zero, and reach the quantum-mechanical one, $\hbar^2/4$. The situation is quite similar to the SQ case, so that we can also regard this process as another model to generate the quantum fluctuation in fictitious-time evolution.

4. Conclusion

We have suggested possible origins or mechanisms of the quantum fluctuation, by means of *stochastic* and *microcanonical* quantization methods. We have also showed that a D-dimensional quantum system is equivalent to a (D + 1)-dimensional classical one, within both quantization schemes.

Almost all theories only explain the situation that $\hbar$ is a measure of the fluctuation, but never seek its deeper origins. As far as we know, only the Hayakawa–Tanaka formula [10]

$$\hbar \sim \frac{mc^2\tau}{\sqrt{N}}, \tag{21}$$

might suggest to us a possible reasoning of $\hbar$ within a deeper (cosmological and/or particle-physical) framework, even though we have some questions. This formula was derived under the assumption that (i) the universe is composed of N ($\gg 1$) fundamental particles, with mass m and lifetime τ, (ii) the universe is an open system put in a bigger spacetime world to give $\Delta N/N \sim 1/\sqrt{N}$ and (iii) the quantum-mechanical energy-time uncertainty relation still holds. As for the last one, we remark that, at the least, this prepares a consistency check of modern physics theory and the structure of the universe or, at the most, a doorway to deeper dynamics beyond quantum mechanics. For this reason, the problem of the quantum fluctuation can be ascribed to a (D + 1)-dimensional classical random motion.

We are now interested in the questions: How deeper level of matter or what kind of particle model is enough to discuss a deeper origin of the quantum fluctuation and the quantum correlation? Besides these kinds of problems, we are also interested in another question, as to whether we can find a deeper origin of the quantum fluctuation in a classical chaotic behavior of the surrounding atmosphere. These questions must be related to the rather philosophical question of *reductionism*. We have to say that quantum mechanics is still developing.

One may also expect to find a doorway to new future dynamics beyond quantum mechanics, in recent cosmic ray observations anomalies of high energy particle distributions, the nonvanishing neutrino mass, and so on.

Acknowledgement

The author is very grateful to M. Kanenaga for his help.

References

1. Schrödinger, E.: *Sitzungsb. Preuss. Akad. W.* (1931), 144; *Ann. Inst. H. Poincaré* **2** (1932), 267; *J. Metadier, Comptes Rendus* **193** (1931), 1173; *R. Fürth, Z. Phys.* **81** (1933), 143.
2. Von Neumann, J.: *Mathematische Grundlagen der Quantenmechanik*, Springer, Berlin, 1932.
3. Pauli, W.: *Writings on Physics and Philosophy* (English version) translated by R. Schlapp, Springer, Berlin, 1994. Its Japanese version translated by K. Okano was published by Springer, Tokyo in 1998.
4. Bohm, D.: *Phys. Rev.* **85** (1952), 166;
   Nelson, E.: *Phys. Rev.* **150** (1966), 1079.
5. Parisi, G. and Wu, Y.: *Sci. Sin.* (1981), 483;
   Namiki, M.: *Stochastic Quantization*, Springer, Heidelberg, 1992;
   Namiki, M. and Okano, K. (eds): Stochastic quantization, *Suppl. Progr. Theoret. Phys.* No. 111 (1993).
6. Suzuki, M.: *Progr. Theoret. Phys.* **56** (1976), 1454.
7. Callaway, D. J. E. and Rahman, A.: *Phys. Rev. Lett.* **49** (1982), 613; *Phys. Rev. D* **28** (1983), 1506.
8. Namiki, M. and Kanenaga, M.: *Phys. Lett. A* **249** (1998), 13.
9. Nakazato, H., Namiki, M., Pascazio, S. and Yamanaka, Y.: *Phys. Lett. A* **222** (1996), 130.
10. Hayakawa, S.: *Progr. Theoret. Phys.* **21** (1959), 324;
    Hayakawa, S. and Tanaka, H.: *Progr. Theoret. Phys.* **23** (1961), 858.

Acta Applicandae Mathematicae **63:** 283–291, 2000.

Complex White Noise and Coherent State Representations

NOBUAKI OBATA
Graduate School of Mathematics, Nagoya University, Nagoya, 464-8602, Japan.
e-mail: obata@math.nagoya-u.ac.jp

(Received: 12 April 1999)

Abstract. An infinite-dimensional extension of a coherent state representation is discussed within the framework of white noise calculus. The exponential vectors (unnormalized coherent states) and the complex Gaussian integral play a role in such representations of a white noise function and of a white noise operator.

Mathematics Subject Classifications (2000): 46F25, 46G20, 60H40, 81R30.

Key words: complex white noise, white noise operator, coherent state, exponential vector, inverse S-transform.

Introduction

The white noise analysis, launched by T. Hida [6] in 1975, has recently developed considerably into an infinite-dimensional analogue of Schwartz-type distribution theory on a *real* Gaussian space with many applications (see [8, 13, 14] and references cited therein). However, the *complex* white noise has received less attention though introduced at a quite early stage in the study of white noise. So far, only a few results are known in connection with, for example, group symmetry emerging in Gaussian analysis [7, 9], an infinite-dimensional analogue of Bargmann spaces [12, 18], and so forth.

In a previous paper [15], we initiated a study pertaining to representations of states and operators by means of coherent states (exponential vectors) and the complex white noise. A motivation of this line of research comes from the well-known fact that quantum mechanical density operators are represented as integrals over projection operators $|z\rangle\langle z|$, where $|z\rangle$ is the (normalized) coherent state. In fact, this is based on the resolution of the identity

$$I = \frac{1}{\pi}\int_{\mathbf{C}} |z\rangle\langle z|\, \mathrm{d}^2 z. \tag{1}$$

In this paper, we show that (1) admits an infinite-dimensional extension where the Euclidean measure $\mathrm{d}^2 z$ on $\mathbf{C}$ is replaced with the complex Gaussian measure. Moreover, we derive integral representations of white noise distributions and of white noise operators in terms of coherent states and the complex white noise. The corresponding results in the case of finite degrees of freedom are well known, see, e.g. [5, 10, Chapter I, 17]. During the discussion, we reproduce the inversion formula for the S-transform which already has appeared implicitly in Kondratiev [11], see also [1, Chapter 2.5]. Further investigation is now in progress.

1. Complex Gaussian Space

We start with a Gelfand triple:

$$E \subset H \subset E^*, \tag{2}$$

where H is a real Hilbert space, E is a nuclear space densely and continuously embedded in H, and E^* is the dual space of E. As usual, we assume that E is constructed from a positive selfadjoint operator A on H satisfying the particular properties described in [14, Chapter 3]. The canonical bilinear form on $E^* \times E$, which is compatible to the inner product of H, is denoted by the common symbol $\langle \cdot, \cdot \rangle$. The norm of H is denoted by $|\cdot|_0$ and the topology of E is given by the family of norms $|\xi|_p = |A^p \xi|_0$, $p \in \mathbf{R}$. Let $\mu_{1/2}$ denote the Gaussian measure on E^* with variance $1/2$ which is defined by the characteristic function:

$$e^{-|\xi|_0^2/4} = \int_{E^*} e^{i\langle x, \xi \rangle} \mu_{1/2}(dx), \qquad \xi \in E.$$

The unique existence is guaranteed by the Bochner–Minlos theorem.

We consider the complexifications

$$E_{\mathbf{C}} = E + iE \quad \text{and} \quad E_{\mathbf{C}}^* = E^* + iE^*.$$

The canonical bilinear form $\langle \cdot, \cdot \rangle$ is uniquely extended to a complex bilinear form on $E_{\mathbf{C}}^* \times E_{\mathbf{C}}$. Based on the topological isomorphism $E_{\mathbf{C}}^* \cong E^* \times E^*$, we introduce a probability measure $\nu = \mu_{1/2} \times \mu_{1/2}$ on $E_{\mathbf{C}}^*$ by

$$\nu(dz) = \mu_{1/2}(dx)\mu_{1/2}(dy), \qquad z = x + iy \in E_{\mathbf{C}}^*.$$

The probability space $(E_{\mathbf{C}}^*, \nu)$ is called the *complex Gaussian (white noise) space*. Notice that we have changed the variance of ν from the previous one [15] in order to keep consistency with the standard notation [7, Chapter 6]. The following formulas are useful:

$$\int_{E^*} e^{\langle x, \xi \rangle} \mu_{1/2}(dx) = e^{\langle \xi, \xi \rangle/4}, \qquad \xi \in E_{\mathbf{C}}, \tag{3}$$

$$\int_{E_{\mathbf{C}}^*} e^{\langle \bar{z}, \xi \rangle + \langle z, \eta \rangle} \nu(dz) = e^{\langle \xi, \eta \rangle}, \qquad \xi, \eta \in E_{\mathbf{C}}, \tag{4}$$

where $\bar{z} = x - iy$ for $z = x + iy \in E^* + iE^*$.

Let $\Gamma(H_{\mathbf{C}})$ be the Boson Fock space over $H_{\mathbf{C}}$. Then, through the celebrated Wiener–Itô–Segal isomorphism, we have

$$L^2(E^*, \mu_{1/2}) \cong \Gamma(H_{\mathbf{C}}), \tag{5}$$

where the isometric isomorphism is uniquely determined by the correspondence

$$\psi_\xi(x) \equiv \mathrm{e}^{\sqrt{2}\langle x,\xi\rangle - \langle \xi,\xi\rangle/2} \leftrightarrow \phi_\xi \equiv \left(1, \xi, \frac{\xi^{\otimes 2}}{2!}, \dots, \frac{\xi^{\otimes n}}{n!}, \dots\right), \quad \xi \in E_{\mathbf{C}}. \tag{6}$$

The above ϕ_ξ is called an *exponential vector* or a *coherent state* (up to a normalizing factor). By definition

$$\langle\langle \phi_\xi, \phi_\eta \rangle\rangle = \mathrm{e}^{\langle \xi, \eta\rangle}, \quad \xi, \eta \in E_{\mathbf{C}}, \tag{7}$$

where $\langle\langle \cdot, \cdot \rangle\rangle$ is the canonical bilinear form on $\Gamma(H_{\mathbf{C}})$.

2. CKS-Space

Let $\{\alpha(n)\}_{n=0}^\infty$ be a positive sequence satisfiying the *CKS-condition*, i.e.,

(A1) $\alpha(0) = 1$ and $\gamma \equiv \sup \alpha^{-1}(n) < \infty$;

(A2) the associated generating function

$$G_\alpha(t) = \sum_{n=0}^\infty \frac{\alpha(n)}{n!} t^n$$

is entire holomorphic, i.e., has an infinite radius of convergence;

(A3) $\widetilde{G}_\alpha(t) = \displaystyle\sum_{n=0}^\infty t^n \frac{n^{2n}}{n!\alpha(n)} \left\{ \inf_{s>0} \frac{G_\alpha(s)}{s^n} \right\}$

has a positive radius of convergence.

In general, given such a sequence $\{\alpha(n)\}_{n=0}^\infty$, we define a *weighted Fock space* over a Hilbert space H by

$$\Gamma_\alpha(H) = \left\{ (f_n);\ f_n \in H^{\widehat{\otimes} n},\ \|(f_n)\|^2 \equiv \sum_{n=0}^\infty n!\alpha(n) \,|\, f_n \,|^2 < \infty \right\}.$$

By definition, the usual Fock space $\Gamma(H)$ is obtained as a special case of $\alpha(n) \equiv 1$.

We now go back to (2). For $p \in \mathbf{R}$, let E_p be the Hilbert space obtained by completing E with respect to the norm $|\cdot|_p$ and define

$$\Gamma_\alpha(E) = \operatorname*{proj\,lim}_{p\to\infty} \Gamma_\alpha(E_p).$$

The topology is given by the family of norms

$$\| \phi \|_{p,+}^2 = \sum_{n=0}^{\infty} n!\alpha(n) \, | \, f_n \, |_p^2 , \quad \phi = (f_n).$$

We say that $\Gamma_\alpha(E)$ is the *Fock space over E associated with* $\{\alpha(n)\}$. The dual space of $\Gamma_\alpha(E)$ is easily described. The space $\Gamma_{1/\alpha}(E_{-p})$ is defined in a similar manner as above, the norm of which is given by

$$\| \Phi \|_{-p,-}^2 = \sum_{n=0}^{\infty} \frac{n!}{\alpha(n)} \, | \, F_n \, |_{-p}^2 , \quad \Phi = (F_n).$$

By a standard argument, it is proved that

$$\Gamma_\alpha(E)^* \cong \operatorname*{ind\,lim}_{p\to\infty} \Gamma_{1/\alpha}(E_{-p}),$$

where $\Gamma_\alpha(E)^*$ carries the strong dual topology. Finally, taking the complexification, we obtain a chain of Fock spaces:

$$\Gamma_\alpha(E_{\mathbf{C}}) \subset \Gamma_\alpha(H_{\mathbf{C}}) \subset \Gamma(H_{\mathbf{C}}) \subset \Gamma_\alpha(H_{\mathbf{C}})^* \subset \Gamma_\alpha(E_{\mathbf{C}})^*.$$

It is known that $\Gamma_\alpha(E_{\mathbf{C}})$ is nuclear. Dropping the index α for simplicity, we come to

$$\mathcal{W} \equiv \Gamma_\alpha(E_{\mathbf{C}}) \subset \Gamma(H_{\mathbf{C}}) \subset \Gamma_\alpha(E_{\mathbf{C}})^* \equiv \mathcal{W}^*, \tag{8}$$

which is called the *Cochran–Kuo–Sengupta space* [4]. The particular cases of $\alpha(n) = 1$ and $\alpha(n) = (n!)^\beta$ with $0 \leqslant \beta < 1$ are respectively called the *Hida–Kubo–Takenaka space* and the *Kondratiev–Streit space*, see also [13]. The canonical bilinear form on $\mathcal{W}^* \times \mathcal{W}$ is denoted also by $\langle\!\langle \cdot, \cdot \rangle\!\rangle$. Then we have

$$\langle\!\langle \Phi, \phi \rangle\!\rangle = \sum_{n=0}^{\infty} n! \, \langle F_n, f_n \rangle , \quad \Phi = (F_n) \in \mathcal{W}^*, \ \phi = (f_n) \in \mathcal{W}.$$

In view of (5) and (8) we obtain

$$\mathcal{W} \otimes \mathcal{W} \subset L^2(E^*, \mu_{1/2}) \otimes L^2(E^*, \mu_{1/2}) \subset (\mathcal{W} \otimes \mathcal{W})^*. \tag{9}$$

On the other hand, identifying a function on $E_{\mathbf{C}}^*$ with one on $E^* \times E^*$ in such a way that

$$\phi \otimes \psi(x + iy) = \phi(x)\psi(y), \quad x, y \in E^*, \ \phi, \psi \in \mathcal{W},$$

we have the isomorphism $L^2(E^*, \mu_{1/2}) \otimes L^2(E^*, \mu_{1/2}) \cong L^2(E_{\mathbf{C}}^*, \nu)$. Finally, we denote by

$$\mathcal{D} \subset L^2(E_{\mathbf{C}}^*, \nu) \subset \mathcal{D}^* \tag{10}$$

the Gelfand triple corresponding to (9).

3. Resolution of the Identity via Coherent States

For any $z \in E_{\mathbf{C}}^*$ the same formula as in (6) is applied to defining an exponential vector $\phi_z \in \mathcal{W}^*$. We then put

$$Q_z\phi = \langle\langle \phi_{\bar{z}}, \phi \rangle\rangle \phi_z, \quad \phi \in \mathcal{W}.$$

As is easily verified, $Q_z \in \mathcal{L}(\mathcal{W}, \mathcal{W}^*)$ and its symbol is given by

$$\widehat{Q}_z(\xi, \eta) = \langle\langle Q_z\phi_\xi, \phi_\eta \rangle\rangle = \langle\langle \phi_{\bar{z}}, \phi_\xi \rangle\rangle \langle\langle \phi_z, \phi_\eta \rangle\rangle = \mathrm{e}^{\langle \bar{z}, \xi \rangle + \langle z, \eta \rangle}. \tag{11}$$

Note also that both maps $z \mapsto \phi_z \in \mathcal{W}^*$ and $z \mapsto Q_z \in \mathcal{L}(\mathcal{W}, \mathcal{W}^*)$ are continuous.

THEOREM 3.1. *It holds that*

$$I = \int_{E_{\mathbf{C}}^*} Q_z \, \nu(\mathrm{d}z), \tag{12}$$

where the integral is understood through the canonical bilinear form on $\mathcal{W}^* \times \mathcal{W}$.

Proof. It follows by a standard argument that there exists an operator $\Xi \in \mathcal{L}(\mathcal{W}, \mathcal{W}^*)$ such that

$$\langle\langle \Xi\phi, \psi \rangle\rangle = \int_{E_{\mathbf{C}}^*} \langle\langle Q_z\phi, \psi \rangle\rangle \, \nu(\mathrm{d}z), \quad \phi, \psi \in \mathcal{W},$$

for a precise argument see the proof of Lemma 5.1 below. Then, by using formulas (11) and (4), the symbol of Ξ is computed as follows:

$$\begin{aligned}\widehat{\Xi}(\xi, \eta) &= \langle\langle \Xi\phi_\xi, \phi_\eta \rangle\rangle = \int_{E_{\mathbf{C}}^*} \langle\langle Q_z\phi_\xi, \phi_\eta \rangle\rangle \, \nu(\mathrm{d}z) \\ &= \int_{E_{\mathbf{C}}^*} \mathrm{e}^{\langle \bar{z}, \xi \rangle + \langle z, \eta \rangle} \, \nu(\mathrm{d}z) = \mathrm{e}^{\langle \xi, \eta \rangle} = \langle\langle \phi_\xi, \phi_\eta \rangle\rangle.\end{aligned}$$

Then the uniqueness of an operator symbol implies that $\Xi = I$. (General results on operator symbols in CKS-space are obtained in [2, 3]. See also [14].) □

4. Decomposition of White Noise Functions via Coherent States

Let $\xi \in E_{\mathbf{C}}$ and consider the exponential function $z \mapsto \mathrm{e}^{\langle z, \xi \rangle}$, $z \in E_{\mathbf{C}}^*$. Writing $z = x + iy$, we observe

$$\begin{aligned}\mathrm{e}^{\langle z, \xi \rangle} &= \mathrm{e}^{\langle x, \xi \rangle} \mathrm{e}^{\langle y, i\xi \rangle} \\ &= \mathrm{e}^{\sqrt{2}\langle x, \xi/\sqrt{2} \rangle - \langle \xi/\sqrt{2}, \xi/\sqrt{2} \rangle/2} \mathrm{e}^{\sqrt{2}\langle y, i\xi/\sqrt{2} \rangle - \langle i\xi/\sqrt{2}, i\xi/\sqrt{2} \rangle/2} \\ &= \psi_{\xi/\sqrt{2}}(x) \psi_{i\xi/\sqrt{2}}(y).\end{aligned}$$

Therefore $\mathrm{e}^{\langle \cdot, \xi \rangle} = \psi_{\xi/\sqrt{2}} \otimes \psi_{i\xi/\sqrt{2}}$ belongs to $\mathcal{W} \otimes \mathcal{W} = \mathcal{D}$.

THEOREM 4.1. *For any $w \in \mathcal{D}^*$ there exists a unique $\Phi \in \mathcal{W}^*$ such that*

$$\langle\langle \Phi, \phi_\xi \rangle\rangle = \langle\langle w, e^{\langle \cdot, \xi \rangle} \rangle\rangle. \tag{13}$$

Proof. By the characterization theorem for the S-transform in CKS-space [4]. □

In view of $\langle\langle \phi_z, \phi_\xi \rangle\rangle = e^{\langle z, \xi \rangle}$ and expression (13), we naturally adopt a formal integral expression:

$$\Phi = \int_{E_{\mathbf{C}}^*} w(z)\phi_z \, \nu(dz). \tag{14}$$

It is not difficult to see that any $\Phi \in \mathcal{W}^*$ admits an expression as in (14). On the other hand, it is noteworthy that (14) gives the inversion formula for the S-transform. Originally, the S-transform of $\Phi \in \mathcal{W}^*$ is defined by

$$S\Phi(\xi) = \langle\langle \Phi, \phi_\xi \rangle\rangle, \qquad \xi \in E_{\mathbf{C}},$$

and, hence, $S\Phi$ is a function on $E_{\mathbf{C}}$. However, if $\phi \in \mathcal{W}$, the S-transform $S\phi$ is naturally extended to a function of $E_{\mathbf{C}}^*$ by

$$S\phi(z) = \langle\langle \phi_z, \phi \rangle\rangle, \qquad z \in E_{\mathbf{C}}^*.$$

THEOREM 4.2. *For $\phi \in \mathcal{W}$, it holds that*

$$\phi = \int_{E_{\mathbf{C}}^*} S\phi(\bar{z})\phi_z \, \nu(dz). \tag{15}$$

Proof. Write ψ for the right-hand side of (15). It is easy to see that $\psi \in \mathcal{W}$ and the S-transform is computed as

$$\begin{aligned} S\psi(\xi) &= \int_{E_{\mathbf{C}}^*} S\phi(\bar{z}) S\phi_z(\xi) \nu(dz) \\ &= \int_{E_{\mathbf{C}}^*} \langle\langle \phi_{\bar{z}}, \phi \rangle\rangle \langle\langle \phi_z, \phi_\xi \rangle\rangle \nu(dz) = \int_{E_{\mathbf{C}}^*} \langle\langle Q_z \phi, \phi_\xi \rangle\rangle \nu(dz), \end{aligned}$$

where we used (11). It then follows from Theorem 3.1 that the last expression is equal to $\langle\langle \phi, \phi_\xi \rangle\rangle = S\phi(\xi)$. Then by the uniqueness of S-transform we conclude that $\phi = \psi$ as desired. □

A coherent state representation of a white noise distribution as in (14) is not unique due to the famous overcompleteness of the exponential vectors, see the forthcoming paper [16].

5. Diagonal Coherent State Representations of Operators

We go back to the operator symbol of Q_z, see (11). Given $\xi, \eta \in E_{\mathbf{C}}$ we put

$$q_{\xi,\eta}(z) = \langle\!\langle Q_z\phi_\xi, \phi_\eta \rangle\!\rangle = \mathrm{e}^{\langle \bar{z}, \xi\rangle + \langle z, \eta\rangle}, \quad z \in E_{\mathbf{C}}^*. \tag{16}$$

Then, for $z = x + iy$, we obtain

$$\begin{aligned} q_{\xi,\eta}(x+iy) &= \mathrm{e}^{\langle x, \xi+\eta\rangle}\mathrm{e}^{\langle y, i(-\xi+\eta)\rangle} \\ &= \mathrm{e}^{\langle \xi, \eta\rangle}\psi_{(\xi+\eta)/\sqrt{2}}(x)\psi_{i(-\xi+\eta)/\sqrt{2}}(y). \end{aligned} \tag{17}$$

Therefore $q_{\xi,\eta} = \mathrm{e}^{\langle \xi, \eta\rangle}\psi_{(\xi+\eta)/\sqrt{2}} \otimes \psi_{i(-\xi+\eta)/\sqrt{2}}$ and belongs to $\mathcal{W} \otimes \mathcal{W} \cong \mathcal{D}$.

LEMMA 5.1. *For $w \in \mathcal{D}^*$ there is a unique operator $\Xi \in \mathcal{L}(\mathcal{W}, \mathcal{W}^*)$ such that*

$$\langle\!\langle \Xi\phi_\xi, \phi_\eta \rangle\!\rangle = \langle\!\langle w, q_{\xi,\eta} \rangle\!\rangle, \quad \xi, \eta \in E_{\mathbf{C}}. \tag{18}$$

Proof. By definition we have

$$\begin{aligned} &\left|\langle\!\langle w, q_{\xi,\eta}\rangle\!\rangle\right|^2 \\ &\quad = \left|\mathrm{e}^{\langle \xi, \eta\rangle}\langle\!\langle w, \psi_{(\xi+\eta)/\sqrt{2}} \otimes \psi_{i(-\xi+\eta)/\sqrt{2}}\rangle\!\rangle\right|^2 \\ &\quad \leqslant \mathrm{e}^{2|\langle \xi, \eta\rangle|} \, \| w \|_{-p}^2 \, \| \psi_{(\xi+\eta)/\sqrt{2}} \|_p^2 \| \psi_{i(-\xi+\eta)/\sqrt{2}} \|_p^2 \\ &\quad = \mathrm{e}^{2|\langle \xi, \eta\rangle|} \, \| w \|_{-p}^2 \, G_\alpha\big(| (\xi+\eta)/\sqrt{2} |_p^2\big) G_\alpha\big(| i(-\xi+\eta)/\sqrt{2} |_p^2\big), \end{aligned}$$

where $p \geqslant 0$ is chosen as $\| w \|_{-p} < \infty$. Then, after usual estimates fitting the generating function G_α, we conclude with the basis on characterization theorem of operator symbols (e.g., [2, 14, §4.4]) that the right-hand side of (18) is the symbol of an operator $\Xi \in \mathcal{L}(\mathcal{W}, \mathcal{W}^*)$. □

The operator Ξ defined as in (18) is written in a formal integral:

$$\Xi = \int_{E_{\mathbf{C}}^*} w(z)Q_z \, \nu(\mathrm{d}z) \tag{19}$$

and is called a *diagonal coherent state representation.*

THEOREM 5.2. *Every operator in $\mathcal{L}(\mathcal{W}, \mathcal{W}^*)$ admits a diagonal coherent state representation.*

Proof. Given $\Xi \in \mathcal{L}(\mathcal{W}, \mathcal{W}^*)$ we consider

$$\Theta(\xi, \eta) = \langle\!\langle \Xi\phi_{(\xi+i\eta)/\sqrt{2}}, \phi_{(\xi-i\eta)/\sqrt{2}}\rangle\!\rangle \mathrm{e}^{-\langle \xi+i\eta, \xi-i\eta\rangle/2}, \quad \xi, \eta \in E_{\mathbf{C}}.$$

It then follows from the characterization theorem of operator symbols that there exists $W \in \mathcal{L}(\mathcal{W}, \mathcal{W}^*)$ such that

$$\Theta(\xi, \eta) = \langle\!\langle W\phi_\xi, \phi_\eta \rangle\!\rangle, \quad \xi, \eta \in E_{\mathbf{C}}.$$

Then, changing the parameters, we have

$$\langle\langle \Xi\phi_\xi, \phi_\eta \rangle\rangle = \langle\langle W\phi_{(\xi+\eta)/\sqrt{2}}, \phi_{i(-\xi+\eta)/\sqrt{2}} \rangle\rangle e^{\langle \xi, \eta \rangle}, \quad \xi, \eta \in E_{\mathbf{C}}.$$

In view of the canonical isomorphism $(\mathcal{W} \otimes \mathcal{W})^* \cong \mathcal{L}(\mathcal{W}, \mathcal{W}^*)$, we choose $w \in (\mathcal{W} \otimes \mathcal{W})^*$ such that

$$\langle\langle \Xi\phi_\xi, \phi_\eta \rangle\rangle = \langle\langle w, \phi_{(\xi+\eta)/\sqrt{2}} \otimes \phi_{i(-\xi+\eta)/\sqrt{2}} \rangle\rangle e^{\langle \xi, \eta \rangle}. \tag{20}$$

Now, taking (6) into account, we consider the functional realization of the right-hand side of (20). As is seen already in (16) and (17), we have

$$\begin{aligned} &\psi_{(\xi+\eta)/\sqrt{2}}(x)\psi_{i(-\xi+\eta)/\sqrt{2}}(y)e^{\langle \xi, \eta \rangle} \\ &\quad = q_{\xi,\eta}(x+iy) = \langle\langle Q_z\phi_\xi, \phi_\eta \rangle\rangle, \quad z = x+iy. \end{aligned}$$

Hence (20) is written in a formal integral

$$\langle\langle \Xi\phi_\xi, \phi_\eta \rangle\rangle = \int_{E^*_{\mathbf{C}}} w(z) \langle\langle Q_z\phi_\xi, \phi_\eta \rangle\rangle \, \nu(\mathrm{d}z), \tag{21}$$

which means that Ξ admits a diagonal coherent state representation as in (19). □

For $z \in E_{\mathbf{C}}$ we have $\phi_z \in \mathcal{W}$ and $Q_z \in \mathcal{L}(\mathcal{W}, \mathcal{W})$. Taking a complete orthonormal basis $\{f_i\} \subset \mathcal{W}$ of $\Gamma(H_{\mathbf{C}})$, we observe

$$\mathrm{Tr}\, Q_z = \sum_i \langle\langle \overline{f_i}, Q_z f_i \rangle\rangle = \sum_i \langle\langle \phi_{\bar{z}}, f_i \rangle\rangle \langle\langle \overline{f_i}, \phi_z \rangle\rangle = \langle\langle \phi_{\bar{z}}, \phi_z \rangle\rangle = e^{\langle \bar{z}, z \rangle}.$$

Recall that $\widehat{Q}_z(\xi, \eta) = e^{\langle \bar{z}, \xi \rangle + \langle z, \eta \rangle}$ by (11). Then, applying (4), we have

$$\int_{E^*_{\mathbf{C}}} \widehat{Q}_z(\bar{\xi}, \xi) \, \nu(\mathrm{d}\xi) = \int_{E^*_{\mathbf{C}}} e^{\langle \bar{z}, \bar{\xi} \rangle + \langle z, \xi \rangle} \, \nu(\mathrm{d}\xi) = e^{\langle \bar{z}, z \rangle}.$$

Consequently,

$$\mathrm{Tr}\, Q_z = \int_{E^*_{\mathbf{C}}} \widehat{Q}_z(\bar{\xi}, \xi) \, \nu(\mathrm{d}\xi), \quad z \in E_{\mathbf{C}}.$$

This formula is extended to a larger class of operators. Further detailed discussion will appear elsewhere.

References

1. Berezansky, Y. M. and Kondratiev, Y. G.: *Spectral Methods in Infinite-Dimensional Analysis*, Kluwer Acad. Publ., Dordrecht, 1995.
2. Chung, D. M., Ji, U. C. and Obata, N.: Higher powers of quantum white noises in terms of integral kernel operators, *Infin. Dimens. Anal. Quantum Probab. Relat. Top.* **1** (1998), 533–559.
3. Chung, D. M., Ji, U. C. and Obata, N.: Normal-ordered white noise differential equations II: Regularity properties of solutions, In: B. Grigolionis *et al.* (eds), *Probability Theory and Mathematical Statistics*, VSP BV and TEV Ltd., 1999, pp. 157–174.

4. Cochran, W. G., Kuo, H.-H. and Sengupta, A.: A new class of white noise generalized functions, *Infinite Dimen. Anal. Quantum Probab.* **1** (1998), 43–67.
5. Glauber, R. J.: Coherent and incoherent states of the radiation field, *Phys. Rev.* **131** (1963), 2766–2788.
6. Hida, T.: *Analysis of Brownian Functionals*, Carleton Math. Lect. Notes 13, Carleton University, Ottawa, 1975.
7. Hida, T.: *Brownian Motion*, Springer, New York, 1980.
8. Hida, T., Kuo, H.-H., Potthoff, J. and Streit, L.: *White Noise*, Kluwer Acad. Publ., Dordrecht, 1993.
9. Jondral, F.: Some remarks about generalized functionals of complex white noise, *Nagoya Math. J.* **81** (1981), 113–122.
10. Klauder, J. R. and Skagerstam, B.-S.: *Coherent States*, World Scientific, Singapore, 1985.
11. Kondratiev, Y. G.: Nuclear spaces of entire functions in problems of infinite-dimensional analysis, *Soviet Math. Dokl.* **22** (1980), 588–592.
12. Kubo, I. and Yokoi, Y.: Generalized functions and functionals in fluctuation analysis, In: T. Hida (ed.), *Mathematical Approach to Fluctuations II*, World Scientific, Singapore, 1995, pp. 203–230.
13. Kuo, H.-H.: *White Noise Distribution Theory*, CRC Press, Boca Raton, 1996.
14. Obata, N.: *White Noise Calculus and Fock Space*, Lecture Notes in Math. 1577, Springer, New York, 1994.
15. Obata, N.: A note on Hida's whiskers and complex white noise, In: H. Heyer and J. Marion (eds), *Analysis on Infinite-Dimensional Lie Groups and Algebras*, World Scientific, Singapore, 1998, pp. 321–336.
16. Obata, N.: Inverse S-transform, Wick product and overcompleteness of exponential vectors, Preprint, 2000.
17. Sudarshan, E. C. G.: Equivalence of semiclassical and quantum mechanical descriptions of statistical light beams, *Phys. Rev. Lett.* **10** (1963), 277–279.
18. Yokoi, Y.: Simple setting for white noise calculus using Bargmann space and Gauss transform, *Hiroshima Math. J.* **25** (1995), 97–121.

Acta Applicandae Mathematicae **63:** 293–306, 2000.

Complexity in Dynamics and Computation

Dedicated to Professor T. Hida for his 70th birthday

MASANORI OHYA
Department of Information Sciences, Science University of Tokyo, Noda City, Chiba, 278 Japan

(Received: 28 October 1999)

Abstract. A new description of chaos in both classical and quantum dynamical systems is discussed in the context of information dynamics, which is called the chaos degree. The algorithm computing this degree is shown. Quantum algorithm solving the SAT problem, on of the NP complete problems, is studied and it is discussed that the SAT problem can be solved in polynomial time when a certain mixture of two orthogonal vectors is physically detected.

Mathematics Subject Classifications (2000): 81P68, 11Y16, 81Q50.

Key words: complexity, information dynamics, chaos degree, SAT problem, quantum algorithm.

1. Introduction

In the theory of complexity, there are two main approaches, one of which is due to the chaotic aspects of natural or nonnatural phenomena and the other is to ask the steps needed to work out a certain job-like computation. The former is strongly related to finding a quantity measuring chaos of dynamical sysyems, and this aspect has been discussed within the context of information dynamics (ID for short) [20, 33, 34, 38, 40], with new descriptions of complexity and chaos. The latter is focused on the research of computational complexity with computer.

In this paper, we review the fundamentals of the complexity in ID and explain the background to constructing the entropic chaos degree both in classical and quantum systems and, further, we discuss the computational complexity in a quantum algorithm based on the work with Masuda [43] to solve the NP complete problem.

2. Complexity and Chaos Degree in Information Dynamics

Several mathematical tools exist to describe chaotic aspects of natural or nonnatural phenomena such as entropy and dynamical entropy, Chaitin's complexity, Lyapunov's exponent, fractal dimensions, bifurcation, ergodicity, multiplicity [2–5, 10, 12, 18, 19, 24, 27, 35, 48].

The author proposed using information dynamics in 1991 to synthesize the dynamics of state change and the complexity of a system, and it is applied to several different fields such as quantum physics, fractal theory, quantum information, and genetics [20].

The quantity measuring chaos in dynamical systems was defined by means of two complexities in ID and was called the *chaos degree*. In particular, among several chaos degrees, the entropic chaos degree was introduced in [40] , and applied to the logistic map [40] and other dynamical maps [22] to study their chaotic behavior.

Here we briefly explain the concept of the complexity of ID in a simplified version (see [20, 33] for details).

Let $(\mathcal{A}, \mathfrak{S}, \alpha(G))$ be an input (or initial) system and $(\overline{\mathcal{A}}, \overline{\mathfrak{S}}, \overline{\alpha}(\overline{G}))$ be an output (or final) system. Here $\mathcal{A}$ is the set of all objects to be observed and $\mathfrak{S}$ is the set of all means for measurement of $\mathcal{A}$, $\alpha(G)$ is a certain evolution of system. Often we have $\mathcal{A} = \overline{\mathcal{A}}$, $\mathfrak{S} = \overline{\mathfrak{S}}$, $\alpha = \overline{\alpha}$. Therefore we claim that

[giving a mathematical structure to input and output triples
$\equiv$ Having a theory].

For instance, when $\mathcal{A}$ is the set $M(\Omega)$ of all measurable functions on a measurable space $(\Omega, \mathcal{F})$ and $\mathfrak{S}(\mathcal{A})$ is the set $P(\Omega)$ of all probability measures on Ω, we have the usual probability theory by which the classical dynamical system is described. When $\mathcal{A} = B(\mathcal{H})$, the set of all bounded linear operators on a Hilbert space $\mathcal{H}$, and $\mathfrak{S}(\mathcal{A}) = \mathfrak{S}(\mathcal{H})$, the set of density operators on $\mathcal{H}$, we have a quantum dynamical system.

Once the input and output systems are set, the situation of the input system is described by a state, an element of $\mathfrak{S}$, and the change of the state is expressed by a mapping from $\mathfrak{S}$ to $\overline{\mathfrak{S}}$, called *a channel*, Λ^*: $\mathfrak{S} \to \overline{\mathfrak{S}}$ (sometimes $\mathfrak{S} \to \mathfrak{S}$). The channel Λ^* describes the dynamics of the system when $\mathcal{A} = \overline{\mathcal{A}}$, so that Λ^* depends on a certain parameter such as time and it may be equal to α. The details of the channels and their uses in physics and quantum communications are discussed in [1, 29, 44].

Moreover, there exist two complexities in ID which are axiomatically given as:

Let $(\mathcal{A}_t, \mathfrak{S}_t, \alpha^t(G^t))$ be the total system of $(\mathcal{A}, \mathfrak{S}, \alpha)$ and $(\overline{\mathcal{A}}, \overline{\mathfrak{S}}, \overline{\alpha})$, and let $C(\varphi) \in [0, \infty]$ be the complexity of a state φ and $T(\varphi; \Lambda^*) \in [0, \infty]$ be the transmitted complexity associated with the state change $\varphi \to \Lambda^*\varphi$. These complexities, C and T, are the quantities satisfying the following conditions:

(i) For any $\varphi \in \mathfrak{S}$, $C(\varphi) \geqslant 0$, $T(\varphi; \Lambda^*) \geqslant 0$.

(ii) For any orthogonal bijection j: $\operatorname{ex} \mathfrak{S} \to \operatorname{ex} \mathfrak{S}$ (the set of all extreme points in $\mathfrak{S}$), $C(j(\varphi)) = C(\varphi)$, $T(j(\varphi); \Lambda^*) = T(\varphi; \Lambda^*)$.

(iii) For $\Phi \equiv \varphi \otimes \psi \in \mathfrak{S}_t$, $C(\Phi) = C(\varphi) + C(\psi)$.

(iv) For any state φ and a channel Λ^*, $0 \leqslant T(\varphi; \Lambda^*) \leqslant C(\varphi)$.

(v) For the identity map 'id' from $\mathfrak{S}$ to $\mathfrak{S}$, $T(\varphi; \mathrm{id}) = C(\varphi)$.

DEFINITION 2.1. Information Dynamics (ID) is defined by

$$(\mathcal{A}, \mathfrak{S}, \alpha(G); \overline{\mathcal{A}}, \overline{\mathfrak{S}}, \overline{\alpha}(\overline{G});\ \Lambda^*;\ C(\varphi), T(\varphi;\ \Lambda^*))$$

and some relations R among them.

Thus, within the framework of ID, we have to

(i) determine $\mathcal{A}, \mathfrak{S}, \alpha(G); \overline{\mathcal{A}}, \overline{\mathfrak{S}}, \overline{\alpha}(\overline{G})$ mathematically,
(ii) choose Λ^* and R, and
(iii) define $C(\varphi), T(\varphi; \Lambda^*)$.

Information dynamics can be applied to the study of chaos in the following ways:

DEFINITION 2.2. (1) ψ is more chaotic than φ, as seen from the reference system $\mathcal{S}$ if $C(\psi) \geqslant C(\varphi)$.

(2)When φ changes to $\Lambda^*\varphi$, the degree of chaos associated to this state change (dynamics) Λ^* is given by

$$D(\varphi; \Lambda^*) = \inf\left\{\int C(\Lambda^* w)\, \mathrm{d}\mu;\ \mu \in M(\varphi)\right\},$$

where $\varphi = \int w\, \mathrm{d}\mu$ is a maximal extremal decomposition of φ equipped with a certain topology in the state space $\mathfrak{S}$ and $M(\varphi)$ is the set of such measures. In some cases such that Λ^* is linear, this chaos degree $D(\varphi; \Lambda^*)$ can be written as $C(\Lambda^*\varphi) - T(\varphi; \Lambda^*)$. A dynamical system is chaotic iff $D(\varphi; \Lambda^*) > 0$ and stable iff $D(\varphi; \Lambda^*) = 0$.

This quantity is called the chaos degree in the sequel.

In ID, several different topics can be treated from a common standpoint [20].

3. Entropic Complexity and Chaos Degree

Although several complexities exist [38], one of the most fundamental pairs of C and T in quantum systems is von Neumann entropy and mutual entropy, where C and T are modified to formulate entropic complexities such as ε-entropy (ε-entropic complexity) [25, 32, 34, 35], Kolmogorov–Sinai-type dynamical entropy (entropic complexity) [3, 24, 27].

The concept of entropy was introduced and developed to study irreversible behavior, symmetry breaking, the amount of information transmission, chaotic properties of states, etc. Here we first show that quantum entropy and quantum mutual entropy are examples of our complexities C and T, respectively.

Let ρ be a state described by a density operator on a Hilbert space $\mathcal{H}$. The entropy of the state ρ was introduced by von Neumann [28, 44] as

$$S(\rho) = -\mathrm{tr}\, \rho \log \rho.$$

If $\rho = \sum_k p_k E_k$ is the Schatten decomposition (i.e., p_k is the eigenvalue of ρ and E_k is the one-dimensional projection associated with p_k, this decomposition is not unique unless every eigenvalue is nondegenerated) of ρ, then the von Neumann entropy takes the Shannon form

$$S(\rho) = -\sum_k p_k \log p_k,$$

because $\{p_k\}$ is a probability distribution. Therefore, von Neumann entropy contains Shannon entropy as a special case.

For two states $\rho, \sigma \in \mathfrak{S}(\mathcal{H})$, the relative entropy [6, 49, 50] is defined by

$$S(\rho, \sigma) = \begin{cases} \operatorname{tr} \rho(\log \rho - \log \sigma) & (\rho \ll \sigma), \\ +\infty & (\text{otherwise}), \end{cases}$$

where $\rho \ll \sigma$ means that $\operatorname{tr} \sigma A = 0 \Rightarrow \operatorname{tr} \rho A = 0$ for any $A \geqslant 0$.

Let Λ^*: $(\mathcal{H}) \to \overline{\mathfrak{S}}(\overline{\mathcal{H}})$ be a channel and define the compound state by

$$\theta_E = \sum_k p_k E_k \otimes \Lambda^* E_k,$$

which expresses the correlation between the initial state ρ and the final state $\Lambda^*\rho$ for a linear (affine) channel [30, 31]. The mutual entropy [30, 32] for a state $\rho \in \mathfrak{S}(\mathcal{H})$ and a channel Λ^*, the amount of information transmitted from ρ to $\Lambda^*\rho$, is given by

$$\begin{aligned} I(\rho; \Lambda^*) &= \sup\{S(\theta_E, \rho \otimes \Lambda^*\rho); \{E_k\}\} \\ &= \sup\left\{\sum_k p_k S(\Lambda^* E_k, \Lambda^*\rho); \{E_k\}\right\}, \end{aligned} \tag{3.1}$$

where the supremum is taken over all Schatten decompositions. The above entropy and mutual entropy become a pair of our two complexities according to the following facts:

(1) The fundamental inequality of the Shannon type [30, 37]:

$$0 \leqslant I(\rho; \Lambda^*) \leqslant \min\{S(\rho), S(\Lambda^*\rho)\},$$

because of

$$S(\Lambda^* E_k, \Lambda^*\rho) = S(\Lambda^*\rho) - \sum_k p_k S(\Lambda^* E_k) \leqslant S(\Lambda^*\rho)$$

and the monotonicity [44, 49] of the relative entropy: $S(\Lambda^* E_k, \Lambda^*\rho) \leqslant S(E_k, \rho)$.

(2) $I(\rho; \text{id}) = S(\rho)$, which is proved as follows:

$$\begin{aligned} I(\rho; \text{id}) &= \sup\left\{\sum_k p_k S(E_{k,}\rho); \{E_k\}\right\} \\ &= \sup\left\{\sum_k p_k(-S(E_k) - E_k \log \rho); \{E_k\}\right\} = S(\rho), \end{aligned}$$

because of $S(E_k) = 0$.

Thus, the quantum entropy and the quantum mutual entropy satisfy all conditions of the complexity and the transmitted complexity, respectively, $C(\rho) = S(\rho)$, $T(\rho; \Lambda^*) = I(\rho; \Lambda^*)$.

In Shannon's communication theory in classical systems, ρ is a probability distribution $p = (p_k) = \sum_k p_k \delta_k$ and Λ^* is a transition probability $(t_{i,j})$, so that the Schatten decomposition of ρ is unique and the compound state of ρ and its output $\overline{\rho}$ $(\equiv \overline{p} = (\overline{p}_i) = \Lambda^* p)$ is the joint distribution $r = (r_{i,j})$ with $r_{i,j} \equiv t_{i,j} p_j$. Then the above complexities C and T become the Shannon entropy and mutual entropy, respectively

$$C(p) = S(p) = -\sum_k p_k \log p_k,$$

$$T(p; \Lambda^*) = I(p; \Lambda^*) = \sum_{i,j} r_{i,j} \log \frac{r_{i,j}}{p_j \overline{p}_i}.$$

We can construct several other types of entropic complexities [38]. For instance, one pair of the complexities is

$$T(\rho; \Lambda^*) = \sup\left\{ \sum_k p_k S(\Lambda^* \rho_k, \Lambda^* \rho);\ \rho = \sum_k p_k \rho_k \right\}, \quad C(\rho) = T(\rho; \mathrm{id}),$$

where $\rho = \sum_k p_k \rho_k$ is a finite decomposition of ρ and the supremum is taken over all such finite decompositions.

When the channel Λ^* is linear, since

$$S(\Lambda^* \rho) = -\mathrm{tr}\, \Lambda^* \rho \log \Lambda^* \rho = -\mathrm{tr}\left(\sum_n p_n \Lambda^* E_n \log \Lambda^* \rho \right)$$

for any Schatten decomposition $\{E_n\}$ of ρ and (3.1), we have

$$\begin{aligned} &D(\rho; \Lambda^*) \\ &\quad = C(\Lambda^* \rho) - T(\rho; \Lambda^*) \\ &\quad = S(\Lambda^* \rho) - I(\rho; \Lambda^*) \\ &\quad = S(\Lambda^* \rho) - \sup\left\{ \mathrm{tr}\left(\sum_n p_n \Lambda^* E_n (\log \Lambda^* E_n - \log \Lambda^* \rho) \right);\ \{E_n\} \right\} \\ &\quad = \inf\left\{ \sum_n p_n S(\Lambda^* E_n);\ \{E_n\} \right\}. \end{aligned} \tag{3.2}$$

The above quantity $D(\rho; \Lambda^*)$ is interpreted as the complexity produced through the channel Λ^*. We apply this quantity $D(\rho; \Lambda^*)$ to study quantum chaos, even when the channel describing the dynamics is not linear. $D(\rho; \Lambda^*)$ is called the entropic chaos degree in the sequel of this paper, which is a special case of the general chaos degree in Definition 2.2.

In order to contain more general dynamics such as in continuous systems, we define the entropic chaos degree in C*-algebraic terninology. This setting will not be used in the sequel application, but for mathematical completeness we will discuss the C*-algebraic setting.

Let $(\mathcal{A}, \mathfrak{S})$ be an input C* system and $(\overline{\mathcal{A}}, \overline{\mathfrak{S}})$ be an output C* system; namely, $\mathcal{A}$ is a C* algebra with unit I and $\mathfrak{S}$ is the set of all states on $\mathcal{A}$. We assume $\overline{\mathcal{A}} = \mathcal{A}$ for simplicity. For a weak* compact convex subset $\mathcal{S}$ (called the *reference space*) of $\mathfrak{S}$, take a state φ from the set $\mathcal{S}$ and let $\varphi = \int_{\mathcal{S}} \omega \, \mathrm{d}\mu_{\varphi}$ be an extremal orthogonal decomposition of φ in $\mathcal{S}$, which describes the degree of mixture of φ in the reference space $\mathcal{S}$. The measure μ_{φ} is not uniquely determined unless $\mathcal{S}$ is the Schoque simplex, so that the set of all such measures is denoted by $M_{\varphi}(\mathcal{S})$. The entropic chaos degree with respect to $\varphi \in \mathcal{S}$ and a channel Λ^* is defined by

$$D^{\mathcal{S}}(\varphi; \Lambda^*) \equiv \inf\left\{ \int_{\mathcal{S}} S^{\mathcal{S}}(\Lambda^*\varphi) \, \mathrm{d}\mu_{\varphi}; \mu_{\varphi} \in M_{\varphi}(\mathcal{S}) \right\}, \tag{3.3}$$

where $S^{\mathcal{S}}(\Lambda^*\varphi)$ is the mixing entropy of a state φ in the reference space $\mathcal{S}$ [36]. The (3.3) is again a special case of one defined in Definition 2.2. When $\mathcal{S} = \mathfrak{S}$, $D^{\mathcal{S}}(\varphi; \Lambda^*)$ is simply written as $D(\varphi; \Lambda^*)$. This $D^{\mathcal{S}}(\varphi; \Lambda^*)$ contains the classical chaos degree and the quantum one (3.2). The classical entropic chaos degree is the case where $\mathcal{A}$ is Abelian and φ is the probability distribution of an orbit generated by the dynamics (channel) Λ^*; $\varphi = \sum_k p_k \delta_k$, where δ_k is the delta measure such that

$$\delta_k(j) \equiv \begin{cases} 1 & (k = j), \\ 0 & (k \neq j). \end{cases}$$

Then the classical entropic chaos degree is

$$D_c(\varphi; \Lambda^*) = \sum_k p_k S(\Lambda^*\delta_k) \tag{3.4}$$

with the Shannon entropy S.

4. Algorithm of Chaos Degree

Algorithmically, the chaos degrees D_c and D_q for classical and quantum dynamics are set as follows: the dynamics of a state is given by a channel $F^*(F_t^*)$ or a mapping $F(F_t)$ on $I \equiv [a, b]^{\mathrm{N}} \subset \mathbf{R}^{\mathrm{N}}$ or a certain Hilbert space $\mathcal{H}$, for instance, $\varphi_t = F_t^* \varphi_0$, $\mathrm{d}x/\mathrm{d}t = F(x)$ with $x \in I$ or $\mathcal{H}$. For a state $\varphi^{(n)}$ at the time (steps) n after a certain time (steps) m, let Λ^* be the channel properly defined by a given dynamics F^* or F. Note that in some cases $\Lambda^* = F^*$, but generally $\Lambda^* \neq F^*$. We will briefly discuss how to compute the entropic chaos degrees for classical dynamics and for quantum dynamics.

(1) *Classical Chaos Degree* D_c: For a map F on $I \equiv [a, b]^{\mathrm{N}} \subset \mathbf{R}^{\mathrm{N}}$ with $x_{n+1} = F(x_n)$ (a difference equation), let $I \equiv \bigcup_k A_k$ be a finite partition with

$A_i \cap A_j = \emptyset$ $(i \neq j)$. The state $\varphi^{(n)}$ of the orbit determined by the difference equation is defined by the probabilty distribution $(p_i^{(n)})$, that is, $\varphi^{(n)} = \sum_i p_i^{(n)} \delta_i$, where for an initial value $x \in I$ and the characteristic function 1_A

$$p_i^{(n)} \equiv \frac{1}{m+1} \sum_{k=n}^{m+n} 1_{A_i}(F^k x). \tag{4.1}$$

When the initial value x is distributed due to a measure υ on I, the above $p_i^{(n)}$ is given as

$$p_i^{(n)} \equiv \frac{1}{m+1} \int_I \sum_{k=n}^{m+n} 1_{A_i}(F^k x)\, d\upsilon. \tag{4.2}$$

The joint distribution $(p_{ij}^{(n,n+1)})$ between the time n and $n+1$ is defined by

$$p_{ij}^{(n,n+1)} \equiv \frac{1}{m+1} \sum_{k=n}^{m+n} 1_{A_i}(F^k x) 1_{A_j}(F^{k+1} x) \tag{4.3}$$

or

$$p_{ij}^{(n,n+1)} \equiv \frac{1}{m+1} \int_I \sum_{k=n}^{m+n} 1_{A_i}(F^k x) 1_{A_j}(F^{k+1} x)\, d\upsilon. \tag{4.4}$$

Then the channel Λ_n^* at n is determined by

$$\Lambda_n^* \equiv \left(\frac{p_{ij}^{(n,n+1)}}{p_i^{(n)}} \right) \Longrightarrow \varphi^{(n+1)} = \Lambda_n^* \varphi^{(n)}, \tag{4.5}$$

and the chaos degree is given by

$$D_c(\varphi^{(n)}; \Lambda_n^*) = \sum_i p_i^{(n)} S(\Lambda_n^* \delta_k) = \sum_{i,j} p_{ij}^{(n,n+1)} \log \frac{p_i^{(n)}}{p_{ij}^{(n,n+1)}}. \tag{4.6}$$

This classical chaos degree was applied to several dynamical maps such as logistic map, Baker's transformation, and Tinkerbel map, and it could explain their chaotic characters [22, 40]. Our chaos degree has several merits compared with such measures such as the Lyapunov exponent.

(2) *Quantum chaos degree* D_q: Here we explain the entropic chaos degree of a quantum system described by a density operator. Let F^* be a channel sending a state to a state and ρ be an intial state. After time n, the state is $F^{*n}\rho$, whose Schatten decomposition is denoted by $\sum_k \lambda_k^{(n)} E_k^{(n)}$. Then define a channel Λ_m^* on $\otimes_1^m \mathcal{H}$ by

$$\Lambda_m^* \sigma = F^* \sigma \otimes \cdots \otimes F^m \sigma, \quad \sigma \in \mathfrak{S}(\mathcal{H}), \tag{4.7}$$

from which the entropic chaos degree (3.3) for the dynamics $\digamma^*$ is written as

$$D_q(\rho;\Lambda_m^*) = \inf\left\{\frac{1}{m}\sum_k \lambda_k^{(n)} S\left(\Lambda_m^* E_k^{(n)}\right);\ \left\{E_k^{(n)}\right\}\right\}, \tag{4.8}$$

where the supremum is taken over all Schatten decompositions of $\digamma^{*n}\rho$.

The quantum entropic chaos degree is applied for the analysis of the quantum spin system [21] and quantum Baker-type transformation [23], and we could measure the chaos of these systems.

5. Computational Complexity

When we solve a problem with the input size n, such as a sequence composed of n letters of 0 and 1, under a certain algorithm, and the time (steps) to solve this problem by computer is in polynominal order of the size n, the algorithm is called a 'good' algorithm and the problem is said to belong to the P (polynominal) class. Such a problem is one to be recognized in polynominal time by a deterministic Turing machine. On the other hand, the problem to be recognized in polynominal time by a nondeterministic Turing machine is called an *NP problem*. A NP problem can also be understood as the problem whose solution cannot be obtained in polynominal time, but a candidacy of the solutions can be examined in polynominal time to be a real solution of this problem or not. One of fundamental problems of computational complexity is whether or not there exists an algorithm to solve the NP problem in polynominal time; namely, NP=P or not. It is known [45] that the most difficult NP problems exist in the NP class, called the NP complete problem, and they are all equivalent. There are several NP complete problems such as the SAT (satisfiability) problem, Salesman problem, and Knapsack problem.

In [43], we showed that the SAT problem can be solved in polynonimal time if a superposition of two orthogonal vectors is detected experimentally. We will explain the basic part of this result in the sequel sections.

6. SAT Problem

Let $X \equiv \{x_1, \ldots, x_n\}$ be a set. Then x_k and its negation $\overline{x}_k (k = 1, 2, \ldots, n)$ are called literals and the set of all such literals is denoted by $X' = \{x_1, \overline{x}_1, \ldots, x_n, \overline{x}_n\}$. The set of all subsets of X' is denoted by $\mathcal{F}(X')$ and an element $C \in \mathcal{F}(X')$ is called a clause. We take a truth assignment to all variables x_k. If we can assign the truth value to at least one element of C, then C is called satisfiable. When C is satisfiable, the truth value $t(C)$ of C is regarded as true, otherwise that of C is false. Take the truth values as 'true $\leftrightarrow$ 1, false $\leftrightarrow$ 0'. Then

$$C \text{ is satisfiable iff } t(C) = 1.$$

Let $L = \{0, 1\}$ be a Boolean lattice with the usual join $\vee$ and meet $\wedge$, and $t(x)$ be the truth value of a literal x in X. Then the truth value of a clause C is written as $t(C) \equiv \bigvee_{x \in C} t(x)$. Moreover, the set $\mathcal{C}$ of all clauses $C_j (j = 1, 2, \ldots, m)$ is called satisfiable iff the meet of all truth values of C_j is 1; $t(\mathcal{C}) \equiv \bigwedge_{j=1}^{m} t(C_j) = 1$. Thus, the SAT problem is written as follows:

DEFINITION 6.1. SAT Problem: Given a set $X \equiv \{x_1, \ldots, x_n\}$ and a set $\mathcal{C} = \{C_1, C_2, \ldots, C_m\}$ of clauses, determine whether $\mathcal{C}$ is satisfiable or not.

That is, this problem is to ask whether there exists a truth assignment to make $\mathcal{C}$ satisfiable.

It is known [45] in usual algorithm that it is polynomial time to check the satisfiability only when a specific truth assignment is given, but we cannot determine the satisfiability in polynomial time when an assignment is not specified.

7. Quantum Algorithm of SAT

Let 0 and 1 of the Boolean lattice L be denoted by the vectors

$$|0\rangle \equiv \begin{pmatrix} 1 \\ 0 \end{pmatrix} \quad \text{and} \quad |1\rangle \equiv \begin{pmatrix} 0 \\ 1 \end{pmatrix}$$

in the Hilbert space $\mathbf{C}^2$, respectively. That is, the vector $|0\rangle$ corresponds to falseness and $|1\rangle$ to truth.

As we explained in the previous section, an element $x \in X$ can be denoted by 0 or 1, so by $|0\rangle$ or $|1\rangle$. In order to describe a clause C with, at most, n length by a quantum state, we need the n-tuple tensor product Hilbert space $\mathcal{H} \equiv \otimes_1^n \mathbf{C}^2$. For instance, in the case of $n = 2$, given $C = \{x_1, x_2\}$ with an assignment $x_1 = 0$ and $x_2 = 1$, the corresponding quantum state vector is $|0\rangle \otimes |1\rangle$, so that the quantum state vector describing C is generally written by $|C\rangle = |x_1\rangle \otimes |x_2\rangle \in \mathcal{H}$ with $x_k = 0$ or 1 $(k = 1, 2)$.

The quantum computation is performed by a unitary gate constructed from several fundamental gates such as Not gate, Controlled-Not gate, Controlled-Controlled Not gate [15, 42]. Once

$$X \equiv \{x_1, \ldots, x_n\} \quad \text{and} \quad \mathcal{C} = \{C_1, C_2, \ldots, C_m\}$$

are given, the SAT is to find the vector $|f(\mathcal{C})\rangle \equiv \bigwedge_{j=1}^{m} \bigvee_{x \in C_j} t(x)$, where $t(x)$ is $|0\rangle$ or $|1\rangle$ when $x = 0$ or 1, respectively, and $t(x) \wedge t(y) \equiv t(x \wedge y)$, $t(x) \vee t(y) \equiv t(x \vee y)$.

We consider the quantum algorithm for the SAT problem. Since we have n variables $x_k (k = 1, \ldots, n)$ and a quantum computation produces some dust bits, the assignments of the n variables and the dusts are represented by n qubits and l qubits in the Hilbert space $\bigotimes_1^n \mathbf{C}^2 \bigotimes_1^l \mathbf{C}^2$. Moreover, the resulting state vector $|f(\mathcal{C})\rangle$ should be added, so that the total Hilbert space is

$$\mathcal{H} \equiv \otimes_1^n \mathbf{C}^2 \otimes_1^l \mathbf{C}^2 \otimes \mathbf{C}^2.$$

Let us start the quantum computation of the SAT problem from an initial vector $|v_0\rangle \equiv \otimes_1^n|0\rangle \otimes_1^l |0\rangle \otimes |0\rangle$ when $\mathcal{C}$ contains n Boolean variables $x_1, \ldots, x_n$. We apply the discrete Fourier transformation denoted by

$$U_F \equiv \otimes_1^n \frac{1}{\sqrt{2}} \begin{pmatrix} 1 & 1 \\ 1 & -1 \end{pmatrix}$$

to the part of the Boolean variables of the vector $|v_0\rangle$, then the resulting state vector becomes

$$|v\rangle \equiv U_F \otimes_1^{l+1} I|v_0\rangle = \frac{1}{\sqrt{2^n}} \otimes_1^n (|0\rangle + |1\rangle) \otimes_1^l |0\rangle \otimes |0\rangle,$$

where I is the identity matrix in $\mathbf{C}^2$. This vector can be written as

$$|v\rangle = \frac{1}{\sqrt{2^n}} \sum_{x_1,\cdots,x_n=0}^{1} \otimes_{j=1}^n |x_j\rangle \otimes_1^l |0\rangle \otimes |0\rangle.$$

Now, we perform the quantum computation to check the satisfiability, which will be done by a unitary operator U_f properly constructed by unitary gates. Then, after the computation by U_f, the vector $|v\rangle$ goes to

$$\begin{aligned} |v_f\rangle &\equiv U_f|v\rangle = \frac{1}{\sqrt{2^n}} \sum_{x_1,\ldots,x_n=0}^{1} U_f \otimes_{j=1}^n |x_j\rangle \otimes_1^l |0\rangle \otimes |0\rangle \\ &= \frac{1}{\sqrt{2^n}} \sum_{x_1,\ldots,x_n=0}^{1} \otimes_{j=1}^n |x_j\rangle \otimes_{i=1}^l |y_i\rangle \otimes |f(x_1, \ldots, x_n)\rangle, \end{aligned}$$

where $f(x_1, \ldots, x_n) \equiv f(\mathcal{C})$ because $\mathcal{C}$ contains $x_1, \ldots x_n$, and $|y_i\rangle$ are the dust bits produced by the computation. As we will explain in an example below, the unitary operator U_f is concretely constructed.

For the case

$$X = \{x_1, x_2, x_3\} \quad \text{and} \quad \mathcal{C} = \{\{x_1\}, \{x_2, x_3\}, \{x_1, \overline{x}_3\}, \{\overline{x}_1, \overline{x}_2, x_3\}\},$$

the resulting state $|f(x_1, x_2, x_3)\rangle$ is written as

$$\begin{aligned} &|f(x_1, x_2, x_3)\rangle \\ &\quad = |x_1\rangle \wedge (|x_2\rangle \vee |x_3\rangle) \wedge (|x_1\rangle \vee |\overline{x}_3\rangle) \wedge (|\overline{x}_1\rangle \vee |\overline{x}_2\rangle \vee |x_3\rangle). \end{aligned}$$

In quantum computation, it is not necessary to substitute all values of x_j ($j = 1, 2, 3$) as the classical computation; we have only to use a unitary operator U_f for the computation of $|f(x_1, x_2, x_3)\rangle$. This unitary operator U_f is constructed as follows: Let U_{NOT}, U_{CN} and U_{CCN} be the Not gate on $\mathbf{C}^2$, the Controlled-Not gate on $\mathbf{C}^2 \otimes \mathbf{C}^2$ and the Controlled-Controlled-Not gate on $\mathbf{C}^2 \otimes \mathbf{C}^2 \otimes \mathbf{C}^2$, respectively, which are given by

$$\begin{aligned}
U_{\text{NOT}} &= |0\rangle\langle 1| + |1\rangle\langle 0|, \\
U_{\text{CN}} &= |0\rangle\langle 0| \otimes I + |1\rangle\langle 1| \otimes (|0\rangle\langle 1| + |1\rangle\langle 0|), \\
U_{\text{CCN}} &= |0\rangle\langle 0| \otimes |0\rangle\langle 0| \otimes I + |0\rangle\langle 0| \otimes |1\rangle\langle 1| \otimes I + \\
&\quad + |1\rangle\langle 1| \otimes |0\rangle\langle 0| \otimes I + |1\rangle\langle 1| \otimes |1\rangle\langle 1| \otimes (|0\rangle\langle 1| + |1\rangle\langle 0|).
\end{aligned}$$

Then the unitary operator U_f is determined by combining the above three unitaries as

$$U_f \equiv U_{36} U_{35} \cdots U_2 U_1,$$

where, for instance,

$$\begin{aligned}
U_1 &\equiv |0\rangle\langle 0| \otimes_1^{23} I + |1\rangle\langle 1| \otimes_1^2 I \otimes (|0\rangle\langle 1| + |1\rangle\langle 0|) \otimes_1^{20} I, \\
U_2 &\equiv I \otimes |0\rangle\langle 0| \otimes_1^{22} I + I \otimes |1\rangle\langle 1| \otimes_1^2 I \otimes (|0\rangle\langle 1| + |1\rangle\langle 0|) \otimes_1^{19} I, \\
U_3 &\equiv \otimes_1^2 I \otimes |0\rangle\langle 0| \otimes_1^{21} I + \otimes_1^2 I \otimes |1\rangle\langle 1| \otimes_1^2 I \otimes (|0\rangle\langle 1| + |1\rangle\langle 0|) \otimes_1^{18} I
\end{aligned}$$

and other $U_4, \ldots, U_{36}$ are similarly constructed. In this case, we need 20 dust bits, so that U_f is operated on the Hilbert space $\otimes_1^{24} \mathbf{C}^2$.

Starting from the initial vector $|v_0\rangle \equiv \otimes_1^3 |0\rangle \otimes_1^{20} |0\rangle \otimes |0\rangle$, the final vector is

$$\begin{aligned}
|v_f\rangle &= U_{36} \cdots U_1 U_F |v_0\rangle = U_f U_F |v_0\rangle \\
&= \frac{1}{\sqrt{2^3}} (|0,0,0,0,1,1,0,1,0,1,0,0,0,1,1,0,0,0,1,1,1,0,1;0\rangle \\
&\quad + |1,0,0,1,1,1,0,0,0,1,1,0,0,1,1,1,0,0,1,1,1,0,1;0\rangle \\
&\quad + |0,1,0,0,0,1,1,1,0,1,0,1,0,1,1,0,1,0,1,1,1,0,1;0\rangle \\
&\quad + |0,0,1,0,1,0,1,1,1,0,0,0,0,0,1,0,1,0,0,1,0,0,0;0\rangle \\
&\quad + |1,1,0,1,0,1,1,0,0,1,1,1,1,1,0,1,1,1,1,1,0,1,0;0\rangle \\
&\quad + |1,0,1,1,1,0,1,0,1,1,1,0,0,0,1,1,1,1,1,1,1,1,1;1\rangle \\
&\quad + |0,1,1,0,0,0,1,1,1,0,0,1,0,0,1,0,1,0,0,1,0,0,0;0\rangle \\
&\quad + |1,1,1,1,0,0,1,0,1,1,1,1,1,0,1,1,1,1,1,1,1,1,1;1\rangle),
\end{aligned}$$

where we used the notation

$$\begin{aligned}
&|x_1, x_2, x_3, y_1, \ldots, y_{20}; f(x_1, x_2, x_3)\rangle \\
&\quad \equiv \otimes_{j=1}^3 |x_j\rangle \otimes_{i=1}^{20} |y_i\rangle \otimes |f(x_1, x_2, x_3)\rangle.
\end{aligned}$$

Go back to general discussion. The final step to check the satisfiability of $\mathfrak{C}$ is to apply the projection $E \equiv \otimes_1^{n+l} I \otimes |1\rangle\langle 1|$ to the state $|v_f\rangle$, mathematically equivalent , to compute the value $\langle v_f | E | v_f \rangle$. If the vector $E|v_f\rangle$ exists or the value $\langle v_f | E | v_f \rangle$ is not 0, then we conclude that $\mathfrak{C}$ is satisfiable. The value of $\langle v_f | E | v_f \rangle$ corresponds to that of the ramdom algorithm and it may be very small, so it may or may not be obtained in polynomial time. Remark: This difficulty can be overcome by applying a chaos dynamics in recent paper [51]. Let us consider an operator V_θ, given by

$$V_\theta \equiv \otimes_1^{n+l}(A|0\rangle\langle 0| + B|1\rangle\langle 1|) \otimes e^{i\theta f(\mathbb{C})} I,$$

and applying it to the vector $|v_f\rangle$, where

$$A \equiv \frac{1}{\sqrt{2}}\begin{pmatrix} 1 & 1 \\ 1 & -1 \end{pmatrix} \quad \text{and} \quad B \equiv \frac{1}{\sqrt{2}}\begin{pmatrix} 1 & 1 \\ -1 & 1 \end{pmatrix}$$

and θ is a certain constant describing the phase of the vector $|f(\mathbb{C})\rangle$. The resulting vector is the superposition of two vectors with some constants α, β such as

$$V_\theta|v_f\rangle = \frac{1}{\sqrt{2^{n+l}}} \otimes_1^{n+l} (|0\rangle + |1\rangle) \otimes (\alpha|0\rangle + \beta e^{i\theta}|1\rangle),$$

one of which is polarized with θ and the other is nonpolarized. The existence of this mixture of two vectors $|0\rangle$ and $e^{i\theta}|1\rangle$ is recognized the SAT problem can be solved in the polynominal order of n [43, 51].

References

1. Accardi, L. and Ohya, M.: Compound channels, transition expectations and liftings, *Appl. Math. Optim.* **39** (1999), 33–59.
2. Accardi, L., Ohya, M. and Watanabe, N.: Dynamical entropy through Markov chain, *Open Systems Inform. Dynam.* **4**(1) (1997), 71–87.
3. Accardi, L., Ohya, M. and Watanabe, N.: Note on quantum dynamical entropy, *Rep. Math. Phys.* **38** (1996), 457–469.
4. Alicki, R.: *Quantum Geometry of Noncommutative Bernoulli Shifts*, Banach Center Publications, 1991.
5. Akashi, S.: The asymptotic behavior of ε-entropy of a compact positive operator, *J. Math. Anal. Appl.* **153** (1990), 250–257.
6. Araki, H.: Relative entropy for states of von Neumann algebras, *Publ. RIMS Kyoto Univ.* **11** (1976), 809–833.
7. Alligood, K. T., Sauer, T. D. and Yorke, J. A.: *Chaos – An Introduction to Dynamical Systems-*, Textbooks in Math. Sci., Springer, New York, 1996.
8. Benatti, F.: *Deterministic Chaos in Infinite Quantum Systems,* Trieste Notes in Phys., Springer, New York, 1993.
9. Billingsley, P.: *Ergodic Theory and Information,* Wiley, NewYork, 1965.
10. Connes, A., Narnhofer, H. and Thirring, W.: Dynamical entropy of C*-algebras and von Neumann algebras, *Comm. Math. Phys.* **112** (1987), 691–719.
11. Connes, A. and Størmer, E.: Entropy for automorphisms of II_1 von Neumann algebras, *Acta Math.* **134** (1975), 289–306.
12. Devaney, R. L.: *An Introduction to Chaotic Dynamical Systems*, Benjamin, New York, 1986.
13. Deutsch, D.: Quantum theory, the Church–Turing principle and the universal quantum computer, *Proc. Royal Soc. London A* **400** (1985), 97–117.
14. Deutsch, D. and Jozsa, R.: Rapid solution of problems by quantum computation, *Proc. Royal Soc. London A* **439** (1992), 553–558.

15. Ekert, A. and Jozsa, R.: Quantum computation and Shor's factoring algorithm, *Rev. Modern Phys.* **68**(3) (1996), 733–753.
16. Emch, G. G.: Positivity of the K-entropy on non-Abelian K-flows, *Z. Wahrscheinlichkeitstheorie Verw. Gebiete* **29** (1974), 241–252.
17. Feymann, R.: Quantum mechanical computer, *Optics News* **11** (1985), 11–20.
18. Hasegawa, H.: Dynamical formulation of quantum level ststistics, *Open Systems Inform. Dynam.* **4** (1997), 359–377.
19. Hida, T.: Complexity in white noise analysis, *The Second International Conference on Quantum Information*, Meijyou Univ., 1999.
20. Ingarden, R. S., Kossakowski, A. and Ohya, M.: *Information Dynamics and Open Systems*, Kluwer Acad. Publ., Dordrecht, 1997.
21. Inoue, K., Kossakowski, A. and Ohya, M.: A description of quantum chaos and its application to spin systems, SUT Preprint.
22. Inoue, K., Ohya, M. and Sato, K.: Application of chaos degree to some dynamical systems, *Chaos, Solitons and Fractals* **11** (2000), 1377–1385.
23. Inoue, K., Ohya, M. and Volovich, I. V.: Semiclassical properties and chaos degree for the quantum bakers map, submitted.
24. Kossakowski, A., Ohya, M. and Watanabe, N.: Quantum dynamical entropy for completely positive operator, *Infin. Dimens. Anal. Quantum Probab. Relat. Top.* **2**(1) (1999), 267–282.
25. Matsuoka, T. and Ohya, M.: Fractal dimensions of states and its application to Ising model, *Rep. Math. Phys.* **36** (1995), 365–379.
26. Misiurewicz, M.: Absolutely continuous measutres for certain maps of interval, *Publ. Math. IHES* **53** (1981), 17–51.
27. Muraki, N. and Ohya, M.: Entropy functionals of Kolmogorov Sinai type and their limit theorems, *Lett. Math. Phys.* **36** (1996), 327–335.
28. von Neumann, J.: Die *Mathematischen Grundlagen der Quantenmechanik*, Springer, Berlin, 1932.
29. Ohya, M.: Quantum ergodic channels in operator algebras, *J. Math. Anal. Appl.* **84** (1981), 318–327.
30. Ohya, M.: On compound state and mutual information in quantum information theory, *IEEE Trans. Inform. Theory* **29** (1983), 770–774.
31. Ohya, M.: Note on quantum probability, *Lett. Nuovo Cimento* **38** (1983), 402–406.
32. Ohya, M.: Some aspects of quantum information theory and their applications to irreversible processes, *Rep. Math. Phys.* **27** (1989), 19–47.
33. Ohya, M.: Information dynamics and its application to optical communication processes, In: *Lecture Notes in Phys.* 378, Springer, New York (1991) pp. 81–92.
34. Ohya, M.: Fractal dimension of states, *Quantum Probab. Related Topics* **6** (1991), 359–369.
35. Ohya, M.: State change, complexity and fractal in quantum systems, *Quantum Comm. Measure.* **2** (1995), 309–320.
36. Ohya, M.: Entropy transmission in C*-dynamical systems, *J. Math. Anal. Appl.* **100** (1984), 222–235.
37. Ohya, M.: Fundamentals of quantum mutual entropy and capacity, *Open Systems Inform. Dynam.* **6**(1) (1999), 69–78.
38. Ohya, M.: Complexity and fractal dimensions for quantum states, *Open Systems Inform. Dynam.* **4** (1997), 141–157.
39. Ohya, M.: Applications of information dynamics to genome sequences, SUT Preprint.
40. Ohya, M.: Complexities and their applications to characterization of chaos, *Internat. J. Theoret. Phys.* **37**(1) (1998), 495–505.
41. Ohya, M.: Foundation of entropy, complexity and fractal in quantum systems, *Internat. Congr. of Probability Toward 2000*, pp. 263–286.
42. Ohya, M.: *Mathematical Foundation of Quantum Computer*, Maruzen Publ. Company, 1999.

43. Ohya, M. and Masuda, A.: NP problem in quantum algorithm, *Open Systems Infom. Dynam.* **7**(1) (2000), 33–39.
44. Ohya, M. and Petz, D.: *Quantum Entropy and Its Use,* Springer, New York, 1993.
45. Papadimitriou, C. H.: *Computational Complexity*, Addison-Wesley, Reading, Mass., 1995.
46. Shaw, R.: Strange attractors, chaotic behavior and information flow, *Z. Naturforsch* **36.a** (1981), 80–112.
47. Shor, P. W.: Algorithm for quantum computation: Discrete logarithm and factoring algorithm, *Proc. 35th Annual IEEE Sympos. Foundation of Computer Science*, 1994, 124–134.
48. Toda, M.: Crisis in chaotic scattering of a highly excited van der Waals complex, *Phys. Rev. Lett.* **74**(14) (1995), 2670–2673.
49. Uhlmann, A.: Relative entropy and the Wigner–Yanase–Dyson–Lieb concavity in interpolation theory, *Comm. Math. Phys.* **54** (1977), 21–32.
50. Umegaki, H.: Conditional expectations in an operator algebra IV (entropy and information), *Kodai Math. Sem. Rep.* **14** (1962), 59–85.
51. Ohya, M. and Volovich, I. V.: Quantum computing, NP-complete problems and chaotic dynamics, quant-ph/9912100, 1999.

Acta Applicandae Mathematicae **63:** 307–322, 2000.

On the Theory of KM_2O-Langevin Equations for Stationary Flows (2): Construction Theorem *

Dedicated to Professor Takeyuki Hida on his 70th birthday

YASUNORI OKABE
Department of Mathematical Engineering and Information Physics, Graduate School and Faculty of Engineering, University of Tokyo, Hongo, Bunkyo-ku 113, Japan

(Received: 8 January 1999)

Abstract. The aim of the present paper is to construct the KM_2O-Langevin matrix $\mathcal{LM}([\mathbf{X}, \mathbf{Y}])$ directly from the correlation matrix function R by using (DDT), (FDT) and (PAC) as an algorithm.

Mathematics Subject Classification (2000): 60H10.

Key words: Langevin equations, stationary flows.

1. Introduction

In a previous paper [2], we introduced the notion of stationarity for the pair $[\mathbf{X}, \mathbf{Y}]$ of two d-dimensional flows $\mathbf{X} = (X(n); 0 \leqslant n \leqslant N)$ and $\mathbf{Y} = (Y(\ell); -N \leqslant \ell \leqslant 0)$ moving in a metric vector space and then characterized the stationary property by the fluctuation-dissipation theorem ((DDT) and (FDT)) which states that there exist certain relations among the KM_2O-Langevin matrix $\mathcal{LM}([\mathbf{X}, \mathbf{Y}])$ associated with the pair $[\mathbf{X}, \mathbf{Y}]$ of flows (Theorem 4.2 in [2]). Furthermore, we have obtained a relation that holds between the KM_2O-Langevin matrix $\mathcal{LM}([\mathbf{X}, \mathbf{Y}])$ and the correlation matrix function R associated with the stationary pair $[\mathbf{X}, \mathbf{Y}]$ of flows (Theorem 6.1 in [2]). We call this relation (PAC). We have found that Burg's relation plays an important role in the characterization theorem stated above (Theorem 6.3 in Section 6 and Claims 8–18 in Section 7 of [2]).

The aim of the present paper is to construct the KM_2O-Langevin matrix $\mathcal{LM}([\mathbf{X}, \mathbf{Y}])$ directly from the correlation matrix function R by using (DDT), (FDT) and (PAC) as an algorithm. The point in this procedure lies in that we can use (PAC) only after showing that the KM_2O-Langevin fluctuation matrices defined according to (FDT) become regular. Therefore, we have to show the latter

* This research was partially supported by Grant-in-Aid for Science Research (B) No. 07459007, No. 10440026, No. 10554001 and Grant-in-Aid for Exploratory Research No. 08874007, the Ministry of Education, Science, Sports and Culture, Japan and by Promotion Work for Creative Software, Information-Technology Promotion Agency, Japan.

statement. As stated in [2], we proved a construction theorem for a stationary solution of the forward KM_2O-Langevin equation under an additional condition that this statement holds (Theorem 6.1 in [1]).

Now let us state the contents of this paper. In Section 2 we will recall the results obtained in [2]. Apart from the stationary pair of flows, in Section 3 we will treat any positive definite $d \times d$ matrix function R defined on a finite set $\{-N, -N+1, \ldots, N-1, N\}$ with the Toeplitz condition and construct a KM_2O-Langevin matrix $\mathcal{LM}(R)$ associated with the matrix function R according to (DDT), (FDT) and (PAC). In Section 4 we will construct a stationary pair $[\mathbf{X}, \mathbf{Y}]$ of flows with the matrix function R its correlation matrix function by solving the KM_2O-Langevin equation describing the time evolution for the pair $[\mathbf{X}, \mathbf{Y}]$ of flows. As stated in [2], there was a gap in the proof of Burg's relation in [1] that is needed for the proof of stationarity of the solution to the forward KM_2O-Langevin equation. We shall eliminate this gap by using the characterization theorem for stationarity proved in [2].

In a forthcoming paper [3], we will discuss the problem of obtaining all stationary extensions of a given stationary flow $\mathbf{X} = (X(n);\ 0 \leqslant n \leqslant N)$ from the point of view of the fluctuation-fluctuation theorem and then apply it to an extension problem for a given positive definite matrix function defined on a finite set $\{-N, -N+1, \ldots, N-1, N\}$ with the Toeplitz condition.

2. The Stationary Pair of Flows

We will recall the results obtained in [2]. Let $[\mathbf{X}, \mathbf{Y}]$ be any pair of flows $\mathbf{X} = (X(n); 0 \leqslant n \leqslant N)$ and $\mathbf{Y} = (Y(\ell); -N \leqslant \ell \leqslant 0)$ with the following independence conditions (2.1) and (2.2) moving in a metric vector space W:

$$\left\{X_j(n); 1 \leqslant j \leqslant d, 0 \leqslant n \leqslant N-1\right\} \text{ is linearly independent in } W, \tag{2.1}$$

$$\left\{Y_j(-n); 1 \leqslant j \leqslant d, 0 \leqslant n \leqslant N-1\right\} \text{ is linearly independent in } W. \tag{2.2}$$

By the innovation method, we have introduced the forward (resp. backward) KM_2O-Langevin fluctuation flow $\nu_+ = (\nu_+(n); 0 \leqslant n \leqslant N)$ (resp. $\nu_- = (\nu_-(\ell);$ $-N \leqslant \ell \leqslant 0)$).

From theorems 2.1 and 2.2 in [2], we know that the KM_2O-Langevin fluctuation flows $\nu_\pm$ are orthogonal in the following sense:

$$(\nu_\pm(\pm m), {}^t\nu_\pm(\pm n)) = \delta_{mn} V_\pm(m) \quad (0 \leqslant m, n \leqslant N). \tag{2.3$_\pm$}$$

We call the matrix function $V_+ = (V_+(n); 0 \leqslant n \leqslant N)$ (resp. $V_- = (V_-(n); 0 \leqslant n \leqslant N)$) the forward (resp. backward) KM_2O-Langevin fluctuation matrix function associated with the flow $\mathbf{X}$ (resp. $\mathbf{Y}$).

The time evolution of the flow $\mathbf{X}$ (resp. $\mathbf{Y}$) is govered by the following forward (resp. backward) KM_2O-Langevin equaton (2.5$_+$) (resp. (2.5$_-$)) with initial condition (2.4$_+$) (resp. (2.4$_-$)):

$$X(0) = \nu_+(0), \tag{2.4$_+$}$$

$$X(n) = -\sum_{k=0}^{n-1} \gamma_+(n,k)X(k) + \nu_+(n) \quad (1 \leqslant n \leqslant N) \tag{2.5$_+$}$$

and

$$Y(0) = \nu_-(0), \tag{2.4$_-$}$$

$$Y(-n) = -\sum_{k=0}^{n-1} \gamma_-(n,k)Y(-k) + \nu_-(-n) \quad (1 \leqslant n \leqslant N). \tag{2.5$_-$}$$

We define the KM$_2$O-Langevin matrix $\mathcal{LM}([\mathbf{X},\mathbf{Y}])$ associated with the pair $[\mathbf{X},\mathbf{Y}]$ of flows by

$$\mathcal{LM}([\mathbf{X},\mathbf{Y}]) \equiv \left\{\gamma_+(n,k), \gamma_-(n,k), V_+(m), V_-(m); 0 \leqslant k < n \leqslant N, 0 \leqslant m \leqslant N\right\}. \tag{2.6}$$

In particular, we define $\delta_\pm(n)$ by

$$\delta_\pm(n) \equiv \gamma_\pm(n,0) \quad (1 \leqslant n \leqslant N). \tag{2.7$_\pm$}$$

We say that the pair $[\mathbf{X},\mathbf{Y}]$ of flows has a stationary property if there exists a matrix function R: $\{-N, -N+1, \ldots, N-1, N\} \to M(d;\mathbf{R})$, said to be the correlation matrix function associated with the pair $[\mathbf{X},\mathbf{Y}]$ of flows, such that for any m, n $(0 \leqslant m, n \leqslant N)$,

$$(X(m), {}^tX(n)) = R(m-n), \tag{2.8$_+$}$$

$$(Y(-m), {}^tY(-n)) = R(-m+n). \tag{2.8$_-$}$$

Now, we shall treat the case where the pair $[\mathbf{X},\mathbf{Y}]$ of flows has a stationary property with independent conditions. We note that the KM$_2$O-Langevin fluctuation matrices $V_\pm(n)$ are regular $(0 \leqslant n \leqslant N-1)$ (Lemma 6.2 in [2]). We recall three fundamental relations (DDT), (FDT) and (PAC):

(DDT) Dissipation-Dissipation Theorem:

$$\gamma_\pm(n,k) = \gamma_\pm(n-1,k-1) + \delta_\pm(n)\gamma_\mp(n-1,n-k-1) \quad (1 \leqslant k < n \leqslant N).$$

(FDT) Fluctuation-Dissipation Theorem:

(i) $V_\pm(n) = (I - \delta_\pm(n)\delta_\mp(n))V_\pm(n-1) \quad (1 \leqslant n \leqslant N)$,
(ii) $V_+(n-1)\,{}^t\delta_-(n) = \delta_+(n)V_-(n-1) \quad (1 \leqslant n \leqslant N)$,
(iii) $V_+(n)\,{}^t\delta_-(n) = \delta_+(n)V_-(n) \quad (1 \leqslant n \leqslant N)$.

(PAC) Formula for KM_2O-Langevin partial correlation matrix function:

$$\delta_\pm(n+1) = -\Big\{R(\pm(n+1)) + \sum_{k=0}^{n-1} \gamma_\pm(n,k)R(\pm(k+1))\Big\}V_\mp(n)^{-1}$$
$$(0 \leqslant n \leqslant N-1).$$

For each n $(0 \leqslant n \leqslant N)$, we define a subset $\mathcal{LM}(n)$ of $\mathcal{LM}([\mathbf{X}, \mathbf{Y}])$ by

$$\mathcal{LM}(n) \equiv \{\gamma_+(m,k), \gamma_-(m,k), V_+(\ell), V_-(\ell); 0 \leqslant k < m \leqslant n, 0 \leqslant \ell \leqslant n\}.$$

In particular, we note that $\mathcal{LM}(0) = \{V_+(0), V_-(0)\} = \{R(0)\}$.

We see from (DDT) and (FDT) that for each n $(1 \leqslant n \leqslant N)$, the matrices $\gamma_\pm(n,k)$ $(0 \leqslant k \leqslant n-1)$ and $V_\pm(n)$ can be obtained from the matrices $\delta_\pm(n)$. On the other hand, we find from (PAC) that for each n $(1 \leqslant n \leqslant N)$, the matrices $\delta_\pm(n)$ can be determined by the system $\mathcal{LM}(n-1)$ and the matrices $R(m)$ $(0 \leqslant m \leqslant n)$.

3. An Algorithm for Constructing the KM_2O-Langevin Matrix from the Correlation Matrix Function

As stated in Section 1, the aim of this section is to establish the relations (DDT), (FDT) and (PAC) stated in Section 2 as an algorithm for constructing the KM_2O-Langevin matrix directly from the correlation matrix function. The point in this procedure is to show that the KM_2O-Langevin fluctuation matrices defined according to (FDT) become regular.

Apart from the stationary pair of flows, we will treat only the correlation matrix function. For that purpose, let us given any $M(d;\mathbf{R})$-valued function $R = (R(n); |n| \leqslant N)$ defined on the set $\{-N, -N+1, \ldots, N-1, N\}$ such that

$$R(-n) = {}^tR(n) \qquad (0 \leqslant n \leqslant N). \tag{3.1}$$

We define for each integer n $(1 \leqslant n \leqslant N+1)$ two $nd \times nd$ block matrices $T_\pm(n)$ by

$$T_\pm(n) = \begin{pmatrix} R(0) & R(\pm1) & \ldots & R(\pm(n-1)) \\ R(\mp1) & R(0) & \ldots & R(\pm(n-2)) \\ \vdots & \vdots & \ddots & \vdots \\ R(\mp(n-1)) & R(\mp(n-2)) & \ldots & R(0) \end{pmatrix}. \tag{3.2$_\pm$}$$

It is to be noted that

$$T_+(1) = T_-(1) = R(0), \tag{3.3}$$

$${}^tT_\pm(n) = T_\pm(n) \quad (1 \leqslant n \leqslant N+1). \tag{3.4$_\pm$}$$

In the sequal below, we will treat the case where for each n $(1 \leqslant n \leqslant N)$, $T_\pm(n)$ are positive definite matrices and $T_\pm(N+1)$ are nonnegative definite matrices:

$$(T_{\pm}(n)\xi, \xi) \geqslant 0, \quad \text{for any } \xi \in \mathbf{R}^{nd} \ (1 \leqslant n \leqslant N+1), \tag{3.5$_\pm$}$$

$$T_{\pm}(n) \in \mathrm{GL}(nd; \mathbf{R}) \quad (1 \leqslant n \leqslant N). \tag{3.6$_\pm$}$$

We call condition (3.5$_\pm$) together with condition (3.6$_\pm$) the *Toeplitz condition*.

Step 1. We define two $d \times d$ matrices $V_{\pm}(0)$ and a system $\mathcal{LM}(R; 0)$ by

$$V_{\pm}(0) \equiv R(0), \tag{3.7$_\pm$}$$

$$\mathcal{LM}(R; 0) \equiv \{V_{+}(0), V_{-}(0)\}. \tag{3.8$_0$}$$

Immediately from (3.3), (3.5$_\pm$) and (3.6$_\pm$), we see that $V_{\pm}(0)$ are positive definite matrices. According to (PAC), therefore, we can define two $d \times d$ matrices $\delta_{\pm}(1)$ by

$$\delta_{\pm}(1) \equiv -R(\pm 1)V_{\mp}(0)^{-1}. \tag{3.9$_\pm$}$$

According to (DDT) and (FDT), furthermore, we define four $d \times d$ matrices $\gamma_{\pm}(1,0)$, $V_{\pm}(1)$ and a system $\mathcal{LM}(R; 1)$ by

$$\gamma_{\pm}(1,0) \equiv \delta_{\pm}(1), \tag{3.10$_\pm$}$$

$$V_{\pm}(1) \equiv (I - \delta_{\pm}(1)\delta_{\mp}(1))V_{\pm}(0), \tag{3.11$_\pm$}$$

$$\begin{aligned}\mathcal{LM}(R; 1) \equiv \{&\gamma_{+}(m,k), \gamma_{-}(m,k), V_{+}(\ell), V_{-}(\ell);\\ &0 \leqslant k < m \leqslant 1, 0 \leqslant \ell \leqslant 1\}.\end{aligned} \tag{3.8$_1$}$$

By a direct calculation, we have

LEMMA 3.1.

(i) $R(\pm 1) = -\delta_{\pm}(1)V_{\mp}(0) = -\delta_{\pm}(1)R(0)$,
(ii) $V_{\pm}(1) = R(0) + \delta_{\pm}(1)R(\mp 1)$,
(iii) $\delta_{+}(1)V_{-}(0) = V_{+}(0)\,{}^{t}\delta_{-}(1)$,
(iv) $\delta_{+}(1)V_{-}(1) = V_{+}(1)\,{}^{t}\delta_{-}(1)$.

Step 2. For any fixed n_0 $(2 \leqslant n_0 \leqslant N)$, we assume that we can construct a system $\mathcal{LM}(R; n_0 - 1) \equiv \{\gamma_{+}(m,k), \gamma_{-}(m,k), V_{+}(\ell), V_{-}(\ell); 0 \leqslant k < m \leqslant n_0 - 1, 0 \leqslant \ell \leqslant n_0 - 1\}$ whose elements satisfy the following properties for any m $(1 \leqslant m \leqslant n_0 - 1)$:

($a_m^{\pm}$) $V_{\pm}(m-1)$ are positive definite matrices,

($b_m^{\pm}$) $\delta_{\pm}(m) = -\left\{R(\pm m) + \sum_{k=0}^{m-2} \gamma_{\pm}(m-1,k)R(\pm(k+1))\right\}V_{\mp}(m-1)^{-1}$,

($c_m^{\pm}$) $\gamma_{\pm}(m,k) = \gamma_{\pm}(m-1,k-1) + \delta_{\pm}(m)\gamma_{\mp}(m-1,m-k-1) \quad (1 \leqslant \forall k \leqslant m-1)$,

$(d_m^{\pm})$ $\quad V_{\pm}(m) = (I - \delta_{\pm}(m)\delta_{\mp}(m))V_{\pm}(m-1),$

$(e_m^{\pm})$ $\quad V_{\pm}(m) = \sum_{k=0}^{m-1} \gamma_{\pm}(m,k)R(\mp(m-k)) + R(0),$

$(f_m^{\pm})$ $\quad R(\pm(m-\ell))) = -\sum_{k=0}^{m-1} \gamma_{\pm}(m,k)R(\pm(k-\ell)) \quad (0 \leqslant \forall \ell \leqslant m-1),$

(g_m) $\quad \sum_{k=0}^{m-1} R(k+1)\,{}^t\gamma_{-}(m,k) = \sum_{k=0}^{m-1} \gamma_{+}(m,k)R(k+1).$

We note that $(b_m^{\pm})$, $(c_m^{\pm})$ and $(d_m^{\pm})$ correspond to (PAC), (DDT) and (FDT), respectively. Further, it is to be noted that (g_m) corresponds to Burg's relation.

It follows from Lemma 3.1 that the case for $n_0 = 2$ holds. Further, by the assumption of induction, we can use $(a_{n_0-1}^{\pm})$ to define two $d \times d$ matrices $\delta_{\pm}(n_0)$ by

$(b_{n_0}^{\pm})$ $\quad \delta_{\pm}(n_0) \equiv -\left\{ R(\pm n_0) + \sum_{k=0}^{n_0-2} \gamma_{\pm}(n_0-1,k)R(\pm(k+1)) \right\} V_{\mp}(n_0-1)^{-1}.$

Furthermore, we define $d \times d$ matrices $\gamma_{\pm}(n_0,k)$ and $V_{\pm}(n_0)$ $(0 \leqslant k < n_0)$ by

$(c_{n_0}^{\pm})$ $\quad \gamma_{\pm}(n_0,k) \equiv \gamma_{\pm}(n_0-1,k-1) + \delta_{\pm}(n_0)\gamma_{\mp}(n_0-1,n_0-k-1) \quad (1 \leqslant k < n_0),$

$(d_{n_0}^{\pm})$ $\quad V_{\pm}(n_0) \equiv (I - \delta_{\pm}(n_0)\delta_{\mp}(n_0))V_{\pm}(n_0-1)$

and construct a system $\mathcal{LM}(R;n_0)$ by

$$\begin{aligned}\mathcal{LM}(R;n_0) \equiv\ & \{\gamma_{+}(m,k), \gamma_{-}(m,k), V_{+}(\ell), V_{-}(\ell); \\ & 0 \leqslant k < m \leqslant n_0, 0 \leqslant \ell \leqslant n_0\}.\end{aligned} \tag{3.8$_{n_0}$}$$

In the sequel, we shall show that the elements of $\mathcal{LM}(R;n_0)$ satisfy four properties $(a_{n_0}^{\pm})$, $(e_{n_0}^{\pm})$, $(f_{n_0}^{\pm})$ and (g_{n_0}).

Step 3. At first, we shall prove the properties $(a_{n_0}^{\pm})$. For that purpose, we shall prepare

LEMMA 3.2. *For any point x_0 in $\mathbf{R}^d$, we put $x_k^{\pm} \equiv {}^t\gamma_{\pm}(n_0-1,n_0-1-k)x_0$ $(1 \leqslant k \leqslant n_0-1)$ and $\xi_{\pm} \equiv {}^t(x_0, x_1^{\pm}, \ldots, x_{n_0-1}^{\pm})$.*

Then, we have

$$(T_{\pm}(n_0)\xi_{\pm}, \xi_{\pm}) = (V_{\pm}(n_0-1)x_0, x_0).$$

Proof. We shall prove only the plus part, because the minus part is similarly proved.

By virtue of the assumption of induction, we can make a matrix representation of n_0 numbers of statements in $(e^+_{n_0-1})$ and $(f^+_{n_0-1})$ for $\ell = 0, 1, \ldots, n_0 - 2$ to get

$$\Gamma_+(n_0-1)T_+(n_0) = \begin{pmatrix} V_+(n_0-1) & 0 & \ldots & 0 \\ {}^tR(1) & & & \\ \vdots & & T_+(n_0-1) & \\ \vdots & & & \\ {}^tR(n_0-1) & & & \end{pmatrix}, \tag{3.12}$$

where the block matrices $\Gamma_\pm(n_0-1)$ is given by

$$\Gamma_\pm(n_0-1) \equiv \begin{pmatrix} I & \gamma_\pm(n_0-1, n_0-2) & \gamma_\pm(n_0-1, n_0-3) & \ldots & \gamma_\pm(n_0-1, 0) \\ 0 & I & 0 & \ldots & 0 \\ \vdots & 0 & \ddots & \ddots & \vdots \\ \vdots & \vdots & \ddots & \ddots & 0 \\ 0 & 0 & \ldots & 0 & I \end{pmatrix}. \tag{3.13}$$

Moreover, by using $(f^+_{n_0-1})$ for $\ell = 0, 1, \ldots, n_0 - 2$ again, we obtain

$$\Gamma_+(n_0-1)T_+(n_0)\,{}^t\Gamma_+(n_0-1) = \begin{pmatrix} V_+(n_0-1) & 0 & \ldots & 0 \\ 0 & & & \\ \vdots & & T_+(n_0-1) & \\ \vdots & & & \\ 0 & & & \end{pmatrix}. \tag{3.14}$$

For any given point x_0 in $\mathbf{R}^d$, we define a point η in $\mathbf{R}^{n_0 d}$ by $\eta \equiv {}^t(x_0, 0, \ldots, 0)$. Then, by caluculating the value of $(\Gamma_+(n_0-1)T_+(n_0)\,{}^t\Gamma_+(n_0-1)\eta, \eta)$ and noting ${}^t\Gamma_+(n_0-1)\eta = \xi_+$, we can show from (3.14) that Lemma 3.2 holds. □

LEMMA 3.3. *The properties* $(a^\pm_{n_0})$ *hold.*

Proof. By (3.5_+) and Lemma 3.2, we find that the matrix $V_+(n_0-1)$ is symmetric and nonnegative definite. Moreover, we see from (3.6_+), (3.12) and (3.13) that the matrix $V_+(n_0-1)$ is positive definite. In a similar way, we can show that the matrix $V_-(n_0-1)$ is positive definite. Therefore, we have Lemma 3.3. □

Step 4. The aim of Step 4 is to show $(e^\pm_{n_0})$ and $(f^\pm_{n_0})$.

LEMMA 3.4.

(i) *The properties* $(e^\pm_{n_0})$ *hold.*
(ii) *The properties* $(f^\pm_{n_0})$ *hold.*

Proof. We shall prove $(e^+_{n_0})$. By $(c^+_{n_0})$, $(b^-_{n_0})$ and $(e^+_{n_0-1})$. We have

$$\begin{aligned}
&\text{the right-hand side of } (e^+_{n_0})\\
&= \delta_+(n_0)\left\{ {}^tR(n_0) + \sum_{k=1}^{n_0-1} \gamma_-(n_0-1, k-1)\, {}^tR(k)\right\} + \\
&\quad + \sum_{k=1}^{n_0-1} \gamma_+(n_0-1, n_0-1-k)\, {}^tR(k) + R(0)\\
&= (I - \delta_+(n_0)\delta_-(n_0))V_+(n_0-1),
\end{aligned}$$

which with $(d^+_{n_0})$ implies that $(e^+_{n_0})$ holds. In a similar way, we have $(e^-_{n_0})$.

Next, we shall prove $(f^+_{n_0})$. We have only to prove

$$R(n_0-\ell) = -\sum_{k=0}^{n_0-1} \gamma_+(n_0, k)R(k-\ell) \quad (0 \leqslant \forall \ell \leqslant n_0-1). \tag{3.15}$$

When $\ell = 0$, it follows from $(c^+_{n_0})$, $(e^-_{n_0-1})$ and $(b^+_{n_0})$ that

$$\begin{aligned}
&\text{the right-hand side of (3.15)}\\
&= -\delta_+(n_0)\left\{ R(0) + \sum_{k=1}^{n_0-1} \gamma_-(n_0-1, n_0-1-k)R(k)\right\} - \\
&\quad - \sum_{k=1}^{n_0-1} \gamma_+(n_0-1, k-1)R(k)\\
&= -\delta_+(n_0)V_-(n_0-1) - \sum_{k=1}^{n_0-1} \gamma_+(n_0-1, k-1)R(k) = R(n_0)
\end{aligned}$$

and so (3.15) holds for $\ell = 0$. Let ℓ be any integer such that $1 \leqslant \ell \leqslant n_0 - 1$. By $(f^+_{n_0-1})$,

$$\begin{aligned}
&\text{the left-hand side of (3.15)}\\
&= R(n_0 - 1 - (\ell - 1))\\
&= -\sum_{k=1}^{n_0-2} \gamma_+(n_0-1, k)R(k-(\ell-1)).
\end{aligned} \tag{3.16}$$

On the other hand, $(c^+_{n_0})$ implies that

$$\begin{aligned}
&\text{the right-hand side of (3.15)}\\
&= -\delta_+(n_0)\left\{ R(-\ell) + \sum_{k=1}^{n_0-1} \gamma_-(n_0-1, n_0-k-1)R(k-\ell)\right\} - \\
&\quad - \sum_{k=1}^{n_0-1} \gamma_+(n_0-1, k-1)R(k-\ell).
\end{aligned}$$

Further, by noting that

$$R(-\ell) = {}^tR(n_0 - 1 - (n_0 - 1 - \ell)) \quad \text{and}$$
$$R(k-\ell) = {}^tR(k - (n_0 - 1 - \ell)),$$

we find from $(f^-_{n_0-1})$ that

$$R(-\ell) + \sum_{k=1}^{n_0-1} \gamma_-(n_0-1, n_0-k-1)R(k-\ell) = 0.$$

Therefore,

$$\begin{aligned}&\text{the right-hand side of (3.15)}\\ &\quad = -\sum_{k=0}^{n_0-2} \gamma_+(n_0-1,k)R(k+1-\ell),\end{aligned}$$

which with (3.16) implies $(f^+_{n_0})$. The property $(f^-_{n_0})$ is also similarly proved. □

Step 5. Let n_0 be the natural number fixed previously such that $2 \leqslant n_0 \leqslant N$. In Step 5, we will show (g_{n_0}). In order to analyse the structure of the statements (g_m) $(1 \leqslant m \leqslant n_0)$, we define $d \times d$ matrices A_m $(1 \leqslant m \leqslant n_0)$ by

$$A_m \equiv \sum_{k=0}^{m-1} R(k+1)\,{}^t\gamma_-(m,k) - \sum_{k=0}^{m-1} \gamma_+(m,k)R(k+1). \tag{3.17$_m$}$$

LEMMA 3.5.

(i) $R(\pm m) = -\delta_\pm(m)V_\mp(m-1) - \sum_{k=0}^{m-2} \gamma_\pm(m-1,k)R(\pm(k+1))$ $(1 \leqslant m \leqslant n_0)$,

(ii) $R(m) = -V_+(m-1)\,{}^t\delta_-(m) - \sum_{k=0}^{m-2} R(k+1)\,{}^t\gamma_-(m-1,k)$ $(1 \leqslant m \leqslant n_0)$,

(iii) $A_m = \delta_+(m+1)V_-(m) - V_+(m)\,{}^t\delta_-(m+1)$ $(0 \leqslant m \leqslant n_0 - 1)$.

Proof. Since $(a^\pm_m)$ hold for any m $(0 \leqslant m \leqslant n_0 - 1)$ and $(b^\pm_m)$ hold for any m $(1 \leqslant m \leqslant n_0)$, we have (i). Since it follows from $(a^+_{n_0-1})$ that ${}^tV_+(m) = V_+(m)$ $(0 \leqslant m \leqslant n_0 - 1)$, we can see that (ii) follows from (i). By (i) and (ii), we have (iii). □

We shall show a kinds of commutation relations that hold between $\delta_\pm(\star)$ and $V_\pm(\star)$.

LEMMA 3.6.

$$\delta_+(m)V_-(m-1) = V_+(m-1)\,{}^t\delta_-(m) \quad (1 \leqslant m \leqslant n_0).$$

Proof. From (ii) and (iii) in Lemma 3.1, Lemma 3.6 holds for $m = 1$.

Since it follows from the assumption of induction that (g_m) holds for any m $(1 \leqslant m \leqslant n_0 - 1)$, we find that $A_m = 0$ for any m $(1 \leqslant m \leqslant n_0 - 1)$ and Lemma 3.5(iii) implies Lemma 3.6. □

LEMMA 3.7. *$V_{\pm}(n_0)$ are symmetric matrices.*

Proof. By Lemma 3.6 and (d_{n_0}), we see that $V_{\pm}(n_0) = V_{\pm}(n_0 - 1) - \delta_{\pm}(n_0)$ $V_{\mp}(n_0)^t \delta_{\pm}(n_0)$, which with $(a^{\pm}_{n_0})$ imply Lemma 3.7. □

LEMMA 3.8.

$$\delta_+(m)V_-(m) = V_+(m)\,{}^t\delta_-(m) \quad (1 \leqslant m \leqslant n_0).$$

Proof. Since $(a^{\pm}_m)$ hold for any m $(0 \leqslant m \leqslant n_0 - 1)$ and $(b^{\pm}_m)$, $(d^{\pm}_m)$ hold for any m $(1 \leqslant m \leqslant n_0)$, we find from Lemma 3.6 that for any m $(1 \leqslant m \leqslant n_0)$,

$$\begin{aligned} \delta_+(m)V_-(m) &= V_+(m-1)(I - {}^t\delta_-(m)\,{}^t\delta_+(m))\,{}^t\delta_-(m) \\ &= {}^tV_+(m)\,{}^t\delta_-(m). \end{aligned}$$

Therefore, it follows from Lemma 3.7 that Lemma 3.8 holds. □

By taking the same procedure as Step 4 in the proof of Lemma 4.3 in [1], we can decompose $d \times d$ matrices A_m with respect to $d \times d$ matrices $\delta_{\pm}(m)$.

LEMMA 3.9. *Let m be any fixed natural number such that $2 \leqslant m \leqslant n_0$. Then, we have*

$$A_m = \delta_+(m)\mathrm{I}_m\,{}^t\delta_-(m) + \delta_+(m)\mathrm{II}_m + \mathrm{III}_m\,{}^t\delta_-(m) + \mathrm{IV}_m,$$

where

$$\mathrm{I}_m = -V_-(m-1)\,{}^t\delta_+(m-1) + \delta_-(m-1)V_+(m-1),$$

$$\begin{aligned} \mathrm{II}_m &= -R(1) - \sum_{j=1}^{m-2} \gamma_-(m-1, m-1-j)R(j+1) + \\ &\quad + \delta_-(m-1)\left\{\sum_{j=0}^{m-2} R(j+1)\,{}^t\gamma_-(m-1, j)\right\} - \\ &\quad - V_-(m-1)\,{}^t\gamma_-(m-1, m-2), \end{aligned}$$

$$\begin{aligned} \mathrm{III}_m &= R(1) + \sum_{j=1}^{m-2} R(j+1)\,{}^t\gamma_+(m-1, m-1-j) - \\ &\quad - \left\{\sum_{j=0}^{m-2} \gamma_+(m-1, j)R(j+1)\right\}{}^t\delta_+(m-1) + \\ &\quad + \gamma_+(m-1, m-2)V_+(m-1), \end{aligned}$$

$$\begin{aligned}\mathrm{IV}_m &= -\sum_{j=0}^{m-3}\gamma_+(m-1,j)R(j+2) - \left\{\sum_{j=0}^{m-3}\gamma_+(m-1,j)R(j+1)\right\}\times \\ &\quad \times {}^t\gamma_-(m-1,m-2) + \sum_{j=0}^{m-3}R(j+2)\,{}^t\gamma_-(m-1,j) + \\ &\quad + \gamma_+(m-1,m-2)\left\{\sum_{j=0}^{m-3}R(j+1)\,{}^t\gamma_-(m-1,j)\right\}.\end{aligned}$$

By using Lemma 3.9, we shall show

LEMMA 3.10. *The property* (g_{n_0}) *holds.*

Proof. Immediately from Lemma 3.8, it follows that $\mathrm{I}_m = 0$ for any m $(2 \leqslant m \leqslant n_0)$. By using $(c^{\pm}_{m-1})$, (d^{-}_{m-1}) and (g_{m-1}), we see that for any m $(2 \leqslant m \leqslant n_0)$,

$$\begin{aligned}\mathrm{II}_m &= -(I - \delta_-(m-1)\delta_+(m-1))\Bigg\{R(1) + \\ &\quad + \sum_{j=1}^{m-2}\gamma_-(m-2,m-2-j)R(j+1) + V_-(m-2)\,{}^t\gamma_-(m-1,m-2)\Bigg\}.\end{aligned}$$

On the other hand, noting from Lemmas 3.6 and 3.7 that

$$-\delta_-(m-2)\delta_+(m-1)V_+(m-2) + V_-(m-2)\,{}^t\delta_+(m-2)\,{}^t\delta_-(m-1) = 0,$$

we can apply (b^{+}_{m-1}) and (c^{-}_{m-1}) to get

$$\begin{aligned}&R(1) + \sum_{j=1}^{m-2}\gamma_-(m-2,m-2-j)R(j+1) + V_-(m-2)\,{}^t\gamma_-(m-1,m-2) \\ &= -\mathrm{II}_{m-1}.\end{aligned}$$

Hence, we find that $\mathrm{II}_m = (I - \delta_-(m-1)\delta_+(m-1))\mathrm{II}_{m-1}$ and so $\mathrm{II}_m = (I - \delta_-(m-1)\delta_+(m-1))(I - \delta_-(m-2)\delta_+(m-2))\cdots(I - \delta_-(2)\delta_+(2))\mathrm{II}_2$. By a direct calculation, we can see from (i) and (iii) in Lemma 3.1 that $\mathrm{II}_2 = 0$. Thus, we conclude that $\mathrm{II}_m = 0$ for any m $(2 \leqslant m \leqslant n_0)$.

In a similar way, we can show that $\mathrm{III}_m = 0$ for any m $(2 \leqslant m \leqslant n_0)$.

Noting that for any j $(0 \leqslant j \leqslant m-3)$, $j+2 = m-1-(m-3-j)$ and $0 \leqslant m-3-j \leqslant (m-1)-1$, we can apply the case $\ell = m-3-j$ in (f^{+}_{m-1}) to see that for any j $(0 \leqslant j \leqslant m-3)$,

$$\begin{aligned}R(j+1) &= R(m-1-(m-3-j)) \\ &= -\gamma_+(m-1,m-2)R(j+1) - \\ &\quad - \sum_{k=0}^{m-3}\gamma_+(m-1,k)R(j-(m-3-k)).\end{aligned}$$

Hence, we have

$$\begin{aligned}
&\sum_{j=0}^{m-3} R(j+2)\,{}^t\gamma_-(m-1,j) \\
&\quad = -\gamma_+(m-1,m-2)\left\{\sum_{j=0}^{m-3}\sum_{j=0}^{m-3} R(j+1)\,{}^t\gamma_-(m-1,j)\right\} - \\
&\quad\quad - \sum_{k=0}^{m-3}\gamma_+(m-1,k)\left\{\sum_{j=0}^{m-3} R(j-(m-3-k))^t\gamma_-(m-1,j)\right\}.
\end{aligned}$$

In a similar way, we can note $R(j+2) = {}^t({}^tR(j+2))$ and use (f^+_{m-1}) to obtain

$$\begin{aligned}
&\sum_{j=0}^{m-3} \gamma_+(m-1,j)R(j+2) \\
&\quad = -\left\{\sum_{j=0}^{m-3}\gamma_+(m-1,j)R(j+1)\right\}{}^t\gamma_-(m-1,m-2) - \\
&\quad\quad - \sum_{j=0}^{m-3}\gamma_+(m-1,j)\left\{\sum_{k=0}^{m-3} R(j-(m-3-k))\,{}^t\gamma_-(m-1,j)\right\}.
\end{aligned}$$

Therefore, we can conclude that $\mathrm{IV}_m = 0$ for any m $(2 \leqslant m \leqslant n_0)$.

Thus we have proved that $A_m = 0$ for any m $(2 \leqslant m \leqslant n_0)$, which gives Lemma 3.10. □

Step 6. We have now arrived at a position to state the following Theorem 3.1.

THEOREM 3.1. *There exists a unique system* $\mathcal{LM}(R) = \{\gamma_+(n,k), \gamma_-(n,k), V_+(\ell), V_-(\ell); 0 \leqslant k < n \leqslant N, 0 \leqslant \ell \leqslant N\}$ *of* $d \times d$ *matrices such that*

(i) $V_\pm(\ell)$ *are positive definite matrices* $(0 \leqslant \ell \leqslant N-1)$,

(ii) $V_\pm(N)$ *are nonnegative definite matrices,*

(iii) $\delta_\pm(n) = -\left\{R(\pm n) + \sum_{k=0}^{n-2}\gamma_\pm(n-1,k)R(\pm(k+1))\right\}V_\mp(n-1)^{-1}$ $(1 \leqslant n \leqslant N)$,

(iv) $\gamma_\pm(n,k) = \gamma_\pm(n-1,k-1) + \delta_\pm(n)\gamma_\mp(n-1,n-k-1)$ $(1 \leqslant k < n \leqslant N)$,

(v) $V_\pm(n) = (I - \delta_\pm(n)\delta_\mp(n))V_\pm(n-1)$ $(1 \leqslant n \leqslant N)$,

(vi) $\delta_+(n+1)V_-(n) = V_+(n)\,{}^t\delta_-(n+1)$ $(0 \leqslant n \leqslant N-1)$,

(vii) $\delta_+(n+1)V_-(n+1) = V_+(n+1)\,{}^t\delta_-(n+1)$ $(0 \leqslant n \leqslant N-1)$.

Proof. By Lemmas 3.4 and 3.10, we find that $(a_m^{\pm})$, $(b_m^{\pm})$, $(c_m^{\pm})$, $(d_m^{\pm})$, $(e_m^{\pm})$, $(f_m^{\pm})$ and (g_m) hold for any m $(1 \leqslant m \leqslant N)$, which implies that all statements except for (ii) in Theorem 3.1 hold. Finally, we shall show (ii) in Theorem 3.1. Since $(e_N^{\pm})$ and $(f_N^{\pm})$ hold, we can take the same procedure as in Lemma 3.3 in Step 3 to see that Lemma 3.2 holds for $n_0 = N + 1$. Therefore, we find from $(3.5_{\pm})$ that the matrices $V_{\pm}(N)$ are nonnegative definite, that is, (ii) in Theorem 3.1 hold. □

We call the system $\mathcal{LM}(R)$ in Theorem 3.1 *a KM$_2$O-Langevin matrix associated with the matrix function R*.

4. A Construction Theorem for Stationary Pair of Flows

Under the same situation as in Section 3, let $R = (R(n); |n| \leqslant N)$ be any $M(d; \mathbf{R})$-valued function defined on the set $\{-N, -N+1, \ldots, N-1, N\}$ satisfying Toeplitz condition $(3.5_{\pm})$ and $(3.6_{\pm})$ with (3.1). Moreover, we shall consider any metric vector space W with an inner product $(*, \star)$ over real field $\mathbf{R}$. The aim of this section is to construct a stationary pair $[\mathbf{X}, \mathbf{Y}]$ of two d-dimensional flows $\mathbf{X} = (X(n); 0 \leqslant n \leqslant N)$ and $\mathbf{Y} = (Y(\ell); -N \leqslant \ell \leqslant 0)$ on the vector space W such that the function R becomes the correlation matrix function of the pair $[\mathbf{X}, \mathbf{Y}]$ of flows.

For that purpose, we assume that there exist two systems $\Xi_{\pm} = \{\xi_{jn}^{\pm}; 1 \leqslant j \leqslant d, 0 \leqslant n \leqslant N\}$ of vectors in the vector space W such that

$$(\xi_{jm}^{\pm}, {}^t\xi_{kn}^{\pm}) = \delta_{jk}\delta_{mn} \quad (1 \leqslant j, k \leqslant d, 0 \leqslant m, n \leqslant N). \tag{4.1$_\pm$}$$

Then we construct $2(N + 1)$ d-dimensional vectors $\xi_{\pm}(\pm n)$ $(0 \leqslant n \leqslant N)$ in W by

$$\xi_{\pm}(\pm n) \equiv {}^t(\xi_{1n}^{\pm}, \xi_{2n}^{\pm}, \ldots, \xi_{dn}^{\pm}) \quad (0 \leqslant n \leqslant N). \tag{4.2$_\pm$}$$

It follows that

$$(\xi_{\pm}(\pm m), {}^t\xi_{\pm}(\pm n)) = \delta_{mn} I \quad (0 \leqslant n \leqslant N). \tag{4.3$_\pm$}$$

Let $\mathcal{LM}(R) = \{\gamma_+(n,k), \gamma_-(n,k), V_+(\ell), V_-(\ell); 0 \leqslant k < n \leqslant N, 0 \leqslant \ell \leqslant N\}$ be the KM$_2$O-Langevin matrix associated with the matix function R constructed in Theorem 3.1. Let $W_{\pm}(n)$ be any $d \times d$ matrices $(0 \leqslant n \leqslant N)$ such that

$$V_{\pm}(n) = W_{\pm}(n)\, {}^tW_{\pm}(n) \quad (0 \leqslant n \leqslant N). \tag{4.4$_\pm$}$$

Then we define two d-dimensional flows $\nu_+ = (\nu_+(n); 0 \leqslant n \leqslant N)$ and $\nu_- = (\nu_-(\ell); -N \leqslant \ell \leqslant 0)$ on W by

$$\nu_{\pm}(\pm n) \equiv W_{\pm}(n)\xi_{\pm}(\pm n) \quad (0 \leqslant n \leqslant N). \tag{4.5$_\pm$}$$

It follows that

$$(\nu_\pm(\pm m),\ {}^t\nu_\pm(\pm n)) = \delta_{mn} V_\pm(n) \quad (0 \leqslant n \leqslant N). \tag{4.6$_\pm$}$$

We note from Theorem 3.1 that the matrices $V_\pm(N)$ are not necessarily regular differently from the matrices $V_\pm(n)$ $(0 \leqslant n \leqslant N-1)$.

By using these flows ν_+ and ν_-, we can inductively construct a pair $[\mathbf{X}, \mathbf{Y}]$ of the d-dimensional flows $\mathbf{X} = (X(n); 0 \leqslant n \leqslant N)$ and $\mathbf{Y} = (Y(\ell); -N \leqslant \ell \leqslant 0)$ on W by

$$\begin{aligned} X(0) &\equiv \nu_+(0), \\ X(n) &\equiv -\sum_{k=0}^{n-1} \gamma_+(n,k) X(k) + \nu_+(n) \quad (1 \leqslant n \leqslant N), \end{aligned} \tag{4.7$_+$}$$

$$\begin{aligned} Y(0) &\equiv \nu_-(0), \\ Y(-n) &\equiv -\sum_{k=0}^{n-1} \gamma_-(n,k) Y(-k) + \nu_-(-n) \quad (1 \leqslant n \leqslant N). \end{aligned} \tag{4.7$_-$}$$

LEMMA 4.1.

(i) $\mathbf{M}_0^n(\mathbf{X}) = \mathbf{M}_0^n(\nu_+)$ $(0 \leqslant n \leqslant N)$,
$\mathbf{M}_{-n}^0(\mathbf{Y}) = \mathbf{M}_{-n}^0(\nu_-)$ $(0 \leqslant n \leqslant N)$,
(ii) $P_{\mathbf{M}_0^{n-1}(\mathbf{X})} X(n) = -\sum_{k=0}^{n-1} \gamma_+(n,k) X(k)$ $(1 \leqslant n \leqslant N)$,
$P_{\mathbf{M}_{-n+1}^0(\mathbf{Y})} Y(-n) = -\sum_{k=0}^{n-1} \gamma_-(n,k) Y(-k)$ $(1 \leqslant n \leqslant N)$.

Proof. We shall prove only the part of the flow $\mathbf{X}$. Immediately, we have (i) from (4.7$_+$). Hence, we find from (i) and (4.6$_+$) that for each n $(1 \leqslant n \leqslant N)$, the d components of the vector $\nu_+(n)$ are orthogonal to the subspace $\mathbf{M}_0^{N-1}(\mathbf{X})$. Therefore, we have (ii) from (4.7$_+$). □

Moreover, we shall show that the pair $[\mathbf{X}, \mathbf{Y}]$ of flows satisfies the following independence condition.

LEMMA 4.2.

(i) $\{X_j(n); 1 \leqslant j \leqslant d, 0 \leqslant n \leqslant N-1\}$ *is linearly independent in* W.
(ii) $\{Y_j(\ell); 1 \leqslant j \leqslant d, -N+1 \leqslant \ell \leqslant 0\}$ *is linearly independent in* W.

Proof. We shall prove only (i), since (ii) is similarly proved. Let c_{jn} $(1 \leqslant j \leqslant d, 0 \leqslant n \leqslant N-1)$ be any real numbers such that $\sum_{j=1}^d \sum_{n=0}^{N-1} c_{jn} X_j(n) = 0$. Putting $c_n \equiv (c_{1n}, c_{2n}, \ldots, c_{dn})$, we see that $\sum_{n=0}^{N-1} c_n X(n) = 0$. By substituting (4.7$_+$) into the above equation, we obtain

$$\sum_{k=0}^{N-2} \left(\sum_{n=k+1}^{N-1} c_n \gamma_+(n,k) \right) X(k) = \sum_{n=0}^{N-1} c_n \nu_+(n). \tag{4.8}$$

We note from Lemma 4.1(i) that the left-hand side of (4.8) belongs to the subspace $\mathbf{M}_0^{N-2}(\mathbf{X})$ and the d components of the vector $\nu_+(N-1)$ are orthogonal to the subspace $\mathbf{M}_0^{N-2}(\mathbf{X})$. By taking the inner product of the both-hand sides with the vector $\nu_+(N-1)$, we find from (4.6$_+$) that $c_{N-1}V_+(N-1)=0$. Since the matrix $V_+(N-1)$ is regular, we see that $c_{N-1}=0$. It then follows that Equation (4.8) becomes

$$\sum_{k=0}^{N-3}\left(\sum_{n=k+1}^{N-1} c_n\gamma_+(n,k)\right)X(k)=\sum_{n=0}^{N-2} c_n\nu_+(n).$$

By repeating the same procedure, we can conclude from (i) in Theorem 3.1 that $c_n=0$ $(0\leqslant n\leqslant N-2)$, which proves (i). □

After the above preparations, we shall prove the following theorem.

THEOREM 4.1. *The pair* $[\mathbf{X},\mathbf{Y}]$ *of* d*-dimensional flows* $\mathbf{X}=(X(n);0\leqslant n\leqslant N)$ *and* $\mathbf{Y}=(Y(\ell);-N\leqslant\ell\leqslant 0)$ *has stationary property with* R *its correlation matrix function.*

Proof. By using Lemmas 4.1 and 4.2, we find that the d-dimensional flow ν_+ (resp. ν_-) the forward (resp. the backward) KM$_2$O-Langevin fluctuation flow associated with the d-dimensional flow $\mathbf{X}=(X(n);0\leqslant n\leqslant N)$ (resp. $\mathbf{Y}=(Y(\ell);-N\leqslant\ell\leqslant 0)$) and the system $\mathcal{LM}(R)=\{\gamma_+(n,k),\gamma_-(n,k),V_+(\ell),V_-(\ell);0\leqslant k<n\leqslant N,0\leqslant\ell\leqslant N\}$ becomes the KM$_2$O-Langevin matrix associated with the pair $[\mathbf{X},\mathbf{Y}]$ of flows. Threfore, by virtue of Theorem 3.1, we can apply the characterization theorem in [2] for the pair $[\mathbf{X},\mathbf{Y}]$ to see that the pair $[\mathbf{X},\mathbf{Y}]$ has stationary property. □

By inverting the time evolution of d-dimensional flow $\mathbf{X}=(X(n);0\leqslant n\leqslant N)$ in Theorem 4.1, we define a d-dimensional flow $\mathbf{X}_-=(X_-(\ell);-N\leqslant\ell\leqslant 0)$ by

$$X_-(-n)\equiv X(N-n). \tag{4.9}$$

Immediately from Theorem 4.1, we have

THEOREM 4.2. *The pair* $[\mathbf{X},\mathbf{X}_-]$ *of* d*-dimensional flows* $\mathbf{X}=(X(n);0\leqslant n\leqslant N)$ *and* $\mathbf{X}_-=(X_-(\ell);-N\leqslant\ell\leqslant 0)$ *has stationary property with* R *its correlation matrix function.*

We shall show that the flow $\mathbf{X}_-$ satisfies the independence condition. By virtue of (i) in Theorem 4.1, we can apply a standard method to obtain the following Lemma 4.3.

LEMMA 4.3. *There exists a unique unitary operator* U *from* $\mathbf{M}_0^{N-1}(\mathbf{X})$ *to* $\mathbf{M}_1^{N}(\mathbf{X})$ *such that*

$$U(X_j(n))=X_j(n+1)\quad(1\leqslant j\leqslant d,0\leqslant n\leqslant N-1).$$

By using Lemma 4.3 and Lemma 4.2(i), we can show

LEMMA 4.4. *$\{X_j(n); 1 \leqslant j \leqslant d, 1 \leqslant n \leqslant N\}$ is linearly independent in W.*

By combining Theorem 4.2 with Lemma 4.2(i) and Lemma 4.4, we have

THEOREM 4.3. *The KM_2O-Langevin matrix $\mathcal{LM}([\mathbf{X}, \mathbf{X}_-])$ associated with the stationary pair $[\mathbf{X}, \mathbf{X}_-]$ of flows equals to the KM_2O-Langevin matrix $\mathcal{LM}(R)$ associated with the matrix function R, which has been constructed in Theorem* 3.1.

Let $\nu_-(\mathbf{X}_-) = (\nu_-(\mathbf{X}_-)(\ell); -N \leqslant \ell \leqslant 0)$ be the backward KM_2O-Langevin flow associated with the flow $\mathbf{X}_-$. Then the backward KM_2O-Langevin equation associated with the flow $\mathbf{X}_-$ reads as

$$X_-(-n) = -\sum_{k=0}^{n-1} \gamma_-(n,k) X_-(-k) + \nu_-(\mathbf{X}_-)(-n) \quad (0 \leqslant n \leqslant N). \tag{4.10}$$

By noting (4.9), we can rewrite (4.10) into

$$X(N-n) = -\sum_{k=0}^{n-1} \gamma_-(n,k) X(N-k) + \nu_-(\mathbf{X}_-)(-n) \quad (0 \leqslant n \leqslant N). \tag{4.11}$$

References

1. Okabe, Y.: On a stochastic difference equation for the multi-dimensional weakly stationary process with discrete time, In: M. Kashiwara and T. Kawai (eds), *Prospect of Algebraic Analysis*, Academic Press, Tokyo, 1988, pp. 601–645.
2. Okabe, Y.: On the theory of KM_2O-Langevin equations for stationary flows (I): Characterization theorem, To appear in *J. Math. Soc. Japan.*
3. Okabe, Y. and Matsuura, M.: On the theory of KM_2O-Langevin equations for stationary flows (III): Extension theorem, To be submitted to *Hokkaido Math. J.*

Acta Applicandae Mathematicae **63:** 323–332, 2000.

Vector Bundle-Valued Poisson and Cauchy Kernel Functions on Classical Domains

K. OKAMOTO, M. TSUKAMOTO and K. YOKOTA
Department of Mathematics, Faculty of Science and Technology, Meijo University, Japan

(Received: 8 January 1999)

Abstract. In this paper we investigate the vector bundle-valued Cauchy and Poisson kernel functions. We compute explicitly matrix-valued eigenfunctions of an invariant differential operator on the classical domain of Type I. Furthermore, a special choice of a vector bundle gives us a matrix-valued Cauchy and Poisson kernel function which satisfies the matrix-valued Laplacian operator on the classical domain of Type I.

Mathematics Subject Classifications (2000): 22E46, 60J50, 65N25.

Key words: Lie group, group representation, homogeneous vector bundle, harmonic analysis, classical domain, Poisson kernel, Cauchy kernel, eigenfunctions of Laplacian operator, boudary theory, eigenfunctions of invariant differential operators.

Introduction

In [2], Hua gave explicit formulas of the Poisson kernel functions and Cauchy kernel functions on classical domains.

In our previous paper [6], we put forward an idea of how to obtain generalized formulas which include not only Poisson kernel functions but also Cauchy kernel functions. Moreover, we carried out explicit computations for each type of classical domain and proved that our formulas, as special cases, give rise to the same Cauchy and Poisson kernel functions as those given by Hua.

Our idea in [6] was based on the group-theoretic method in such a way that, instead of the space of ordinary functions, we considered the space of sections of the homogeneous line bundles on the classical domain and the Shilov boundary.

In this paper, we generalize this result further to the vector bundle-valued Cauchy and Poisson kernel functions.

In [2], Hua showed that for the classical domain of Type I, the Laplacian operator Δ is obtained by the trace of a certain matrix-valued differential operator $\widehat{\Delta}$. Moreover, he proved that the Poisson kernel function $\mathscr{P}$ satisfies not only the equation $\Delta\mathscr{P} = 0$ but also the equation $\widehat{\Delta}\mathscr{P} = \mathbf{0}$ (where $\mathbf{0}$ denotes the zero matrix), so that the Poisson kernel function satisfies the system of differential equations.

In this paper, we explain another feature of this matrix-valued differential operator $\widehat{\Delta}$. Namely, $\widehat{\Delta}$ can be regarded as an invariant differential operator on a certain class of vector bundles over the classical domain of Type I and the vector-bundle valued Cauchy and Poisson kernel functions can be realized as matrix valued eigenfunctions of $\widehat{\Delta}$. Furthermore, a special choice of a vector bundle gives us a matrix valued Cauchy and Poisson kernel function $\mathcal{K}$ which satisfies the differential equation $\widehat{\Delta}\mathcal{K} = \mathbf{0}$.

Some of the results of this paper can be obtained by the similar computation as that in [6], so that instead of duplicating the computation we indicate only the difference and new results.

1. Homogeneous Vector Bundles and Poisson–Cauchy Kernels

Let G be a connected noncompact semisimple Lie group admitting a finite-dimensional faithful representation. Let K be a maximal compact subgroup of G. We suppose that G/K is Hermitian symmetric.

We denote by G_c and K_c the complexifications of G and K, respectively. Then G/K is realized as the G-orbit of the origin U of the generalized flag manifold G_c/U. Let $P_+K_cP_-$ $(U = K_cP_-)$ be the Harish-Chandra decomposition. Then we can prove that $GU \subset P_+U$ and that there exists the unique bounded domain $\boldsymbol{D}$ in the Lie algebra $\mathfrak{p}_+$ of P_+ such that $GU = (\exp \boldsymbol{D})U$.

For any w in GU, we denote by $z(w)$ (resp. $u(w)$) the unique element of $\mathfrak{p}_+$ (resp. U) such that $w = (\exp z(w))u(w)$. For any $g \in G$ and $z \in \boldsymbol{D}$, we denote by $g[z]$ the unique element of $\boldsymbol{D}$ such that $g(\exp z)U = (\exp g[z])U$. Then we have the canonical onto-isomorphisms:

$$\begin{array}{ccccc} G/K & \cong & GU/U & \cong & \boldsymbol{D} \\ \Cup & & \Cup & & \Cup \\ gK & \longmapsto & gU & \longmapsto & g[0] = z. \end{array}$$

Moreover, G acts on $G/K \cong GU/U \cong \boldsymbol{D}$ from the following commutative diagram:

For any $g_1 \in G$,

$$\begin{array}{ccccc} G/K & \cong & GU/U & \cong & \boldsymbol{D} \\ \Cup & & \Cup & & \Cup \\ gK & \longmapsto & gU & \longmapsto & g[0] = z \\ \downarrow & & \downarrow & & \downarrow \\ g_1gK & \longmapsto & g_1gU & \longmapsto & g_1g[0] = g_1[z] \\ \Cap & & \Cap & & \Cap \\ G/K & \cong & GU/U & \cong & \boldsymbol{D}. \end{array}$$

We fix a point $\mu U \in G_c/U$ such that μU belongs to the boundary of GU/U and that the G-orbit of μU is compact. Then the G-orbit $G\mu U/U$ is the Shilov boundary of the bounded domain $GU/U \cong G/K$ and the isotropy subgroup at the point μU of G_c/U is a maximal parabolic subgroup of G which we denote by P.

Put $u_o = z(\mu)$ and $\mu_o = \exp u_o$. Then clearly we get $\mu U = \mu_o U = (\exp u_o)U$ which implies that $G \cap \mu U \mu^{-1} = G \cap \mu_o U \mu_o^{-1} = P$.

Put

$$\check{S} = \{u \in \mathfrak{p}_+;\ \exp uU \in G\mu_o U/U\}.$$

Then $\check{S}$ is the Shilov boundary of $\boldsymbol{D}$.

For any $g \in G$ and $u \in \check{S}$, we denote by $g[u]$ the unique element of $\check{S}$ such that $g(\exp u)U = (\exp g[u])U$. Then we have the canonical onto-isomorphisms:

$$\begin{array}{ccccc} G/P & \cong & G\mu_o U/U & \cong & \check{S} \\ \Cup & & \Cup & & \Cup \\ gP & \longmapsto & g\mu_o U & \longmapsto & g[u_o] = u. \end{array}$$

Moreover, G acts on $G/P \cong G\mu_o U/U \cong \check{S}$ from the following commutative diagram:

For any $g_1 \in G$,

$$\begin{array}{ccccc} G/P & \cong & G\mu_o U/U & \cong & \check{S} \\ \Cup & & \Cup & & \Cup \\ gP & \longmapsto & g\mu_o U & \longmapsto & g[u_o] = u \\ \downarrow & & \downarrow & & \downarrow \\ g_1 gP & \longmapsto & g_1 g\mu_o U & \longmapsto & g_1 g[u_o] = g_1[u] \\ \Cap & & \Cap & & \Cap \\ G/P & \cong & G\mu_o U/U & \cong & \check{S}. \end{array}$$

Let $P = MAN$ be the Langlands decomposition of P and let $\mathfrak{m}$ (resp. $\mathfrak{a}$, $\mathfrak{n}$) be the Lie algebras corresponding to M (resp. A, N). Then MA is the centralizer of A in G and N is a normal subgroup of P. Moreover, $\mathfrak{m}$ is the orthogonal complement of $\mathfrak{a}$ in the centralizer of $\mathfrak{a}$ in $\mathfrak{g}$ with respect to the Killing form.

We denote by ad the adjoint representation of $\mathfrak{g}$ on $\mathfrak{g}$ so that for any $X \in \mathfrak{g}$ we have $\mathrm{ad}(X)(Y) = [X, Y]$ $(Y \in \mathfrak{g})$. Let ρ be the linear form on $\mathfrak{a}$ defined by $\rho(H) = \frac{1}{2}\mathrm{Tr}(\mathrm{ad}(H)|_{\mathfrak{n}})$, where $\mathrm{ad}(H)|_{\mathfrak{n}}$ denotes the restriction of $\mathrm{ad}(H)$ to $\mathfrak{n}$.

Let M be a $\boldsymbol{C}^\infty$ manifold and V a finite-dimensional vector space. We denote by $\boldsymbol{C}^\infty(M, V)$ the space of all V-valued $\boldsymbol{C}^\infty$ functions on M.

For any positive integer k, we denote by $M_k(\boldsymbol{C})$, $\mathrm{GL}(k, \boldsymbol{C})$, $\mathrm{SL}(k, \boldsymbol{C})$ and $\mathrm{U}(k)$ the set of all (k, k) complex matrices, complex nonsingular matrices, complex matrices of determinant one and unitary matrices, respectively.

Let τ be a representation of K_c on a finite-dimensional vector space V. Then τ is uniquely extended to a representation of U which is trivial on P_-. We regard the complex Lie group G_c as the principal fiber bundle over the homogeneous space G_c/U. We denote by $\widetilde{E}_\tau$ the $\boldsymbol{C}^\infty$ vector bundle over G_c/U associated to τ. Then the space of all $\boldsymbol{C}^\infty$ sections of $\widetilde{E}_\tau$ is identified with

$$\boldsymbol{C}^\infty(\widetilde{E}_\tau) = \{h \in \boldsymbol{C}^\infty(G_c, V);\ h(wu) = \tau(u)^{-1}h(w)\ (w \in G_c,\ u \in U)\}.$$

We denote by E_τ the restriction of $\widetilde{E}_\tau$ to the open submanifold $G/K \cong GU/U$ of G_c/U. Then the space of all C^∞ sections of E_τ is identified with

$$C^\infty(E_\tau) = \{h \in C^\infty(GU, V); h(wu) = \tau(u)^{-1}h(w)\ (w \in GU, u \in U)\}.$$

Let η be a representation of U such that the restriction of η to K coincides with the restriction of τ to K. We denote by $\widetilde{L}_\eta$ the C^∞ vector bundle on G_c/U associated to η. Then the space of all C^∞ sections of $\widetilde{L}_\eta$ is identified with

$$C^\infty(\widetilde{L}_\eta) = \{h \in C^\infty(G_c, V); h(wu) = \eta(u)^{-1}h(w)\ (w \in G_c,\ u \in U)\}.$$

We denote by L_η the restriction of $\widetilde{L}_\eta$ to the compact submanifold $G/P \cong G\mu_o U/U$ of G_c/U. Then the space of all C^∞ sections of L_η is identified with

$$\begin{aligned} &C^\infty(L_\eta) \\ &\quad = \{h \in C^\infty(G\mu_o U, V); h(wu) = \eta(u)^{-1}h(w)\ (w \in G\mu_o U, u \in U)\}. \end{aligned}$$

We define the C^∞ character ξ of P by

$$\xi(p) = \eta(\mu_o^{-1} p \mu_o) \quad (p \in P).$$

Put

$$\begin{aligned} C^\infty(G, V)_\tau &= \{f \in C^\infty(G, V); f(gk) = \tau(k)^{-1}f(g)\ (g \in G,\ k \in K)\}, \\ C^\infty(G, V)_\xi &= \{\phi \in C^\infty(G, V); \phi(gp) = \xi(p)^{-1}\phi(g)\ (g \in G,\ p \in P)\}. \end{aligned}$$

Then we obtain the following four onto-isomorphisms:

$$\begin{aligned} C^\infty(E_\tau) &\ni h \longmapsto f \in C^\infty(G, V)_\tau, \quad f(g) = h(g) \quad (g \in G), \\ C^\infty(E_\tau) &\ni h \longmapsto F \in C^\infty(\boldsymbol{D}, V), \quad F(z) = h(\exp z) \quad (z \in \boldsymbol{D}), \\ C^\infty(L_\eta) &\ni \psi \longmapsto \phi \in C^\infty(G, V)_\xi, \quad \phi(g) = \psi(g\mu_o) \quad (g \in G), \\ C^\infty(L_\eta) &\ni \psi \longmapsto \Phi \in C^\infty(\check{\boldsymbol{S}}, V), \quad \Phi(u) = \psi(\exp u) \quad (u \in \check{\boldsymbol{S}}). \end{aligned}$$

Now, we define the vector bundle-valued Poisson–Cauchy transform

$$\mathcal{P}_{\tau,\eta} : C^\infty(G, V)_\xi \ni \phi \longmapsto f \in C^\infty(G, V)_\tau$$

by

$$f(g) = \int_K \tau(k)\phi(gk)\, \mathrm{d}k \quad (g \in G).$$

For any $g \in G$, we put

$$\begin{aligned} &\kappa(g) = km, \quad H(g) = \log a, \quad n(g) = n, \\ &(g = kman,\ k \in K,\ m \in M,\ a \in A,\ n \in N). \end{aligned}$$

Then there exists a linear form λ on $\mathfrak{a}$ such that $\xi(a) = \mathrm{e}^{\lambda(\log a)}$.

We define

$$\omega(km) = \tau(k)\xi(m) \quad (k \in K,\ m \in M).$$

The computation of kernel functions is almost the same as in [6]. However, we must be careful about the order of the multiplication of the operators. Unless both τ and ξ are characters the formula of the generalized Poisson–Cauchy kernel function in [6] is not correct. The right formula is given by the following theorem:

THEOREM. *The generalized Poisson–Cauchy transform is given by*

$$\begin{aligned} F(z) &= \int_K \mathrm{e}^{(\lambda-2\rho)(H(g^{-1}k))}\tau(u(g))\omega(\kappa(g^{-1}k))\tau(k)^{-1}\Phi(u)\,\mathrm{d}u \\ &= \int_K \mathcal{K}_{\tau,\eta}(z,u)\Phi(u)\,\mathrm{d}u, \end{aligned}$$

where

$$\mathcal{K}_{\tau,\eta}(z,u) = \mathrm{e}^{(\lambda-2\rho)(H(g^{-1}k))}\tau(u(g))\omega(\kappa(g^{-1}k))\tau(k)^{-1}$$
$$(z = g[0],\ u = k[u_o]).$$

2. Invariant Differential Operators

We keep the notation of the previous section.

For any $g \in G$ and $h \in \boldsymbol{C}^\infty(E_\tau)$, we define

$$(\pi_\tau(g)h)(w) = h(g^{-1}w) \quad (w \in GU).$$

Then π_τ is a representation of G on $\boldsymbol{C}^\infty(E_\tau)$. Let P be a differential operator on $\boldsymbol{C}^\infty(E_\tau)$ which satisfies the following commutative diagram:

For any $g \in G$,

$$\begin{array}{ccc} \boldsymbol{C}^\infty(E_\tau) & \stackrel{P}{\longmapsto} & \boldsymbol{C}^\infty(E_\tau) \\ \pi_\tau(g)\downarrow & & \pi_\tau(g)\downarrow \\ \boldsymbol{C}^\infty(E_\tau) & \stackrel{P}{\longmapsto} & \boldsymbol{C}^\infty(E_\tau). \end{array}$$

Then we call this differential operator an invariant differential operator (with respect to the representation π_τ). Using the following isomorphism, we realize the vector bundle-valued Poisson–Cauchy kernel functions by the matrix-valued Poisson–Cauchy kernel functions:

$$\begin{array}{ccc} \boldsymbol{C}^\infty(E_\tau) & \longmapsto & \boldsymbol{C}^\infty(\boldsymbol{D}, V) \\ \Cup & & \Cup \\ h & \longmapsto & F \end{array}$$

where

$$F(z) = h(\exp z) \quad (z \in \boldsymbol{D}).$$

We define a representation T_τ of G on $\boldsymbol{C}^\infty(\boldsymbol{D}, V)$ by the following commutative diagram:

For any $g \in G$,

$$\begin{array}{ccc} \boldsymbol{C}^\infty(E_\tau) & \longmapsto & \boldsymbol{C}^\infty(\boldsymbol{D}, V) \\ \pi_\tau(g) \downarrow & & T_\tau(g) \downarrow \\ \boldsymbol{C}^\infty(E_\tau) & \longmapsto & \boldsymbol{C}^\infty(\boldsymbol{D}, V). \end{array}$$

The above-mentioned isomorphism induces a differential operator $P(z, D_z)$ on $\boldsymbol{C}^\infty(\boldsymbol{D}, V)$ which satisfies the following commutative diagram:

$$\begin{array}{ccc} \boldsymbol{C}^\infty(E_\tau) & \overset{P}{\longmapsto} & \boldsymbol{C}^\infty(E_\tau) \\ \downarrow & & \downarrow \\ \boldsymbol{C}^\infty(\boldsymbol{D}, V) & \overset{P(z,D_z)}{\longmapsto} & \boldsymbol{C}^\infty(\boldsymbol{D}, V). \end{array}$$

Now, we explain the close relation between vector bundle-valued Poisson–Cauchy kernel functions and invariant differential operators. To show this relation explicitly, we confine our consideration to the special case: $G = \mathrm{SU}(2, 2)$.

In this case the bounded symmetric domain is defined by

$$\boldsymbol{D} = \left\{ z \in M_2(\boldsymbol{C}); z^* z \ll I_2 \right\},$$

where

$$I_2 = \begin{pmatrix} 1 & 0 \\ 0 & 1 \end{pmatrix}.$$

The Silov boundary of $\boldsymbol{D}$ is given by

$$\check{\boldsymbol{S}} = \left\{ u \in M_2(\boldsymbol{C}); u^* u = I_2 \right\}.$$

We define

$$G = \mathrm{SU}(2, 2) = \left\{ g \in \mathrm{SL}(4, \boldsymbol{C}); g^* I_{2,2} g = I_{2,2} \right\},$$

where

$$I_{2,2} = \begin{pmatrix} I_2 & 0 \\ 0 & -I_2 \end{pmatrix},$$

$$K = \left\{ \begin{pmatrix} k_1 & 0 \\ 0 & k_2 \end{pmatrix} \in \mathrm{SL}(4, \boldsymbol{C}); k_1 \in \mathrm{U}(2), k_2 \in \mathrm{U}(2) \right\},$$

$$M = \left\{ \begin{pmatrix} m_1 & m_2 \\ m_2 & m_1 \end{pmatrix} \in G; \begin{array}{c} m_1, m_2 \in M_2(\boldsymbol{C}) \\ |\det(m_1 + m_2)| = 1 \end{array} \right\},$$

$$A = \left\{ \begin{pmatrix} \mathrm{ch}\, t\ I_2 & \mathrm{sh}\, t\ I_2 \\ \mathrm{sh}\, t\ I_2 & \mathrm{ch}\, t\ I_2 \end{pmatrix}; t \in \boldsymbol{R} \right\},$$

$$N = \left\{ \begin{pmatrix} I_2 + \xi & -\xi \\ \xi & I_2 - \xi \end{pmatrix}; \xi^* = -\xi,\ \xi \in M_2(\boldsymbol{C}) \right\},$$

$$P = MAN.$$

Then we have $G = KMAN$.
Further, we define

$$G_c = \mathrm{SL}(4, \boldsymbol{C}),$$

$$K_c = \left\{ \begin{pmatrix} \alpha & 0 \\ 0 & \delta \end{pmatrix} \in \mathrm{SL}(4, \boldsymbol{C}); \alpha \in \mathrm{GL}(2, \boldsymbol{C}), \delta \in \mathrm{GL}(2, \boldsymbol{C}) \right\},$$

$$P_+ = \left\{ \begin{pmatrix} I_2 & z \\ 0 & I_2 \end{pmatrix}; z \in M_2(\boldsymbol{C}) \right\},$$

$$P_- = \left\{ \begin{pmatrix} I_2 & 0 \\ w & I_2 \end{pmatrix}; w \in M_2(\boldsymbol{C}) \right\},$$

$$U = K_c P_-,$$

$$\mu = \begin{pmatrix} \dfrac{i}{1+i} I_2 & \dfrac{1}{1+i} I_2 \\ \dfrac{-i}{1+i} I_2 & \dfrac{1}{1+i} I_2 \end{pmatrix}, \qquad \mu_o = \begin{pmatrix} I_2 & I_2 \\ 0 & I_2 \end{pmatrix}.$$

Then we get

$$G \cap \mu U \mu^{-1} = G \cap \mu_o U \mu_o^{-1} = P.$$

The action of G on $\boldsymbol{D}$ is given by

$$\boldsymbol{D} \ni z \longmapsto g[z] = (az + b)(cz + d)^{-1} \in \boldsymbol{D}, \quad g = \begin{pmatrix} a & b \\ c & d \end{pmatrix} \in G.$$

We have the canonical onto-isomorphisms:

$$\begin{array}{ccccc} G/K & \cong & GU/U & \cong & \boldsymbol{D} \\ \Cup & & \Cup & & \Cup \\ gK & \longmapsto & gU & \longmapsto & z = g[0] = bd^{-1}. \end{array}$$

Moreover, the action of G on $\check{\boldsymbol{S}}$ is given by

$$\check{\boldsymbol{S}} \ni u \longmapsto g[u] = (au + b)(cu + d)^{-1} \in \check{\boldsymbol{S}}, \quad g = \begin{pmatrix} a & b \\ c & d \end{pmatrix} \in G.$$

Fix $u_o = I_2 \in \check{S}$. Then we have the canonical onto-isomorphisms:

$$\begin{array}{ccccc} G/P & \cong & G\mu_o U/U & \cong & \check{S} \\ \Cup & & \Cup & & \Cup \\ gP & \longmapsto & g\mu_o U & \longmapsto & u = g[u_o] = (a+b)(c+d)^{-1}. \end{array}$$

Let τ be the trivial representation of U:

$$\tau\colon U \ni \begin{pmatrix} \alpha & 0 \\ \zeta & \delta \end{pmatrix} \longmapsto 1 \in \boldsymbol{C}^*.$$

For any $s \in \boldsymbol{C}$, we define a character η_s of U and a character ξ_s of P by

$$\eta_s\colon U \ni \begin{pmatrix} \alpha & 0 \\ \zeta & \delta \end{pmatrix} \longmapsto |\det(\delta)|^s \in \boldsymbol{C}^*,$$

$$\xi_s\colon P \ni p \longmapsto \eta_s(\mu_o^{-1} p \mu_o) \in \boldsymbol{C}^*.$$

Then the generalized Poisson–Cauchy kernel function is given by

$$\mathcal{P}_s(z, u) = \left(\frac{\det(I_2 - z^* z)}{|\det(I_2 - u^* z)|^2} \right)^{2-s/2}.$$

We define

$$\partial_z = \begin{pmatrix} \dfrac{\partial}{\partial_{z_{11}}} & \dfrac{\partial}{\partial_{z_{21}}} \\ \dfrac{\partial}{\partial_{z_{12}}} & \dfrac{\partial}{\partial_{z_{22}}} \end{pmatrix}, \quad (\partial_z)^* = \begin{pmatrix} \dfrac{\partial}{\partial_{\bar{z}_{11}}} & \dfrac{\partial}{\partial_{\bar{z}_{12}}} \\ \dfrac{\partial}{\partial_{\bar{z}_{21}}} & \dfrac{\partial}{\partial_{\bar{z}_{22}}} \end{pmatrix},$$

$$\partial_w = \begin{pmatrix} \dfrac{\partial}{\partial_{w_{11}}} & \dfrac{\partial}{\partial_{w_{12}}} \\ \dfrac{\partial}{\partial_{w_{21}}} & \dfrac{\partial}{\partial_{w_{22}}} \end{pmatrix}, \quad (\partial_w)^* = \begin{pmatrix} \dfrac{\partial}{\partial_{\bar{w}_{11}}} & \dfrac{\partial}{\partial_{\bar{w}_{21}}} \\ \dfrac{\partial}{\partial_{\bar{w}_{12}}} & \dfrac{\partial}{\partial_{\bar{w}_{22}}} \end{pmatrix},$$

$$\widehat{\Delta}_z = (I_2 - z^* z)\partial_z \cdot (I_2 - z z^*) \cdot (\partial_z)^*,$$

$$\widehat{\Delta}_w = (I_2 - w^* w)\partial_w \cdot (I_2 - w w^*) \cdot (\partial_w)^*,$$

where $\cdot \sim \cdot$ means that the part $\sim$ should not be differentiated.

Then one can show that the Laplacian for the bounded domain $\boldsymbol{D}$ is given by $\Delta_z = \mathrm{Tr}(\widehat{\Delta}_z)$, and that the Poisson kernel function $\mathcal{P}_0$ satisfies the differential equation

$$\Delta_z \mathcal{P}_0(z, u) = 0 \quad (u \in \check{S}).$$

Moreover, the generalized Poisson kernel function $\mathcal{P}_s$ is an eigenfunction of Δ_z:

$$\Delta_z \mathcal{P}_s(z, u) = \frac{s(s-4)}{4} \mathcal{P}_s(z, u) \quad (u \in \check{S}).$$

Furthermore, the generalized Poisson kernel function $\mathcal{P}_s$ satisfies the system of differential equations:

$$\widehat{\Delta}_z \mathcal{P}_s(z,u) = \frac{s(s-4)}{4}\mathcal{P}_s(z,u) \quad (u \in \check{\boldsymbol{S}}).$$

(In the case of $s = 0$, this was proved by Hua [2].)

Now, we fix any representation σ of $\mathrm{GL}(2, \boldsymbol{C})$ on $\boldsymbol{C}^n$. We define representations τ_σ, $\eta_{\sigma,s}$ of U and a representation $\xi_{\sigma,s}$ of P by

$$\tau_\sigma : U \ni \begin{pmatrix} \alpha & 0 \\ \zeta & \delta \end{pmatrix} \longmapsto \sigma(\delta) \in \mathrm{GL}(n, \boldsymbol{C}),$$

$$\eta_{\sigma,s} : U \ni \begin{pmatrix} \alpha & 0 \\ \zeta & \delta \end{pmatrix} \longmapsto |\det(\delta)|^s \sigma(\delta) \in \mathrm{GL}(n, \boldsymbol{C}),$$

$$\xi_{\sigma,s} : P \ni p \longmapsto \eta_{\sigma,s}(\mu_o^{-1} p \mu_o) \in \mathrm{GL}(n, \boldsymbol{C}).$$

Then, for any $g \in G$ and $F \in \boldsymbol{C}^\infty(\boldsymbol{D}, \boldsymbol{C}^n)$, we have

$$(T_{\tau_\sigma}(g)F)(z) = \sigma(cz+d)^{-1} F\left((az+b)(cz+d)^{-1}\right)$$

$$\left(g^{-1} = \begin{pmatrix} a & b \\ c & d \end{pmatrix},\ z \in \boldsymbol{D}\right).$$

Let $P(z, \boldsymbol{D}_z)$ be an invariant differential operator on $\boldsymbol{C}^\infty(\boldsymbol{D}, \boldsymbol{C}^n)$, so that the following commutative diagram holds:

$$\begin{array}{ccc} \boldsymbol{C}^\infty(\boldsymbol{D}, \boldsymbol{C}^n) & \overset{P(z, D_z)}{\longmapsto} & \boldsymbol{C}^\infty(\boldsymbol{D}, \boldsymbol{C}^n) \\ T_{\tau_\sigma}(g) \downarrow & & T_{\tau_\sigma}(g) \downarrow \\ \boldsymbol{C}^\infty(\boldsymbol{D}, \boldsymbol{C}^n) & \overset{P(w, D_w)}{\longmapsto} & \boldsymbol{C}^\infty(\boldsymbol{D}, \boldsymbol{C}^n), \end{array}$$

where

$$g^{-1} = \begin{pmatrix} a & b \\ c & d \end{pmatrix} \quad \text{and} \quad w = (az+b)(cz+d)^{-1}.$$

Then we get the following equation:

$$P(w, D_w) = \sigma(cz+d) P(z, D_z) \sigma(cz+d)^{-1},$$

where

$$g^{-1} = \begin{pmatrix} a & b \\ c & d \end{pmatrix} \quad \text{and} \quad w = (az+b)(cz+d)^{-1}.$$

Conversely, one can prove that if $P(z, D_z)$ satisfies this equation, $P(z, D_z)$ is an invariant differential operator on $\boldsymbol{C}^\infty(\boldsymbol{D}, \boldsymbol{C}^n)$.

On the other hand, it is easy to check that

$$\widehat{\Delta}_w = ((cz+d)^*)^{-1}\widehat{\Delta}_z(cz+d)^*,$$

where

$$g^{-1} = \begin{pmatrix} a & b \\ c & d \end{pmatrix} \quad \text{and} \quad w = (az+b)(cz+d)^{-1}.$$

In the case $n = 2$, this shows that if $\sigma(\delta)(\delta)^*$ ($\delta \in \mathrm{GL}(2, \boldsymbol{C})$) is a scalar operator, $\widehat{\Delta}_z$ is an invariant differential operator.

Finally, we choose the special representation σ defined by

$$\sigma(\delta) = (\overline{\det(\delta)})^{-2}(\delta^*)^{-1} \quad (\delta \in \mathrm{GL}(2, \boldsymbol{C})).$$

Then $\widehat{\Delta}_z$ is an invariant differential operator.

Moreover, by an easy computation, we obtain

$$\mathcal{K}_{\tau_\sigma,\eta_s}(z,u) = \left(\frac{\det(I_2 - z^*z)}{|\det(I_2 - u^*z)|^2}\right)^{2-s/2}\left(\frac{\det(I_2 - z^*z)}{\det(I_2 - u^*z)}\right)^{-2}(I_2 - z^*z)(I_2 - u^*z)^{-1}.$$

Furthermore, we can prove the following proposition:

PROPOSITION. *Put*

$$\sigma(\delta) = (\overline{\det(\delta)})^{-2}(\delta^*)^{-1} \quad (\delta \in \mathrm{GL}(2, \boldsymbol{C})).$$

Then $\widehat{\Delta}_z$ is an invariant differential operator with respect to the representation T_{τ_σ}, and the matrix-valued Poisson–Cauchy kernel function $\mathcal{K}_{\tau_\sigma,\eta_s}(z,u)$ satisfies the following differential equation:

$$\widehat{\Delta}_z\mathcal{K}_{\tau_\sigma,\eta_s}(z,u) = \frac{s(s-4)}{4}\mathcal{K}_{\tau_\sigma,\eta_s}(z,u) \quad (u \in \check{\boldsymbol{S}}).$$

We will discuss the general case else where.

References

1. Helgason, S.: *Differential Geometry and Symmetric Spaces*, Academic Press, New York, 1962.
2. Hua, L. K.: *Harmonic Analysis of Functions of Several Complex Variables in the Classical Domains*, Amer. Math. Soc. Transl. 6, Amer. Math. Soc., Providence, 1963.
3. Okamoto, K.: *Harmonic Analysis on Homogeneous Vector Bundles*, Lecture Notes in Math. 266, Springer, New York, 1971, pp. 255–271.
4. Narasimhan, M. S. and Okamoto, K.: An analogue of the Borel–Weil–Bott theorem for Hermitian symmetric pairs of non-compact type, *Ann. of Math.* **91** (1970), 486–511.
5. Kashiwara, M., Kowata, A., Minemura, K., Okamoto, K., Oshima, T. and Tanaka, M.: Eigenfunctions of invariant differential operators on a symmetric space, *Ann. of Math.* **107** (1978), 1–39.
6. Okamoto, K., Tsukamoto, M. and Yokota, K.: Generalized Poisson and Cauchy kernel functions on classical domains, *Japan. J. Math.* **26** (2000), 51–103.
7. Warner, G.: *Harmonic Analysis of Semi-Simple Lie Groups*, Springer, New York, 1973.

Acta Applicandae Mathematicae **63:** 333–347, 2000.

On Differential Operators in White Noise Analysis

Dedicated to Professor Takeyuki Hida on the occasion of his 70th birthday

JÜRGEN POTTHOFF
Lehrstuhl für Mathematik V, Fakultät für Mathematik und Informatik, Universität Mannheim, D–68131 Mannheim, Germany

(Received: 5 February 1999)

Abstract. White noise analysis is formulated on a general probability space which is such that (1) it admits a standard Brownian motion, and (2) its σ-algebra is generated by this Brownian motion (up to completion). As a special case, the white noise probability space with time parameter being the half-line is worked out in detail. It is shown that the usual differential operators can be defined on the smooth, finitely based functions of at most exponential growth via the chain rule, without supposing the existence of a linear structure (or translations) on the underlying probability space.

Mathematics Subject Classifications (2000): 60B05, 60B11, 60H07, 60H40.

Key words: white noise analysis, Malliavin calculus, differential operators.

1. Introduction

In [DP97], T. Deck, G. Våge, and the author began the attempt to formulate T. Hida's white noise analysis on a general probability space. One of the reasons that motivated the article [DP97] was that there are important applications – such as in nonlinear filtering – in which a Brownian motion and its noise are *given*, and it seems unnatural, if not wrong, to require that it is realized on a fixed probability space, such as the white noise space. Another reason was the wish to unify the bases of white noise analysis and the Malliavin calculus. It turned out that indeed at least two of the main features of white noise analysis can be formulated within a general framework without any loss: the theory of generalized random variables, and the calculus of differential operators.

In the above-mentioned paper, however, the problem of showing that the differential operators defined there are well-defined was left open, and the main purpose of the present paper is to give a proof of this fact.

Our starting point here will be a general probability space which carries a Brownian motion whose time parameter domain is the half-line $\mathbb{R}_+$. The only additional assumption is that the underlying σ-algebra is (up to completion) generated by the Brownian motion. This entails that the S-transform, which is one of the basic tools of white noise analysis, is injective.

The framework and basic notions will be established in Section 2. In Section 3 one realization of the framework via the white noise space with time parameter domain $\mathbb{R}_+$ is provided in a rather detailed fashion, because this realization seems not to be so well known. Also, I reproduce there a result by S. Albeverio and M. Röckner [AR90] on quasi-invariant measures on Suslin spaces. This result implies that on a Suslin space with a (nontrivial) measure ν which is quasi-invariant with respect to translations of a dense linear subspace, a ν-class of random variables can have at most one continuous representative. This fact will be essential for the proof of the well-definedness of the differential operators which is given in Section 4.

2. Framework

Let $(\Omega, \mathcal{B}, P)$ a complete probability space which satisfies the following conditions:

(H.1) There exists a standard Brownian motion $(B_t,\ t \in \mathbb{R}_+)$ on $(\Omega, \mathcal{B}, P)$;
(H.2) The P-completion of $\sigma(B_t,\ t \in \mathbb{R}_+)$ is equal to $\mathcal{B}$.

It is well known, e.g., [RY91], Lemma V.3.1, that condition (H.2) implies that the algebra generated by the (real, imaginary or complex) exponential functions in the variables $B_{t_1}, \ldots, B_{t_n}$, for $n \in \mathbb{N}$, $t_1, \ldots, t_n \in \mathbb{R}_+$, is dense in $L^2(P)$. In fact, both statements are equivalent, and it is obvious that the same holds just as well for the algebra of polynomials.

Let us mention two realizations of these assumptions:

(1) Wiener space: $\Omega = C_0(\mathbb{R}_+)$, P is Wiener measure, $\mathcal{B}$ is the completion of the σ-algebra generated by the cylinder sets, and $B_t(\omega) = \omega(t)$ for $\omega \in \Omega$.
(2) White noise over the half line, which is discussed in detail in Section 3.

Let $(\mathcal{F}_t^0,\ t \in \mathbb{R}_+)$ be the filtration generated by $(B_t,\ t \in \mathbb{R}_+)$. Denote by $(\mathcal{F}_t,\ t \in \mathbb{R}_+)$ its standard μ-augmentation: if $\mathcal{N}$ is the ideal of the μ-null-sets in $\mathcal{B}$, $\mathcal{F}_t := \mathcal{F}_t^0 \Delta \mathcal{N}$, $t \in \mathbb{R}_+$. Then $(\mathcal{F}_t,\ t \in \mathbb{R}_+)$ is right-continuous because $(B_t,\ t \in \mathbb{R}_+)$ is strongly Markov (cf. Proposition 2.7.7 in [KS88]). Actually, $(\mathcal{F}_t,\ t \in \mathbb{R}_+)$ is continuous (e.g., [KS88], Corollary 2.7.8), and for all $t, s \in \mathbb{R}_+$ with $t > s$, $B_t - B_s$ is independent of $\mathcal{F}_s$ (e.g., [KS88], Theorem 2.7.9), i.e., $(B_t,\ t \in \mathbb{R}_+)$ is a Brownian motion relative to $(\mathcal{F}_t,\ t \in \mathbb{R}_+)$.

Consider the linear mapping

$$\begin{aligned} X\colon C_c^\infty(\mathbb{R}_+) &\longrightarrow L^2(P) \\ f &\longmapsto X_f, \end{aligned}$$

where $X_f := \int_{\mathbb{R}_+} f(t)\,\mathrm{d}B_t$. A trivial application of the Itô isometry shows that this mapping is continuous, if we give $C_c^\infty(\mathbb{R}_+)$ the norm $|\cdot|_2$ of $L^2(\mathbb{R}_+) \equiv L^2(\mathbb{R}_+, \mathcal{B}(\mathbb{R}_+), \lambda)$, where λ is the Lebesgue measure. Since $C_c^\infty(\mathbb{R}_+)$ is dense in $L^2(\mathbb{R}_+)$, this mapping extends to a continuous linear mapping $\widehat{X}$ from $L^2(\mathbb{R}_+)$ into

$L^2(P)$. If $f \in L^2(\mathbb{R}_+)$ is not in $C_c^\infty(\mathbb{R}_+)$, we mean by X_f any representative of the P-class $\widehat{X}_f$. It is clear, that the family $(X_f,\ f \in L^2(\mathbb{R}_+))$ forms a centered Gaussian family with covariance given by the inner product of $L^2(\mathbb{R}_+)$.

I think it is appropriate to call the mapping $\widehat{X}$ *(Gaussian) white noise*, and an essential part of white noise analysis is really the analysis of this mapping and its extensions. Consider the Schwartz space $\mathcal{S}(\mathbb{R}_+)$ over the half-line as defined in [DP97], cf. also Section 3, and the dense continuous embedding $\mathcal{S}(\mathbb{R}_+) \subset L^2(\mathbb{R}_+)$. $\mathcal{S}(\mathbb{R}_+)$ is by construction nuclear, and therefore we have the canonically associated Gel'fand–Hida triple (e.g., [KL96])

$$(\mathcal{S}_+) \subset L^2(P) \subset (\mathcal{S}_+)^*,$$

where $(\mathcal{S}_+)^*$ is a space of generalized random variables. Therefore, we may consider now $\widehat{X}$ as a mapping from $C_c^\infty(\mathbb{R}_+)$ into $(\mathcal{S}_+)^*$, and it has a continuous linear extension to the Schwartz space of tempered distributions $\mathcal{S}'(\mathbb{R}_+)$ over the half-line (cf. Section 3). In particular, we have for $t \in \mathbb{R}_+$, $\widehat{X}_{\delta_t} \in (\mathcal{S}_+)^*$ (δ_t is the Dirac distribution at t), and this is *white noise at time t*. It is customary to denote this generalized random variable by $\dot{B}_t$, and it is not hard to see that it is indeed the time derivative of Brownian motion at time t, the derivative being taken in the strong topology of $(\mathcal{S}_+)^*$.

Let $Z \in L^2(P)$. Define its S-transform SZ as a function on $C_c^\infty(\mathbb{R}_+)$ by

$$SZ(f) := \mathrm{e}^{-1/2|f|_2^2}\,\mathbb{E}\big(Z\,\mathrm{e}^{X_f}\big), \quad f \in C_c^\infty(\mathbb{R}_+).$$

Among other purposes, the factor in front of the expectation serves to normalize the S-transform so that $S1 = 1$.

By what has been said above, it is clear that for $Z \in L^2(P)$, SZ has a continuous extension to $L^2(\mathbb{R}_+)$, and we shall not distinguish the two mappings in the sequel.

The S-transform was first introduced in white noise analysis by I. Kubo and S. Takenaka [KT80], and it is closely related to the Segal–Bargman transform on Fock space, cf., e.g., [KL96].

The S-transform is very useful for a number of reasons. Two are:

(i) It 'diagonalizes' the Itô integral [DP97], and thereby allows to handle and extend 'multiplication by Gaussian (white) noise' very efficiently.
(ii) Spaces of generalized random variables, like $(\mathcal{S}_+)^*$, can be characterized via the S-transform in a way (e.g., [KL96] and literature quoted there) which is quite convenient for theoretical work as well as applications.

3. Schwartz Spaces and White Noise on the Half-Line

In this section we construct a white noise probability space over the half-line $\mathbb{R}_+$, which – according to Section 2 – is to be interpreted as the domain of the time parameter. The construction and properties are very similar to those of the usual white noise probability space (e.g., [Hi80, HK93]).

Let $C^2_\infty(\mathbb{R}_+)$ denote the set of twice continuously differentiable functions f on $(0, +\infty)$, which are such that

(i) f and its first two derivatives vanish rapidly (i.e., faster than any power) at infinity, and
(ii) the first two right derivatives of f exist at 0, and they coincide with $f'(0+)$ and $f''(0+)$, respectively.

Consider the following differential operator

$$Af(t) := -\frac{1}{2}\frac{\mathrm{d}}{\mathrm{d}t}\, t\, \frac{\mathrm{d}}{\mathrm{d}t} f(t) + \frac{1}{4}\, t\, f(t), \qquad t \in \mathbb{R}_+,$$

acting on $C^2_\infty(\mathbb{R}_+)$. It is obvious that A is symmetric on the dense subspace $C^2_\infty(\mathbb{R}_+)$ of $L^2(\mathbb{R}_+)$. (As usual, we talk about the elements of this Hilbert space as if they were functions. There will be no danger of confusion.) Introduce the set of Laguerre functions

$$l_k(t) := \mathrm{e}^{-t/2}\, L_k(t), \quad t \in \mathbb{R}_+,\ k \in \mathbb{N}_0,$$

where L_k is the Laguerre polynomial of order $k \in \mathbb{N}_0$ (e.g., [Sa77]). It is well-known that $(l_k,\ k \in \mathbb{N}_0)$ is a complete orthonormal system in $L^2(\mathbb{R}_+)$. The Laguerre functions l_k obviously belong to $C^2_\infty(\mathbb{R}_+)$, and an elementary calculation shows that they are the eigenfunctions of A:

$$A\, l_k = \left(k + \frac{1}{2}\right) l_k, \quad k \in \mathbb{N}_0.$$

Therefore it is plain (for example by an application of Nelson's analytic vector theorem [RS75]), that A is essentially self-adjoint on $C^2_\infty(\mathbb{R}_+)$, and its unique self-adjoint extension has domain

$$\mathcal{D}(A) := \left\{ f \in L^2(\mathbb{R}_+);\ f = \sum_{k \in \mathbb{N}_0} f_k\, l_k \text{ with } \sum_{k \in \mathbb{N}_0} f_k^2\, k^2 < +\infty \right\}.$$

We see that on the half-line A plays a role which is very similar to the one of the Hamiltonian of the harmonic oscillator on the full line, and the Laguerre functions play the role of the Hermite functions. Now we can proceed as in the case of the full line (e.g., [Hi80, RS72, Si71]). Define

$$\mathcal{S}(\mathbb{R}_+) := \bigcap_{n \in \mathbb{N}_0} \mathcal{D}(A^n),$$

and equip this space as usual with the projective limit topology defined by the Hilbert spaces $\mathcal{D}(A^n)$. Due to the spectral properties of A, it is obvious that $\mathcal{S}(\mathbb{R}_+)$ is a nuclear countable Hilbert space, and as such a Fréchet and a Montel space (e.g., [Tr67]). In [DP97] an elementary argument was given which shows that the functions f in $\mathcal{S}(\mathbb{R}_+)$ are infinitely differentiable on $(0, +\infty)$, and that f together with all its derivatives vanishes rapidly at infinity. It is also straightforward to check that all right derivatives of $f \in \mathcal{S}(\mathbb{R}_+)$ exist at zero, and that they coincide with the right limits of the derivatives of f at 0. That is, $f \in \mathcal{S}(\mathbb{R}_+)$ can be considered as the restriction of a function in $\mathcal{S}(\mathbb{R})$ to $\mathbb{R}_+$.

We denote the dual of $\mathcal{S}(\mathbb{R}_+)$ by $\mathcal{S}'(\mathbb{R}_+)$, and call it *Schwartz space of tempered distributions over the half-line*.

Let us equip $\mathcal{S}'(\mathbb{R}_+)$ with some of the various common topologies, and discuss its topological properties. The arguments below are all completely standard, and given here for the convenience of the interested reader.

If we give $\mathcal{S}'(\mathbb{R}_+)$ the strong topology (i.e., the topology of uniform convergence on bounded subsets of $\mathcal{S}(\mathbb{R}_+)$), then $\mathcal{S}'(\mathbb{R}_+)$ is reflexive since $\mathcal{S}(\mathbb{R}_+)$ is Montel (e.g., [Tr67], p. 376, Corollary). For example, from the Lemma, p. 166, in [RS72] we can conclude that the strong and the Mackey topology coincide. (The Mackey topology is the topology of uniform convergence on the weakly compact, convex subsets of $\mathcal{S}(\mathbb{R}_+)$, and it is the finest dual topology on $\mathcal{S}'(\mathbb{R}_+)$, i.e., the finest locally convex Hausdorff topology on $\mathcal{S}'(\mathbb{R}_+)$ so that $\mathcal{S}(\mathbb{R}_+)$ is its dual in this topology. The weak topology on $\mathcal{S}'(\mathbb{R}_+)$ is the topology of pointwise convergence, and it is the coarsest dual topology on $\mathcal{S}'(\mathbb{R}_+)$. The Mackey–Arens theorem (e.g., [RS72]) states that all locally convex dual topologies are between the weak and the Mackey topology. Furthermore, the strong and the Mackey topology on $\mathcal{S}'(\mathbb{R}_+)$ also coincide with the topology of uniform convergence on compact subsets of $\mathcal{S}(\mathbb{R}_+)$ (e.g., [Tr67], p. 357, Proposition 34.5, but it also follows directly from the definition of a Montel space).

Now we can use Theorem 7 in [S73], p. 112, to conclude that $\mathcal{S}'(\mathbb{R}_+)$ equipped with the strong topology is a Lusin space. (One only has to make the trivial remark that the compact-open topology used for $\mathcal{S}'(\mathbb{R}_+) = \mathcal{L}(E, F)$ there, with $E = \mathcal{S}(\mathbb{R}_+)$ and $F = \mathbb{R}$, is just the topology of uniform convergence on compact subsets of $\mathcal{S}(\mathbb{R}_+)$, which had already been identified with the strong topology on $\mathcal{S}'(\mathbb{R}_+)$ above.)

It is evident from the definition of a Lusin space (e.g., [S73], p. 94), that if we equip $\mathcal{S}'(\mathbb{R}_+)$ with any topology weaker than the strong topology, it remains a Lusin space. In particular, $\mathcal{S}'(\mathbb{R}_+)$ equipped with *any* locally convex dual topology is a Lusin space, and we consider this case from now on. As a Lusin space, $\mathcal{S}'(\mathbb{R}_+)$ is also a Suslin space ([S73], p. 96), and it is strongly Lindelöf ([S73], p. 104, Proposition 3), i.e., every open cover of any open subset of $\mathcal{S}'(\mathbb{R}_+)$ has a countable subcover.

Moreover, since $\mathcal{S}'(\mathbb{R}_+)$ is Suslin for any dual topology, Corollary 2, p. 101, in [S73] entails that they all generate the same Borel σ-algebra, which is the σ-algebra generated by the cylinder sets, because the weak topology is a dual topology on $\mathcal{S}'(\mathbb{R}_+)$. Let us denote this σ-algebra by $\mathcal{B}_0$.

Consider a total subset $\mathcal{T}$ in $\mathcal{S}(\mathbb{R}_+)$. As a Fréchet space $\mathcal{S}(\mathbb{R}_+)$ is barreled. Therefore we have the Banach–Steinhaus theorem which implies that on $\mathcal{S}'(\mathbb{R}_+)$ the topology of pointwise convergence on $\mathcal{T}$ coincides with the weak topology.

Consequently, the σ-algebra $\sigma(\mathcal{T}) := \sigma(\langle\cdot, f\rangle; f \in \mathcal{T})$ is equal to $\mathcal{B}_0$. For example, we may choose for $\mathcal{T}$ the set of Laguerre functions.

As usual, we may use Minlos' theorem (e.g., [Hi80]) to introduce the centered Gaussian measure μ on $(\mathcal{S}'(\mathbb{R}_+), \mathcal{B}_0)$ via the relation

$$\int_{\mathcal{S}'(\mathbb{R}_+)} e^{i\langle\omega, f\rangle}\, \mu(d\omega) = e^{-\frac{1}{2}|f|_2^2}, \quad f \in \mathcal{S}(\mathbb{R}_+),$$

where $\langle\cdot,\cdot\rangle$ stands for the dual pairing of $\mathcal{S}'(\mathbb{R}_+)$ and $\mathcal{S}(\mathbb{R}_+)$, and $|\cdot|_2$ for the norm of $L^2(\mathbb{R}_+)$. We shall call μ the *white noise measure on the half line.* It shares almost all properties with the usual white noise measure, and the analysis of $L^2(\mathcal{S}'(\mathbb{R}_+), \mathcal{B}_0, \mu)$ can be done just as in, e.g., [Hi80, HK93]. One aspect, however, should be recorded here for later use, namely that μ is quasi-invariant under the translations by elements in $L^2(\mathbb{R}_+)$ (which we embed in the canonical way into $\mathcal{S}'(\mathbb{R}_+)$ so that it becomes a dense linear subspace). Here is a quick way to see this: give the spaces $\mathcal{D}(A^m)$, $m \in \mathbb{N}$, the norms $|A^m \cdot|_2$ so that they and their duals $\mathcal{D}(A^m)^*$ become Hilbert spaces. Choose $m \in \mathbb{N}$ so that the injection of $\mathcal{D}(A^m)$ into $L^2(\mathbb{R}_+)$ is Hilbert–Schmidt. Notice that Minlos' theorem implies that μ is carried by $\mathcal{D}^*(A^m) \subset \mathcal{S}'(\mathbb{R}_+)$. Thus $(L^2(\mathbb{R}_+), \mathcal{D}^*(A^m), \mu)$ forms an abstract Wiener space, and we can use the well-known Cameron–Martin formula derived, e.g., in [Ku75].

Our next step is to quote a very useful result by S. Albeverio and M. Röckner [AR90].

Consider a Suslin topological vector space E over the reals, equipped with its Borel σ-algebra, denoted again by $\mathcal{B}_0$. Assume that ν is a measure on $(E, \mathcal{B}_0)$ which is not identically zero. Define an open subset V of E as the union of all open subsets which have ν-measure zero. Since by construction V is covered by open subsets of zero measure, the fact that E is strongly Lindelöf (s.a.) tells us, that V has a countable subcover by ν-null sets, and consequently has itself ν-measure zero. Hence V is the largest open subset of E which has zero ν-measure. Albeverio and Röckner define $\operatorname{supp}(\nu) := E \setminus V$.

Let $k \in E$. ν is called k-quasi-invariant if ν is quasi-invariant w.r.t. translations by the elements of the one-dimensional subspace generated by k, i.e., for every $A \in \mathcal{B}_0$, and all $\lambda \in \mathbb{R}$, $\nu(A) = 0$ implies $\nu(A + \lambda k) = 0$. Albeverio and Röckner prove in [AR90] the following

PROPOSITION. *Let $K := \{k \in E;\ \nu$ is k-quasi-invariant$\}$. Then K is a linear space. If K is dense in E then* $\operatorname{supp}(\nu) = E$.

For the convenience of the reader – and with the permission of the authors – I repeat the proof given in [AR90] here:

The fact that K is linear follows directly from the definition of k-quasi-invariance. Assume that K is dense in E, and let $z \in \operatorname{supp}(\nu)$. (Such z exists, otherwise ν would be the zero measure.) We want to show then that every element $z'' \in E$ of the form $z'' = z + \alpha z'$, $z' \in E$, belongs to $\operatorname{supp}(\nu)$, and consequently also every $z'' \in E$. First we show this for $z'' = z + k$ with $k \in K$: Let V_{z+k} be an open neighborhood of $z + k$. Then $V_z := V_{z+k} - k$ is an open neighborhood of z. Thus we must have $\nu(V_z) > 0$, and k-quasi-invariance of ν implies that $\nu(V_{z+k}) > 0$. Hence $z + k \in \operatorname{supp}(\nu)$. Now let $z' \in E$, and choose a net $(k_i,\ i \in I)$ in K which converges to z'. Then $z + \alpha z' = \lim_i (z + \alpha k_i)$. We have $z + \alpha k_i \in \operatorname{supp}(\nu)$ for all $i \in I$. $\operatorname{supp}(\nu)$ is closed, and thus we get $z + \alpha z' \in \operatorname{supp}(\nu)$. □

In other words: if ν is quasi-invariant with respect to the translations of a dense linear subset, then every nonempty open set has nonvanishing ν-measure. Assume now that f and g are continuous mappings from E into a Hausdorff topological space F, which are ν-a.e. equal. If ν is k-quasi-invariant for all k in a dense set K, then they must be equal: The set $\{\omega;\ f(\omega) \neq g(\omega)\}$ is open, and the assumption that it is nonempty leads to a contradiction. For the white noise measure μ on $(\mathcal{S}'(\mathbb{R}_+), \mathcal{B}_0)$ we get therefore the following conclusion.

COROLLARY. *Let f and g be two mappings from $\mathcal{S}'(\mathbb{R}_+)$ into a Hausdorff topological space which are continuous w.r.t. any dual topology on $\mathcal{S}'(\mathbb{R}_+)$, and which are such that $f = g$ μ-a.s. Then $f = g$. In particular, a μ-class of real or complex valued random variables on $(\mathcal{S}'(\mathbb{R}_+), \mathcal{B}_0, \mu)$ can only have one continuous representative.*

Our next task is to introduce a Brownian motion on $(\mathcal{S}'(\mathbb{R}_+), \mathcal{B}_0, \mu)$, and we shall follow the usual constructions. One way to do this is as follows: Consider the linear mapping $\widehat{X}_\cdot\colon \mathcal{S}(\mathbb{R}_+) \to L^2(\mu)$ where $\widehat{X}_f$ is the μ-class of the random variable given by $X_f(\omega) := \langle \omega, f\rangle$ for $f \in \mathcal{S}(\mathbb{R}_+)$, $\omega \in \mathcal{S}'(\mathbb{R}_+)$. It follows directly from the definition of μ that this mapping is continuous, if $\mathcal{S}(\mathbb{R}_+)$ is equipped with the norm of $L^2(\mathbb{R}_+)$. Therefore, it has a unique linear continuous extension to $L^2(\mathbb{R}_+)$, which we continue to denote by $\widehat{X}_\cdot$, and by X_f we mean any representative of $\widehat{X}_f$ if $f \in L^2(\mathbb{R}_+) \setminus \mathcal{S}(\mathbb{R}_+)$. It is clear that $(X_f;\ f \in L^2(\mathbb{R}_+))$ is a centered Gaussian family whose covariance is given by the inner product in $L^2(\mathbb{R}_+)$. Choose for $t \in \mathbb{R}_+$, $f = 1_{[0,t]}$, and set $\widetilde{B}_t := X_{1_{[0,t]}}$. Then it is plain to check, that the process $(\widetilde{B}_t;\ t \in \mathbb{R}_+)$ has the same finite-dimensional distributions as a Brownian motion. Therefore, we can apply the Kolmogorov–Chentsov lemma to find a modification of $(\widetilde{B}_t;\ t \in \mathbb{R}_+)$ with a.s. continuous sample paths. This modification is a standard Brownian motion, and it will be denoted by $(B_t;\ t \in \mathbb{R}_+)$.

Alternatively, we can mimic the Lévy–Ciesielski construction, e.g., [MK69]. To this end, we consider $L^2([0,1])$ (with Lebesgue measure) as embedded into $L^2(\mathbb{R}_+)$. Let $(s_{n,k};\ n \in \mathbb{N}_0,\ k = 1, 3, \dots, 2^{n-1} - 1)$ be the system of Schauder functions on $[0, 1]$. That is, for $t \in \mathbb{R}_+$, $s_{n,k}(t) := \int_0^t f_{n,k}(s)\,ds$, where $f_{n,k}$ is the Haar function of index (n, k), which is $2^{(n-1)/2}$ on the interval $[(k-1)/2^n, k/2^n)$, $-2^{(n-1)/2}$ on $[k/2^n, (k+1)/2^n)$ and zero otherwise. Recall that the system of Haar functions forms a complete orthonormal system on $L^2([0,1])$ (e.g., [MK69]). It is clear that $s_{n,k}$ is continuous on $[0, 1]$.

Now choose any CONS $(e^0_{n,k};\ n \in \mathbb{N}_0,\ k = 1, 3, \dots, 2^{n-1} - 1)$ of $L^2([0, 1])$ in $C^\infty_c([0, 1])$. Let $N_{n,k}(\omega) := \langle \omega, e^0_{n,k} \rangle$. Then the family $\big(N_{n,k};\ n \in \mathbb{N}_0,\ k = 1, 3, \dots, 2^{n-1} - 1\big)$ is an i.i.d. system of everywhere defined standard normal random variables. Define

$$B_t := t\, N_{0,1} + \sum_{n,k} s_{n,k}(t)\, N_{n,k}, \quad t \in [0, 1].$$

Then we are in the same situation as in [MK69], p. 7, and can use the Borel–Cantelli lemma to prove that this series converges a.s. absolutely and *uniformly in* $t \in [0, 1]$. Therefore, the process B_t, $t \in [0, 1]$, has a.s. continuous sample paths. It is obvious that B_t is centered Gaussian, and to check that the covariance of B_s, B_t is $s \wedge t$ is an easy exercise. Thus $(B_t,\ t \in [0, 1])$ is a standard Brownian motion over $[0, 1]$.

For any $N \in \mathbb{N}$ we can now repeat the same construction on the interval $[N, N+1]$ with a CONS $(e^N_{n,k},\ n \in \mathbb{N}_0,\ k = 1, 3, \dots, 2^{n-1} - 1)$ of $L^2([N, N+1])$ in $C^\infty_c([N, N+1])$, and obtain an independent standard Brownian motion indexed by $t \in [N, N+1]$. Gluing these Brownian motions continuously together, we obtain a standard Brownian motion indexed by the half-line.

Let us denote the μ-completion of $\mathcal{B}_0$ by $\mathcal{B}$. (The extension of μ to $\mathcal{B}$ will be denoted by the same symbol.)

Consider our first construction of Brownian motion $(B_t,\ t \in \mathbb{R}_+)$, i.e., by modification of the random variables $\widetilde{B}_t$, $t \in \mathbb{R}_+$. We want to show that the μ-completion $\sigma^\mu(B_t,\ t \in \mathbb{R}_+)$ of $\sigma(B_t,\ t \in \mathbb{R}_+)$ generated by it is equal to $\mathcal{B}$. First we prove the following result which is almost obvious.

LEMMA. *Let $\mathcal{T}$ be any subset of $\mathcal{S}(\mathbb{R}_+)$ which is total in $\mathcal{S}(\mathbb{R}_+)$ with respect to the norm of $L^2(R_+)$ restricted to $\mathcal{S}(\mathbb{R}_+)$. Then the μ-completion of $\sigma(\mathcal{T}) := \sigma(X_f,\ f \in \mathcal{T})$ coincides with $\mathcal{B}$.*

Proof. Since all linear combinations of random variables of the form X_f with $f \in \mathcal{T}$ are $\sigma(\mathcal{T})$-measurable, we may assume without loss of generality that $\mathcal{T}$ is a dense linear subspace of $\mathcal{S}(\mathbb{R}_+)$ with respect to $|\cdot|_2$. Let $f \in \mathcal{S}(\mathbb{R}_+)$, and choose a sequence $(f_n,\ n \in \mathbb{N}_0)$ in $\mathcal{T}$ so that $|f - f_n|_2 \to 0$. Then X_{f_n} converges in mean square to X_f. By choosing a subsequence if necessary, we may assume that X_{f_n} converges to X_f μ-a.s. This implies that X_f has a modification, say

$$\widetilde{X}_f := \limsup_{n \in \mathbb{N}} X_{f_n},$$

which is measurable with respect to the σ-algebra generated by the random variables X_{f_n}, $n \in \mathbb{N}$. In particular, $\widetilde{X}_f$ is $\sigma(\mathcal{T})$-measurable, and therefore X_f is measurable with respect to the μ-completion $\sigma^\mu(\mathcal{T})$ of the σ-algebra $\sigma(\mathcal{T})$. On the other hand, the random variables X_f, $f \in \mathcal{S}(\mathbb{R}_+)$, generate $\mathcal{B}_0$, so that we have the following inclusions

$$\mathcal{B}_0 \subset \sigma^\mu(\mathcal{T}) \subset \mathcal{B}.$$

Consequently $\sigma^{\mu}(\mathcal{T}) = \mathcal{B}$. □

Let $f \in C_c^{\infty}(\mathbb{R}_+)$, and consider the random variable X_f. Define the following sequence

$$X_f^n := \sum_{k=0}^{\infty} f(s_k^n)\left(B_{s_{k+1}^n} - B_{s_k^n}\right), \quad n \in \mathbb{N},$$

where $s_k^n = k/2^n$. Observe that for each $n \in \mathbb{N}$, the series above has only a finite number of terms. It is straightforward to check that X_f^n converges to X_f in mean square sense. By an argument similar to the one in the proof of the lemma, we find that X_f is measurable with respect to $\sigma^{\mu}(B_t,\, t \in \mathbb{R}_+)$. This holds for every f in the subspace $C_c^{\infty}(\mathbb{R}_+)$ of $\mathcal{S}(\mathbb{R}_+)$, which is dense w.r.t. the norm $|\cdot|_2$. Using the lemma with $\mathcal{T} = C_c^{\infty}(\mathbb{R}_+)$ we therefore get the following inclusions:

$$\sigma\left(C_c^{\infty}(\mathbb{R}_+)\right) \subset \sigma^{\mu}(B_t,\, t \in \mathbb{R}_+) \subset \mathcal{B} = \sigma^{\mu}\left(C_c^{\infty}(\mathbb{R}_+)\right),$$

where the last σ-algebra is the μ-completion of $\sigma(C_c^{\infty}(\mathbb{R}_+))$. Hence we can conclude that

$$\sigma^{\mu}(B_t,\, t \in \mathbb{R}_+) = \mathcal{B}.$$

For the Brownian motion defined by the Lévy–Ciesielski construction we obtain the same result – more easily – in a similar way.

Thus we have completed our discussion of a realization of the hypotheses (H.1), (H.2) in the setting of the white noise space over the half-line.

We conclude this section with the following remark. Instead of $\mathcal{S}(\mathbb{R}_+)$ we could as well have chosen the smaller space $\mathcal{S}_0(\mathbb{R}_+)$ which is defined as the closure of $C_c^{\infty}(\mathbb{R}_+)$ in $\mathcal{S}(\mathbb{R})$ (or in $\mathcal{S}(R_+)$, which leads to the same space). It is clear that $\mathcal{S}(\mathbb{R}_+)$ is the subspace of those functions of $\mathcal{S}(\mathbb{R})$ which are supported in $\mathbb{R}_+$ (or those functions in $\mathcal{S}(\mathbb{R}_+)$ which together with all their derivatives vanish at the origin). As a linear subspace of $\mathcal{S}(\mathbb{R})$ or $\mathcal{S}(\mathbb{R}_+)$, $\mathcal{S}_0(\mathbb{R}_+)$ is again nuclear. Therefore, we can repeat the preceding discussion almost word by word, with obvious modifications here and there.

4. Differential Operators

Consider a random variable Y on $(\Omega, \mathcal{B}, P)$. If Ω is a topological vector space (or admits some suitable notion of translations), like in the case of the two realizations mentioned in Section 2, it is obvious how to define directional derivatives. Since in the general case such a structure is not available, we introduce directional derivatives by imitating the chain rule. This has been done in [DP97], but the question whether the derivatives are well-defined had been left open. We start with this question here.

For $n \in \mathbb{N}$, denote by $C_e^\infty(\mathbb{R}^n)$ the space of infinitely often continuously differentiable functions on $\mathbb{R}^n$ which – together with all their derivatives – have at most exponential growth. It will be convenient to denote the ith partial derivative of $f \in C_e^\infty(\mathbb{R}^n)$ at $x \in \mathbb{R}^n$ by $f_{,i}(x)$.

By $C_{\mathrm{fi},e}^\infty(\Omega)$ we mean the space of all random variables Y for which there exist $n \in \mathbb{N}$, $h_1, \ldots, h_n \in C_c^\infty(\mathbb{R}_+)$, and $f \in C_e^\infty(\mathbb{R}^n)$ so that

$$Y = f(X_{h_1}, \ldots, X_{h_n}).$$

Note that $C_{\mathrm{fi},e}^\infty(\Omega) \subset L^p(P)$ for all $p \geqslant 1$, and these inclusions are dense.

LEMMA 1. *Assume that $Y \in C_{\mathrm{fi},e}^\infty(\Omega)$ has two representations of the form*

$$\begin{aligned} Y &= f(X_{h_1}, \ldots, X_{h_n}) \\ &= g(X_{k_1}, \ldots, X_{k_m}) \end{aligned}$$

with n, $m \in \mathbb{N}$, $h_1, \ldots, h_n$, $k_1, \ldots, k_m \in C_c^\infty(\mathbb{R}_+)$, and $f \in C_e^\infty(\mathbb{R}^n)$, $g \in C_e^\infty(\mathbb{R}^m)$. Then for all $u \in L^2(\mathbb{R}_+)$,

$$\begin{aligned} &\sum_{i=1}^n f_{,i}(X_{h_1}, \ldots, X_{h_n})\,(u, h_i)_{L^2(\mathbb{R}_+)} \\ &\quad = \sum_{j=1}^m g_{,j}(X_{k_1}, \ldots, X_{k_m})\,(u, k_j)_{L^2(\mathbb{R}_+)}, \qquad P\text{-a.s.} \end{aligned}$$

Proof. By assumption, we have for all $l \in C_c^\infty(\mathbb{R}_+)$,

$$\big(f(X_{h_1}, \ldots, X_{h_n}) - g(X_{k_1}, \ldots, X_{k_m})\big)\,\mathrm{e}^{X_l} = 0,$$

and in particular the law of this random variable is the Dirac measure ε_0 at 0.

Now consider the Gaussian random variables $Z_{h_1}, \ldots, Z_{h_n}, Z_{k_1}, \ldots, Z_{k_m}, Z_l$ in the white noise realization on $(\mathcal{S}'(\mathbb{R}_+), \mathcal{B}, \mu)$ of Section 3. That is, $Z_\eta(\omega) = \langle \omega, \eta \rangle$ for $\eta \in \mathcal{S}(\mathbb{R}_+)$, $\omega \in \mathcal{S}'(\mathbb{R}_+)$. The random variables $Z_{h_1}, \ldots, Z_l$ have the same joint distribution as $X_{h_1}, \ldots, X_l$. Therefore also the random variable

$$\big(f(Z_{h_1}, \ldots, Z_{h_n}) - g(Z_{k_1}, \ldots, Z_{k_m})\big)\,\mathrm{e}^{Z_l}$$

has ε_0 as its law. Since this is true for every $l \in C_c^\infty(\mathbb{R}_+)$, this implies that the S-transform of

$$f(Z_{h_1}, \ldots, Z_{h_n}) - g(Z_{k_1}, \ldots, Z_{k_m})$$

is zero. Because of the injectivity of the S-transform we conclude that

$$f(Z_{h_1}, \ldots, Z_{h_n}) = g(Z_{k_1}, \ldots, Z_{k_m}), \quad \mu\text{-a.s.}$$

By construction, both sides of the last equality are weakly continuous on $\mathcal{S}'(\mathbb{R}_+)$. The corollary of Section 3 entails that these two random variables are equal *everywhere* on $\mathcal{S}'(\mathbb{R}_+)$. Hence we may compute the directional derivative of both sides in direction $u \in L^2(\mathbb{R}_+) \subset \mathcal{S}'(\mathbb{R}_+)$ as follows:

$$\begin{aligned} D_u\, f(Z_{h_1}, \ldots, Z_{h_n}) &= \sum_{i=1}^{n} f_{,i}(Z_{h_1}, \ldots, Z_{h_n})\,(u, h_i)_{L^2(\mathbb{R}_+)} \\ &= D_u\, g(Z_{k_1}, \ldots, Z_{k_m}) \\ &= \sum_{j=1}^{m} g_{,j}(Z_{k_1}, \ldots, Z_{k_m})\,(u, k_j)_{L^2(\mathbb{R}_+)}, \end{aligned}$$

because for $\eta \in \mathcal{S}(\mathbb{R}_+)$, ω, $u \in \mathcal{S}'(\mathbb{R}_+)$,

$$\begin{aligned} D_u\, Z_\eta(\omega) &= \left.\frac{\mathrm{d}}{\mathrm{d}\lambda} \langle \omega + \lambda u, \eta \rangle \right|_{\lambda=0} \\ &= (u, \eta)_{L^2(\mathbb{R}_+)}. \end{aligned}$$

Consequently, we have for all $l \in C_c^\infty(\mathbb{R}_+)$ that

$$\left(\sum_{i=1}^{n} f_{,i}(Z_{h_1}, \ldots, Z_{h_n})\,(u, h_i)_{L^2(\mathbb{R}_+)} - \right. \\ \left. - \sum_{j=1}^{m} g_{,j}(Z_{k_1}, \ldots, Z_{k_m})\,(u, k_j)_{L^2(\mathbb{R}_+)} \right) \mathrm{e}^{Z_l}$$

has ε_0 as its law. Now we go back to the original probability space, i.e., replace the Gaussian random variables Z_{h_1}, etc., by X_{h_1}, etc. Then we obtain that the S-transform of

$$\sum_{i=1}^{n} f_{,i}(X_{h_1}, \ldots, X_{h_n})\,(u, h_i)_{L^2(\mathbb{R}_+)} - \sum_{j=1}^{m} g_{,j}(X_{k_1}, \ldots, X_{k_m})\,(u, k_j)_{L^2(\mathbb{R}_+)}$$

is zero, and hence this random variable is P-a.s. zero. □

The preceding lemma justifies the following

DEFINITION. Let $Y \in C^\infty_{\mathrm{fi},e}(\Omega)$ be given by

$$Y = f(X_{h_1}, \ldots, X_{h_n}),$$

with $n \in \mathbb{N}$, $h_1, \ldots, h_n \in C_c^\infty(\mathbb{R}_+)$, and $f \in C_e^\infty(\mathbb{R}^n)$. For all $u \in L^2(\mathbb{R}_+)$, the random variable in $C^\infty_{\mathrm{fi},e}(\Omega)$ defined by

$$D_u Y := \sum_{i=1}^{n} f_{,i}(X_{h_1}, \ldots, X_{h_n})\,(u, h_i)_{L^2(\mathbb{R}_+)}$$

is called the *derivative of* Y *in direction* u.

For every $u \in L^2(\mathbb{R}_+)$, D_u is a densely defined operator on $L^2(P)$. Thus it has an adjoint D_u^* with domain $\mathcal{D}(D_u^*)$. It is not very difficult to check that $C^\infty_{\text{fi},e}(\Omega)$ is a subset of $\mathcal{D}(D_u^*)$, and that on $Y \in C^\infty_{\text{fi},e}(\Omega)$ it acts as

$$D_u^* Y = X_u Y - D_u Y.$$

In particular, D_u is closable on $L^2(P)$, and we denote its closure by $(D_u, \mathcal{D}(D_u))$. At this point we can repeat all arguments, as, e.g., given in [HK93], to show that actually

$$\mathcal{D}(D_u) = \mathcal{D}(D_u^*) = \mathcal{D}(X_u \cdot) = \mathcal{D}(N^{1/2}),$$

where N is the number operator, which can – for example – be defined on the core $C^\infty_{\text{fi},e}(\Omega)$ by the formula

$$N = \sum_{k \in \mathbb{N}} D_{e_k}^* D_{e_k},$$

with a CONS $(e_k, k \in \mathbb{N})$ of $L^2(\mathbb{R}_+)$.

The following result will be useful.

LEMMA 2. *For every $u \in L^2(\mathbb{R}_+)$, D_u commutes with the S-transform in the sense that for all $Y \in \mathcal{D}(N^{1/2})$, $f \in L^2(\mathbb{R}_+)$,*

$$S(D_u Y)(f) = D_u(SY)(f),$$

where the right-hand side is the usual Gâteaux derivative of a function on $L^2(\mathbb{R}_+)$.

Proof. Compute

$$\begin{aligned} S(D_u Y)(f) &= \mathbb{E}\big(D_u Y : e^{X_f} :\big) \\ &= \mathbb{E}\big(Y D_u^* : e^{X_f} :\big) \\ &= \mathbb{E}\big(Y \left(X_u - (u, f)\right) : e^{X_f} :\big), \end{aligned}$$

where we have set

$$: e^{X_f} : = e^{X_f - 1/2 |f|_2^2}.$$

On the other hand, we have

$$\begin{aligned} D_u(SY)(f) &= \frac{\partial}{\partial \lambda} (SY)(f + \lambda u) \Big|_{\lambda=0} \\ &= \frac{\partial}{\partial \lambda} \mathbb{E}\big(Y : e^{X_{f+\lambda u}} :\big) \Big|_{\lambda=0} \\ &= \frac{\partial}{\partial \lambda} \mathbb{E}\big(Y e^{X_f + \lambda X_u - 1/2 |f + \lambda u|_2^2}\big) \Big|_{\lambda=0} \\ &= \mathbb{E}\big(Y \left(X_u - (f, u)\right) : e^{X_f} :\big), \end{aligned}$$

where the interchange of the derivative w.r.t. λ and the expectation is readily justified, e.g., by an application of the dominated convergence theorem. □

Consider again the formula for $D_u Y$ with $Y \in C^\infty_{\mathrm{fi},e}(\Omega)$. We may write it as follows

$$D_u Y = \int_{\mathbb{R}_+} (\partial_t Y)\, u(t)\, \mathrm{d}t,$$

where we have put

$$\partial_t Y := \sum_{i=1}^{n} f_{,i}(X_{h_1}, \ldots, X_{h_n})\, h_i(t).$$

The integral above could of course be interpreted pointwise on Ω, but we shall take in the sense of an $L^2(P)$-valued Pettis integral (e.g., [HP57]). Therefore we have for all $Z \in L^2(P)$ the relation

$$\mathbb{E}\left(Z \int_{\mathbb{R}_+} (\partial_t Y)\, u(t)\, \mathrm{d}t\right) = \int_{\mathbb{R}_+} \mathbb{E}\big(Z\, (\partial_t Y)\big)\, u(t)\, \mathrm{d}t.$$

Choosing in particular $Z = \exp(X_f - \frac{1}{2}|f|_2^2)$ with $f \in C_c^\infty(\mathbb{R}_+)$, we find that

$$S(D_u Y)(f) = \int_{\mathbb{R}_+} S(\partial_t Y)(f)\, u(t)\, \mathrm{d}t.$$

We are interested to compute the S-transform of $\partial_t Y$. To this end, let us prove first the following result

LEMMA 3. *For all $f \in C_c^\infty(\mathbb{R}_+)$, $Y \in C^\infty_{\mathrm{fi},e}(\Omega)$, the mapping*

$$u \mapsto D_u(SY)(f)$$

from $L^2(\mathbb{R}_+)$ into $\mathbb{R}$ is linear and continuous.

Proof. That for all $Y \in C^\infty_{\mathrm{fi},e}(\Omega)$, the mapping $u \mapsto D_u Y$ is linear and continuous from $L^2(\mathbb{R}_+)$ into $L^2(P)$, is obvious from the definition. On the other hand, the S-transform of a random variable is its $L^2(P)$-inner product with $\exp(X_f - \frac{1}{2}|f|_2^2)$, and hence it is clear that this is a continuous linear map from $L^2(P)$ into $\mathbb{R}$. Thus $u \mapsto S(D_u Y)(f)$ is linear and continuous from $L^2(\mathbb{R}_+)$ into $\mathbb{R}$. The proof is finished by an application of Lemma 2. □

From Lemma 3 we conclude that for given $Y \in C^\infty_{\mathrm{fi},e}(\Omega)$, $f \in C_c^\infty(\mathbb{R}_+)$, there exists an element in $L^2(\mathbb{R}_+)$ which we denote by

$$t \mapsto \frac{\delta}{\delta f(t)} SY(f)$$

so that for all $u \in L^2(\mathbb{R}_+)$,

$$D_u(SY)(f) = \int_{\mathbb{R}_+} \frac{\delta}{\delta f(t)} SY(f)\, u(t)\, \mathrm{d}t.$$

$\delta\, SY(f)/\delta f(t)$ is also called the *Fréchet functional derivative of SY at f*. If we use again the fact that D_u and S commute, we arrive at the following (slightly informal) intertwining relation for ∂_t and S

$$\partial_t = S^{-1} \frac{\delta}{\delta f(t)} S,$$

which is essentially T. Hida's original definition of ∂_t in the ground-breaking paper [Hi75].

Acknowledgement

It is a pleasure to thank Thomas Deck and Gjermund Våge for many stimulating discussions. The author is very grateful to S. Albeverio and M. Röckner for their advice and their permission to reproduce here one of their results.

References

[AR90] Albeverio, S. and Röckner, M.: New developments in the theory and applications of Dirichlet forms, In: S. Albeverio *et al.* (eds), *Stochastic Processes, Physics and Geometry*, World Scientific, Singapore, 1990.

[DP97] Deck, T., Potthoff, J. and Våge, G.: A review of white noise analysis from a probabilistic standpoint, *Acta Appl. Math.* **48** (1997), 91–112.

[Hi75] Hida, T.: *Analysis of Brownian Functionals*, Carleton Math. Lecture Notes, No. 13, 1975.

[Hi80] Hida, T.: *Brownian Motion*, Springer, New York, 1980.

[HK93] Hida, T., Kuo, H.-H., Potthoff, J. and Streit, L.: *White Noise – An Infinite Dimensional Calculus*, Kluwer Acad. Publ., Dordrecht, 1993.

[HP57] Hille, E. and Phillips, R. S.: *Functional Analysis and Semigroups*, Amer. Math. Soc. Colloq. Publ., Amer. Math. Soc., Providence, 1957.

[KS88] Karatzas, I. and Shreve, S. E.: *Brownian Motion and Stochastic Calculus*, Springer, New York, 1988.

[KL96] Kondratiev, Yu. G., Leukert, P., Potthoff, J., Streit, L. and Westerkamp, W.: Generalized functionals on Gaussian spaces – the characterization theorem revisited, *J. Funct. Anal.* **141** (1996), 301–318.

[KT80] Kubo, I. and Takenaka, S.: Calculus on Gaussian white noise I, *Proc. Japan Acad.* **56** (1980), 376–380.

[Ku75] Kuo, H.-H.: *Gaussian Measures in Banach Spaces*, Springer, New York, 1975.

[MK69] McKean, H. P., Jr.: *Stochastic Integrals*, Academic Press, New York, 1969.

[RS72] Reed, M. and Simon, B.: *Methods of Modern Mathematical Physics I: Functional Analysis*, Academic Press, New York, 1972.

[RS75] Reed, M. and Simon, B.: *Methods of Modern Mathematical Physics II: Fourier Analysis, Self-Adjointness*, Academic Press, New York, 1975.

[RY91] Revuz, D. and Yor, M.: *Continuous Martingales and Brownian Motion*, Springer, New York, 1991.

[Sa77] Sansone, G.: *Orthogonal Functions*, Robert Krieger Publ. Co., 1977.
[Si71] Simon, B.: Distributions and their Hermite expansions, *J. Math. Phys.* **12** (1971), 140–148.
[S73] Schwartz, L.: *Radon Measures on Arbitrary Topological Vector Spaces and Cylindrical Measures*, Oxford Univ. Press, London, 1973.
[Tr67] Trèves, F.: *Topological Vector Spaces, Distributions and Kernels*, Academic Press, New York, 1967.

Acta Applicandae Mathematicae **63:** 349–361, 2000.

Stochastic Differentiation – A Generalized Approach

MYLAN REDFERN
Department of Mathematics, University of Southern Mississippi, Hattiesburg, MS 39406-5045, U.S.A. e-mail: m.redfern@usm.edu

(Received: 8 January 1999)

Abstract. The space $(\mathcal{D}^*)$ of Wiener distributions allows a natural Pettis-type stochastic calculus. For a certain class of generalized multiparameter processes $X\colon \mathbb{R}^N \to (\mathcal{D}^*)$ we prove several differentiation rules (Itô formulas); these processes can be anticipating. We then apply these rules to some examples of square integrable Wiener functionals and look at the integral versions of the resulting formulas.

Mathematics Subject Classifications (2000): 60G20, 60H05, 60H40.

Key words: Wiener distributions, stochastic differentiation, generalized stochastic integrals, multiparameter processes.

1. Introduction

In [BR3] a generalized theory of multiparameter stochastic integration was developed. The framework for this theory is the white noise space $(\mathcal{S}^*, \mu)$ where $\mathcal{S}^* = \mathcal{S}^*(\mathbb{R}^N)$ is the space of tempered distributions on $\mathbb{R}^N$ with Gaussian measure μ given by Minlos' Theorem. The stochastic integrals which were studied appear naturally in integral versions of certain differentiation formulas. One such formula is given in the following theorem from [BR2]:

THEOREM 1.1. *Suppose $\phi^1, \phi^2, \dots, \phi^p\colon \mathbb{R}^N \to K_1$ are differentiable K_1-valued processes with covariance matrix $A(t) = \{A^{ab}(t)\} = \{\langle \phi^a(t), \phi^b(t)\rangle\}$ which is differentiable and invertible at each point t in $\mathbb{R}^N$. Suppose that F is $C_b^2(\mathbb{R}^p)$ and let $F(\phi)\colon \mathbb{R}^N \to (\mathcal{D}^*)$ be the corresponding generalized function of the random variable $\phi(t) = (\phi^1(t), \dots, \phi^p(t))$. Then $F(\phi)$ is differentiable, and*

$$\frac{\partial}{\partial t_i} F(\phi) = (D_i F)(\phi) + \frac{\partial F}{\partial x_a}(\phi) \widehat{\otimes} \frac{\partial \phi^a}{\partial t_i}.$$

Here D_i is the operator

$$D_i = \frac{\partial}{\partial t_i} + \frac{1}{2} A_i^{ab} \frac{\partial^2}{\partial x_a \partial x_b}$$

with $A_i^{ab} = \partial A^{ab}/\partial t_i$, and there is implied summation over repeated indices.

The purpose of this paper is to discuss and extend this formula and give some examples. In the above theorem each $\phi^a(t)$ is in K_1, i.e. is the Wiener integral of a function $f(t)$ in $L^2(\mathbb{R}^N)$. We will extend this result to the case where $\phi^a(t)$ is the sum of Wick products of such Wiener integrals, see Theorem 3.1 and Corollary 3.1. We then obtain, in Corollary 3.2, a differentiation formula for products of corresponding generalized functions. The function F in the above theorem can be replaced by $F \in \mathcal{S}^*(\mathbb{R}^p)$ but we do not need such generality here.

2. Background

In this section we will set some notation and briefly discuss the space of Wiener distributions and explain integration and differentiation in this space. For the complete development see [BR2].

Our approach exploits the Fock space structure of $(L^2) \equiv L^2(\mathcal{S}^*)$ and has its origins in the works of T. Hida, see [Hi1, Hi2]. There is the well-known Wiener–Itô decomposition $(L^2) = \bigoplus_{n=0}^{\infty} K_n$ where K_n is the space of n-tuple Wiener integrals of functions in $L_n^2 \equiv L^2((\mathbb{R}^N)^n)$. Let $I_n(f_n) = I_n(\hat{f}_n)$ represent the n-tuple Wiener integral of f_n and $\hat{}$ represent symmetrization. Then (L^2) is isomorphic to the Fock space $\hat{L}^2 \equiv \bigoplus_{n=0}^{\infty} \sqrt{n!}\hat{L}_n^2$.

Densely embedded in L_n^2 is the space $\mathcal{D}_n \equiv C_c^{\infty}((\mathbb{R}^N)^n)$. As a linear operator I_n restricts to $\hat{\mathcal{D}}_n$ and by duality extends to $\hat{\mathcal{D}}^*$. We denote $(\mathcal{D}_n) \equiv I_n(\hat{\mathcal{D}}_n)$ and $(\mathcal{D}_n^*) \equiv I_n(\hat{\mathcal{D}}_n^*)$. We define $\mathcal{D} \equiv \coprod_{n=0}^{\infty} \mathcal{D}_n$, the coproduct with the inductive limit topology, so that $\mathcal{D}^* = \prod_{n=0}^{\infty} \hat{\mathcal{D}}_n^*$. We call $(\mathcal{D}) \equiv \coprod_{n=0}^{\infty}(\mathcal{D}_n)$ the space of Wiener test functionals and $(\mathcal{D}^*) \equiv \prod_{n=0}^{\infty}(\mathcal{D}_n^*)$ the space of Wiener distributions. For $X = \sum_{n=0}^{\infty} X_n = \sum_{n=0}^{\infty} I_n(T_n)$ in $(\mathcal{D}^*)$, the dual pairing between $(\mathcal{D}^*)$ and $(\mathcal{D})$ is

$$\langle X, \Phi \rangle = \sum_{n=0}^{\infty} \langle X_n, \Phi_n \rangle = \sum_{n=0}^{\infty} n! \langle T_n, \xi_n \rangle,$$

where $\Phi = \sum_{n=0}^{\infty} \Phi_n = \sum_{n=0}^{\infty} I_n(\xi_n)$ is a Wiener test functional. We see that $(\mathcal{D}^*)$ is the set of formal series $\{X = \sum_{n=0} I_n(T_n) | T_n \in \mathcal{D}^*\}$. This space contains the space $(\mathcal{S})^*$ of Hida distributions, see [HKPS, Ku1], or [Ku2].

For $X_n = \sum_{n=0}^{\infty} I_n(T_n)$ and $Y = \sum_{n=0}^{\infty} I_n(U_n)$ in $(\mathcal{D}^*)$ there is the natural product $\hat{\otimes}$ given by

$$X \hat{\otimes} Y = \sum_{n=0}^{\infty} \left(\sum_{r=0}^{n} I_n(T_r \hat{\otimes} U_{n-r}) \right).$$

Here $T_n \hat{\otimes} U_{n-r}$ is the symmetric tensor product of the distributions T_r and U_{n-r}.

For X: $\mathbb{R}^N \to (\mathcal{D}^*)$ the derivative $\partial X / \partial t_i$ and the integral $\int X(t)\, dt$ are elements of $(\mathcal{D}^*)$ which satisfy:

$$\left\langle \frac{\partial X}{\partial t_i}, \Phi \right\rangle = \frac{\partial}{\partial t_i} \langle X(t), \Phi \rangle, \tag{1}$$

$$\left\langle \int X(t)\,\mathrm{d}t, \Phi \right\rangle = \int \langle X(t), \Phi \rangle\,\mathrm{d}t \tag{2}$$

for all $\Phi \in (\mathcal{D})$. It is in the sense of (1) that the conclusion of Theorem 1.1 holds.

Because of the structure of the test function space, there is the product rule:

$$\frac{\partial}{\partial t_i}(\zeta \mathbin{\widehat{\otimes}} \theta) = \frac{\partial \zeta}{\partial t_i} \mathbin{\widehat{\otimes}} \theta + \zeta \mathbin{\widehat{\otimes}} \frac{\partial \theta}{\partial t_i}. \tag{3}$$

Also, $\theta = \sum_{n=0}^{\infty} \theta_n$ is differentiable (integrable) if and only if each component is differentiable (integrable) in which case

$$\frac{\partial \theta}{\partial t_i} = \sum_{n=0}^{\infty} \frac{\partial \theta_n}{\partial t_i}, \tag{4}$$

$$\int \theta(t)\,\mathrm{d}t = \sum_{n=0}^{\infty} \int \theta_n(t)\,\mathrm{d}t. \tag{5}$$

As an example illustrating the ideas in the theorem we offer the following.

EXAMPLE 2.1. For $F \in C_b^2(\mathbb{R})$ and $W(t) = I(1_{[0,t)})$, $t = (t_1, \ldots, t_N) \in \mathbb{R}^N$, the N-parameter Wiener process,

$$\frac{\partial}{\partial t_i} F(W(t)) = \frac{1}{2}\frac{\partial}{\partial t_i}(t_1 t_2 \cdots t_N) F''(W(t)) + F'(W(t)) \mathbin{\widehat{\otimes}} \frac{\partial W}{\partial t_i}(t).$$

Thus, in particular, for $N = 2$ and $F \in C_b^4(\mathbb{R})$,

$$\begin{aligned}
\frac{\partial^2}{\partial t_1 \partial t_2} F(W(t)) \;=\; & \frac{1}{2} F''(W(t)) + \frac{1}{4} t_1 t_2 F^{(4)}(W(t)) + \\
& + \frac{1}{2} F'''(W(t)) \mathbin{\widehat{\otimes}} \left(t_1 \frac{\partial W}{\partial t_1}(t) + t_2 \frac{\partial W}{\partial t_2}(t) \right) + \\
& + F''(W(t)) \mathbin{\widehat{\otimes}} \frac{\partial W}{\partial t_1}(t) \mathbin{\widehat{\otimes}} \frac{\partial W}{\partial t_2}(t) + \\
& + F'(W(t)) \mathbin{\widehat{\otimes}} \frac{\partial^2 W}{\partial t_1 \partial t_2}(t).
\end{aligned}$$

It is interesting to consider the integrals that appear in the integral version of the above formula:

$$\begin{aligned}
\Delta_{[0,1]^2} F(W(t)) \;=\; & \frac{1}{2} \int_{[0,1]^2} F''(W(t))\,\mathrm{d}t + \frac{1}{4} \int_{[0,1]^2} t_1 t_2 F^{(4)}(W(t))\,\mathrm{d}t + \\
& + \frac{1}{2} \int_{[0,1]^2} F'''(W(t)) \mathbin{\widehat{\otimes}} \left(t_1 \frac{\partial W}{\partial t_1}(t) + t_2 \frac{\partial W}{\partial t_2}(t) \right) \mathrm{d}t + \\
& + \int_{[0,1]^2} F''(W(t)) \mathbin{\widehat{\otimes}} \frac{\partial W}{\partial t_1}(t) \mathbin{\widehat{\otimes}} \frac{\partial W}{\partial t_2}(t)\,\mathrm{d}t + \\
& + \int_{[0,1]^2} F'(W(t)) \mathbin{\widehat{\otimes}} \frac{\partial^2 W}{\partial t_1 \partial t_2}(t)\,\mathrm{d}t.
\end{aligned}$$

In general, these integrals are defined as elements of $(\mathcal{D}^*)$ but for suitable functions F, they reduce to well-known stochastic integrals. In this example,

$$\int_{[0,1]^2} F'(W(t)) \widehat{\otimes} \frac{\partial^2 W}{\partial t_1 \partial t_2}(t)\,\mathrm{d}t$$

is the two-parameter Skorokhod integral of $F'(W(t))$: $\int_{[0,1]^2} F'(W(t))\mathrm{d}W(t)$. Furthermore, the integral

$$\int_{[0,1]^2} F''(W(t)) \widehat{\otimes} \frac{\partial W}{\partial t_1}(t) \widehat{\otimes} \frac{\partial W}{\partial t_2}(t)\,\mathrm{d}t$$

reduces to a Wong–Zakai integral, see [WZ]. This example first appeared in [BR1] and further discussion can be found there.

In [BR3] generalized stochastic integrals of processes $X\colon \mathbb{R}^N \to (\mathcal{D}^*)$ are defined as weak integrals $\int X(t) \widehat{\otimes} D(t)$ with respect to a density $D(t)$. The approach is to define integrals first on an elementary class, analogous to Stieltjes integrals, but using Wick multiplication. Thus suppose $Y\colon \mathrm{T} \to (\mathcal{D}^*)$, where $\mathrm{T} \equiv [0,1]^N$, is a given process and $X\colon \mathrm{T} \to (\mathcal{D}^*)$ is a finitely-valued, weakly measurable process of the form

$$X(t) = \sum_{j=1}^{l} v^j 1_{B_j}(t),$$

where $v^j \in (\mathcal{D}^*)$ and $B_j = [b^j, c^j] \equiv \{x = (x_1, \ldots, x_N) \in \mathrm{T} \mid b_i^j \leqslant x_i \leqslant c_i^j\} \subseteq \mathrm{T}$, is a rectangle. We assume that $B_i \cap B_j = \emptyset$ for $i \neq j$. Then the integral of X is defined by

$$\mathcal{I}^Y(X) = \sum_{j=1}^{l} v^j \widehat{\otimes} \triangle_{B_j} Y,$$

where $\triangle_{B_j} Y \in (\mathcal{D}^*)$ is the increment of Y over the rectangle B_j. Here, $\triangle_{[b,c]} Y \equiv Y(t)|_{b_1}^{c_1} \cdots |_{b_N}^{c_N} = [Y(c_1, t_2, \ldots, t_N) - Y(b_1, t_2, \ldots, t_N)]|_{b_2}^{c_2} \cdots |_{b_N}^{c_N} = etc.$ If Y is suitably differentible then one has

$$\triangle_{[c,d]} Y = \int_{[c,d]} \frac{\partial^N Y}{\partial t_1 \cdots t_N}(t)\,\mathrm{d}t.$$

Thus,

$$\begin{aligned}
\mathcal{I}^Y(X) &= \sum_{j=1}^{l} v^j \widehat{\otimes} \int 1_{B_j}(t) \frac{\partial^N Y}{\partial t_1 \cdots \partial t_N}(t)\,\mathrm{d}t \\
&= \sum_{j=1}^{l} \int v^j \widehat{\otimes} \left(1_{B_j}(t) \frac{\partial^N Y}{\partial t_1 \cdots \partial t_N}(t) \right) \mathrm{d}t \\
&= \int X(t) \widehat{\otimes} \frac{\partial^N Y}{\partial t_1 \cdots \partial t_N}(t)\,\mathrm{d}t.
\end{aligned}$$

This gives a stochastic integral

$$\mathcal{I}^Y\colon L^1\left(\mathrm{T}, (\mathcal{D}^*), \frac{\partial^N Y}{\partial t_1 \cdots \partial t_N}(t)\,\mathrm{d}t\right) \to (\mathcal{D}^*)$$

with domain consisting of those processes $X\colon \mathrm{T} \to (\mathcal{D}^*)$ for which $t \to X(t) \,\widehat{\otimes}\, \frac{\partial^N Y}{\partial t_1 \cdots \partial t_N}(t)$ is weakly integrable.

3. The Extended Itô Formula and Examples

As mentioned in the introduction, our first goal is to extend Theorem 1.1 from $F(\phi)$ where $\phi(t) = (\phi^1(t), ..., \phi^p(t))$ with $\phi^a(t) \in K_1$ to $F(\phi)$ with $\phi(t) = (\phi^1(t), ..., \phi^p(t))$ and $\phi^a(t) \in K_{m_a}$ for various values of m_a. We then will look at $F(X)$ where X is a stochastic integral. We would like to have a general formula for any $\phi^a(t) \in (L^2)$ but so far we have been unable to obtain such a formula.

Note. We will use the convention of implied summation on repeated indices in the following theorem and its two corollaries.

THEOREM 3.1. *Let $\phi^a = \phi_1^a \,\widehat{\otimes} \cdots \widehat{\otimes}\, \phi_{m_a}^a$ where $\phi_j^a\colon \mathbb{R}^N \to K_1$ is differentiable for $j = 1, \ldots, m_a$. Suppose the covariance matrix $A(t) = \{A_{rs}^{ab}(t)\} = \{\langle \phi_r^a(t), \phi_s^b(t)\rangle\}$ is differentiable and invertible at each $t \in \mathbb{R}^N$. Define $\widetilde{\phi_r^a} \equiv \widehat{\bigotimes}_{i \neq r} \phi_i^a$ and let $\phi = (\phi^1, \ldots, \phi^p)$. Assume that $F \in C_b^2(\mathbb{R}^p)$. Then*

$$\frac{\partial}{\partial t_j} F(\phi) = \frac{\partial \phi_s^a}{\partial t_j} \,\widehat{\otimes} \left(\widetilde{\phi_s^a} \frac{\partial F}{\partial x_a}(\phi)\right) + \frac{1}{2} \frac{\partial}{\partial t_j} \langle \phi_r^a, \phi_s^b \rangle \widetilde{\phi_r^a} \widetilde{\phi_s^b} \frac{\partial^2 F}{\partial x_a \partial x_b}(\phi).$$

Proof. For $\phi^a = \phi_1^a \,\widehat{\otimes} \cdots \widehat{\otimes}\, \phi_{m_a}^a$ let $\phi_j^a(t) = I(f_j^a(t))$ where $f_j^a(t) \in L^2(\mathbb{R}^N)$. Let $S \subset \{1, \ldots, m_a\}$ with $|S| = 2k$. Define $\langle f_1^a, \ldots, f_{m_a}^a \rangle_S \equiv \sum \langle f_{i_1}^a, f_{j_1}^a \rangle \cdots \langle f_{i_k}^a, f_{j_k}^a \rangle$, the sum being over all pairs $\{i_1, j_1\}, \ldots, \{i_k, j_k\}$ on S and $\langle \cdot, \cdot \rangle$ indicating the $L^2(\mathbb{R}^N)$ inner product. Then there is the formula, see [BR2, Si],

$$\phi^a = \sum_{k=0}^{[\frac{m_a}{2}]} (-1)^k \sum_{|S|=2k} \langle f_1^a, \ldots, f_{m_a}^a \rangle_S \prod_{j \notin S} \phi_j^a. \tag{6}$$

We see then that $\phi^a = P^a(\phi_1^a, \ldots, \phi_{m_a}^a)$ where $P^a(x_1, \ldots, x_{m_a})$ is a polynomial. Note that by the product rule (3)

$$\frac{\partial \phi^a}{\partial t_j} = \frac{\partial P^a}{\partial x_k}(\phi_1^a, \ldots, \phi_{m_a}^a) \,\widehat{\otimes}\, \frac{\partial \phi_k^a}{\partial t_j}. \tag{7}$$

Now, by Theorem 1.1,

$$\frac{\partial \phi^a}{\partial t_j} = \frac{\partial}{\partial t_j} P^a(\phi_1^a, \ldots, \phi_{m_a}) \tag{8}$$

$$= \frac{\partial P^a}{\partial t_j}(\phi_1^a, \dots, \phi_{m_a}^a) + \frac{\partial P}{\partial x_k}(\phi_1^a, \dots, \phi_{m_a}^a) \widehat{\otimes} \frac{\partial \phi_k^a}{\partial t_j} + \tag{9}$$

$$+ \frac{\partial}{\partial t_j}\langle \phi_k^a, \phi_m^b \rangle \frac{\partial^2 P}{\partial x_k \partial x_m}(\phi_1, \dots, \phi_{m_a}^a). \tag{10}$$

From the above displayed Equations (7)–(10), we have that

$$\frac{\partial P^a}{\partial t_j}(\phi_1^a, \dots, \phi_{m_a}^a) + \frac{1}{2}\frac{\partial}{\partial t_j}\langle \phi_k^a, \phi_m^a \rangle \frac{\partial^2 P}{\partial x_k \partial x_m}(\phi_1^a, \dots, \phi_{m_a}^a) = 0. \tag{11}$$

For $\phi = (\phi^1, \dots, \phi^p)$,where $\phi^a = \phi_1^a \widehat{\otimes} \cdots \widehat{\otimes} \phi_{m_a}^a$, $1 \leqslant a \leqslant p$, we have polynomials $P^1, \dots, P^p$ such that

$$\phi = (P^1(\phi_1^1, \dots, \phi_{m_1}^1), \dots, P^p(\phi_1^p, \dots, \phi_{m_p}^p)).$$

Hence, we can apply Theorem 1.1 to $F(\phi)$ to get

$$\begin{aligned}
\frac{\partial}{\partial t_j}F(\phi) &= \frac{\partial F}{\partial x^a}(\phi)\frac{\partial P^a}{\partial x_s^a}(\phi_1^a, \dots, \phi_{m_a}^a) \widehat{\otimes} \frac{\partial \phi_s^a}{\partial t_j} + \\
&\quad + \frac{\partial F}{\partial t_j}(\phi) + \frac{1}{2}\frac{\partial}{\partial t_j}\langle \phi_s^a, \phi_l^b \rangle \frac{\partial^2 F}{\partial x_s^a \partial x_l^b}(\phi) \\
&= \frac{\partial F}{\partial x^a}(\phi)\frac{\partial P^a}{\partial x_s^a}(\phi_1^a, \dots, \phi_{m_a}^a) \widehat{\otimes} \frac{\partial \phi_s^a}{\partial t_j} + \frac{\partial F}{\partial x^a}(\phi)\frac{\partial P^a}{\partial t_j}(\phi_1^a, \dots, \phi_{m_a}^a) + \\
&\quad + \frac{1}{2}\frac{\partial}{\partial t_j}\langle \phi_s^a, \phi_l^a \rangle \frac{\partial^2 P^a}{\partial x_s^a \partial x_l^a}(\phi_1^a, \dots, \phi_{m_a}^a)\frac{\partial F}{\partial x^a}(\phi) + \\
&\quad + \frac{1}{2}\frac{\partial}{\partial t_j}\langle \phi_s^a, \phi_l^b \rangle \frac{\partial P^a}{\partial x_s^a}(\phi_1^a, \dots, \phi_{m_a}^a)\frac{\partial P^b}{\partial x_l^b}(\phi_1^b, \dots, \phi_{m_b}^b)\frac{\partial^2 F}{\partial x^a \partial x^b}(\phi).
\end{aligned}$$

Upon incorporating (11) and observing that $\frac{\partial P^a}{\partial x_s^a}(\phi_1^a, \dots, \phi_{m_a}^a) = \widetilde{\phi_s^a}$, we get the conclusion of the theorem. □

COROLLARY 3.1. *Let* $\phi^a = \phi_0^a + \cdots + \phi_{m_a}^a$ *where* $\phi_k^a = \phi_{k1}^a \widehat{\otimes} \cdots \widehat{\otimes} \phi_{km_k}^a$, *with* $\phi_{kl}^i(t) \in K_1$, *and* $\widetilde{\phi_{kl}^a} \equiv \widehat{\bigotimes}_{j \neq l}\phi_{kj}^a$. *Set* $X = (\phi^1, \dots, \phi^p)$ *then*

$$\frac{\partial}{\partial t_i}F(X) = \frac{\partial \phi_{sn}^a}{\partial t_j} \widehat{\otimes} \left(\widetilde{\phi_{sn}^a} \frac{\partial F}{\partial x_a}(X) \right) + \frac{1}{2}\frac{\partial}{\partial t_j}\langle \phi_{rn}^a, \phi_{sk}^b \rangle \widetilde{\phi_{rn}^a}\widetilde{\phi_{sk}^b}\frac{\partial^2 F}{\partial x_a \partial x_b}(X).$$

COROLLARY 3.2. *Suppose that* $F \in C_b^2(\mathbb{R}^p)$*and* $G \in C_b^2(\mathbb{R}^q)$ *with* $X = (\phi^1, \dots, \phi^p)$, $Y = (\psi^1, \dots, \psi^q)$ *as in Corollary* 3.1. *Then*

$$\begin{aligned}
\frac{\partial}{\partial t_j}[F(X)G(Y)] &= \frac{\partial \phi_{sn}^a}{\partial t_i} \widehat{\otimes} \left(\widetilde{\phi_{sn}^a} \frac{\partial F}{\partial x_a}(X)G(Y) \right) + \\
&\quad + \frac{\partial \psi_{sn}^a}{\partial t_j} \widehat{\otimes} \left(\widetilde{\psi_{sn}^a} F(X) \frac{\partial G}{\partial y_a}(Y) \right) +
\end{aligned}$$

$$+\frac{1}{2}\frac{\partial}{\partial t_j}\langle\phi_{sn}^a,\phi_{rk}^b\rangle\widetilde{\phi_{sn}^a}\widetilde{\phi_{rk}}\frac{\partial^2 F}{\partial x_a\partial x_b}(X)G(Y)+$$

$$+\frac{1}{2}\frac{\partial}{\partial t_j}\langle\psi_{sn}^a,\psi_{rk}^b\rangle\widetilde{\psi_{sn}^a}\widetilde{\psi_{rk}^b}F(X)\frac{\partial^2 G}{\partial y_a\partial y_b}(Y)+$$

$$+\frac{\partial}{\partial t_j}\langle\phi_{sn}^a,\psi_{rk}^b\rangle\widetilde{\phi_{sn}^a}\widetilde{\psi_{rk}^b}\frac{\partial F}{\partial x_a}(X)\frac{\partial G}{\partial y_b}(Y).$$

We can now apply the above results to the case where X is a particular type of stochastic integral. Let $T=[0,1]^N$ and let $t\in T$. Suppose that $G\colon T\to K_n$ is a simple process of the form

$$G(s)=\sum_{a=1}^{p}v^a 1_{[b^a,c^a)}(s),$$

where $\{[b^a,c^a)\}$ is a partition of $[0,1]^N$ into congruent rectangles and $v^a=v_1^a\,\widehat{\otimes}\cdots\widehat{\otimes}\,v_n^a$ with $v_j^a=I(f_j^a)$ for $f_j^a\in L^2(\mathbb{R}^N)$. Let $X(t)=\int_{[0,t)}G(s)\,\mathrm{d}W(s)$ be the Skorokhod integral of G. In our setup we have the following:

$$\begin{aligned}X(t)&=\int_{[0,t)}G(s)\,\widehat{\otimes}\,\frac{\partial^N W}{\partial t_1\cdots\partial t_N}(s)\,\mathrm{d}s\\&=\sum_{a=1}^{p}v^a\,\widehat{\otimes}\int_{[0,t)}1_{[b^a,c^a)}(s)\frac{\partial^N W}{\partial t_1\cdots\partial t_N}(s)\,\mathrm{d}s.\end{aligned}$$

EXAMPLE 3.1. Take $N=1$ so that $B(t)=W(t)$ is Brownian motion. Assume that F is $C_b^2(\mathbb{R})$. Then in $(\mathcal{D}^*)$

$$\begin{aligned}X(t)&=\int_0^t G(s)\,\widehat{\otimes}\,\dot{B}(s)\,\mathrm{d}s\\&=\sum_{a=1}^{l-1}v^a\,\widehat{\otimes}\,\Delta_{[b^a,c^a)}B+v^l\,\widehat{\otimes}\,(B(t)-B(b^l)),\end{aligned}$$

where $t\in[b^l,c^l)$. To simplify notation, let $\Delta_a B\equiv\Delta_{[b^a,c^a)}B$ and recall that $v_j^a=I(f_j^a)$ where $f_j^a\in L^2(\mathbb{R})$. Applying Theorem 3.1 gives

$$\begin{aligned}&\frac{\mathrm{d}}{\mathrm{d}t}F(X(t))\\&\quad=F'(X(t))v^l\,\widehat{\otimes}\,\frac{\mathrm{d}}{\mathrm{d}t}(B(t)-B(b^l)+\\&\qquad+\frac{1}{2}F''(X(t))\left[\sum_{a=1}^{l-1}\sum_{j=1}^{n}(\widetilde{v_j^a}\,\widehat{\otimes}\,\Delta_a B)v^l\frac{\mathrm{d}}{\mathrm{d}t}\langle v_j^a,B(t)-B(b^l)\rangle+\right.\end{aligned}$$

$$+ (\widetilde{v_j^l} \widehat{\otimes} (B(t) - B(b^l)))v^l \frac{\mathrm{d}}{\mathrm{d}t}\langle v_j^l, B(t) - B(b^l)\rangle \Bigg] +$$

$$+ \frac{1}{2} F''(X(t))(v^l)^2 \frac{\mathrm{d}}{\mathrm{d}t}\langle B(t) - B(b^l), B(t) - B(b^l)\rangle$$

$$= F'(X(t))v^l \widehat{\otimes} \dot{B}(t) +$$

$$+ \frac{1}{2} F'' X(t) \Bigg[\sum_{a=1}^{l-1} \sum_{j=1}^{n} (\widetilde{v_j^a} \widehat{\otimes} \Delta_a B) v^l \left(\frac{\mathrm{d}}{\mathrm{d}t} \int_{b^l}^{t} f_j^a(s)\,\mathrm{d}s \right) +$$

$$+ \widetilde{v_j^a} \widehat{\otimes} (B(t) - B(b^l) v^l \frac{\mathrm{d}}{\mathrm{d}t} \int_0^t f_l^a(s)\,\mathrm{d}s \Bigg] + \frac{1}{2} F''(X(t))(v^l)^2.$$

Notice that the sum: $\sum_{a=1}^{l-1} \sum_{j=1}^{n} (\widetilde{v_j^a} \widehat{\otimes} \Delta_a B) f_j^a(t) + \sum_{j=1}^{n} (\widetilde{v_j^l} \widehat{\otimes} \Delta_l B) f_j^l(t)$ can be written as $\int_0^t D_t(G(s)) \widehat{\otimes} \dot{B}(s)\,\mathrm{d}s$ where D_t is the derivative operator, see [Nu], which, for $F = \sum_{m=0}^{\infty} I_m(f_m)$ in $L^2(\mathcal{S}^*)$, with $f_m \in \hat{L}^2(\mathbb{R}^m)$ and $\sum_{m=1}^{\infty} mm! \parallel f_m \parallel_{L^2(\mathbb{R}^m)} < \infty$, is defined by $D_t F = \sum_{m=1}^{\infty} m I_{m-1}(f_m(t))$. Hence, when we look at the integral version of this formula, we have

$$\begin{aligned} F(X(1)) - F(X(0)) &= \int_0^1 F'(X(t))G(t) \widehat{\otimes} \dot{B}(t)\,\mathrm{d}t + \\ &\quad + \frac{1}{2} \int_0^1 F''(X(t))(G(t))^2\,\mathrm{d}t + \\ &\quad + \frac{1}{2} \int_0^1 F''(X(t)) \left(\int_0^t D_t G(s) \widehat{\otimes} \dot{B}(s)\,\mathrm{d}s \right) G(t)\,\mathrm{d}t. \end{aligned}$$

If G is nonanticipating the integral involving $D_t G$ is zero and one has the usual Itô formula. If G is anticipating then the above formula is a simple case of the formula in Theorem 6.1 of [NP].

In the next example we will assume the usual partial ordering on $\mathbb{R}^2$: $s \leqslant t$ if $s_1 \leqslant t_1$ and $s_2 \leqslant t_2$.

EXAMPLE 3.2. Here take $N = 2$ and $F \in C_b^4(\mathbb{R})$. Assume that G is previsible: v^a is measurable with respect to $\mathcal{F}_{b^a}$ where $\mathcal{F}_{b^a} \equiv \sigma\{W_s | s \leqslant b^a\}$. Let $t \in [b^l, c^l)$. Then

$$\begin{aligned} X(t) &= \int_0^t G(s) \widehat{\otimes} \frac{\partial^2 W}{\partial t_1 \partial t_2}(s)\,\mathrm{d}s = \sum_{a<l} v^a \widehat{\otimes} \Delta_{[b^a, c^a)} W + v^l \widehat{\otimes} \Delta_{[b^l, t)} W + \\ &\quad + \sum_{\substack{a \\ b^a = (b_1^a, b_2^l), b_1^a < b_1^l}} v^a \widehat{\otimes} \Delta_{[b^a, (c_1^a, t_2))} W + \sum_{\substack{a \\ b^a = (b_1^l, b_2^a), b_2^a < b_2^l}} v^a \widehat{\otimes} \Delta_{[b^a, (t_1, c_2^a))} W. \end{aligned}$$

Thus by Theorem 3.1

$$
\begin{aligned}
&\frac{\partial}{\partial t_1} F(X(t)) \\
&= F'(X(t))v^l \,\widehat{\otimes}\, \frac{\partial}{\partial t_1}\Delta_{[b^l,t)}W + \\
&+ \sum_{b^a=(b_1^l,b_2^a),\, b_2^a<b_2^l}^{a} F'(X(t))v^a \,\widehat{\otimes}\, \frac{\partial}{\partial t_1}\Delta_{[b^a,(t_1,c_2^a))}W + \\
&+ \frac{1}{2}\sum_{j=1}^{n} \sum_{b^r=(b_1^l,b_2^r),\, b_1^a<b_1^l}^{a,r} F''(X(t))(\widetilde{v_j^a} \,\widehat{\otimes}\, \Delta_{[b^a,c^a)}W)v^r \int_{b_2^r}^{c_2^r} f_j^a(t_1,s_2)\,\mathrm{d}s_2 + \\
&+ \frac{1}{2}\sum_{b^a=(b_1^l,b_2^a),\, b_2^a<b_2^l}^{a} F''(X(t))(v^a)^2(c_2^a-b_2^a) + \frac{1}{2}F''(X(t))(v^l)^2(t_2-b_2^l).
\end{aligned}
$$

Since G is previsible, $\int_{b_2^r}^{c_2^r} f_j^a(t_1,s_2)\,\mathrm{d}s_2 = 0$ in the double sum above. To simplify the notation let us relable $c^l \equiv t$. Hence,

$$
\begin{aligned}
\frac{\partial}{\partial t_1} F(X(t)) \;=\; & \sum_{b^a=(b_1^l,b_2^a),\, b_2^a\leqslant b_2^l}^{a} F'(X(t))v^a \,\widehat{\otimes}\, \frac{\partial}{\partial t_1}\Delta_{[b^a,(t_1,c_2^a))}W + \\
&+ \frac{1}{2}\sum_{b^a=(b_1^l,b_2^a),\, b_2^a\leqslant b_2^l}^{a} F''(X(t))(v^a)^2(c_2^a-b_2^a).
\end{aligned}
$$

Now, applying Corollary 3.2

$$
\begin{aligned}
&\frac{\partial}{\partial t_1\partial t_2} F(X(t)) \\
&= F'(X(t))v^l \,\widehat{\otimes}\, \frac{\partial}{\partial t_1\partial t_2}\Delta_{[b^l,t)}W + \\
&+ \sum_{b^a=(b_1^l,b_2^a),\, b_2^a\leqslant b_2^l}^{a} \frac{\partial}{\partial t_2}[F'(X(t))v^a] \,\widehat{\otimes}\, \frac{\partial}{\partial t_1}\Delta_{[b^a,(t_1,c_2^a))}W + \\
&+ \frac{1}{2}\sum_{b^a=(b_1^l,b_2^a),\, b_2^a\leqslant b_2^l}^{a} \frac{\partial}{\partial t_2}[F''(X(t))(v^a)^2(c_2^a-b_2^a)] \\
&= F'(X(t))v^l \,\widehat{\otimes}\, \frac{\partial^2}{\partial t_1\partial t_2}W(t) + \\
&+ \sum_{b^a=(b_1^l,b_2^a),\, b_2^a\leqslant b_2^l}^{a} \Bigg(\sum_{b^r=(b_1^r,b_2^l),\, b_1^r\leqslant b_1^l}^{r} \Bigg[\frac{\partial}{\partial t_2}\Delta_{[b^r,(c_1^r,t_2))}W \,\widehat{\otimes}\, v^r v^a F''(X(t)) +
\end{aligned}
$$

$$+\frac{1}{2}(c_1^r - b_1^r)(v^r)^2 v^a F'''(X(t))\Big]\Big) \mathbin{\widehat{\otimes}} \frac{\partial}{\partial t_1}\triangle_{[b^a,(t_1,c_2^a))}W +$$

$$+\frac{1}{2}\sum_{\substack{a\\ b^a=(b_1^l,b_2^a),b_2^a\leqslant b_2^l}}\Bigg(\sum_{\substack{r\\ b^r=(b_1^r,b_2^l),b_1^r\leqslant b_1^l}}\Bigg[\frac{\partial}{\partial t_2}\triangle_{[b^r,(c_1^r,t_2))}W \mathbin{\widehat{\otimes}}$$

$$\mathbin{\widehat{\otimes}} v^r(v^a)^2 F'''(X(t))(c_2^a - b_2^a) +$$

$$+\frac{1}{2}(c_1^r - b_1^r)(v^r)^2(v^a)^2 F^{(4)}(X(t))(c_2^a - b_2^a)\Big]\Big) + \frac{1}{2}F''(X(t))(v^l)^2.$$

In the following formula we will assume that the conditions on the indices a and r in the double sums are those given in the above formula.

$$\begin{aligned}
&\frac{\partial^2}{\partial t_1\partial t_2}F(X(t))\\
&\quad = F'(X(t))v^l \mathbin{\widehat{\otimes}} \frac{\partial^2 W}{\partial t_1\partial t_2}(t) +\\
&\qquad + \sum_a\sum_r v^r v^a F''(X(t)) \mathbin{\widehat{\otimes}} \frac{\partial}{\partial t_1}\triangle_{[b^a,(t_1,c_2^a))}W \mathbin{\widehat{\otimes}} \frac{\partial}{\partial t_2}\triangle_{[b^r,(c_1^r,t_2))}W +\\
&\qquad + \frac{1}{2}\sum_a\sum_r (c_1^r - b_1^r)(v^r)^2 v^a F'''(X(t)) \mathbin{\widehat{\otimes}} \frac{\partial}{\partial t_1}\triangle_{[b^a,(t_1,c_2^a))}W +\\
&\qquad + \frac{1}{2}\sum_a\sum_r v^r(v^a)^2(c_2^a - b_2^a)F'''(X(t)) \mathbin{\widehat{\otimes}} \frac{\partial}{\partial t_2}\triangle_{[b_1^r,(c_1^r,t_2))}W +\\
&\qquad + \frac{1}{4}\sum_a\sum_r (c_1^r - b_1^r)(v^r)^2(v^a)^2(c_2^a - b_2^a)F^{(4)}(X(t)) +\\
&\qquad + \frac{1}{2}F''(X(t))(v^l)^2.
\end{aligned}$$

Since G is a simple process, the double sums in this formula are integrals:

$$\begin{aligned}
&\frac{\partial^2}{\partial t_1\partial t_2}F(X(t))\\
&\quad = F'(X(t))v^l \mathbin{\widehat{\otimes}} \frac{\partial^2 W}{\partial t_1\partial t_2}(t) + \frac{1}{2}F''(X(t))(v^l)^2 +\\
&\qquad + \int_0^{t_2}\int_0^{t_1} G(s_1,t_2)G(t_1,s_2)F''(X((s_1,t_2)\vee(t_1,s_2)) \mathbin{\widehat{\otimes}}\\
&\qquad \mathbin{\widehat{\otimes}} \frac{\partial^2 W}{\partial t_1\partial t_2}(t_1,s_2) \mathbin{\widehat{\otimes}} \frac{\partial^2 W}{\partial t_1\partial t_2}(s_1,t_2)\,\mathrm{d}s_1\,\mathrm{d}s_2 +\\
&\qquad + \frac{1}{2}\int_0^{t_1}\int_0^{t_2} G(s_1,t_2)^2 G(t_1,s_2)F'''(X((s_1,t_2)\vee(t_1,s_2)) \mathbin{\widehat{\otimes}}\\
&\qquad \mathbin{\widehat{\otimes}} \frac{\partial^2 W}{\partial t_1\partial t_2}(t_1,s_2)\,\mathrm{d}s_2\,\mathrm{d}s_1 +
\end{aligned}$$

$$+\frac{1}{2}\int_0^{t_2}\int_0^{t_1} G(s_1,t_2)G(t_1,s_2)^2F'''(X((s_1,t_2)\vee(t_1,s_2))\mathbin{\widehat{\otimes}}$$
$$\mathbin{\widehat{\otimes}}\frac{\partial^2 W}{\partial t_1\partial t_2}(s_1,t_2)\,\mathrm{d}s_1\,\mathrm{d}s_2+$$
$$+\frac{1}{4}\int_0^{t_2}\int_0^{t_1} G(s_1,t_2)^2G(t_1,s_2)^2F^{(4)}(X(t))F''(X(t))\,\mathrm{d}s_1\,\mathrm{d}s_2.$$

If we now look at the integral version of this formula we have

$$\triangle_T F(X(t))$$
$$=\int_T F'(X(t))G(t)\mathbin{\widehat{\otimes}}\frac{\partial^2 W}{\partial t_1\partial t_2}(t)\,\mathrm{d}t+\frac{1}{2}\int_T F''(X(t))G(t)^2\,\mathrm{d}t+ \tag{12}$$
$$+\int_{T^2} 1_{[0,t)}(s)G(s_1,t_2)G(t_1,s_2)F''(X((s_1,t_2)\vee(t_1,s_2))\mathbin{\widehat{\otimes}}$$
$$\mathbin{\widehat{\otimes}}\frac{\partial^2 W}{\partial t_1\partial t_2}(t_1,s_2)\mathbin{\widehat{\otimes}}\frac{\partial^2 W}{\partial t_1\partial t_2}(s_1,t_2)\,\mathrm{d}s_1\,\mathrm{d}s_2\,\mathrm{d}t+ \tag{13}$$
$$+\frac{1}{2}\int_{T^2} 1_{[0,t)}(s)G(s_1,t_2)^2G(t_1,s_2)F'''(X((s_1,t_2)\vee(t_1,s_2))\mathbin{\widehat{\otimes}}$$
$$\mathbin{\widehat{\otimes}}\frac{\partial^2 W}{\partial t_1\partial t_2}(t_1,s_2)\,\mathrm{d}s_2\,\mathrm{d}s_1\,\mathrm{d}t+ \tag{14}$$
$$+\frac{1}{2}\int_{T^2} 1_{[0,t)}(s)G(s_1,t_2)G(t_1,s_2)^2F'''(X((s_1,t_2)\vee(t_1,s_2))\mathbin{\widehat{\otimes}}$$
$$\mathbin{\widehat{\otimes}}\frac{\partial^2 W}{\partial t_1\partial t_2}(s_1,t_2)\,\mathrm{d}s_1\,\mathrm{d}s_2\,\mathrm{d}t+ \tag{15}$$
$$+\frac{1}{4}\int_{T^2} 1_{[0,t)}(s)G(s_1,t_2)^2G(t_1,s_2)^2F^{(4)}(X((s_1,t_2)\vee$$
$$(t_1,s_2))\,\mathrm{d}s_1\,\mathrm{d}s_2\,\mathrm{d}t. \tag{16}$$

The first integral on the right-hand side is the two-parameter Skorokhod integral: $\int_T G(t)F'(X(t))\,\mathrm{d}W(t)$. Some of the double integrals in the above formula can be reformulated so that they, too, can be recognized as ordinary stochastic integrals. To do this we need to define the set D:

$$D=\{(s,t)\in[0,1]^2 \mid (s_1<t_1,s_2>t_2)\text{ or }(t_1<s_1,t_2>s_2)\}.$$

This set is just the set of pairs of unordered points in $[0,1]^2$. In the following calculations, remember that these integrals are elements of $(\mathcal{D}^*)$ and the reformulations require looking at how they act on the elements of $\mathcal{D}$. For more details see [BR3].

First look at (13). If we change the order of integration: $\mathrm{d}s_1\,\mathrm{d}s_2\,\mathrm{d}t_1\,\mathrm{d}t_2\to\mathrm{d}s_1\,\mathrm{d}t_2\,\mathrm{d}t_1\,\mathrm{d}s_2$, and then relable: $t_2\to s_2$, $s_2\to t_2$, we get

$$\int_{T^2} 1_{[0,t_1)}(s_1)1_{[0,s_2)}(t_2)G(s)G(t)F''(X(s\vee t))\mathbin{\widehat{\otimes}}\frac{\partial^2 W}{\partial t_1\partial t_2}(t)\mathbin{\widehat{\otimes}}\frac{\partial^2 W}{\partial t_1\partial t_2}(s)\,\mathrm{d}t\,\mathrm{d}s.$$

This is easily seen to be

$$\frac{1}{2}\int_{T^2} 1_D(s,t)G(s)G(t)F''(X(s\vee t))\,\widehat{\otimes}\,\frac{\partial^2 W}{\partial t_1\partial t_2}(t)\,\widehat{\otimes}\,\frac{\partial^2 W}{\partial t_1\partial t_2}(s)\,\mathrm{d}t\,\mathrm{d}s.$$

We call this a multiple Skorokhod integral and in this case, as mentioned in Example 2.1, it is a generalization of a Wong–Zakai integral. In fact, here it is just

$$\frac{1}{2}\int_{T^2} 1_D(s,t)G(s)G(t)F''(X(s\vee t))\,\mathrm{d}W(t)\,\mathrm{d}W(s).$$

Making similar adjustments, the sum (14) + (15) can be reformulated as the following element of $(\mathcal{D}^*)$:

$$\begin{aligned}&\frac{1}{2}\int_{T^2} 1_{[0,t_1)}(s_1)1_{[0,s_2)}(t_2)G(s_1,s_2)^2G(t_1,t_2)F'''(X(s\vee t))\,\widehat{\otimes}\,\frac{\partial^2 W}{\partial t_1\partial t_2}(t)\,\mathrm{d}t\,\mathrm{d}s\,+\\&\quad+\frac{1}{2}\int_{T^2} 1_{[0,t_1)}(s_1)1_{[0,s_2)}(t_2)G(s_1,s_2)G(t_1,t_2)^2F'''(X(s\vee t))\,\widehat{\otimes}\\&\quad\quad\widehat{\otimes}\,\frac{\partial^2 W}{\partial t_1\partial t_2}(s)\,\mathrm{d}s\,\mathrm{d}t.\end{aligned}$$

Reversing the order in the second integral and then combining we have

$$\frac{1}{2}\int_{T^2} 1_D(s,t)G(s)G(t)^2F'''(X(s\vee t))\,\widehat{\otimes}\,\frac{\partial^2 W}{\partial t_1\partial t_2}(t)\,\mathrm{d}t\,\mathrm{d}s.$$

We call this a Skorokhod–Lebesgue integral. In this case it is the ordinary mixed integral:

$$\frac{1}{2}\int_{T^2} 1_D(s,t)G(s)G(t)^2F'''(X(s\vee t))\,\mathrm{d}W(t)\,\mathrm{d}s.$$

With the relabling $t_2\to s_2$, $s_2\to t_2$ the integral (16) becomes

$$\frac{1}{4}\int_{T^2} 1_{[0,s_2)}(t_2)1_{[0,t_1)}(s_1)G(s)^2G(t)^2F^{(4)}(X(s\vee t))\,\mathrm{d}s\,\mathrm{d}t.$$

Thus, in more familiar notation, the above formula becomes

$$\begin{aligned}\Delta_T F(X) \;=\;& \int_T F'(X(t))G(t)\,\mathrm{d}W(t)+\frac{1}{2}\int_{T^2} G(t)^2F''(X(t))\,\mathrm{d}t\,+\\&+\frac{1}{2}\int_{T^2} 1_D(s,t)G(s)G(t)F''(X(s\vee t))\,\mathrm{d}W(t)\,\mathrm{d}W(s)\,+\\&+\frac{1}{2}\int_{T^2} 1_D(s,t)G(s_1,s_2)G(t_1,t_2)^2F'''(X(s\vee t)\,\mathrm{d}W(t)\,\mathrm{d}s\,+\\&+\frac{1}{4}\int_{T^2} 1_{[0,s_2)}(t_2)1_{[0,t_1)}(s_1)G(s)^2G(t)^2F^{(4)}(X(s\vee t))\,\mathrm{d}s\,\mathrm{d}t.\end{aligned}$$

Work related to Example 3.2 can be found in [JS], where the methods of Malliavin Calculus are applied.

As illustrated in the above examples, using our distributional approach, one can naturally generate differentiation formulas for certain classes of simple processes. We have found this to be an advantage in the multiparameter setting.

References

[BR1] Betounes, D. and Redfern, M.: A generalized Itô formula for N-dimensional time, In: T. Hida, H.-H. Kuo, J. Potthoff and L. Streit (eds), *White Noise Analysis – Mathematics and Applications*, World Scientific, Singapore, 1990, 337–343.

[BR2] Betounes, D. and Redfern, M.: Wiener distributions and white noise analysis, *Appl. Math. Optim.* **26** (1992), 63–93.

[BR3] Betounes, D. and Redfern, M.: Stochastic integrals for nonprevisible, multiparameter processes, *Appl. Math. Optim.* **28** (1993), 197–223.

[Hi1] Hida, T.: *Analysis of Brownian Functionals*, Carleton Mathematical Lecture Notes 13 (1975).

[Hi2] Hida, T.: Generalized multiple Wiener integrals, *Proc. Japan Acad. Ser. A Math. Sci.* **54** (1978), 55–58.

[HKPS] Hida, T., Kuo, H.-H., Potthoff, J. and Streit, L.: *White Noise, An Infinite Dimensional Calculus*, Kluwer Acad. Publ., Dordrecht, 1993.

[JS] Jolis, M. and Sanz, M.: On generalized multiple stochastic integrals and multiparameter anticipative calculus, In: *Lecture Notes in Math.* 1444, Springer, New York, 1988, 141–182.

[Ku1] Kuo, H.-H.: Lectures on white noise analysis, *Soochow J. Math.* **18** (1992), 229–300.

[Ku2] Kuo, H.-H.: *White Noise Distribution Theory*, CRC Press, Boca Raton, 1996.

[Nu] Nualart, D.: *The Malliavin Calculus and Related Topics*, Springer, New York, 1995.

[NP] Nualart, D. and Pardoux, E.: Stochastic calculus with anticipating integrands, *Probab. Theory Related Fields* **78** (1988), 535–581.

[Si] Simon, B.: *The $P(\phi)_2$ Euclidean (Quantum) Field Theory*, Princeton University Press, 1974.

[WZ] Wong, E. and Zakai, M.: Martingales and stochastic integrals for processes with a multidimensional parameter, *Z. Wahrsch. Verw. Gebiete* **29** (1974), 109–122.

Acta Applicandae Mathematicae **63:** 363–373, 2000.

A Stochastic Process Generated by the Lévy Laplacian

KIMIAKI SAITÔ
Department of Mathematics, Meijo University, Nagoya 468-8502, Japan

(Received: 14 August 1999)

Abstract. In this paper we give a stochastic process generated by the Lévy Laplacian in the white noise analysis with a characterization of the Laplacian.

Mathematics Subject Classification (2000): 60H40.

Key words: Lévy Laplacian, stable processes.

1. Introduction

The Lévy Laplacian was introduced by P. Lévy in [21]. T. Hida [6] introduced this Laplacian to his theory of white noise functionals. It has been studied by many authors (see [1–3, 7, 9, 10, 16–18, 22, 25–28, etc.]).

In a previous paper [29] we extend the Lévy Laplacian to a self-adjoint operator densely defined on a Hilbert space and also gave a relation between the Laplacian and an infinite-dimensional Ornstein–Uhlenbeck process with some operator.

The purpose of this paper is to give a stochastic process generated by the Lévy Laplacian acting on white noise distributions.

The paper is organized as follows. In Section 2 we summarize some basic definitions and results in the white noise analysis. In Section 3 we discuss a self-adjointness of the Lévy Laplacian following our previous paper [29] and give a characterization of the Laplacian in terms of rotation invariant. In last section we give an infinite-dimensional stochastic process generated by the Lévy Laplacian using stable processes.

2. Preliminaries

In this section we assemble some basic notations of white noise analysis following [9, 14, 17, 23]. White noise analysis can be regarded as a distribution theory on an infinite-dimensional space. We take this space to be the space $E^* \equiv \mathcal{S}'(\mathbf{R})$ of tempered distributions with the standard Gaussian measure μ such that

$$\int_{E^*} \exp\{i\langle x, \xi\rangle\}\,\mathrm{d}\mu(x) = \exp\left(-\frac{1}{2}|\xi|_0^2\right), \quad \xi \in E \equiv \mathcal{S}(\mathbf{R}),$$

where $\langle \cdot, \cdot \rangle$ is the canonical bilinear form on $E^* \times E$ and $|\cdot|_0$ is the $L^2(\mathbf{R})$-norm. Let $A = -(d/du)^2 + u^2 + 1$. This is a densely defined self-adjoint operator on $L^2(\mathbf{R})$ and there exists an orthonormal basis $\{e_\nu; \nu \geqslant 0\} \subset E$ for $L^2(\mathbf{R})$ such that $Ae_\nu = 2(\nu+1)e_\nu$. We define the norm $|\cdot|_p$ by $|f|_p = |A^p f|_0$ for $f \in E$ and $p \in \mathbf{R}$, and let E_p be the completion of E with respect to the norm $|\cdot|_p$. Then the dual space E'_p of E_p is the same as E_{-p} (see [12]). We denote the complexifications of $L^2(\mathbf{R})$, E and E_p by $L^2_{\mathbf{C}}(\mathbf{R})$, $E_{\mathbf{C}}$ and $E_{\mathbf{C},p}$, respectively.

The space $(L^2) = L^2(E^*, \mu)$ of complex-valued square-integrable functionals defined on E^* admits the well-known Wiener–Itô decomposition: $(L^2) = \bigoplus_{n=0}^\infty H_n$, where H_n is the space of multiple Wiener integrals of order $n \in \mathbf{N}$ and $H_0 = \mathbf{C}$. Let $L^2_{\mathbf{C}}(\mathbf{R})^{\hat{\otimes} n}$ denote the n-fold symmetric tensor product of $L^2_{\mathbf{C}}(\mathbf{R})$. If $\varphi \in (L^2)$ is represented by $\varphi = \sum_{n=0}^\infty \mathbf{I}_n(f_n)$, $f_n \in L^2_{\mathbf{C}}(\mathbf{R})^{\hat{\otimes} n}$, then the (L^2)-norm $\|\varphi\|_0$ is given by

$$\|\varphi\|_0 = \left(\sum_{n=0}^\infty n! |f_n|_0^2 \right)^{1/2},$$

where $|\cdot|_0$ means also the norm of $L^2_{\mathbf{C}}(\mathbf{R})^{\hat{\otimes} n}$.

For $p \in \mathbf{R}$, let $\|\varphi\|_p = \|\Gamma(A)^p \varphi\|_0$, where $\Gamma(A)$ is the second quantization operator of A. If $p \geqslant 0$, let $(E)_p$ be the domain of $\Gamma(A)^p$. If $p < 0$, let $(E)_p$ be the completion of (L^2) with respect to the norm $\|\cdot\|_p$. Then $(E)_p$, $p \in \mathbf{R}$, is a Hilbert space with the norm $\|\cdot\|_p$. It is easy to see that for $p > 0$, the dual space $(E)_p^*$ of $(E)_p$ is given by $(E)_{-p}$. Moreover, for any $p \in \mathbf{R}$, we have the decomposition $(E)_p = \bigoplus_{n=0}^\infty H_n^{(p)}$, where $H_n^{(p)}$ is the completion of $\{\mathbf{I}_n(f); f \in E_{\mathbf{C}}^{\hat{\otimes} n}\}$ with respect to $\|\cdot\|_p$. Here $E_{\mathbf{C}}^{\hat{\otimes} n}$ is the n-fold symmetric tensor product of $E_{\mathbf{C}}$. We also have $H_n^{(p)} = \{\mathbf{I}_n(f); f \in E_{\mathbf{C},p}^{\hat{\otimes} n}\}$ for any $p \in \mathbf{R}$, where $E_{\mathbf{C},p}^{\hat{\otimes} n}$ is also the n-fold symmetric tensor product of $E_{\mathbf{C},p}$. The norm $\|\varphi\|_p$ of $\varphi = \sum_{n=0}^\infty \mathbf{I}_n(f_n) \in (E)_p$ is given by

$$\|\varphi\|_p = \left(\sum_{n=0}^\infty n! |f_n|_p^2 \right)^{1/2}, \qquad f_n \in E_{\mathbf{C},p}^{\hat{\otimes} n},$$

where the norm of $E_{\mathbf{C},p}^{\hat{\otimes} n}$ is denoted also by $|\cdot|_p$.

The projective limit space (E) of spaces $(E)_p$, $p \in \mathbf{R}$ is a nuclear space. The inductive limit space $(E)^*$ of spaces $(E)_p$, $p \in \mathbf{R}$ is nothing but the dual space of (E). The space $(E)^*$ is called the space of *Hida distributions* or *generalized white noise functionals*. We denote by $\langle\langle \cdot, \cdot \rangle\rangle$ the canonical bilinear form on $(E)^* \times (E)$. Then we have

$$\langle\langle \Phi, \varphi \rangle\rangle = \sum_{n=0}^\infty n! \langle F_n, f_n \rangle$$

for any $\Phi = \sum_{n=0}^\infty \mathbf{I}_n(F_n) \in (E)^*$ and $\varphi = \sum_{n=0}^\infty \mathbf{I}_n(f_n) \in (E)$, where the canonical bilinear form on $(E_{\mathbf{C}}^{\hat{\otimes} n})^* \times (E_{\mathbf{C}}^{\hat{\otimes} n})$ is denoted also by $\langle \cdot, \cdot \rangle$.

Since $\exp\langle\cdot,\xi\rangle \in (E)$, the S- *transform* is defined on $(E)^*$ by

$$S[\Phi](\xi) = \exp\left(-\frac{1}{2}\langle\xi,\xi\rangle\right)\langle\langle\Phi, \exp\langle\cdot,\xi\rangle\rangle\rangle, \quad \xi \in E_{\mathbf{C}}.$$

3. The Lévy Laplacian as a Self-Adjoint Operator

Let Φ be in $(E)^*$. The S-transform $S[\Phi]$ of Φ is *Fréchet differentiable*, i.e.

$$S[\Phi](\xi+\eta) = S[\Phi](\xi) + S[\Phi]'(\xi)(\eta) + \mathrm{o}(\eta),$$

where $\mathrm{o}(\eta)$ means that there exists $p \geqslant 0$ depending on ξ such that $\mathrm{o}(\eta)/|\eta|_p \to 0$ as $|\eta|_p \to 0$. If the first variation $S[\Phi]'(\xi)(\eta)$ is expressed in the form $S[\Phi]'(\xi)(\eta) = \int_T F(\xi;t)\eta(t)\,\mathrm{d}t$ for every $\eta \in E$ by using the generalized function $F(\xi;\cdot)$ and $F(\cdot;t)$ is in $S[(E)^*]$, then the Hida derivative $\partial_t\Phi$ of Φ is defined as an element in $(E)^*$ whose S-transform is given by $F(\cdot;t)$.

We fix a finite interval T in $\mathbf{R}$. Take an orthonormal basis $\{\zeta_n\}_{n=0}^{\infty} \subset E$ for $L^2(T)$ satisfying the equally dense and uniform boundedness property (see [9, 17, 18, 22, 28, etc.]). Let $\mathcal{D}_L$ denote the set of all $\Phi \in (E)^*$ such that the limit

$$\widetilde{\Delta}_L S[\Phi](\xi) = \lim_{N\to\infty}\frac{1}{N}\sum_{n=0}^{N-1} S[\Phi]''(\xi)(\zeta_n,\zeta_n)$$

exists for any $\xi \in E_{\mathbf{C}}$ and is in $S[(E)^*]$. The Lévy Laplacian Δ_L is defined by $\Delta_L\Phi = S^{-1}\widetilde{\Delta}_L S\Phi$ for $\Phi \in \mathcal{D}_L$. We denote the set of all functionals $\Phi \in \mathcal{D}_L$ such that $S[\Phi](\eta) = 0$ for all $\eta \in E$ with $\mathrm{supp}(\eta) \subset T^c$ by $\mathcal{D}_L^T$. The operator is formally expressed in the form $\Delta_L = 1/|T|\int_{\mathrm{T}}\partial_t^2(\mathrm{d}t)^2$.

A white noise functional

$$\Phi = \int_{\mathbf{R}^n} f(u_1,\ldots,u_n) : \mathrm{e}^{ia_1x(u_1)}\cdots\mathrm{e}^{ia_nx(u_n)} : \mathbf{du} \in \mathcal{D}_L^T, \tag{3.1}$$

$$f \in L^1_{\mathbf{C}}(\mathbf{R})^{\hat{\otimes}n} \cap L^2_{\mathbf{C}}(\mathbf{R})^{\hat{\otimes}n},\ a_k \in \mathbf{R},\ k = 1,2,\ldots,n,$$

is equal to

$$\int_{T^n} f(u_1,\ldots,u_n) : \mathrm{e}^{ia_1x(u_1)}\cdots\mathrm{e}^{ia_nx(u_n)} : \mathbf{du}$$

and the S-transform $S[\Phi]$ of Φ is given by

$$S[\Phi](\xi) = \int_{T^n} f(\mathbf{u})\mathrm{e}^{ia_1\xi(u_1)}\cdots\mathrm{e}^{ia_n\xi(u_n)}\,\mathbf{du}. \tag{3.2}$$

This functional is important as an eigenfunction of the operator Δ_L. In fact, we have the following result:

THEOREM 1 [29]. *A white noise functional* Φ *as in* (3.1) *satisfies the equation*

$$\Delta_L \Phi = -\frac{1}{|T|} \sum_{k=1}^{n} a_k^2 \Phi. \tag{3.3}$$

We put

$$\mathbf{D}_n = \left\{ \int_{T^n} f(\mathbf{u}) : \prod_{\nu=1}^{n} \mathrm{e}^{ix(u_\nu)} : \mathbf{du} \in \mathcal{D}_L^T;\ f \in L^1_{\mathbf{C}}(\mathbf{R})^{\hat{\otimes} n} \cap L^2_{\mathbf{C}}(\mathbf{R})^{\hat{\otimes} n} \right\}$$

for each $n \in \mathbf{N} \cup \{0\}$. Then $\mathbf{D}_n$ is a linear subspace of $(E)_{-p}$ for any $p \geqslant 1$, and Δ_L is a linear operator from $\mathbf{D}_n$ into itself such that $\|\Delta_L \Phi\|_{-p} = n/|T| \|\Phi\|_{-p}$ for any $\Phi \in \mathbf{D}_n$. We define a space $\overline{\mathbf{D}}_n$ by the completion of $\mathbf{D}_n$ in $(E)_{-p}$ with respect to $\|\cdot\|_{-p}$. Then for each $n \in \mathbf{N} \cup \{0\}$, $\overline{\mathbf{D}}_n$ becomes a Hilbert space with the inner product of $(E)_{-p}$. For each $n \in \mathbf{N} \cup \{0\}$, the operator Δ_L becomes a continuous linear operator from $\overline{\mathbf{D}}_n$ into itself satisfying

$$\|\Delta_L \Phi\|_{-p} = \frac{n}{|T|} \|\Phi\|_{-p} \quad \text{for any } \Phi \in \overline{\mathbf{D}}_n.$$

The operator Δ_L is a self-adjoint operator on $\overline{\mathbf{D}}_n$ for each $n \in \mathbf{N} \cup \{0\}$.

Put $\alpha_N(n) = \sum_{\ell=0}^{N} (n/|T|)^{2\ell}$ and define a space $\mathbf{E}_{-p,N}$ by

$$\mathbf{E}_{-p,N} = \left\{ \sum_{n=0}^{\infty} \Phi_n \in (E)^*;\ \sum_{n=0}^{\infty} \alpha_N(n) \|\Phi_n\|_{-p}^2 < \infty, \right.$$
$$\left. \Phi_n \in \overline{\mathbf{D}}_n,\ n = 0, 1, 2, \ldots \right\}$$

with the norm $|||\cdot|||_{-p,N}$ given by

$$|||\Phi|||_{-p,N} = \left(\sum_{n=0}^{\infty} \alpha_N(n) \|\Phi_n\|_{-p}^2 \right)^{1/2}, \quad \Phi = \sum_{n=0}^{\infty} \Phi_n \in \mathbf{E}_{-p,N}$$

for each $N \in \mathbf{N} \cup \{0\}$ and $p \geqslant 1$. Then for any $N \geqslant 1$ and $p \geqslant 1$, $\mathbf{E}_{-p,N}$ is in $(E)_{-p}$ and is a Hilbert space with respect to the norm $|||\cdot|||_{-p,N}$.

Put $\mathbf{E}_{-p,\infty} = \bigcap_{N=1}^{\infty} \mathbf{E}_{-p,N}$ with the projective limit topology. Then we have the following inclusion relations:

$$\mathbf{E}_{-p,\infty} \subset \cdots \subset \mathbf{E}_{-p,N+1} \subset \mathbf{E}_{-p,N} \subset \cdots \subset \mathbf{E}_{-p,1} \subset (E)_{-p}.$$

The space $\mathbf{E}_{-p,\infty}$ includes $\overline{\mathbf{D}}_n$ for any $n \in \mathbf{N} \cup \{0\}$. The operator Δ_L becomes a continuous linear operator defined on $\mathbf{E}_{-p,2}$ into $\mathbf{E}_{-p,1}$ satisfying $|||\Delta_L \Phi|||_{-p,N} \leqslant |||\Phi|||_{-p,N+1}$, $N = 1, 2, 3, \ldots$, $\Phi \in \mathbf{E}_{-p,\infty}$. With these properties, we have the following:

THEOREM 2 [29]. *The operator Δ_L is a self-adjoint operator densely defined on $\mathbf{E}_{-p,N}$ for each $N \geqslant 1$ and $p \geqslant 1$.*

Put

$$\mathcal{E}_{-p} = \Big\{ \Phi = \sum_{n=0}^{\infty} \mathbf{I}_n(f_n) \in (E)_{-p};\ S[\Phi](e^{i\xi}) \equiv \sum_{n=0}^{\infty} \langle (e^{i\xi})^{\otimes n}, f_n \rangle \text{ exists in } S[\mathbf{E}_{-p,\infty}] \Big\}$$

and define an operator K on $\mathcal{E}_{-p}$ by

$$K[\Phi] = S^{-1}[S[\Phi](e^{i\xi})].$$

The operator K implies a relation between Δ_L and the number operator $\mathcal{N}$ on $(E)^*$ given by

$$\mathcal{N}\Phi = \sum_{n=0}^{\infty} n\mathbf{I}_n(f_n) \quad \text{for } \Phi = \sum_{n=0}^{\infty} \mathbf{I}_n(f_n) \in (E)^*.$$

PROPOSITION 3 [29]. *For any $\Phi \in \mathcal{E}_{-p}$ we have*

$$\Delta_L K[\Phi] = -\frac{1}{|T|} K[\mathcal{N}[\Phi]].$$

Put

$$[E]_{q,N} = \Big\{ \varphi = \sum_{n=0}^{\infty} \mathbf{I}_n(f_n) \in (E);\ \sum_{n=0}^{\infty} \alpha_N(n) e^{n^2/2} |f_n|_q^2 < \infty,\ \operatorname{supp}(f_n) \subset T, n = 0, 1, 2, \ldots \Big\}$$

for $q > 0$ and $N \in \mathbf{N}$. Define a space $\overline{[E]_{q,N}}$ by the completion of $[E]_{q,N}$ with respect to the norm $\|\cdot\|_{\overline{[E]_{q,N}}}$ given by

$$\|\varphi\|_{\overline{[E]_{q,N}}} = \left(\sum_{n=0}^{\infty} \alpha_N(n) e^{n^2/2} |f_n|_q^2 \right)^{1/2}$$

for $\varphi = \sum_{n=0}^{\infty} \mathbf{I}_n(f_n) \in (E)^*$. Then $\overline{[E]_{q,N}}$ is a Hilbert space with norm $\|\cdot\|_{\overline{[E]_{q,N}}}$. It is easily checked that $\overline{[E]_{q,N}} \subset (E)_q$ for any $q > 0$. Put $\overline{[E]_{\infty,N}} = \bigcap_{q>0} \overline{[E]_{q,N}}$ with the projective limit topology and also put $\overline{[E]_{\infty,\infty}} = \bigcap_{N \geqslant 1} \overline{[E]_{\infty,N}}$ with the projective limit topology. Then we have the following:

PROPOSITION 4. *Let $p \geqslant 1$. Then the operator K is a continuous linear operator from $\overline{[E]_{\infty,\infty}}$ into $\mathbf{E}_{-p,\infty}$.*

Proof. Let $p \geqslant 1$. Then for each $N \in \mathbf{N}$ we can calculate the norm $|||K[\varphi]|||_{-p,N}^2$ of $K[\varphi]$ for $\varphi = \sum_{n=0}^{\infty} \mathbf{I}_n(f_n) \in \overline{[E]_{\infty,\infty}}$ as follows:

$$\begin{aligned} |||K[\varphi]|||_{-p,N}^2 &= \sum_{n=0}^{\infty} \alpha_N(n) || \langle : (\mathrm{e}^{ix})^{\otimes n} :, f_n \rangle ||_{-p}^2 \\ &\leqslant \sum_{n=0}^{\infty} \alpha_N(n) \sum_{\ell=0}^{\infty} \ell! \sum_{k_1,\ldots,k_\ell=0}^{\infty} \prod_{j=1}^{\ell} (2k_j+2)^{-2p} \times \\ &\quad \times \left| \sum_{|\nu|=\ell} \frac{1}{\nu!} \langle F_\nu, e_{k_1} \otimes \cdots \otimes e_{k_\ell} \rangle \right|^2, \end{aligned}$$

where $\nu = (\nu_1, \ldots, \nu_n) \in \mathbf{N} \cup \{0\}$, $|\nu| = \nu_1 + \cdots + \nu_n$ and $F_\nu = \int_{\mathbf{R}^n} f(\mathbf{u}) \delta_{u_1}^{\hat{\otimes}\nu_1} \hat{\otimes} \cdots \hat{\otimes} \delta_{u_n}^{\hat{\otimes}\nu_n} \, \mathrm{d}\mathbf{u}$. Since there exists $q > 0$ such that

$$\begin{aligned} &\sum_{k_1,\ldots,k_\ell=0}^{\infty} \prod_{j=1}^{\ell} (2k_j+2)^{-2p} \left| \sum_{|\nu|=\ell} \frac{1}{\nu!} \langle F_\nu, e_{k_1} \otimes \cdots \otimes e_{k_\ell} \rangle \right|^2 \\ &\quad \leqslant |f_n|_q^2 n^{2\ell} \left(\sum_{|\nu|=\ell} \frac{1}{\nu!} \right)^2 \left(\sum_{k=0}^{\infty} (2k+2)^{-2p} |e_k|_{-q}^2 \right)^\ell, \end{aligned}$$

we get that

$$\begin{aligned} |||K[\varphi]|||_{-p,N}^2 &\leqslant \sum_{n=0}^{\infty} \alpha_N(n) \mathrm{e}^{n^2 \sum_{k=0}^{\infty} (2k+2)^{-2(p+q)}} |f_n|_q^2 \\ &\leqslant \sum_{n=0}^{\infty} \alpha_N(n) \mathrm{e}^{n^2/2} |f_n|_q^2. \end{aligned}$$

This is nothing but the inequality:

$$|||K[\varphi]|||_{-p,N} \leqslant ||\varphi||_{\overline{[E]_{q,N}}}.$$

Thus the proof is completed. □

Let κ be in $(E^*)^{\hat{\otimes}2}$ such that $\Xi_{1,1}(\kappa)\varphi = \int_{\mathbf{R}^2} \kappa(u,v) \partial_u^* \partial_v \varphi \, \mathrm{d}u \, \mathrm{d}v$ exists in $\mathcal{E}_{-p}$ for all $\varphi \in \mathcal{E}_{-p} \cap (E)$. (See [11].) Define an operator $\mathbf{L}(\kappa)$ on $K[\mathcal{E}_{-p} \cap (E)]$ by $\mathbf{L}(\kappa) K[\varphi] = K[\Xi_{1,1}\varphi]$. The differntial operator D_y, $y \in E^*$, on (E) is defined by

$$D_y \varphi \equiv \sum_{n=0}^{\infty} n \mathbf{I}_{n-1}(\langle y, f_n \rangle)$$

for $\varphi(x) = \sum_{n=0}^{\infty} \mathbf{I}_n(f_n) \in (E)$. Then the operator D_y, $y \in E^*$, is a continuous linear operator from (E) into itself. We can see that for any $\xi \in E$, D_ξ is extended to a continuous linear operator $\widetilde{D}_\xi$ from $(E)^*$ into itself. For any $y \in E^*$, the adjoint operator D_y^* of D_y is also a continuous linear operator from $(E)^*$ into itself.

In order to characterize the Lévy Laplacian we use an operator

$$R_{\xi,\eta} = D_\xi^* \widetilde{D}_\eta - D_\eta^* \widetilde{D}_\xi$$

for $\xi, \eta \in E$ introduced in [15]. Let $E|_T = \{\xi \in E;\ \mathrm{supp}(\xi) \subset T\}$. Then it can be checked that $R_{\xi,\eta}[[E]_{\infty,\infty}] \subset \mathcal{E}_{-p} \cap (E)$ for all $\xi, \eta \in E|_T$ by Proposition 4. Therefore we can define an operator $\overline{R}_{\xi,\eta}$ on $K[\overline{[E]_{\infty,\infty}}]$ by $\overline{R}_{\xi,\eta} K[\varphi] = K[R_{\xi,\eta}\varphi]$ for all ξ, η in $E|_T$. As in [15, Theorem 12.41] we can prove that if $[R_{\xi,\eta}, \Xi_{1,1}(\kappa)] = 0$ on $\overline{[E]_{\infty,\infty}}$ holds for all $\xi, \eta \in E|_T$, then $\Xi_{1,1}(\kappa)$ is equal to $\mathcal{N}$ up to a constant. Consequently, by Proposition 3 we have the following:

THEOREM 5. *Let κ be a generalized function in $(E^*)^{\hat{\otimes}2}$ such that $\Xi_{1,1}\varphi$ exists in $\mathcal{E}_{-p}$ for all $\varphi \in \mathcal{E}_{-p} \cap (E)$. If $[\overline{R}_{\xi,\eta}, \mathbf{L}(\kappa)]K[\varphi] = 0$ for all $\xi, \eta \in E|_T$ and $\varphi \in \overline{[E]_{\infty,\infty}}$, then there exists a constant c such that $\mathbf{L}(\kappa) = c\Delta_L$.*

4. A Stochastic Process Generated by the Lévy Laplacian

We define a semi-group $\{G_t;\ t \geqslant 0\}$ on $\mathbf{E}_{-p,\infty}$ by

$$G_t\Phi = \sum_{n=0}^{\infty} \mathrm{e}^{-\frac{nt}{|T|}} \Phi_n$$

for $\Phi = \sum_{n=0}^{\infty} \Phi_n \in \mathbf{E}_{-p,\infty}$, $\Phi_n \in \overline{\mathbf{D}}_n$, $n = 0, 1, 2, \ldots$. Then the infinitesimal generator of $\{G_t;\ t \geqslant 0\}$ is given by Δ_L. (See [29].)

Let $\{X_t^\alpha;\ t \geqslant 0\}$ be a strictly stable process with the characteristic function of X_t^α given by

$$E[\mathrm{e}^{izX_t^\alpha}] = \mathrm{e}^{-t|z|^\alpha}$$

for any $0 < \alpha \leqslant 2$. Take a smooth function $\eta_T \in E$ with $\eta_T(u) = 1/|T|$ on T. Then we can construct stochastic processes generated by powers of the Lévy Laplacian as follows.

THEOREM 6. *Let F be the S-transform of a white noise functional in $\mathbf{E}_{-p,\infty}$. Then it holds that*

$$\mathrm{e}^{-t(-\widetilde{\Delta}_L)^\alpha} F(\xi) = E[F(\xi + X_t^\alpha \eta_T)], \quad \xi \in E$$

for any $0 < \alpha \leqslant 2$. In particular, the equalities hold:

$$\mathrm{e}^{t\widetilde{\Delta}_L} F(\xi) = E[F(\xi + X_t^1 \eta_T)];$$

$$\mathrm{e}^{-t\widetilde{\Delta}_L^2} F(\xi) = E[F(\xi + B_{2t}\eta_T)],$$

where $\{B_t;\ t \geqslant 0\}$ is a standard Brownian motion.

Proof. Put $F(\xi) = \int_{T^n} f(\mathbf{u})e^{i\xi(u_1)} \cdots e^{i\xi(u_n)}\,d\mathbf{u}$ with $f \in L^1_{\mathbf{C}}(T)^{\hat{\otimes} n} \cap L^2_{\mathbf{C}}(T)^{\hat{\otimes} n}$. Then we have

$$\begin{aligned} E[F(\xi + X^\alpha_t \eta_T)] &= \int_{T^n} f(\mathbf{u})e^{i\xi(u_1)} \cdots e^{i\xi(u_n)} E[e^{i\frac{n}{|T|}X^\alpha_t}]\,d\mathbf{u} \\ &= e^{-t(\frac{n}{|T|})^\alpha} F(\xi) = e^{-t(-\widetilde{\Delta}_L)^\alpha} F(\xi). \end{aligned}$$

Let $F = \sum_{n=0}^\infty F_n \in S[\mathbf{E}_{-p,\infty}]$. Then for any $n \in \mathbf{N} \cup \{0\}$, F_n is expressed in the following form:

$$F_n(\xi) = \lim_{N\to\infty} \int_{T^n} f^{[N]}(\mathbf{u})e^{i\xi(u_1)} \cdots e^{i\xi(u_n)}\,d\mathbf{u},$$

where $(f^{[N]})_N$ is a sequence of functions in $L^1_{\mathbf{C}}(\mathbf{R})^{\hat{\otimes} n} \cap L^2_{\mathbf{C}}(\mathbf{R})^{\hat{\otimes} n}$. Hence, we have

$$\begin{aligned} &\sum_{n=0}^\infty E[|F_n(\xi + X^\alpha_t \eta_T)|] \\ &\quad = \sum_{n=0}^\infty E\left[\lim_{N\to\infty}\left|\int_{T^n} f^{[N]}(\mathbf{u})e^{i\xi(u_1)} \cdots e^{i\xi(u_n)} e^{iX^\alpha_t \eta_T(u_1)} \cdots e^{iX^\alpha_t \eta_T(u_n)}\,d\mathbf{u}\right|\right] \\ &\quad = \sum_{n=0}^\infty \lim_{N\to\infty}\left|\int_{T^n} f^{[N]}(\mathbf{u})e^{i\xi(u_1)} \cdots e^{i\xi(u_n)}\,d\mathbf{u}\right| \\ &\quad = \sum_{n=0}^\infty |F_n(\xi)|. \end{aligned}$$

Since $F_n \in S[\mathbf{E}_{-p,\infty}]$, there exists some $\Phi_n \in \mathbf{E}_{-p,\infty}$ such that $F_n = S[\Phi_n]$ for any n. By the Schwarz inequality, we see that

$$\sum_{n=0}^\infty |F_n(\xi)| \leqslant \sum_{n=0}^\infty \|\Phi_n\|_{-p}\|\varphi_\xi\|_p < \infty, \quad \xi \in E,$$

where $\varphi_\xi(x) = :\exp\{\langle x, \xi\rangle\}:$. Therefore by the continuity of $e^{-t(-\widetilde{\Delta}_L)^\alpha}$ we get that

$$\begin{aligned} E[F(\xi + X^\alpha_t \eta_T)] &= \sum_{n=0}^\infty E[F_n(\xi + X^\alpha_t \eta_T)] \\ &= \sum_{n=0}^\infty e^{-t(-\widetilde{\Delta}_L)^\alpha} F_n(\xi) \\ &= e^{-t(-\widetilde{\Delta}_L)^\alpha} F(\xi). \end{aligned}$$

Thus we obtain the assertion. □

Equivalently, an integral expression of $e^{-t(-\widetilde{\Delta}_L)^\alpha}$ is given as follows:

Remark 7. Let F be the S-transform of a generalized white noise fnctional in $\mathbf{E}_{-p,\infty}$. Then it holds that

$$e^{-t(-\widetilde{\Delta}_L)^\alpha} F(\xi) = \int_{\mathbf{R}} \int_E F(\xi + t^{1/\alpha} \cdot u \cdot x) \, d\delta_{\eta_T}(x) \, dm_\alpha(u),$$

where δ_{η_T} is the Dirac delta measure on E at η_T and m_α denotes a stable distribution characterized by

$$\int_{\mathbf{R}} e^{izu} \, dm_\alpha(u) = e^{-|z|^\alpha}.$$

For any $\Phi \in (E)^*$ and $\eta \in E$, the translation $\Phi_\eta(\cdot) = \Phi(\cdot + \eta)$ in $(E)^*$ of Φ by η is defined by $S[\Phi_\eta](\xi) = S[\Phi](\xi + \eta)$, $\xi \in E$. Then we can translate Theorem 6 to be in words of white noise functionals. (See [15].)

COROLLARY 8. *Let Φ be a white noise functional in $\mathbf{E}_{-p,\infty}$. Then it holds that*

$$e^{-t(-\Delta_L)^\alpha} \Phi(x) = E[\Phi(x + X_t^\alpha \eta_T)].$$

In particular, the equality holds:

$$G_t \Phi(x) = E[\Phi(x + X_t^1 \eta_T)].$$

Acknowledgements

This paper was written based on the author's talk in the Volterra International School 'White Noise Approach to Classical and Quantum Stochastic Calculi and Quantum Probability', Trento, Italy, July 19–23, 1999. Thc author would like to express his thanks to the organizers, Professor L. Accardi and Professor I. V. Volovich, for the invitation to this stimulating school and for their warm hospitality and support. This work was supported, in part, by the Research Project 'Quantum Information Theoretical Approach to Life Science' for the Academic Frontier in Science promoted by the Ministry of Education in Japan and was also supported, in part, by JMESSC Grant-in-Aid for Scientific Research (C)(2) 11640139. The author is grateful for their supports.

References

1. Accardi, L., Gibilisco, P. and Volovich, I. V.: The Lévy Laplacian and the Yang–Mills equations, *Rend. Accad. Lincei* (1993).
2. Accardi, L. and Smolyanov, O. G.: Trace formulae for Levy–Gaussian measures and their application, *Proc. IIAS Workshop 'Mathematical Approach to Fluctuations'*, Vol. II, World Scientific, Singapore, 1995, pp. 31–47.

3. Chung, D. M., Ji, U. C. and Saitô, K.: Cauchy problems associated with the Lévy Laplacian in white noise analysis, *Infin. Dimens. Anal. Quantum Probab. Relat. Top.* **2**(1) (1999), 131–153.
4. Gross, L.: Abstract Wiener spaces, *Proc. 5th Berkeley Symp. Math. Stat. Probab.* Vol. 2, Univ. Calif. Press, Berkeley, 1965, pp. 31–42.
5. Gross, L.: Potential theory on Hilbert space, *J. Funct. Anal.* **1** (1967), 123–181.
6. Hida, T.: *Analysis of Brownian Functionals*, Carleton Math. Lecture Notes 13, Carleton University, Ottawa, 1975.
7. Hida, T.: A role of the Lévy Laplacian in the causal calculus of generalized white noise functionals, In: S. Cambanis *et al.* (eds), *Stochastic Processes, A Festschrift in Honour of G. Kallianpur*, Springer-Verlag, New York, 1992.
8. Hida, T., Kuo, H.-H. and Obata, N.: Transformations for white noise functionals, *J. Funct. Anal.* (1990).
9. Hida, T., Kuo, H.-H., Potthoff, J. and Streit, L.: *White Noise: An Infinite Dimensional Calculus*, Kluwer Acad. Publ., Dordrecht, 1993.
10. Hida, T. and Saitô, K.: White noise analysis and the Lévy Laplacian, In: S. Albeverio *et al.* (eds), *Stochastic Processes in Physics and Engineering*, D. Reidel Publ., Dordrecht, 1988, pp. 177–184.
11. Hida, T., Obata, N. and Saitô, K.: Infinite dimensional rotations and Laplacian in terms of white noise calculus, *Nagoya Math. J.* **128** (1992), 65–93.
12. Itô, K.: Stochastic analysis in infinite dimensions, In: *Proc. International Conference on Stochastic Analysis*, Evanston, Acad. Press, 1978, pp. 187–197.
13. Kubo, I.: A direct setting of white noise calculus, In: *Stochastic Analysis on Infinite Dimensional Spaces*, Pitman Res. Notes in Math. Ser. 310, Longman, Harlow, 1994, pp. 152–166.
14. Kubo, I. and Takenaka, S.: Calculus on Gaussian white noise I, II, III and IV, *Proc. Japan Acad. Ser. A Math. Sci.* **56** (1980), 376–380; **56** (1980), 411–416; **57** (1981), 433–436; **58** (1982), 186–189.
15. Kuo, H.-H.: Lectures on white noise calculus, *Soochow J. Math.* (1991).
16. Kuo, H.-H.: On Laplacian operators of generalized Brownian functionals, In: Lecture Notes in Math. 1203, Springer-Verlag, New York, 1986, pp. 119–128.
17. Kuo, H.-H.: *White Noise Distribution Theory*, CRC Press, Boca Raton, 1996.
18. Kuo, H.-H., Obata, N. and Saitô, K.: Lévy Laplacian of generalized functions on a nuclear space, *J. Funct. Anal.* **94** (1990), 74–92.
19. Lee, Y.-J.: Integral transforms of analytic functions on abstract Wiener spaces, *J. Funct. Anal.* **47** (1983), 153–164.
20. Lee, Y.-J.: Unitary operators on the space of L^2-functions over abstract Wiener spaces, *Soochow J. Math.* **13** (1987), 165–174.
21. Lévy, P.: *Leçons d'analyse fonctionnelle*, Gauthier-Villars, Paris, 1922.
22. Obata, N.: A characterization of the Lévy Laplacian in terms of infinite dimensional rotation groups, *Nagoya Math. J.* **118** (1990), 111–132.
23. Obata, N.: *White Noise Calculus and Fock Space*, Lecture Notes in Math. 1577, Springer-Verlag, New York, 1994.
24. Potthoff, J. and Streit, L.: A characterization of Hida distributions, *J. Funct. Anal.* **101** (1991), 212–229.
25. Saitô, K.: Itô's formula and Lévy's Laplacian I and II, *Nagoya Math. J.* **108** (1987), 67–76; **123** (1991), 153–169.
26. Saitô, K.: A group generated by the Lévy Laplacian and the Fourier–Mehler transform, In: *Stochastic Analysis on Infinite Dimensional Spaces*, Pitman Research Notes in Mathematics Series 310, Longman, Harlow, 1994, pp. 274–288.
27. Saitô, K.: A (C_0)-group generated by the Lévy Laplacian, *J. Stochast. Anal. Appl.* **16**(3) (1998), 567–584.

28. Saitô, K.: A (C_0)-group generated by the Lévy Laplacian II, *Infinite Dimensional Analysis, Quantum Probab. Related Topics* **1**(3) (1998), 425–437.
29. Saitô, K. and Tsoi, A. H.: The Lévy Laplacian as a self-adjoint operator, *Proc. First Internat. Conf. Quantum Information*, World Scientific, Singapore, 1999.
30. Yosida, K.: *Functional Analysis*, 3rd edn, Springer-Verlag, New York, 1971.

Acta Applicandae Mathematicae **63:** 375–384, 2000.

Recurrence-Transience for Self-similar Additive Processes Associated with Stable Distributions

KEN-ITI SATO[1] and KOUJI YAMAMURO[2]
[1]*Hachiman-yama 1101-5-103, Tenpaku-ku, Nagoya, 468-0074, Japan*
[2]*Aichi Konan College, Takaya-cho, Konan, 483-8086, Japan*

(Received: January 1999)

Abstract. For any stable distribution μ on the line, recurrence-transience of the selfsimilar additive process $\{X_t,\ t \geqslant 0\}$ with $\mathcal{L}(X_1) = \mu$ is determined. Comparison with the stable Lévy process $\{Y_t,\ t \geqslant 0\}$ with $\mathcal{L}(Y_1) = \mu$ is made: if μ is not strictly stable, then $\{Y_t\}$ is transient but $\{X_t\}$ is recurrent except the obviously transient case of monotone sample functions.

Mathematics Subject Classification (2000): 60Gxx.

Key words: recurrence-transience, additive processes, stable distributions.

1. Introduction

The problem of recurrence-transience for random walks and Lévy processes has been studied in many papers (see the monographs of Spitzer [10] and Sato [7]). The problem is also studied in the temporally inhomogeneous case in [8, 12], and [13] where selfsimilar additive processes are treated. Those processes were introduced by Sato [5, 6], and then studied by Watanabe [11]. They are connected to self-decomposable distributions; any self-similar additive process has self-decomposable distribution at any fixed time and, conversely, given a self-decomposable distribution μ, we can construct, for any $H > 0$, an H-self-similar additive process such that its distribution at time 1 equals μ. The process is determined by μ and H uniquely in the sense of the relevant law. Their recurrence-transience does not depend on the exponent H. Hence, it depends only on μ. However, the attempt to find a criterion of their recurrence-transience in terms of μ or its characteristic function $\widehat{\mu}(z)$ has not been successful so far. Some sufficient conditions for recurrence and for transience were obtained by Yamamuro [12, 13].

Now assume that we are given a stable distribution μ on the line $\mathbb{R}$. Since stable distributions are self-decomposable, an H-self-similar additive process $\{X_t,\ t \geqslant 0\}$ satisfying $\mathcal{L}(X_1) = \mu$ is associated with μ. Here $\mathcal{L}(X_1)$ denotes the distribution of X_1. In this paper, we determine recurrence-transience of this process $\{X_t\}$. The result is expressed by the index α of stability of μ and by the parameters β and γ in the canonical representation of $\widehat{\mu}(z)$. In the case of $\alpha \geqslant 1$, the result was obtained in [8, 12], and [13], but the case of $0 < \alpha < 1$ has not been treated so far. The distribution μ also induces the stable Lévy process $\{Y_t,\, t \geqslant 0\}$ with $\mathcal{L}(Y_1) = \mu$. Thus, the two processes $\{X_t\}$ and $\{Y_t\}$ are associated, both having distribution μ at time 1. They are identical if and only if μ is strictly stable. Recurrence-transience of the process $\{Y_t\}$ is well known. If μ is not strictly stable, we shall see that $\{X_t\}$ and $\{Y_t\}$ are widely different, that is, except the cases of increasing sample functions a.s. and of decreasing sample functions a.s., $\{X_t\}$ is recurrent and $\{Y_t\}$ is transient. This is an interesting phenomenon.

Here we give the definitions of the notions that have appeared in the preceding two paragraphs. A stochastic process $\{X_t,\, t \geqslant 0\}$ on a probability space $(\Omega, \mathcal{F}, P)$ is self-similar with exponent H, or H-selfsimilar, if $\{X_{at},\, t \geqslant 0\} \stackrel{\mathrm{d}}{=} \{a^H X_t, t \geqslant 0\}$ for every $a > 0$, where $\stackrel{\mathrm{d}}{=}$ denotes the the equality in finite-dimensional distributions.

A distribution μ on $\mathbb{R}$ is self-decomposable if, for every $b > 1$, there is a distribution ρ_b such that $\widehat{\mu}(z) = \widehat{\mu}(b^{-1}z)\widehat{\rho}_b(z)$. A stochastic process $\{X_t,\, t \geqslant 0\}$ is an additive process if it has independent increments, is stochastically continuous, starts at the origin, and has right-continuous sample functions with left limits a.s. It is called a Lévy process if, moreover, it has stationary increments.

A distribution μ is stable with index α if it is infinitely divisible and if, for every $a > 0$, there is $\gamma_a \in \mathbb{R}$ such that

$$\widehat{\mu}(z)^a = \widehat{\mu}(a^{1/\alpha}z)\mathrm{e}^{i\gamma_a z}. \tag{1.1}$$

It is known that α must satisfy $0 < \alpha \leqslant 2$. If, for every $a > 0$, (1.1) holds with $\gamma_a = 0$, then μ is called strictly stable. A stable Lévy process $\{Y_t\}$ with index α is a Lévy process with $\mathcal{L}(Y_1) = \mu$ for some stable distribution μ with index α, i.e., $\{Y_t\}$ is a stable Lévy process with index α if it is a Lévy process such that, for every $a > 0$, there is $\gamma_a \in \mathbb{R}$ satisfying

$$\{Y_{at},\, t \geqslant 0\} \stackrel{\mathrm{d}}{=} \{a^{1/\alpha} Y_t + \gamma_a t,\, t \geqslant 0\}.$$

Likewise, a strictly stable Lévy process is a Lévy process associated with a strictly stable distribution. An additive process $\{X_t\}$ is said to be recurrent if, for every $s \geqslant 0$,

$$P\Big[\liminf_{t\to\infty} |X_{s+t} - X_s| = 0\Big] = 1. \tag{1.2}$$

It is said to be transient if

$$P\Big[\lim_{t\to\infty} |X_t| = \infty\Big] = 1. \tag{1.3}$$

We say that a probability measure μ is trivial if it is a δ-distribution. A stochastic process $\{X_t\}$ is called trivial if $\mathcal{L}(X_t)$ is trivial for every t.

We use the following canonical representation of stable distributions: a nontrivial probability measure μ is stable with index α if and only if $\widehat{\mu}(z)$ has one of the following forms:

$$\alpha = 2, \qquad \widehat{\mu}(z) = \exp[-cz^2 + i\gamma z] \tag{1.4}$$

with $c > 0$ and $\gamma \in \mathbb{R}$;

$$\alpha \in (0, 1) \cup (1, 2),$$
$$\widehat{\mu}(z) = \exp\left[-c|z|^\alpha \left(1 - i\beta \left(\tan \frac{\pi\alpha}{2}\right) \operatorname{sgn} z\right) + i\gamma z\right] \tag{1.5}$$

with $c > 0$, $-1 \leqslant \beta \leqslant 1$, and $\gamma \in \mathbb{R}$;

$$\alpha = 1, \quad \widehat{\mu}(z) = \exp\left[-c|z| \left(1 + i\beta \frac{2}{\pi} (\log|z|) \operatorname{sgn} z\right) + i\gamma z\right] \tag{1.6}$$

with $c > 0$, $-1 \leqslant \beta \leqslant 1$, and $\gamma \in \mathbb{R}$. If $\alpha = 2$ and $\gamma = 0$, then μ essentially induces the Brownian motion, which is the subject of Hida's book [3]. If $\alpha < 2$, then the Lévy measure ν of μ is such that

$$\nu(\mathrm{d}x) = \begin{cases} c_1 x^{-\alpha-1}\,\mathrm{d}x, & \text{on } (0, \infty), \\ c_2 |x|^{-\alpha-1}\,\mathrm{d}x, & \text{on } (-\infty, 0), \end{cases}$$

with some nonnegative constants c_1, c_2 satisfying $c_1 + c_2 > 0$ and we have $\beta = (c_1 - c_2)/(c_1 + c_2)$. Among nontrivial stable distributions of (1.4)–(1.6), strictly stable ones are $\alpha \in (0, 1) \cup (1, 2]$ with $\gamma = 0$; $\alpha = 1$ with $\beta = 0$.

2. Results

The following theorem contains main results of this paper:

THEOREM 2.1. *Let μ be a nontrivial stable distribution with index α on $\mathbb{R}$ having characteristic function* (1.4), (1.5), *or* (1.6). *Let $\{X_t\}$ be the H-self-similar additive process satisfying $\mathcal{L}(X_1) = \mu$. Then the following are true:*

(i) *If $1 \leqslant \alpha \leqslant 2$, then $\{X_t\}$ is recurrent.*
(ii) *Let $0 < \alpha < 1$ and $|\beta| = 1$. If $\beta\gamma \geqslant 0$, then $\{X_t\}$ is transient. If $\beta\gamma < 0$, then $\{X_t\}$ is recurrent.*
(iii) *Let $0 < \alpha < 1$ and $|\beta| < 1$. If $\gamma = 0$, then $\{X_t\}$ is transient. If $\gamma \neq 0$, then $\{X_t\}$ is recurrent.*

COROLLARY 2.2. *Let μ and $\{X_t\}$ be as in Theorem 2.1. Let $\{Y_t\}$ be the stable Lévy process with $\mathcal{L}(Y_1) = \mu$. If μ is not strictly stable and not one-sided, then $\{X_t\}$ is recurrent and $\{Y_t\}$ is transient.*

Here we say that μ is not one-sided if $\mu(-\infty,0) > 0$ and $\mu(0,\infty) > 0$. Note that, a stable distribution μ satisfies $\mu(-\infty,0) = 0$ if and only if $0 < \alpha < 1$, $\beta = 1$, and $\gamma \geqslant 0$. Likewise, a stable distribution μ satisfies $\mu(0,\infty) = 0$ if and only if $0 < \alpha < 1$, $\beta = -1$, and $\gamma \leqslant 0$.

Another consequence of Theorem 2.1 is about recurrent modifications of stable Lévy processes.

COROLLARY 2.3. *Let μ be a nontrivial stable distribution satisfying* (1.4), (1.5), *or* (1.6).

Let $\{Y_t\}$ be the stable Lévy process with $\mathcal{L}(Y_1) = \mu$. Given $b \in \mathbb{R}$, let

$$Z_t^{(b)} = \begin{cases} Y_t - \gamma t + bt^{1/\alpha}, & \text{if } \alpha \neq 1, \\ Y_t - c\beta \frac{2}{\pi} t \log t + bt, & \text{if } \alpha = 1. \end{cases} \tag{2.1}$$

Then $\{Z_t^{(b)}\}$ is recurrent if and only if μ and b satisfy one of the following conditions:

$$1 \leqslant \alpha \leqslant 2; \tag{2.2}$$

$$0 < \alpha < 1, |\beta| = 1, \text{ and } \beta b < 0; \tag{2.3}$$

$$0 < \alpha < 1, |\beta| < 1, \text{ and } b \neq 0. \tag{2.4}$$

Recall that the conditions of recurrence and transience of stable Lévy processes are well known. We give them in the form of a proposition.

PROPOSITION 2.4. *Let $\{Y_t\}$ be the nontrivial stable Lévy process as in Corollary* 2.3.

(i) *Let $1 < \alpha \leqslant 2$. If $\gamma = 0$, then $\{Y_t\}$ is recurrent. If $\gamma \neq 0$, then $\{Y_t\}$ is transient.*
(ii) *Let $\alpha = 1$. If $\beta = 0$, then $\{Y_t\}$ is recurrent. If $\beta \neq 0$, then $\{Y_t\}$ is transient.*
(iii) *Let $0 < \alpha < 1$. Then $\{Y_t\}$ is transient.*

In order to check this proposition, it is enough to use the fact that a Lévy process $\{Y_t\}$ with $\mathcal{L}(Y_1) = \mu$ is recurrent if and only if

$$\limsup_{q \downarrow 0} \int_{-1}^{1} \mathrm{Re}\left(\frac{1}{q - \log \widehat{\mu}(z)}\right) dz = \infty. \tag{2.5}$$

This is a criterion analogous to that of random walks by Chung and Fuchs [1].

3. Proofs

We prepare four lemmas.

LEMMA 3.1. *If $\{X_t\}$ is a self-similar additive process, then it is either recurrent or transient.*

This is Theorem 3.2 of [8].

LEMMA 3.2. *The recurrence-transience of an H-self-similar additive process associated with a self-decomposable distribution μ does not depend on H.*

Proof. Let $\{X_t\}$ and $\{X'_t\}$ be, respectively, H-self-similar and H'-self-similar additive processes satisfying $\mathcal{L}(X_1) = \mathcal{L}(X'_1) = \mu$. Let $\eta = H'/H$. then $\{X'_t\} \stackrel{\mathrm{d}}{=} \{X_{t^\eta}\}$, as remarked on p. 295 of [6]. Hence, if $\{X_t\}$ is recurrent (resp. transient), then $\{X'_t\}$ is recurrent (resp. transient). □

LEMMA 3.3. *Let μ be a nontrivial stable distribution on $\mathbb{R}$. Then μ has a bounded continuous density $p(x)$ on $\mathbb{R}$. Moreover, $p(x) > 0$ on $\mathbb{R}$ if $1 \leqslant \alpha \leqslant 2$ or if $0 < \alpha < 1$ and $|\beta| < 1$; $p(x) > 0$ on (γ, ∞) if $0 < \alpha < 1$ and $\beta = 1$; $p(x) > 0$ on $(-\infty, \gamma)$ if $0 < \alpha < 1$ and $\beta = -1$.*

Proof. Existence of a bounded continuous density $p(x)$ of μ follows from the integrability of $\widehat{\mu}(z)$. Let S be the support of μ. It is known from the general theory of infinitely divisible distributions that $S = \mathbb{R}$ if $1 \leqslant \alpha \leqslant 2$; $S = \mathbb{R}$ if $0 < \alpha < 1$ and $|\beta| < 1$; $S = [\gamma, \infty)$ if $0 < \alpha < 1$ and $\beta = 1$; $S = (-\infty, \gamma]$ if $0 < \alpha < 1$ and $\beta = -1$. Since μ is unimodal by Yamazato's theorem [14], the density $p(x)$ is positive in the interior of the support. Instead of the unimodality, we can also use Sharpe's result [9] that the distribution density $p(t, x)$ of a Lévy process $\{Y_t\}$ on $\mathbb{R}$ is positive in the interior of the support, if $p(t, x)$ is jointly continuous in (t, x). □

The next lemma is the essence of the method of the proof of the recurrence by Yamamuro [13].

LEMMA 3.4. *Let $\{X_t\}$ be a 1-self-similar additive process on $\mathbb{R}$ with $\mathcal{L}(X_1) = \mu$ such that $\widehat{\mu}(z)$ is integrable and*

$$\int_{-\infty}^{\infty} \widehat{\mu}(z)\,\mathrm{d}z > 0. \tag{3.1}$$

Let

$$f(x) = (1 - |x|) \vee 0, \tag{3.2}$$

$$W_n = \left(\sum_{j=1}^{n} j^{-1}\right)^{-1} \sum_{j=1}^{n} f(X_j). \tag{3.3}$$

If $\{W_n, n = 1, 2, \ldots\}$ is uniformly integrable, then $\{X_t\}$ is recurrent.

Proof. Let

$$g(x) = \int_{-\infty}^{\infty} f(x)\mathrm{e}^{izx}\,\mathrm{d}x = (z/2)^{-2}\sin^2(z/2),$$

then

$$f(x) = (2\pi)^{-1} \int_{-\infty}^{\infty} g(z)\mathrm{e}^{izx}\,\mathrm{d}z.$$

Let $b = (2\pi)^{-1} \int_{-\infty}^{\infty} \widehat{\mu}(z)\, \mathrm{d}z$ and $a_n = \sum_{j=1}^{n} j^{-1}$.

Using the 1-self-similarity, we have

$$\begin{aligned} E[f(X_j)] &= E[f(jX_1)] = (2\pi)^{-1} \int \mu(\mathrm{d}x) \int g(z) \mathrm{e}^{ijxz}\, \mathrm{d}z \\ &= (2\pi)^{-1} j^{-1} \int g(j^{-1}z) \widehat{\mu}(z)\, \mathrm{d}z \end{aligned} \tag{3.4}$$

and, hence,

$$|E[W_n] - b| = (2\pi)^{-1} \left| a_n{}^{-1} \sum_{j=1}^{n} j^{-1} \int (g(j^{-1}z) - 1)\, \widehat{\mu}(z)\, \mathrm{d}z \right|.$$

Noticing that $g(j^{-1}z) \to 1$ boundedly as $j \to \infty$, we see that $\int (g(j^{-1}z) - 1)\, \widehat{\mu}(z)\, \mathrm{d}z \to 0$ as $j \to \infty$. Since $a_n \to \infty$, we get

$$\lim_{n\to\infty} E[W_n] = b. \tag{3.5}$$

Suppose that $\{X_t\}$ is not recurrent. Then it is transient by Lemma 3.1. Hence, almost surely, $W_n \to 0$ as $n \to \infty$. Thus, by the uniform integrability, $E[W_n] \to 0$ as $n \to \infty$. This contradicts (3.5), since $b > 0$. □

Proof of Theorem 2.1. If $\alpha = 2$, then the assertion is shown in Proposition 5.1 of [8]. If $1 \leqslant \alpha < 2$, then the assertion is proved in Theorem 1.1 of [13] together with a new proof in the case of $\alpha = 2$. Henceforth, let $0 < \alpha < 1$. By Lemma 3.2 we can assume that $H = 1$.

Let $\{Z_t\}$ be the strictly stable Lévy process with

$$E[\mathrm{e}^{izZ_1}] = \exp\left[-c|z|^{\alpha}\left(1 - i\beta\left(\tan\frac{\pi\alpha}{2}\right)\operatorname{sgn} z\right)\right].$$

Then $\{Z_{t^\alpha} + \gamma t\}$ is a 1-self-similar additive process with distribution μ at time 1. Hence $\{X_t\} \stackrel{\mathrm{d}}{=} \{Z_{t^\alpha} + \gamma t\}$. If $\beta = 1$ (resp. $\beta = -1$), then $\{Z_t\}$ has increasing (resp. decreasing) sample functions a.s. Hence, if $\beta = 1$ (resp. $\beta = -1$) and $\beta\gamma \geqslant 0$, then $\{X_t\}$ has increasing (resp. decreasing) sample functions a.s. and it is transient. If $|\beta| < 1$ and $\gamma = 0$, then $\{X_t\}$ is transient, since $\{Z_t\}$ is transient.

It remains for us to prove the recurrence of $\{X_t\}$ under the following condition (3.6) or (3.7):

$$|\beta| = 1 \text{ and } \beta\gamma < 0, \tag{3.6}$$

$$|\beta| < 1 \text{ and } \gamma \neq 0. \tag{3.7}$$

We shall apply Lemma 3.4. Let $p(x)$ be the bounded continuous density of μ.

First note that $p(0) > 0$ by Lemma 3.3. Since $p(x) = (2\pi)^{-1} \int_{-\infty}^{\infty} \widehat{\mu}(z) \mathrm{e}^{-izx}\, \mathrm{d}z$, this means that (3.1) is satisfied. Define $f(x)$ and W_n by (3.2) and (3.3). The proof will be complete, if we check the uniform integrability of $\{W_n\}$. It sufficcs to show

$$\sup_n E[W_n{}^2] < \infty. \tag{3.8}$$

We use $g(x)$ and a_n in the proof of Lemma 3.4. We have

$$W_n{}^2 = a_n{}^{-2} \sum_{j=1}^{n} f(X_j)^2 + 2a_n{}^{-2} \sum_{j=1}^{n-1} \sum_{k=1}^{n-j} f(X_j) f(X_{j+k}). \tag{3.9}$$

Since $f(X_j)^2 \leqslant f(X_j)$, the expectation of the first term in the right-hand side of (3.9) is of $O(a_n{}^{-1})$ by virtue of (3.4). Since $\{X_t\}$ has independent increments, the expectation of the second term is

$$2a_n{}^{-2} \sum_{j=1}^{n-1} \sum_{k=1}^{n-j} E[f(X_j) h_{j,k}(X_j)], \tag{3.10}$$

where

$$\begin{aligned} h_{j,k}(x) &= E[f(x + X_{j+k} - X_j)] \\ &= (2\pi)^{-1} \int \rho(\mathrm{d}y) \int g(z) \mathrm{e}^{i(x+y)z} \, \mathrm{d}z \\ &= (2\pi)^{-1} \int g(z) \mathrm{e}^{ixz} \widehat{\rho}(z) \, \mathrm{d}z \end{aligned}$$

with the distribution ρ of $X_{j+k} - X_j$. If we consider $\{-X_t\}$ instead of $\{X_t\}$, then β and γ change their signs. Hence, we may and do assume that $\gamma > 0$. We have

$$\widehat{\rho}(z) = \widehat{\mu}((j+k)z)/\widehat{\mu}(jz) = \mathrm{e}^{i\gamma kz} \psi(z),$$

where

$$\psi(z) = \exp[-c((j+k)^\alpha - j^\alpha)|z|^\alpha (1 - i\beta' \mathrm{sgn} z)] \quad \text{with } \beta' = \beta \tan \frac{\pi\alpha}{2}.$$

For simplicity let us write $A = ((j+k)^\alpha - j^\alpha)^{1/\alpha}$. Then

$$\begin{aligned} h_{j,k}(x) &= (2\pi)^{-1} \int_{-\infty}^{\infty} g(z) \mathrm{e}^{i(x+\gamma k)z} \psi(z) \, \mathrm{d}z = (2\pi)^{-1}(I_1 + I_2), \\ I_1 &= \int_0^\infty g(z) \mathrm{e}^{i(x+\gamma k)z} \mathrm{e}^{-c(Az)^\alpha (1-i\beta')} \, \mathrm{d}z, \\ I_2 &= \int_0^\infty g(z) \mathrm{e}^{-i(x+\gamma k)z} \mathrm{e}^{-c(Az)^\alpha (1+i\beta')} \, \mathrm{d}z. \end{aligned}$$

Suppose that $x + \gamma k \neq 0$. Then by integration by parts

$$\begin{aligned} I_1 = (x+\gamma k)^{-1} \Bigg[-i^{-1} - i^{-1} \int_0^\infty \{g'(z) - g(z)c(1 - i\beta')A^\alpha \alpha z^{\alpha-1}\} \times \\ \times \mathrm{e}^{-c(Az)^\alpha (1-i\beta')} \mathrm{e}^{i(x+\gamma k)z} \, \mathrm{d}z \Bigg] \end{aligned}$$

and I_2 is the complex conjugate of I_1. Note that $g'(z) \sim -z/6$ as $z \to 0$ and $g'(z) = O(z^{-2})$ as $z \to \infty$. It follows that

$$|I_1 + I_2| \\ \leqslant 2|x+\gamma k|^{-1}\left(\int_0^\infty |g'(z)|\,dz + (1+\beta'^2)^{1/2} c A^\alpha \alpha \int_0^\infty e^{-c(Az)^\alpha} z^{\alpha-1}\,dz\right).$$

Since $A^\alpha \int_0^\infty e^{-c(Az)^\alpha} z^{\alpha-1}\,dz = \int_0^\infty e^{-cz^\alpha} z^{\alpha-1}\,dz$, we get

$$h_{j,k}(x) = (2\pi)^{-1}|I_1 + I_2| \leqslant \text{const}\ |x+\gamma k|^{-1}, \tag{3.11}$$

where the constant is independent of x, j, and k. Now let N be the greatest integer not exceeding $\gamma^{-1}+1$. For any $k \geqslant N$ we have $\gamma k > 1$ and

$$E[f(X_j)\,h_{j,k}(X_j)] \leqslant \text{const}\ (\gamma k - 1)^{-1} E[f(X_j)] \leqslant \text{const}\ (\gamma k - 1)^{-1} j^{-1}$$

by (3.4) and (3.11). Here we have used the fact that $f(x) = 0$ for $|x| \geqslant 1$. Hence

$$\sum_{j=1}^{n-1} \sum_{N \leqslant k \leqslant n-j} E[f(X_j)\,h_{j,k}(X_j)] \leqslant \text{const}\ a_n^2. \tag{3.12}$$

If $N \geqslant 2$, then take into consideration that

$$\sum_{j=1}^{n-1} \sum_{k=1}^{(N-1)\wedge(n-j)} E[f(X_j)\,h_{j,k}(X_j)] \leqslant \text{const}\ (N-1)a_n \tag{3.13}$$

by $h_{j,k} \leqslant 1$ and by (3.4). We now conclude from (3.12) and (3.13) that (3.10) is bounded. This finishes the proof of (3.8). □

Remark 3.5. In the above, the proof of the recurrence in the case of $1 \leqslant \alpha < 2$ relies on [13]. But, exactly the same argument as above works also when $1 < \alpha < 2$ and $\gamma \neq 0$. A similar proof of the recurrence can be given also in the case of $\alpha = 1$; we have only to replace the proof of the boundedness of (3.10) by the consideration that

$$|\widehat{\rho}(z)| = |\widehat{\mu}((j+k)z)/\widehat{\mu}(jz)| \leqslant e^{-ck|z|}$$

and that

$$h_{j,k}(x) \leqslant (2\pi)^{-1}\int_{-\infty}^{\infty} g(z)|\widehat{\rho}(z)|\,dz \leqslant (\pi c k)^{-1}.$$

If $1 < \alpha < 2$ and $\gamma = 0$, then μ is strictly stable and the recurrence is part of Proposition 2.4.

Proof of Corollary 2.2. Assume that μ is not strictly stable and that μ is not one-sided. If $1 < \alpha \leqslant 2$, then $\gamma \neq 0$ and we get the recurrence of $\{X_t\}$ by Theorem 2.1

and the transience of $\{Y_t\}$ by Proposition 2.4. If $\alpha = 1$, then $\beta \neq 0$ and we get the same conclusion on $\{X_t\}$ and $\{Y_t\}$. If $0 < \alpha < 1$, then $\gamma \neq 0$ and we have either $|\beta| < 1$ or $|\beta| = 1$ and $\beta\gamma < 0$ (since μ is not one-sided), which leads to the same conclusion again. □

Proof of Corollary 2.3. If $\alpha \neq 1$, then, for any $a > 0$,

$$\{Z_{at}^{(b)}\} \stackrel{\mathrm{d}}{=} \{a^{1/\alpha}(Y_t - \gamma t) + ba^{1/\alpha}t^{1/\alpha}\} = \{a^{1/\alpha}Z_t^{(b)}\},$$

that is, $\{Z_t^{(b)}\}$ is a $(1/\alpha)$-self-similar additive process and the characteristic function of $\mathcal{L}(Z_1^{(b)})$ is the right-hand side of (1.4) or (1.5) with γ replaced by b. If $\alpha = 1$, then

$$E[\exp(izZ_t^{(b)})] = \exp\left[-tc|z| - itcz\beta\frac{2}{\pi}\log(t|z|) + itz(\gamma + b)\right],$$

which shows that $\{Z_t^{(b)}\}$ is 1-self-similar and $\mathcal{L}(Z_1^{(b)})$ is stable with index 1. Hence our assertion follows from Theorem 2.1. □

4. Remarks

Remark 4.1. With any self-decomposable distribution μ we can associate a self-similar additive process $\{X_t\}$ and a Lévy process $\{Y_t\}$ such that $\mathcal{L}(X_1) = \mathcal{L}(Y_1) = \mu$. In the case where μ is stable on $\mathbb{R}$, the processes $\{X_t\}$ and $\{Y_t\}$ fall into the following three cases:

(1) both $\{X_t\}$ and $\{Y_t\}$ are recurrent;
(2) both $\{X_t\}$ and $\{Y_t\}$ are transient;
(3) $\{X_t\}$ is recurrent and $\{Y_t\}$ is transient.

This is essentially Corollary 2.2. However, if μ is not stable, then the following case also arises on $\mathbb{R}$:

(4) $\{X_t\}$ is transient and $\{Y_t\}$ is recurrent.

See Proposition 5.2 of [8].

Remark 4.2. Let $\{X_t\}$ be an H-self-similar additive process on $\mathbb{R}$ with $\mathcal{L}(X_1) = \mu$ being stable but not strictly stable. Assume that $1 \leqslant \alpha \leqslant 2$ or that $0 < \alpha < 1$ and $|\beta| < 1$. Then

$$\limsup_{t\to\infty} t^{-H}X_t = \infty \quad \text{a.s.} \tag{4.1}$$

and

$$\liminf_{t\to\infty} t^{-H}X_t = -\infty \quad \text{a.s.} \tag{4.2}$$

Indeed, let $b \in \mathbb{R}$ and let $X'_t = X_t + bt^H$. Here, if $0 < \alpha < 1$, then we assume that $b \neq -\gamma$. Then $\{X'_t\}$ is a recurrent H-self-similar additive process by Theorem 2.1. Since $X_t = X'_t - bt^H$, the recurrence of $\{X'_t\}$ implies $\limsup_{t\to\infty} X_t = \infty$ a.s. and $\liminf_{t\to\infty} X_t = -\infty$ a.s. by the freedom of choice of b. Therefore $\limsup_{t\to\infty} X'_t = \infty$ a.s. and $\liminf_{t\to\infty} X'_t = -\infty$ a. s. Now (4.1) and (4.2) follow from $t^{-H} X_t = t^{-H} X'_t - b$, again by the freedom of choice of b. The facts (4.1), (4.2) are known by Fristedt [2], p. 361, and Pruitt and Taylor [4], p. 322, if we rewrite them into behaviors of Lévy processes as in Corollary 2.3. But our proof reveals the underlying structure for these facts.

If $0 < \alpha < 1$ and $\beta = 1$, then we get (4.1). If $0 < \alpha < 1$ and $\beta = -1$, then we get (4.2). Their proofs are similar.

References

1. Chung, K. L. and Fuchs, W. H.: On the distribution of values of sums of random variables, In: *Four Papers in Probability*, Mem. Amer. Math. Soc. 6, Amer. Math. Soc., Providence, RI, 1951, pp. 1–12.
2. Fristedt, B.: Sample functions of stochastic processes with stationary, independent increments, In: P. Ney and S. Port (eds), *Advances in Probability*, Vol. 3, Marcel Dekker, New York, 1974, pp. 241–396.
3. Hida, T.: *Brownian Motion*, Springer, New York, 1980.
4. Pruitt, W. E. and Taylor, S. J.: The behavior of asymmetric Cauchy processes for large time, *Ann. Probab.* **11** (1983), 302–327.
5. Sato, K.: Distributions of class L and self-similar processes with independent increments, In: T. Hida *et al.* (eds), *White Noise Analysis*, World Scientific, Singapore, 1990, pp. 360–373.
6. Sato, K.: Self-similar processes with independent increments, *Probab. Theory Related Fields* **89** (1991), 285–300.
7. Sato, K.: *Lévy Processes and Infinitely Divisible Distributions*, Cambridge Univ. Press, to appear.
8. Sato, K. and Yamamuro, K.: On selfsimilar and semi-selfsimilar processes with independent increments, *J. Korean Math. Soc.* **35** (1998), 207–224.
9. Sharpe, M.: Zeroes of infinitely divisible densities, *Ann. Math. Statist.* **40** (1969), 1503–1505.
10. Spitzer, F.: *Principles of Random Walk*, Van Nostrand, Princeton, NJ, 1964 (2nd edn, Springer, New York, 1976).
11. Watanabe, Toshiro: Sample function behavior of increasing processes of class L, *Probab. Theory Related Fields* **104** (1996), 349–374.
12. Yamamuro, K.: Transience conditions for self-similar additive processes, *J. Math. Soc. Japan*, to appear.
13. Yamamuro, K.: On recurrence for self-similar additive processes, Preprint 1999.
14. Yamazato, M.: Unimodality of infinitely divisible distribution functions of class L, *Ann. Probab.* **6** (1978), 523–531.

Acta Applicandae Mathematicae **63:** 385–410, 2000.

Semigroup Domination on a Riemannian Manifold with Boundary

Dedicated to Professor Takeyuki Hida on his 70th birthday

ICHIRO SHIGEKAWA
Department of Mathematics, Graduate School of Science, Kyoto University, Kyoto 606-8502, Japan. e-mail: ichiro@kusm.kyoto-u.ac.jp

(Received: 24 December 1998)

Abstract. We discuss the semigroup domination on a Riemannian manifold with boundary. Our main interest is the Hodge–Kodaira Laplacian for differential forms. We consider two kinds of boundary conditions; the absolutely boundary condition and the relative boundary condition. Our main tool is the square field operator. We also develop a general theory of semigroup commutation.

Mathematics Subject Classifications (2000): 60J60, 58G32.

Key words: semigroup domination, Hodge–Kodaira Laplacian, absolute boundary condition, relative boundary condition.

1. Introduction

In this paper, we discuss a theory of semigroup domination. To be precise, let $(X, \mathcal{B}, m)$ be a σ-finite measure space. Suppose we are given a semigroup $\{T_t\}$ on $L^2 = L^2(m)$. We assume that $\{T_t\}$ is positivity preserving. Besides, we are given a semigroup $\{\vec{T}_t\}$ acting on Hilbert space valued square integrable functions. We denote the norm of the Hilbert space by $|\cdot|$. If we have

$$|\vec{T}_t u| \leqslant T_t|u|, \quad \forall u \in L^2, \tag{1.1}$$

we say that the semigroup $\{\vec{T}_t\}$ is dominated by $\{T_t\}$. We are interested in when this inequality holds.

A necessary and sufficient condition for (1.1) is given by the abstract Kato theorem due to B. Simon [18, 19]. Later, E. Ouhabaz [13] gave a necessary and sufficient condition in terms of bilinear form under the sector condition. In my previous paper [16], we discuss this problem within the framework of square field operator. A typical example to which our theorem is applicable is the Hodge–Kodaira Laplacian for differential forms on a Riemannian manifold (see also [10] in this direction).

In this paper, we consider the Hodge–Kodaira Laplacian on a Riemannian manifold with boundary. We consider two kinds of boundary condition: the relative boundary condition and the absolute boundary condition. In this case, we cannot apply the result in [16] and so we generalize the notion of $\vec{\Gamma}$ that corresponds to the Bakry–Emery Γ_2. So far, $\vec{\Gamma}$ is an L^1 function. But, in our formulation, it is no more a function; it is a smooth measure in the sense of Dirichlet form. The positivity of the smooth measure is essential. Using this notion, we give a sufficient condition to (1.1).

We remark that this kind of problem was also discussed by Donnelly and Li [6]. They proved the heat kernel domination. We take a different approach. Méritet [12] also proved the cohomology vanishing theorem.

The organization of this paper is as follows. In Section 2, we give a generalization of $\vec{\Gamma}$ and prepare the general theory for the semigroup domination. To do this, we use the Ouhabaz criterion for the semigroup domination. We also discuss the theory of semigroup commutation in Section 3. Combining the semigroup domination with the semigroup commutation, we can reformulate the Bakry–Emery criterion for the logarithmic Sobolev inequality. In Section 4, we consider the Hodge–Kodaira Laplacian and apply our theory to it. To give a sufficient condition for the domination, the second fundamental form on the boundary is crucial.

2. Square Field Operator and the Contraction Semigroup

In this section, we discuss the contraction semigroup in the framework of the square field operator, which is called '*opérateur carré du champ*' in the French literature. Our main interest is the semigroup domination on a Riemannian manifold with boundary in Section 4, but we prepare a general theory in this section.

Let $(X, \mathcal{B})$ be a measurable space and m be a σ-finite measure. Suppose we are given a strongly continuous contraction symmetric semigroup $\{T_t\}$ on L^2. We further assume that T_t is Markovian, i.e., if $f \in L^2$ satisfies $0 \leqslant f \leqslant 1$, then $0 \leqslant T_t f \leqslant 1$. Here these inequalities hold a.e. But we do not specify 'a.e.' either in the sequel. We denote the generator by A and the associated Dirichlet form by $\mathcal{E}$.

We note that we can regard $\{T_t\}$ as a semigroup on $L^p(m)$, $p \geqslant 1$ by the Riesz–Thorin interpolation theorem. We denote the generator of $\{T_t\}$ on $L^p(m)$ by A_p.

To introduce the square field operator, we assume the following condition (see [3, Chapter 1, §4] for details):

(Γ) For $f, g \in \mathrm{Dom}(A_2)$, we have $f \cdot g \in \mathrm{Dom}(A_1)$.

Under the above assumption, we set

$$\Gamma(f, g) = \tfrac{1}{2}\{A_1(f \cdot g) - A_2 f \cdot g - f \cdot A_2 g\}. \tag{2.1}$$

Here we remark that our definition of Γ is different from that of [3] up to a constant.

Furthermore, we suppose that $\mathcal{E}$ has the local property in the following sense (see [3, Definition I.5.1.2]):

(L) For any real valued function $f \in \mathrm{Dom}(\mathcal{E})$, $F, G \in C_0^\infty(\mathbb{R})$,

$$\operatorname{supp} F \cap \operatorname{supp} G = \emptyset \implies \mathcal{E}(F_0(f), G_0(f)) = 0, \tag{2.2}$$

where $F_0(x) = F(x) - F(0)$, $G_0(x) = G(x) - G(0)$.

The above condition is satisfied as soon as it is satisfied for each element of a dense subset of $\mathrm{Dom}(\mathcal{E})$. In particular, the following identity is most commonly used: For $f, g \in \mathrm{Dom}(\mathcal{E}) \cap L^\infty$,

$$\Gamma(fg, h) = f\Gamma(g, h) + g\Gamma(f, h) \quad \text{for } \forall h \in \mathrm{Dom}(\mathcal{E}). \tag{2.3}$$

In addition, we are given a strongly continuous semigroup $\{\vec{T}_t\}$ on $L^2(m; K)$ where $L^2(m; K)$ is the set of all square integrable K-valued functions, K being a separable Hilbert space. We denote the generator by $\vec{A}$ (or $\vec{A}_2$ to specify the space $L^2(m; K)$). We also assume that $\{\vec{T}_t\}$ is symmetric and is associated with a bilinear form $\vec{\mathcal{E}}$. Of course, $\vec{\mathcal{E}}$ is bounded from below.

Our interest is the following semigroup domination:

$$|\vec{T}_t u| \leqslant T_t|u|, \quad \forall u \in L^2. \tag{2.4}$$

To give a sufficient condition for (2.4), we now define the square field operator for $\{\vec{T}_t\}$. To do this, we imposed the following condition in [16]:

($\vec{\Gamma}_\lambda$) For $u, v \in \mathrm{Dom}(\vec{A}_2)$, we have $(u|v)_K \in \mathrm{Dom}(A_1)$ and there exists $\lambda \in \mathbb{R}$ such that

$$A_1|u|^2 - 2(\vec{A}_2 u|u)_K + 2\lambda|u|^2 \geqslant 0. \tag{2.5}$$

Under the above condition, we define $\vec{\Gamma}$ by

$$\vec{\Gamma}(u, v) = \tfrac{1}{2}\{A_1(u|v)_K - (\vec{A}_2 u|v)_K - (u|\vec{A}_2 v)_K\}. \tag{2.6}$$

The condition ($\vec{\Gamma}_\lambda$) is rather restrictive. In particular, we have to assume that $(u|v)_K \in \mathrm{Dom}(A_1)$ for $u, v \in \mathrm{Dom}(\vec{A}_2)$. We can define $\vec{\Gamma}$ without this condition. In fact, under the above definition, a formal calculation leads to

$$-\mathcal{E}((u, v), f) + \vec{\mathcal{E}}(fu, v) + \vec{\mathcal{E}}(u, fv) = 2\int_X \vec{\Gamma}(u, v) f \, dm. \tag{2.7}$$

In the above expression it is only necessary to assume that $(u, v) \in \mathrm{Dom}(\mathcal{E})$, fu, $fv \in \mathrm{Dom}(\vec{\mathcal{E}})$.

Keeping this observation in mind, we modify the condition ($\vec{\Gamma}_\lambda$) as follows:

($\vec{\Gamma}'_\lambda$-1) For $u, v \in \mathrm{Dom}(\vec{\mathcal{E}}) \cap L^\infty$ and $f \in \mathrm{Dom}(\mathcal{E}) \cap L^\infty$, we have $(u, v) \in \mathrm{Dom}(\mathcal{E})$, fu, $fv \in \mathrm{Dom}(\vec{\mathcal{E}})$ and, moreover, $\mathrm{Dom}(\vec{\mathcal{E}}) \cap L^\infty$ is dense in $\mathrm{Dom}(\vec{\mathcal{E}})$.

($\vec{\Gamma}'_\lambda$-2) There exists $\vec{\Gamma}: \mathrm{Dom}(\vec{\mathcal{E}}) \cap L^\infty \times \mathrm{Dom}(\vec{\mathcal{E}}) \cap L^\infty \to L^1$, a smooth measure σ, and a nonnegative symmetric tensor $\vec{\gamma}$ (i.e. $\vec{\gamma}$ is an $H^* \otimes H^*$-valued function on X) such that

$$\begin{aligned} &-\mathcal{E}((u, v), f) + \vec{\mathcal{E}}(fu, v) + \vec{\mathcal{E}}(u, fv) \\ &= 2\int_X \vec{\Gamma}(u, v) f \, dm + 2\int_X \vec{\gamma}(\tilde{u}, \tilde{v}) \tilde{f} \, d\sigma \\ &\forall f \in \mathrm{Dom}(\mathcal{E}) \cap L^\infty, \quad \forall u, v \in \mathrm{Dom}(\vec{\mathcal{E}}) \cap L^\infty. \end{aligned} \tag{2.8}$$

Here $\tilde{u}$, $\tilde{v}$, $\tilde{f}$ are quasi-continuous modification of u, v, f, respectively. Of course, we have assumed that each element of $\mathrm{Dom}(\vec{\mathcal{E}})$ admits a quasi-continuous modification.

($\vec{\Gamma}'_\lambda$-3) For the real constant λ, it holds that

$$\vec{\Gamma}(u, u) + \lambda |u|^2 \geqslant 0 \quad \text{for } u \in \mathrm{Dom}(\vec{A}_2). \tag{2.9}$$

($\vec{\Gamma}'_\lambda$-4) It holds that

$$\begin{aligned} \vec{\mathcal{E}}(u, v) &= \int_X \vec{\Gamma}(u, v)\, dm + \int_X \vec{\gamma}(\tilde{u}, \tilde{v})\, d\sigma \\ \forall u, v &\in \mathrm{Dom}(\vec{\mathcal{E}}) \cap L^\infty. \end{aligned} \tag{2.10}$$

In the sequel, ($\vec{\Gamma}'_\lambda$) refers to these four conditions. We emphasize that we assumed the *positivity* of $\vec{\gamma}$ and σ. Since $\mathrm{Dom}(\vec{\mathcal{E}}) \cap L^\infty$ is dense in $\mathrm{Dom}(\vec{\mathcal{E}})$, it is easy to see that $\vec{\Gamma}$ is well-defined on $\mathrm{Dom}(\vec{\mathcal{E}}) \times \mathrm{Dom}(\vec{\mathcal{E}})$ as a continuous bilinear mapping into L^1 and (2.10) holds for $u, v \in \mathrm{Dom}(\vec{\mathcal{E}})$.

Next we introduce the following condition on $\vec{\Gamma}$ and $\vec{\mathcal{E}}$. This has already appeared in [16]. It is related to the derivation property of $\vec{\Gamma}$.

($\vec{D}$) For $u, v \in \mathrm{Dom}(\vec{\mathcal{E}}) \cap L^\infty$, $f \in \mathrm{Dom}(\mathcal{E}) \cap L^\infty$, it holds that

$$2f\vec{\Gamma}(u, v) = -\Gamma(f, (v|u)_K) + \vec{\Gamma}(u, fv) + \vec{\Gamma}(fu, v). \tag{2.11}$$

From this, we have

$$2f\vec{\Gamma}(u, u) = -\Gamma(f, |u|^2) + 2\vec{\Gamma}(u, fu). \tag{2.12}$$

Further, by (2.8), we have

$$-\mathcal{E}((u, u), f) + 2\vec{\mathcal{E}}(fu, u) = 2\int_X \vec{\Gamma}(u, u) f\, dm + 2\int_X \vec{\gamma}(\tilde{u}, \tilde{u})\tilde{f}\, d\sigma. \tag{2.13}$$

Substituting (2.12) into this equation, we have

$$\begin{aligned} &-\mathcal{E}((u, u), f) + 2\vec{\mathcal{E}}(fu, u) \\ &= -\int_X \Gamma(f, |u|^2)\, dm + 2\int_X \vec{\Gamma}(fu, u)\, dm + 2\int_X \vec{\gamma}(\tilde{u}, \tilde{u})\tilde{f}\, d\sigma \\ &= -\mathcal{E}((u, u), f) + 2\int_X \vec{\Gamma}(fu, u)\, dm + 2\int_X \vec{\gamma}(\tilde{u}, \tilde{u})\tilde{f}\, d\sigma. \end{aligned}$$

Thus we have

$$\vec{\mathcal{E}}(fu, u) = \int_X \vec{\Gamma}(fu, u)\, dm + \int_X \vec{\gamma}(\tilde{u}, \tilde{u})\tilde{f}\, d\sigma. \tag{2.14}$$

In particular, for $f \geqslant 0$, it holds that

$$\vec{\mathcal{E}}(fu, u) \geqslant \int_X \vec{\Gamma}(fu, u)\, dm. \tag{2.15}$$

We remark that if $\mathbf{1} \in \mathrm{Dom}(\mathcal{E})$, (2.10) follows from (2.14).

Here $\mathbf{1}$ denotes the function identically equal to 1.

We can give a sufficient condition for (2.4) in terms of square field operator. To do this, we need the following Ouhabaz criterion. For the semigroup domination (2.4), the following condition is necessary and sufficient:

(1) If $u \in \mathrm{Dom}(\mathcal{E})$, then $|u| \in \mathrm{Dom}(\mathcal{E})$ and

$$\mathcal{E}(|u|, |u|) \leqslant \vec{\mathcal{E}}(u, u) + \lambda \|u\|_2^2. \tag{2.16}$$

(2) If $u \in \mathrm{Dom}(\vec{\mathcal{E}})$ and $f \in \mathrm{Dom}(\mathcal{E})$ with $0 \leqslant f \leqslant |u|$, then $f \operatorname{sgn} u \in \mathrm{Dom}(\vec{\mathcal{E}})$ and

$$\mathcal{E}(f, |u|) \leqslant \vec{\mathcal{E}}(u, f \operatorname{sgn} u) + \lambda(f, |u|), \tag{2.17}$$

where $\operatorname{sgn} u = u/|u|$.

Now we have the following theorem.

THEOREM 2.1. *Assume conditions* (Γ), (L), $(\vec{\Gamma}_\lambda)$ *and* $(\vec{D})$. *Then, for* $u \in \mathrm{Dom}(\vec{\mathcal{E}})$ *we have* $|u| \in \mathrm{Dom}(\mathcal{E})$ *and*

$$\Gamma(|u|, |u|) \leqslant \vec{\Gamma}(u, u) + \lambda |u|^2. \tag{2.18}$$

In addition, for $f \in \mathrm{Dom}(\mathcal{E}) \cap L^\infty$ *and* $u \in \mathrm{Dom}(\vec{\mathcal{E}}) \cap L^\infty$, *we have*

$$\{\vec{\Gamma}(fu, fu) + \lambda |fu|^2\}^{1/2} \leqslant |f| \{\vec{\Gamma}(u, u) + \lambda |u|^2\}^{1/2} + |u| \Gamma(f, f)^{1/2}. \tag{2.19}$$

Furthermore, we have the following semigroup domination:

$$|\vec{T}_t u| \leqslant \mathrm{e}^{\lambda t} T_t |u| \quad \text{for } u \in L^2(X; K). \tag{2.20}$$

Proof. Many parts are the same as in [16]. We give a proof for completeness. For simplicity, we give a proof in the case $\lambda = 0$. Take $u \in \mathrm{Dom}(\vec{\mathcal{E}}) \cap L^\infty$ and $f \in \mathrm{Dom}(\mathcal{E}) \cap L^\infty$. From the assumption, $v = fu \in \mathrm{Dom}(\vec{\mathcal{E}})$ and we substitute v in (2.11):

$$\vec{\Gamma}(fu, fu) + \vec{\Gamma}(u, |f|^2 u) = 2f \vec{\Gamma}(u, fu) + \Gamma(f, f|u|^2).$$

Hence

$$\begin{aligned}
\vec{\Gamma}(fu, fu) &= -\vec{\Gamma}(u, |f|^2 u) + 2f \vec{\Gamma}(u, fu) + \Gamma(f, f|u|^2) \\
&= -|f|^2 \vec{\Gamma}(u, u) - \frac{1}{2} \Gamma(|f|^2, |u|^2) + f\{2f \vec{\Gamma}(u, u) + \\
&\quad + \Gamma(f, |u|^2)\} + \Gamma(f, f|u|^2) \\
&= -|f|^2 \vec{\Gamma}(u, u) - \frac{1}{2} f \Gamma(f, |u|^2) - \frac{1}{2} f \Gamma(f, |u|^2) + \\
&\quad + 2|f|^2 \vec{\Gamma}(u, u) + f \Gamma(f, |u|^2) + f \Gamma(f, |u|^2) + |u|^2 \Gamma(f, f) \\
&= |f|^2 \vec{\Gamma}(u, u) + \frac{1}{2} f \Gamma(f, |u|^2) + \frac{1}{2} f \Gamma(f, |u|^2) + |u|^2 \Gamma(f, f).
\end{aligned}$$

In particular, if we take $f = |u|^2$, we get

$$\vec{\Gamma}(|u|^2 u, |u|^2 u) = |u|^4 \vec{\Gamma}(u, u) + 2|u|^2 \Gamma(|u|^2, |u|^2). \tag{2.21}$$

On the other hand, substituting $v = u$ and $f = |u|^2$ in (2.11), we have

$$2|u|^2 \vec{\Gamma}(u, u) + \Gamma(|u|^2, |u|^2) = 2\vec{\Gamma}(|u|^2 u, u).$$

Taking square,

$$\begin{aligned}
&4|u|^4 \vec{\Gamma}(u, u)^2 + 4|u|^2 \vec{\Gamma}(u, u)\Gamma(|u|^2, |u|^2) + \Gamma(|u|^2, |u|^2)^2 \\
&\quad = 4\vec{\Gamma}(|u|^2 u, u)^2 \\
&\quad \leqslant 4\vec{\Gamma}(|u|^2 u, |u|^2 u)\vec{\Gamma}(u, u) \quad \text{(by the Schwarz inequality)} \\
&\quad = 4\{|u|^4 \vec{\Gamma}(u, u) + 2|u|^2 \Gamma(|u|^2, |u|^2)\}\vec{\Gamma}(u, u) \quad (\because (2.21)) \\
&\quad = 4|u|^4 \vec{\Gamma}(u, u)^2 + 8|u|^2 \Gamma(|u|^2, |u|^2)\vec{\Gamma}(u, u).
\end{aligned}$$

Thus we have

$$\Gamma(|u|^2, |u|^2) \leqslant 4|u|^2 \vec{\Gamma}(u, u). \tag{2.22}$$

Now for $\varepsilon > 0$, set $\varphi_\varepsilon(t) = \sqrt{t + \varepsilon^2} - \varepsilon$. Then by the derivation property and (2.22),

$$\begin{aligned}
&\Gamma(\varphi_\varepsilon(|u|^2), \varphi_\varepsilon(|u|^2)) \\
&\quad \leqslant \frac{1}{4(|u|^2 + \varepsilon^2)} \Gamma(|u|^2, |u|^2) \\
&\quad \leqslant \frac{4|u|^2}{4(|u|^2 + \varepsilon^2)} \vec{\Gamma}(u, u) \leqslant \vec{\Gamma}(u, u).
\end{aligned}$$

From this we can show that $\{\varphi_\varepsilon(|u|^2)\}_{\varepsilon > 0}$ is a bounded set in $\mathrm{Dom}(\mathcal{E})$. In fact

$$\begin{aligned}
\mathcal{E}(\varphi_\varepsilon(|u|^2), \varphi_\varepsilon(|u|^2)) &= \int_X \Gamma(\varphi_\varepsilon(|u|^2), \varphi_\varepsilon(|u|^2))\, \mathrm{d}m \\
&\leqslant \int_X \vec{\Gamma}(u, u)\, \mathrm{d}m \\
&\leqslant \vec{\mathcal{E}}(u, u).
\end{aligned}$$

Hence we can take a sequence $\{\varepsilon_j\}$ tending to 0 such that $\{\varphi_{\varepsilon_j}(|u|^2)\}_j$ converges weakly in $\mathrm{Dom}(\mathcal{E})$. The limit is $|u|$ since it converges to $|u|$ strongly in L^2. This means that $|u| \in \mathrm{Dom}(\mathcal{E})$. Further, taking a subsequence if necessary, we may assume that the Cesaro mean converges strongly in $\mathrm{Dom}(\mathcal{E})$, i.e.,

$$\frac{1}{n} \sum_{j=1}^{n} \varphi_{\varepsilon_j}(|u|^2) \to |u| \quad \text{strongly in } \mathrm{Dom}(\mathcal{E}).$$

By the continuity of Γ, we have

$$\Gamma\left(\frac{1}{n}\sum_{j=1}^{n}\varphi_{\varepsilon_j}(|u|^2), \frac{1}{n}\sum_{j=1}^{n}\varphi_{\varepsilon_j}(|u|^2)\right) \to \Gamma(|u|, |u|) \quad \text{strongly in } L^1.$$

On the other hand, by the Minkowski inequality, it follows that

$$\begin{aligned}&\Gamma\left(\frac{1}{n}\sum_{j=1}^{n}\varphi_{\varepsilon_j}(|u|^2), \frac{1}{n}\sum_{j=1}^{n}\varphi_{\varepsilon_j}(|u|^2)\right)^{1/2}\\ &\quad\leqslant \frac{1}{n}\sum_{j=1}^{n}\Gamma(\varphi_{\varepsilon_j}(|u|^2), \varphi_{\varepsilon_j}(|u|^2))^{1/2} \leqslant \vec{\Gamma}(u, u)^{1/2}.\end{aligned}$$

Therefore we have

$$\Gamma(|u|, |u|) \leqslant \vec{\Gamma}(u, u). \tag{2.23}$$

In fact, it is enough to take a sequence $\{\varepsilon_n\}$ such that the Cesaro mean of $\{\varphi_{\varepsilon_n}(|u|^2)\}$ converges to $|u|$ in $\mathrm{Dom}(\mathcal{E})$.

Now we return to $\vec{\Gamma}(fu, fu)$:

$$\begin{aligned}\vec{\Gamma}(fu, fu) &= |f|^2\vec{\Gamma}(u, u) + \frac{1}{2}f\Gamma(f, |u|^2) + \frac{1}{2}f\Gamma(f, |u|^2) + |u|^2\Gamma(f, f)\\ &\leqslant |f|^2\vec{\Gamma}(u, u) + |f|\Gamma(|u|^2, |u|^2)^{1/2}\Gamma(f, f)^{1/2} + |u|^2\Gamma(f, f)\\ &\leqslant |f|^2\vec{\Gamma}(u, u) + 2|f||u|\vec{\Gamma}(u, u)^{1/2}\Gamma(f, f)^{1/2} + |u|^2\Gamma(f, f)\\ &= \{|f|\vec{\Gamma}(u, u)^{1/2} + |u|\Gamma(f, f)^{1/2}\}^2,\end{aligned}$$

which shows (2.19).

We show that for $g \in \mathrm{Dom}(\mathcal{E}) \cap L^\infty_+$ and $u \in \mathrm{Dom}(\vec{\mathcal{E}}) \cap L^\infty$ we have

$$\vec{\mathcal{E}}(u, gu) \geqslant \mathcal{E}(g|u|, |u|). \tag{2.24}$$

To see this,

$$\begin{aligned}\vec{\mathcal{E}}(u, gu) &\geqslant \int_X \vec{\Gamma}(u, gu)\,\mathrm{d}m \quad (\because (2.15))\\ &= \int_X\left\{g\vec{\Gamma}(u, u) + \frac{1}{2}\Gamma(g, |u|^2)\right\}dm\\ &\geqslant \int_X\{g\Gamma(|u|, |u|) + |u|\Gamma(g, |u|)\}\,dm\\ &= \int_X \Gamma(g|u|, |u|)\,dm\\ &= \mathcal{E}(g|u|, |u|).\end{aligned}$$

Now we have to check Ouhabaz's condition. Since we have assumed that $\mathrm{Dom}(\vec{\mathcal{E}}) \cap L^\infty$ is dense in $\mathrm{Dom}(\vec{\mathcal{E}})$, for any $u \in \mathrm{Dom}(\vec{\mathcal{E}})$, we take a sequence $\{u_n\} \subseteq \mathrm{Dom}(\vec{\mathcal{E}}) \cap L^\infty$ that converges to u in $\mathrm{Dom}(\vec{\mathcal{E}})$. Set

$$\psi_\varepsilon(t) = \frac{1}{1+\varepsilon t}, \qquad t \geqslant 0.$$

Then $\psi_\varepsilon(|u_n|)|u_n| \in \mathrm{Dom}(\mathcal{E})$ and by (2.19) we have

$$\begin{aligned}
&\vec{\Gamma}(\psi_\varepsilon(|u_n|)u_n, \psi_\varepsilon(|u_n|)u_n)^{1/2} \\
&\quad \leqslant \psi_\varepsilon(|u_n|)\vec{\Gamma}(u_n, u_n)^{1/2} + |u_n|\Gamma(\psi_\varepsilon(|u_n|), \psi_\varepsilon(|u_n|))^{1/2} \\
&\quad \leqslant \psi_\varepsilon(|u_n|)\vec{\Gamma}(u_n, u_n)^{1/2} + |u_n||\psi_\varepsilon'|\Gamma(|u_n|, |u_n|)^{1/2} \\
&\quad \leqslant \left\{\psi_\varepsilon(|u_n|) + \frac{\varepsilon|u_n|}{(1+\varepsilon|u_n|)^2}\right\}\vec{\Gamma}(u_n, u_n)^{1/2} \\
&\quad \leqslant 2\psi_\varepsilon(|u_n|)\vec{\Gamma}(u_n, u_n)^{1/2} \\
&\quad \leqslant 2\vec{\Gamma}(u_n, u_n)^{1/2}.
\end{aligned}$$

Now

$$\begin{aligned}
&\vec{\mathcal{E}}(\psi_\varepsilon(|u_n|)u_n, \psi_\varepsilon(|u_n|)u_n) \\
&\quad = \int_X \vec{\Gamma}(\psi_\varepsilon(|u_n|)u_n, \psi_\varepsilon(|u_n|)u_n)\, \mathrm{d}m + \int_X \vec{\gamma}(\psi_\varepsilon(|u_n|)u_n, \psi_\varepsilon(|u_n|)u_n)\, \mathrm{d}\sigma \\
&\quad \leqslant 4\int_X \vec{\Gamma}(u_n, u_n)\, \mathrm{d}m + 4\int_X \vec{\gamma}(u_n, u_n)\, \mathrm{d}\sigma \\
&\quad = 4\vec{\mathcal{E}}(u_n, u_n).
\end{aligned}$$

Thus we have $\{\psi_\varepsilon(|u_n|)u_n\}$ is $\vec{\mathcal{E}}_1$ bounded and converges to $\psi_\varepsilon(|u|)u$ strongly in L^2 by taking a subsequence if necessary. Therefore we have $\psi_\varepsilon(|u|)u \in \mathrm{Dom}(\vec{\mathcal{E}}) \cap L^\infty$.

For any $f \in \mathrm{Dom}(\mathcal{E})$ and $u \in \mathrm{Dom}(\vec{\mathcal{E}})$ with $0 \leqslant f \leqslant |u|$. We further assume that u is bounded. Set

$$\varphi_\varepsilon(t) = \frac{1}{t+\varepsilon}$$

and

$$g = f\varphi_\varepsilon(|u|) = \frac{f}{|u|+\varepsilon}.$$

Clearly $g \in \mathrm{Dom}(\mathcal{E}) \cap L^\infty$. Now substituting g into (2.24), we have

$$\mathcal{E}\left(|u|, \frac{f}{|u|+\varepsilon}|u|\right) \leqslant \vec{\mathcal{E}}\left(u, \frac{f}{|u|+\varepsilon}u\right).$$

We note that the boundedness of u is necessary to obtain this inequality. Letting $\varepsilon \to 0$, we get the desired result. To do this, we note

$$\begin{aligned}
&\vec{\Gamma}(f\varphi_\varepsilon(|u|)u, f\varphi_\varepsilon(|u|)u)^{1/2} \\
&\quad \leqslant f\varphi_\varepsilon(|u|)\vec{\Gamma}(u, u) + |u|\Gamma(f\varphi_\varepsilon(|u|), f\varphi_\varepsilon(|u|))^{1/2}
\end{aligned}$$

and further

$$\begin{aligned}
&\vec{\Gamma}(f\varphi_\varepsilon(|u|), f\varphi_\varepsilon(|u|))^{1/2}\\
&\quad\leqslant f\Gamma(\varphi_\varepsilon(|u|), \varphi_\varepsilon(|u|))^{1/2} + \varphi_\varepsilon(|u|)\Gamma(f, f)\\
&\quad\leqslant f|\varphi_\varepsilon'(|u|)|\Gamma(|u|, |u|)^{1/2} + \varphi_\varepsilon(|u|)\Gamma(f, f)\\
&\quad\leqslant \varphi_\varepsilon(|u|)\{\Gamma(|u|, |u|)^{1/2} + \Gamma(f, f)\}.
\end{aligned}$$

Combining both of them, we have

$$\begin{aligned}
&\vec{\Gamma}(f\varphi_\varepsilon(|u|)u, f\varphi_\varepsilon(|u|)u)^{1/2}\\
&\quad\leqslant f\varphi_\varepsilon(|u|)\vec{\Gamma}(u, u) + |u|\varphi_\varepsilon(|u|)\{\Gamma(|u|, |u|)^{1/2} + \Gamma(f, f)\}\\
&\quad\leqslant 2\vec{\Gamma}(u, u)^{1/2} + \Gamma(f, f)^{1/2}.
\end{aligned}$$

Now we have $\{f\varphi_\varepsilon(|u|)u; \varepsilon > 0\}$ is bounded in $\mathrm{Dom}(\vec{\mathcal{E}})$. Clearly $f\varphi_\varepsilon(|u|)u \to f \operatorname{sgn} u$ in L^2 and hence we have $f \operatorname{sgn} u \in \mathrm{Dom}(\vec{\mathcal{E}})$. Moreover, we can take a sequence whose Cesaro mean converges strongly in $\mathrm{Dom}(\vec{\mathcal{E}})$ and, hence, we have

$$\vec{\Gamma}(f \operatorname{sgn} u, f \operatorname{sgn} u)^{1/2} \leqslant 2\vec{\Gamma}(u, u)^{1/2} + \Gamma(f, f)^{1/2} \tag{2.25}$$

and

$$\mathcal{E}(|u|, f) \leqslant \vec{\mathcal{E}}(u, f \operatorname{sgn} u). \tag{2.26}$$

Lastly we remove the restriction of the boundedness of u. So now we do not assume that u is bounded. So first we notice that

$$\psi_\varepsilon(|u|)|u| \geqslant \psi_\varepsilon(f)f$$

and both are bounded. We take a sequence ε_j converging to 0 such that

$$\begin{aligned}
u_N &= \frac{1}{N}\sum_{j=1}^{N} \psi_{\varepsilon_j}(|u|)u \to u \quad \text{strongly in } \mathrm{Dom}(\vec{\mathcal{E}}),\\
f_N &= \frac{1}{N}\sum_{j=1}^{N} \psi_{\varepsilon_j}(f)f \to f \quad \text{strongly in } \mathrm{Dom}(\mathcal{E}).
\end{aligned}$$

Now u_N is bounded and $f_N \leqslant |u_N|$ and, hence, by using (2.26), we have

$$\mathcal{E}(|u_N|, f_N) \leqslant \vec{\mathcal{E}}(u_N, f_N \operatorname{sgn} u_N). \tag{2.27}$$

By taking a subsequence if necessary, we may assume that $\{|u_N|\}$ converges to $|u|$ weakly in $\mathrm{Dom}(\mathcal{E})$ and $\{f_N \operatorname{sgn} u_N\}$ converges to $f \operatorname{sgn} u$ weakly in $\mathrm{Dom}(\vec{\mathcal{E}})$. Hence, by taking limit, we have

$$\mathcal{E}(|u|, f) \leqslant \vec{\mathcal{E}}(u, f \operatorname{sgn} u) \tag{2.28}$$

which is the desired result. This completes the proof. □

As before, it is sufficient to assume $(\vec{D})$ for elements of a core. We state it as a proposition. We also include an analogue of (Γ').

PROPOSITION 2.2. *Assume that the assumptions* (Γ) *and* (L) *hold. Furthermore assume that there exist an algebra* $\mathcal{C} \subseteq \mathrm{Dom}(\mathcal{E}) \cap L^\infty$ *and a subspace* $\mathcal{D} \subseteq \mathrm{Dom}(\vec{A}_2) \cap L^\infty$ *such that* $fu \in \mathcal{D}$ *(i.e.,* $\mathcal{D}$ *is a* $\mathcal{C}$*-module),* $(u, v) \in \mathcal{C}$ *for* $f \in \mathcal{C}$, $u, v \in \mathcal{D}$ *and further* $(\vec{\Gamma}'_\lambda)$ *and* $(\vec{D})$ *hold for* $f \in \mathcal{C}$, $u, v \in \mathcal{D}$, *e.g.,* $(\vec{\Gamma}'_\lambda\text{-}2)$ *means that any element in* $\mathcal{D}$ *admits a quasi-continuous modification and* (2.8) *holds. We also suppose that* $\mathcal{C}$ *is a core for* $\mathcal{E}$ *and* $\mathcal{D}$ *is a core for* $\vec{\mathcal{E}}$.

Then $(\vec{\Gamma}_\lambda)$ *and* $(\vec{D})$ *are satisfied for* $f \in \mathrm{Dom}(\mathcal{E}) \cap L^\infty$, $u, v \in \mathrm{Dom}(\vec{\mathcal{E}}) \cap L^\infty$.

Proof. We may suppose $\lambda = 0$. Take $u \in \mathcal{D}$ and $f \in \mathcal{C}$. From the assumption, $v = fu \in \mathcal{D}$. Now by the same argument as in the proof of Theorem 2.1, we have

$$\begin{aligned}\vec{\Gamma}(fu, fu) &= f^2\vec{\Gamma}(u,u) + \frac{1}{2} f\Gamma(f, |u|^2) + \frac{1}{2} f\Gamma(f, |u|^2) + |u|^2\Gamma(f,f)\\ &\leqslant f^2\vec{\Gamma}(u,u) + |f|\Gamma(f,f)^{1/2}\Gamma(|u|^2, |u|^2)^{1/2} + |u|^2\Gamma(f,f).\end{aligned}$$

Clearly this shows that

$$\begin{aligned}&\vec{\mathcal{E}}(fu, fu)\\ &= \int_X \vec{\Gamma}(fu, fu)\,\mathrm{d}m + \int_X \vec{\gamma}(fu, fu)\,\mathrm{d}\sigma\\ &\leqslant \|f\|_\infty^2 \int_X \vec{\Gamma}(u,u)\,\mathrm{d}m + \|f\|_\infty \int_X \Gamma(f,f)^{1/2}\Gamma(|u|^2, |u|^2)^{1/2}\,\mathrm{d}m +\\ &\quad + \|u\|_\infty^2 \int_X \Gamma(f,f)\,\mathrm{d}m + \|f\|_\infty^2 \int_X \vec{\gamma}(u,u)\,\mathrm{d}\sigma\\ &\leqslant \|f\|_\infty^2 \int_X \vec{\Gamma}(u,u)\,\mathrm{d}m + \|f\|_\infty \Big\{\int_X \Gamma(f,f)\,\mathrm{d}m\Big\}^{1/2}\Big\{\int_X \Gamma(|u|^2, |u|^2)\,\mathrm{d}m\Big\}^{1/2} +\\ &\quad + \|u\|_\infty^2 \int_X \Gamma(f,f)\,\mathrm{d}m + \|f\|_\infty^2 \int_X \vec{\gamma}(u,u)\,\mathrm{d}\sigma\\ &= \|f\|_\infty^2 \vec{\mathcal{E}}(u,u) + \|f\|_\infty \mathcal{E}(f,f)^{1/2}\mathcal{E}(|u|^2, |u|^2)^{1/2} + \|u\|_\infty^2 \mathcal{E}(f,f).\end{aligned}$$

We claim that for $f \in \mathrm{Dom}(\mathcal{E}) \cap L^\infty$, $u \in \mathcal{D}$, we have $fu \in \mathrm{Dom}(\vec{\mathcal{E}})$ and (2.11), (2.8) hold. To show this, take any real valued function $f \in \mathcal{C}$. Then, for any C^1-function $F\colon \mathbb{R} \to \mathbb{R}$, we take a sequence of polynomials $\{P_n\}$ such that $P_n \to F$ uniformly on any compact sets up to the first derivative. Since $\vec{\mathcal{E}}(P_n(f)u, P_n(f)u)$ is bounded and $P_n(f)u \to F(f)u$ in L^2. Thus we have $F(f)u \in \mathrm{Dom}(\vec{\mathcal{E}})$ and

$$2F(f)\vec{\Gamma}(u,v) = \Gamma(F(f), (u|v)_K) - \vec{\Gamma}(u, F(f)v) - \vec{\Gamma}(F(f)u, v), \tag{2.29}$$

$$\begin{aligned}&\vec{\mathcal{E}}(F(f)u, F(f)u)\\ &\quad\leqslant 2\|F(f)\|_\infty^2 \vec{\mathcal{E}}(u,u) + 2\|F(f)\|_\infty \mathcal{E}(F(f), F(f))^{1/2}\mathcal{E}(|u|^2, |u|^2)^{1/2} +\\ &\qquad + 2\|u\|_\infty^2 \mathcal{E}(F(f)f, F(f)).\end{aligned}\tag{2.30}$$

Now for any $f \in \mathrm{Dom}(\mathcal{E}) \cap L^\infty$, we take a bounded C^1 function φ such that $\varphi(t) = t$ for $|t| \leqslant \|f\|_\infty$. Let $\{f_n\} \subseteq \mathcal{C}$ be a sequence converging to f in $\mathrm{Dom}(\mathcal{E})$.

Clearly $g_n = \varphi(f_n) \to f$ weakly in $\mathrm{Dom}(\mathcal{E})$ and $\{g_n\}$ is uniformly bounded. Moreover, by (2.30) we have $\sup_n \vec{\mathcal{E}}(g_n u, g_n u) < \infty$. Hence, we can extract a subsequence $\{g_{n_j} u\}$ whose Cesaro mean converges strongly in $\mathrm{Dom}(\vec{\mathcal{E}})$. Together with the fact that $g_n u \to fu$ in L^2, we can see that for $f \in \mathrm{Dom}(\mathcal{E}) \cap L^\infty$ and $u \in \mathcal{D}$, we have $fu \in \mathrm{Dom}(\vec{\mathcal{E}})$ and (2.11) hold.

Since (2.11) holds for $u, v \in \mathcal{D}$ and $f \in \mathrm{Dom}(\mathcal{E}) \cap L^\infty$, we can take $f = |u|^2$. By repeating the argument in the proof of Theorem 2.1, for $u \in \mathcal{D}$ and $f \in \mathrm{Dom}(\mathcal{E}) \cap L^\infty$, we have $fu \in \mathrm{Dom}(\vec{\mathcal{E}})$ and

$$\vec{\Gamma}(fu, fu)^{1/2} \leqslant |f|\vec{\Gamma}(u, u)^{1/2} + |u|\Gamma(f, f)^{1/2}, \tag{2.31}$$

$$\Gamma(|u|, |u|) \leqslant \vec{\Gamma}(u, u). \tag{2.32}$$

Since $\mathcal{D}$ is dense in $\mathrm{Dom}(\vec{\mathcal{E}})$, it is easy to see that (2.32) holds for $u \in \mathrm{Dom}(\vec{\mathcal{E}})$.

Now we can prove (2.11) for $f \in \mathrm{Dom}(\mathcal{E}) \cap L^\infty$ and $u \in \mathrm{Dom}(\vec{\mathcal{E}}) \cap L^\infty$. Take any $u \in \mathcal{D}$. For $\varepsilon > 0$, set

$$\psi_\varepsilon(t) = \frac{1}{1 + \varepsilon t}, \quad t \in [0, \infty).$$

Note that $\psi_\varepsilon' = -(\varepsilon/(1 + \varepsilon t)^2)$. Since $\psi_\varepsilon(|u|) \in \mathrm{Dom}(\mathcal{E})$, we have $\psi_\varepsilon(|u|)u \in \mathrm{Dom}(\vec{\mathcal{E}})$, and

$$\begin{aligned}
&\vec{\Gamma}(f\psi_\varepsilon(|u|)u, f\psi_\varepsilon(|u|)u)^{1/2} \\
&\quad\leqslant |f|\psi_\varepsilon(|u|)\vec{\Gamma}(u, u)^{1/2} + |u|\Gamma(f\psi_\varepsilon(|u|), f\psi_\varepsilon(|u|))^{1/2} \\
&\quad\leqslant |f|\psi_\varepsilon(|u|)\vec{\Gamma}(u, u)^{1/2} + |u|\{\psi_\varepsilon(|u|)^2\Gamma(f, f) + 2\psi_\varepsilon(|u|)\psi_\varepsilon'(|u|)f\Gamma(|u|, f) + \\
&\qquad + f^2\psi_\varepsilon'(|u|)^2\Gamma(|u|, |u|)\}^{1/2} \\
&\quad\leqslant |f|\psi_\varepsilon(|u|)\vec{\Gamma}(u, u)^{1/2} + |u|\{\psi_\varepsilon(|u|)^2\Gamma(f) + \\
&\qquad + 2\psi_\varepsilon(|u|)\psi_\varepsilon'(|u|)|f|\Gamma(|u|, |u|)^{1/2}\Gamma(f, f)^{1/2} + f^2\psi_\varepsilon'(|u|)^2\Gamma(|u|, |u|)\}^{1/2} \\
&\quad\leqslant |f|\psi_\varepsilon(|u|)\vec{\Gamma}(u, u)^{1/2} + |u|\psi_\varepsilon(|u|)\Gamma(f, f)^{1/2} + \\
&\qquad + 2\psi_\varepsilon(|u|)|\psi_\varepsilon'(|u|)||f|\Gamma(|u|, |u|)^{1/2}\Gamma(f, f)^{1/2} + f^2|\psi_\varepsilon'(|u|)|^2\Gamma(|u|, |u|)\}^{1/2} \\
&\quad\leqslant |f|\psi_\varepsilon(|u|)\vec{\Gamma}(u, u)^{1/2} + |u|\psi_\varepsilon(|u|)\Gamma(f, f)^{1/2} + |u||f||\psi_\varepsilon'(|u|)|\vec{\Gamma}(u, u)^{1/2}
\end{aligned}$$

Now for any $v \in \mathrm{Dom}(\vec{\mathcal{E}}) \cap L^\infty$, we take a sequence $\{u_n\} \subseteq \mathcal{D}$ converging to v in $\mathrm{Dom}(\vec{\mathcal{E}})$. Then by the above estimate, we can see that

$$\sup_n \vec{\mathcal{E}}(f\psi_\varepsilon(|u_n|)u_n, f\psi_\varepsilon(|u_n|)u_n) < \infty.$$

Hence, by virtue of the Banach–Saks theorem, we have, by taking a subsequence if necessary, the Cesaro mean of $\{f\psi_\varepsilon(|u_n|)u_n\}$ converges to $f\psi_\varepsilon(|u|)u$ in $\mathrm{Dom}(\vec{\mathcal{E}})$. Moreover, by noting that

$$\vec{\Gamma}\left(\frac{1}{n}\sum_{k=1}^{n} f\psi_\varepsilon(|u_k|)u_k, \frac{1}{n}\sum_{k=1}^{n} f\psi_\varepsilon(|u_k|)u_k\right)^{1/2}$$

$$\leqslant \frac{1}{n}\sum_{k=1}^{n}\vec{\Gamma}(f\psi_\varepsilon(|u_k|)u_k, f\psi_\varepsilon(|u_k|)u_k)^{1/2}$$
$$\leqslant \frac{1}{n}\sum_{k=1}^{n}\{|f|\psi_\varepsilon(|u_k|)\vec{\Gamma}(u_k,u_k)^{1/2} + |u_k|\psi_\varepsilon(|u_k|)\Gamma(f,f)^{1/2} +$$
$$+ |u_k||f||\psi'_\varepsilon(|u_k|)|\vec{\Gamma}(u_k,u_k)^{1/2}\}.$$

Letting $n \to \infty$, we have for $v \in \mathrm{Dom}(\vec{\mathcal{E}}) \cap L^\infty$,

$$\begin{aligned}&\vec{\Gamma}(f\psi_\varepsilon(|u|)v, f\psi_\varepsilon(|v|)v)^{1/2}\\ &\quad\leqslant |f|\psi_\varepsilon(|v|)\vec{\Gamma}(v,v)^{1/2} + |v|\psi_\varepsilon(|v|)\Gamma(f,f)^{1/2} +\\ &\quad\quad + |v||f||\psi'_\varepsilon(|v|)|\vec{\Gamma}(v,v)^{1/2}.\end{aligned}$$

Next, letting $\varepsilon \to 0$, we eventually obtain that $fv \in \mathrm{Dom}(\vec{\mathcal{E}})$ and

$$\vec{\Gamma}(fv,fv)^{1/2} \leqslant |f|\vec{\Gamma}(v,v)^{1/2} + |v|\Gamma(f,f)^{1/2}.$$

Next, we see that (2.11) holds for $u, v \in \mathrm{Dom}(\vec{\mathcal{E}}) \cap L^\infty$ and $f \in \mathrm{Dom}(\mathcal{E}) \cap L^\infty$. To see this, we first note the following identity: for $u, v \in \mathcal{D}$ and $f, \psi \in \mathrm{Dom}(\mathcal{E})$,

$$2f\vec{\Gamma}(\psi u, v) = \Gamma(f, (v|\psi u)_K) - \vec{\Gamma}(\psi u, fv) - \vec{\Gamma}(f\psi u, v).$$

This identity is clear from the assumption if $\psi \in \mathcal{C}$. By an approximating argument, we can see it for $\psi \in \mathrm{Dom}(\mathcal{E}) \cap L^\infty$. Using this, we can repeat the above argument and get the desired result.

So far, we have obtained that $\mathrm{Dom}(\vec{\mathcal{E}}) \cap L^\infty$ is a $\mathrm{Dom}(\mathcal{E}) \cap L^\infty$-module. We can also prove (2.8). Here we remark that every element of $\mathrm{Dom}(\vec{\mathcal{E}})$ admits a quasi-continuous modification. To see this, we note that for $u \in \mathcal{D}$,

$$\begin{aligned}\mathrm{Cap}_1(|u| \geqslant \varepsilon) &\leqslant \frac{1}{\varepsilon^2}\mathcal{E}_1(|u|,|u|)\\ &\leqslant \frac{1}{\varepsilon^2}\int_X \Gamma(|u|,|u|)\,\mathrm{d}m + \int_X |u|^2\,\mathrm{d}m\\ &\leqslant \frac{1}{\varepsilon^2}\int_X \vec{\Gamma}(u,u)\,\mathrm{d}m + \int_X |u|^2\,\mathrm{d}m\\ &\leqslant \frac{1}{\varepsilon^2}\vec{\mathcal{E}}_1(u,u).\end{aligned}$$

The rest is easy by a standard Borel–Cantelli argument.

This completes the proof. □

3. Commutation Relation

In this section, we discuss the commutation relation between two semigroups. Combining this with semigroup domination, we can discuss the logarithmic Sobolev inequality along the Bakry–Emery argument.

Let B and $\hat{B}$ be Banach spaces. We assume that $\hat{B}$ is reflexive. Suppose we are given strongly continuous semigroups $\{T_t\}$ on B and $\{\hat{T}_t\}$ on $\hat{B}$. We denote their generators by A and $\hat{A}$, respectively. We also denote the resolvents of A and $\hat{A}$ by G_α and $\hat{G}_\alpha$, respectively. Further, we are given a closed linear operator D from B to $\hat{B}$ with the dense domain $\mathrm{Dom}(D)$. The dual operator of D is denoted by D^*. We assume that $\mathrm{Dom}(A) \subseteq \mathrm{Dom}(D)$. Under this condition, we have the following theorem.

THEOREM 3.1. *Then the following conditions are equivalent to each other.*

(i) $\mathrm{Dom}(\hat{A}^*) \subseteq \mathrm{Dom}(D^*)$ *and*

$$\langle Au, D^*\theta\rangle = \langle Du, \hat{A}^*\theta\rangle, \quad \forall u \in \mathrm{Dom}(A),\ \forall\theta \in \mathrm{Dom}(\hat{A}^*). \tag{3.1}$$

(ii) *For sufficiently large* α,

$$DG_\alpha u = \hat{G}_\alpha Du, \quad \forall u \in \mathrm{Dom}(D). \tag{3.2}$$

(iii) *For any* $t \geqslant 0$, $T_t\,\mathrm{Dom}(D) \subseteq \mathrm{Dom}(D)$ *and*

$$DT_t u = \hat{T}_t Du, \quad \forall u \in \mathrm{Dom}(D). \tag{3.3}$$

Proof. First we show the implication (i) $\Rightarrow$ (ii). We note that $\{\hat{T}_t^*\}$ is also a strongly continuous semigroup with the generator $\hat{A}^*$ from the reflexivity of $\hat{B}$. From (i), we have, for $u \in \mathrm{Dom}(A)$ and $\theta \in \mathrm{Dom}(\hat{A}^*)$,

$$\langle(\alpha - A)u, D^*\theta\rangle = \langle Du, (\alpha - \hat{A}^*)\theta\rangle.$$

Substituting $u = G_\alpha v$, $\theta = \hat{G}_\alpha^*\xi$, we have

$$\langle v, D^*\hat{G}_\alpha^*\xi\rangle = \langle DG_\alpha v, \xi\rangle.$$

If $v \in \mathrm{Dom}(D)$, it follows

$$\langle \hat{G}_\alpha Dv, \xi\rangle = \langle DG_\alpha v, \xi\rangle$$

which implies (ii).

To show (ii) $\Rightarrow$ (iii), we note that for any integer n,

$$DG_\alpha^n u = \hat{G}_\alpha^n Du$$

which can be easily shown by the induction on n. Since the semigroup can be expresses in terms of the resolvents as

$$T_t u = \lim_{n\to\infty}\left(\frac{n}{t}G_{nt}\right)^n u,$$

we have for $u \in \mathrm{Dom}(D)$,

$$\begin{aligned}\hat{T}_t Du &= \lim_{n\to\infty}\left(\frac{n}{t}\hat{G}_{nt}\right)^n Du \\ &= \lim_{n\to\infty} D\left(\frac{n}{t}G_{nt}\right)^n u.\end{aligned}$$

This, combined with the closeness of D, implies $T_t u \in \mathrm{Dom}(D)$ and $\hat{T}_t Du = DT_t u$.

The implication (iii) $\Rightarrow$ (ii) can be similarly shown by noting

$$G_\alpha u = \int_0^\infty \mathrm{e}^{-\alpha t} T_t u \, \mathrm{d}t.$$

Lastly we show (ii) $\Rightarrow$ (i). From assumption $\mathrm{Dom}(A) \subseteq \mathrm{Dom}(D)$, the closed operator $S = DG_\alpha$ is defined whole space B and hence bounded. Hence, for $u \in \mathrm{Dom}(D)$, $\theta \in \hat{B}^*$, we have

$$\langle u, S^*\theta\rangle = \langle Su, \theta\rangle = \langle DG_\alpha u, \theta\rangle = \langle \hat{G}_\alpha Du, \theta\rangle = \langle Du, \hat{G}_\alpha^* \theta\rangle.$$

Now, setting $\theta = (\alpha - \hat{A}^*)\xi$, $\xi \in \mathrm{Dom}(\hat{A}^*)$, we have

$$\langle u, S^*(\alpha - \hat{A}^*)\xi\rangle = \langle Du, \xi\rangle.$$

From this, it follows $\xi \in \mathrm{Dom}(D^*)$. Thus, for $\xi \in \mathrm{Dom}(\hat{A}^*)$, it holds that

$$\langle Su, (\alpha - \hat{A}^*)\xi\rangle = \langle u, D^*\xi\rangle, \quad \forall u \in \mathrm{Dom}(D). \tag{3.4}$$

Since $\mathrm{Dom}(D)$ is dense in B, the above identity holds for all $u \in B$. In particular, if we set $u = (\alpha - A)v$, $v \in \mathrm{Dom}(A)$, we have

$$\langle S(\alpha - A)v, (\alpha - \hat{A}^*)\xi\rangle = \langle (\alpha - A)v, D^*\xi\rangle.$$

By noting $S(\alpha - A)v = DG_\alpha(\alpha - A)v = Dv$, we get (i). This completes the proof. □

Next we give an sufficient condition for the above theorem. We consider the same situation as in Theorem 3.1. In the sequel, any domain of an operator is equipped with a topology given by the graph norm.

PROPOSITION 3.2. *Suppose that there exist a subspace $\mathfrak{D} \subset \mathrm{Dom}(A)$ satisfying one of the following conditions:*

(i) $A\mathfrak{D} \subseteq \mathrm{Dom}(D)$, $D\mathfrak{D} \subseteq \mathrm{Dom}(\hat{A})$, *$\mathfrak{D}$ is dense in* $\mathrm{Dom}(A)$ *and* $\mathrm{Dom}(\hat{A}^*) \subseteq \mathrm{Dom}(D^*)$, *Further for any $u \in \mathfrak{D}$, it holds that*

$$DAu = \hat{A}Du. \tag{3.5}$$

(ii) $A\mathfrak{D} \subseteq \mathrm{Dom}(D)$, $D\mathfrak{D} \subseteq \mathrm{Dom}(\hat{A})$ *and $(\alpha - A)\mathfrak{D}$ is dense in* $\mathrm{Dom}(D)$ *for sufficiently large α. Further* (3.5) *holds for any $u \in \mathfrak{D}$.*

Then, one of (and hence all) three conditions in Theorem 3.1 *holds.*

Proof. First we assume (i). Then, for any $u \in \mathcal{D}$ and $\theta \in \mathrm{Dom}(\hat{A}^*)$, we have

$$\langle Au, D^*\theta\rangle = \langle DAu, \theta\rangle = \langle \hat{A}Du, \theta\rangle = \langle Du, \hat{A}^*\theta\rangle.$$

Since $\mathcal{D}$ is dense in $\mathrm{Dom}(A)$, we can see that $\langle Au, D^*\theta\rangle = \langle Du, \hat{A}^*\theta\rangle$ holds for $u \in \mathrm{Dom}(A)$ and $\theta \in \mathrm{Dom}(\hat{A}^*)$. Here, we use that D is a bounded operator from $\mathrm{Dom}(A)$ to $\hat{B}$. Hence (i) in Theorem 3.1 holds.

Next assume (ii). From the assumption, we have for $u \in \mathcal{D}$ and $\theta \in \mathrm{Dom}(\hat{A}^*)$,

$$\langle D(\alpha - A)u, \theta\rangle = \langle(\alpha - \hat{A})Du, \theta\rangle = \langle Du, (\alpha - \hat{A}^*)\theta\rangle.$$

If we see $v = (\alpha - A)u \in (\alpha - A)\mathcal{D}$, it follows that

$$\langle Dv, \theta\rangle = \langle DG_\alpha v, (\alpha - \hat{A}^*)\theta\rangle.$$

Now, using the denseness of $(\alpha - A)\mathcal{D}$ in $\mathrm{Dom}(D)$, the above identity holds for all $v \in \mathrm{Dom}(D)$. Further, as was seen in the proof of Theorem 3.1, $S = DG_\alpha$ is a bounded operator on B, we have

$$\langle Dv, \theta\rangle = \langle v, S^*(\alpha - \hat{A}^*)\theta\rangle \quad \forall v \in \mathrm{Dom}(D).$$

Then it follows $\theta \in \mathrm{Dom}(D^*)$ which implies $\mathrm{Dom}(\hat{A}^*) \subseteq \mathrm{Dom}(D^*)$. Hence, this case is reduced to the case (i). □

Now, we take Hilbert spaces H and $\hat{H}$ in place of Banach spaces B and $\hat{B}$. Moreover generators A and $\hat{A}$ are self-adjoint and associated with the closed quadratic forms $\mathcal{E}$ and $\hat{\mathcal{E}}$, respectively. Of course, $\mathcal{E}$ and $\hat{\mathcal{E}}$ are bounded from below. Then we have the following;

PROPOSITION 3.3. *Suppose that there exist a subspace* $\mathcal{D} \subset \mathrm{Dom}(A^{3/2})$ *satisfying the following conditions.* $\mathrm{Dom}(\mathcal{E}) \subseteq \mathrm{Dom}(D)$, $D\mathcal{D} \subseteq \mathrm{Dom}(\hat{A})$, $\mathcal{D}$ *is dense in* $\mathrm{Dom}(A^{3/2})$, *and it holds that*

$$DAu = \hat{A}Du, \quad \forall u \in \mathcal{D}. \tag{3.6}$$

Then, one of (and hence all) three conditions in Theorem 3.1 *holds.*

Proof. From the assumption $\mathcal{D} \subseteq \mathrm{Dom}(A^{3/2})$, we have $(\alpha - A)\mathcal{D} \subseteq \mathrm{Dom}(A^{1/2}) = \mathrm{Dom}(\mathcal{E})$. Moreover $(\alpha - A)\mathcal{D}$ is dense in $\mathrm{Dom}(\mathcal{E})$ by virtue of the denseness of $\mathcal{D}$ in $\mathrm{Dom}(A^{3/2})$. By an argument similar to the proof of Proposition 3.2(ii), we can get the conclusion. □

We can also give another sufficient condition.

PROPOSITION 3.4. *We assume the following conditions:*

$$\begin{aligned}
&\mathcal{D} \subseteq \mathrm{Dom}(A) \subseteq \mathrm{Dom}(\mathcal{E}) \subseteq \mathrm{Dom}(D),\\
&\hat{\mathcal{D}} \subseteq \mathrm{Dom}(\hat{A}) \subseteq \mathrm{Dom}(\hat{\mathcal{E}}) \subseteq \mathrm{Dom}(D^*),\\
&D\,\mathrm{Dom}(A) \subseteq \mathrm{Dom}(\hat{\mathcal{E}}),\\
&D^*\hat{\mathcal{D}} \subseteq \mathrm{Dom}(\mathcal{E}).
\end{aligned}$$

$\mathcal{D}$ is a dense subspace of $\mathrm{Dom}(\mathcal{E})$ *and $\hat{\mathcal{D}}$ is a dense subspace of* $\mathrm{Dom}(\hat{\mathcal{E}})$. *Further, we assume that it holds*

$$(Au, D^*\theta) = (Du, \hat{A}\theta), \quad \text{for } u \in \mathcal{D}, \theta \in \hat{\mathcal{D}}. \tag{3.7}$$

Then, one of (and hence all) three conditions in Theorem 3.1 *holds.*

Proof. Take any $u \in \mathcal{D}$, $\theta \in \hat{\mathcal{D}}$. From the assumption, $D^*\theta \in \mathrm{Dom}(\mathcal{E})$ and hence

$$\mathcal{E}(u, D^*\theta) = (Au, D^*\theta) = (Du, \hat{A}\theta).$$

Now using the denseness of $\mathcal{D}$ in $\mathrm{Dom}(\mathcal{E})$, we have

$$\mathcal{E}(u, D^*\theta) = (Du, \hat{A}\theta), \quad \text{for } u \in \mathrm{Dom}(\mathcal{E}), \theta \in \hat{\mathcal{D}}.$$

In particular, $Du \in \mathrm{Dom}(\hat{\mathcal{E}})$ for $u \in \mathrm{Dom}(\hat{\mathcal{E}})$, we have, for $\theta \in \hat{\mathcal{D}}$

$$(Au, D^*\theta) = \mathcal{E}(u, D^*\theta) = (Du, \hat{A}\theta) = \hat{\mathcal{E}}(Du, \theta).$$

Since $\hat{\mathcal{D}}$ is dense in $\mathrm{Dom}(\hat{\mathcal{E}})$, we obtain

$$(Au, D^*\theta) = \hat{\mathcal{E}}(Du, \theta) \quad \text{for } u \in \mathrm{Dom}(A), \theta \in \mathrm{Dom}(\hat{\mathcal{E}}).$$

Taking θ from $\mathrm{Dom}(\hat{A}) \subseteq \mathrm{Dom}(\mathcal{E})$, it holds that

$$(Au, D^*\theta) = \hat{\mathcal{E}}(Du, \theta) = (Du, \hat{A}\theta), \quad \text{for } u \in \mathrm{Dom}(A), \theta \in \mathrm{Dom}(\hat{A}),$$

which is the desired result. □

Remark 3.1. The condition (3.7) can be replaced by $DAu = \hat{A}Du$ for $u \in \mathcal{D}$. Of course in this case we assume that they are well-defined.

In many applications, $\mathcal{E}$ is given by

$$\mathcal{E}(u, v) = (Du, Dv)_{\hat{H}}. \tag{3.8}$$

Hence the assumption $\mathrm{Dom}(D) \subseteq \mathrm{Dom}(\mathcal{E})$ is automatic, i.e., the identity holds.

Now we formulate the Bakry–Emery criterion for the logarithmic Sobolev inequality in our setting. Let the notations be as before and we impose the following conditions. The Hilbert space H is $L^2(X, m)$ on a measure space (X, m). We assume that m is a *probability* measure. The quadratic form $\mathcal{E}$ is a local Dirichlet form with $\mathbf{1} \in \mathrm{Dom}(\mathcal{E})$ and $\mathcal{E}(\mathbf{1}, \mathbf{1}) = 0$. D is a closed operator (usually first-order differential operator) from $L^2(X, m)$ to the space of L^2 sections of a vector bundle E over X. We assume, for the notational simplicity, that E is a trivial bundle with a fiber K that is a Hilbert space. We denote ∇ instead of D. We assume that the square field operator is given by

$$\Gamma(f, g)(x) = (\nabla f(x), \nabla g(x))_K. \tag{3.9}$$

The Dirichlet form $\mathcal{E}$ is written as

$$\mathcal{E}(f, g) = \int_X (\nabla f(x), \nabla g(x))_K \,\mathrm{d}m. \tag{3.10}$$

Suppose further that we are given another symmetric semigroup $\{\vec{T}_t\}$ on $L^2(m; K)$. We assume the following semigroup domination for which we have given a sufficient condition in the previous section.

$$|\vec{T}_t\theta| \leqslant \mathrm{e}^{-\lambda t} T_t|\theta|. \tag{3.11}$$

Lastly we assume that the following commutativity condition:

$$\vec{T}_t \nabla f = \nabla T_t f. \tag{3.12}$$

As we have seen, we already have sufficient conditions for this.

Under the above conditions, we can show the logarithmic Sobolev inequality for the Dirichlet form $\mathcal{E}$. Before that, let us see the relation to the Bakry–Emery Γ_2-criterion. Suppose the following commutativity condition on a space $\mathcal{D}$,

$$\nabla A f = \vec{A} \nabla f, \quad \forall f \in \mathcal{D}. \tag{3.13}$$

Of course, we have to choose a suitable space $\mathcal{D}$. We ignore the technical detail for a while. We also assume the condition $(\vec{\Gamma}_\lambda)$. Hence $\vec{\Gamma}$ is realized as an L^1 function. Under these conditions,

$$\begin{aligned} 2\vec{\Gamma}(\nabla f, \nabla g) &= A_1(\nabla f, \nabla g) - (\vec{A}\nabla f, \nabla g) - (\nabla f, \vec{A}\nabla g) \\ &= A_1(\nabla f, \nabla g) - (\nabla A f, \nabla g) - (\nabla f, \nabla A g) \\ &= A_1\Gamma(f, g) - \Gamma(Af, g) - \Gamma(f, Ag). \end{aligned}$$

The right-hand side is usually called the Bakry–Emery Γ_2. So our condition (2.9) corresponds to

$$\Gamma_2(f, f) \geqslant \lambda\Gamma(f, f)$$

which is the famous Bakry–Emery criterion.

Now we can state the Bakry–Emery theorem in our formulation as follows:

THEOREM 3.5. *Under assumptions* (3.11) *and* (3.12) *for* $\lambda > 0$, *the following logarithmic Sobolev inequality holds:*

$$\int_X f^2 \log(f^2/\|f\|_2^2)\,\mathrm{d}m \leqslant \frac{2}{\lambda}\mathcal{E}(f, f), \quad \forall f \in \mathrm{Dom}(\mathcal{E}). \tag{3.14}$$

Proof. From our assumption, it holds that

$$|\vec{T}_t\omega| \leqslant \mathrm{e}^{-\lambda t} T_t|\omega|.$$

Therefore, taking $\omega = \nabla f$, we have

$$|\nabla T_t f| \leqslant \mathrm{e}^{-\lambda t} T_t |\nabla f|.$$

Now the rest is the standard argument (see, e.g., [4, Proof of Theorem 6.2.42]). □

4. Riemannian Manifold with Boundary

In this section, we consider a Riemannian manifold with boundary.

Let M be a d-dimensional compact Riemannian manifold with a boundary ∂M. We assume that everything is smooth. As usual, the Hodge–Kodaira Laplacian is defined as follows:

$$\Box = -(\mathrm{d}\delta + \delta\mathrm{d}). \tag{4.1}$$

Here d is the exterior differentiation and δ is its (formal) dual. To specify that it acts p-forms, we denote it by $\Box_{(p)}$. Moreover, we suppose that $\Box$ is defined on a set of all C^∞ differential forms. We always assume that differential forms are always smooth, i.e., C^∞. Later we consider symmetric operators but they are all essentially self-adjoint on smooth differential forms (to be precise, under a suitable boundary condition) and hence it is enough to achieve formal calculation on smooth forms. We denote the pth exterior bundle of T^*M by $\bigwedge^p T^*M$. The set of all smooth differential p-forms, i.e., smooth sections of $\bigwedge^p T^*M$, will be denoted by $A^p(M)$.

Since the manifold M has a boundary, we have to impose boundary condition. Several boundary conditions are known but we consider the two conditions: the absolute boundary condition and the relative boundary condition. To describe boundary condition, we introduce the following local coordinate $(x^1, x^2, \ldots, x^{d-1}, r)$ near the boundary: The local coordinate gives an diffeomorphism between a neighborhood in M and $U \times [0, \varepsilon) \subset \mathbb{R}^{d-1} \times R_+$. We identify the neighborhood in M with $U \times [0, \varepsilon)$ through the local coordinate. We assume that, on the boundary ∂M, $r = 0$ and $\partial/\partial x^j \perp \partial/\partial r$, $j = 1, \ldots, d-1$. We denote the inner normal vector by N, i.e., $N = \partial/\partial r$. Further, we assume that for fixed $(x_0^1, \ldots, x_0^{d-1}) \in U$, $\gamma_t = (x_0^1, \ldots, x_0^{d-1}, t)$ is a geodesic with the velocity 1. For any 1-form θ on M, we can decompose it as

$$\theta = \theta_t + \theta_n \wedge \mathrm{d}r, \tag{4.2}$$

where neither θ_n nor θ_t contains $\mathrm{d}r$. This decomposition is well-defined on the boundary. Using this decomposition, we define

$$B_r(\theta) = \theta_t, \tag{4.3}$$

$$B_a(\theta) = \theta_n. \tag{4.4}$$

Recall that θ_t and θ_n are defined only on ∂M. The suffix r stands for the relative boundary condition and a for the absolute boundary condition. We use these operators to define the boundary condition.

In the sequel, B stands for either B_r or B_a. Now we restrict the domain of $\Box$ according to boundary condition. Set

$$\mathrm{Dom}(\Box^B_{(p)}) = \{\phi \in A^p(M);\ B\phi = 0,\ B(d+\delta)\phi = 0\}. \tag{4.5}$$

By an integration by parts formula for differential forms, we can see that $\Box^B_{(p)}$ is symmetric and moreover it is essentially self-adjoint on $\mathrm{Dom}(\Box^B_{(p)})$. We can think of the associated bilinear form $\mathcal{E}^B_{(p)}$, which is given as follows:

$$\mathrm{Dom}(\mathcal{E}^B_{(p)}) = \{\phi \in A^p(M);\ B\phi = 0\} \tag{4.6}$$

and

$$\mathcal{E}^B_{(p)}(\omega, \eta) = (\mathrm{d}\omega, \mathrm{d}\eta) + (\delta\omega, \delta\eta). \tag{4.7}$$

Let see why this is so. We recall the integration by parts formula for differential forms. Take any $\omega \in A^p(M)$ and $\eta \in A^{p+1}(M)$, then it holds that

$$\int_M (\mathrm{d}\omega, \eta)\,\mathrm{d}m = \int_M (\omega, \delta\eta)\,\mathrm{d}m + \int_{\partial M} \omega \wedge *\eta. \tag{4.8}$$

Here $*$ is the Hodge star operator which sends p-form to $d-p$-form. Now we have

$$\begin{aligned}
-(\Box^{B_a}_{(p)}\omega, \eta) &= -\int_M (\Box^{B_a}_{(p)}\omega, \eta)\,\mathrm{d}m \\
&= \int_M (\mathrm{d}\delta\omega + \delta\,\mathrm{d}\omega, \eta)\,\mathrm{d}m \\
&= \int_M (\delta\omega, \delta\eta)\,\mathrm{d}m + \int_{\partial M} \delta\omega \wedge *\eta\ + \\
&\quad + \int_M (\mathrm{d}\omega, \mathrm{d}\eta)\,\mathrm{d}m - \int_{\partial M} \eta \wedge *\,\mathrm{d}\omega \\
&= \int_M (\delta\omega, \delta\eta)\,\mathrm{d}m + \int_M (d\omega, \mathrm{d}\eta)\,\mathrm{d}m \\
&= \mathcal{E}^{B_a}_{(p)}(\omega, \eta).
\end{aligned}$$

Here $\mathrm{d}\sigma$ is the surface element of ∂M. We used the boundary condition B_a to get $\delta\omega \wedge *\eta = 0$ on ∂M. To see this, $*\eta$ contains $\mathrm{d}r$ since $B_a\eta = 0$. On the other hand, $\delta\omega$ does not contain $\mathrm{d}r$ since $B_a\delta\omega = 0$. Hence $\delta\omega \wedge *\eta$ contains $\mathrm{d}r$. Therefore $\delta\omega \wedge *\eta = 0$ on ∂M since $\mathrm{d}r = 0$ on ∂M. $\eta \wedge *\mathrm{d}\omega$ also vanishes on ∂M.

We can show similar identity for $\Box^{B_r}_{(p)}$.

Next we see the commutation relation. Formally the commutation relation $\mathrm{d}\Box = \Box\mathrm{d}$ holds, i.e., at least in the interior of M. But the boundary condition is involved and we need some arguments.

Take $\omega \in \mathrm{Dom}(\square^{B_a}_{(p)})$, $\eta \in \mathrm{Dom}(\square^{B_a}_{(p+1)})$

$$\begin{aligned}
-\int_M (\square\omega, \delta\eta)\,\mathrm{d}m &= \int_M (\mathrm{d}\delta\omega + \delta\,\mathrm{d}\omega, \delta\eta)\,\mathrm{d}m \\
&= \int_M (\delta\omega + \delta\,\mathrm{d}\omega, \delta\delta\eta)\,\mathrm{d}m + \int_{\partial M} \delta\omega \wedge *\delta\eta + \\
&\quad + \int_M (\mathrm{d}\omega, \mathrm{d}\delta\eta)\,\mathrm{d}m - \int_{\partial M} \delta\eta \wedge *\mathrm{d}w \\
&= \int_M (\mathrm{d}\omega, \mathrm{d}\delta\eta)\,\mathrm{d}m + \int_M (\mathrm{d}\mathrm{d}\omega, \mathrm{d}\eta)\,\mathrm{d}m \\
&= \int_M (\mathrm{d}\omega, \mathrm{d}\delta\eta)\,\mathrm{d}m + \int_M (\mathrm{d}\omega, \delta\,\mathrm{d}\eta)\,\mathrm{d}m + \int_{\partial M} \mathrm{d}\omega \wedge *\mathrm{d}\eta \\
&= -\int_M (\mathrm{d}\omega, \square\eta)\,\mathrm{d}m.
\end{aligned}$$

We used the boundary condition B_a. Absolutely same argument works for B_r.

Now invoking the essential self-adjointness of $\square^{B_a}_{(p)}$ and $\square^{B_r}_{(p)}$, we can get the semigroup commutation. Of course, we must choose the same boundary condition.

We proceed to the issue of semigroup domination. For 0-forms, i.e., scalar functions, the boundary condition B_a corresponds to the Neumann condition: $\partial f/\partial r = 0$ on ∂M. On the other hand, B_r corresponds to the Dirichlet boundary condition: $f = 0$ on ∂M. From now on, for scalar functions, we always consider the Neumann Laplacian $\square^{B_a}_{(0)}$ and we denote it by Δ for simplicity. We also denote the associated Dirichlet form by $\mathcal{E}$, i.e., $\mathcal{E} = \mathcal{E}^{B_a}_{(0)}$. We do not use the Dirichlet form (for scalar functions) with Dirichlet boundary condition. The reason is that the semigroup generated by Hodge–Kodaira Laplacian can not be dominated by the Dirichlet semigroup since the diffusion dies after hitting the boundary.

In order to apply our theorem, we have to check condition $(\vec{\Gamma}_\lambda)$. In particular, to show (2.8), we have to calculate $-\mathcal{E}((\theta, \eta), f) + \vec{\mathcal{E}}(f\theta, \eta) + \vec{\mathcal{E}}(\theta, f\eta)$.

First we note that

$$\begin{aligned}
&-\int_M \Delta(\theta, \eta) f\,\mathrm{d}m \\
&= \int_M \nabla^*\nabla(\theta, \eta) f\,\mathrm{d}m \\
&= \int_M \delta d(\theta, \eta) f\,\mathrm{d}m \\
&= \int_M (d(\theta, \eta), df)\,\mathrm{d}m + \int_{\partial M} f\langle d(\theta, \eta), N\rangle\,\mathrm{d}\sigma \\
&(\because \text{Green–Stokes formula}) \\
&= \mathcal{E}((\theta, \eta), \mathrm{d}f) + \int_{\partial M} f\nabla_N(\theta, \eta)\,\mathrm{d}\sigma.
\end{aligned}$$

Here $d\sigma$ denotes the surface element of ∂M and we recall that N is the inner normal vector. Hence

$$\mathcal{E}((\theta,\eta),f) = -\int_M \Delta(\theta,\eta) f \, dm - \int_{\partial M} f \nabla_N(\theta,\eta)\, d\sigma.$$

On the other hand, it holds that

$$\Delta(\theta,\eta) + (\nabla^*\nabla\theta,\eta) + (\theta,\nabla^*\nabla\eta) = 2(\nabla\theta,\nabla\eta).$$

By the Weitzenböck formula $-\Box^B_{(p)} = \nabla^*\nabla + R_{(p)}$, we have

$$\Delta(\theta,\eta) - (\Box^B_{(p)}\theta,\eta) - (\theta,\Box^B_{(p)}\eta) = 2(\nabla\theta,\nabla\eta) + 2R_{(p)}(\theta,\eta).$$

We do not give the explicit form of $R_{(p)}$, but we just say that $R_{(p)}$ can be written in terms of the curvature. Now by combining these identities, we can obtain the following identity:

$$\begin{aligned} &-\mathcal{E}((\theta,\eta),f) + \mathcal{E}^B_{(p)}(f\theta,\eta) + \mathcal{E}^B_{(p)}(\theta,f\eta) \\ &= 2\int_M (\nabla\theta,\nabla\eta) f\, dm + 2\int_M R_{(p)}(\theta,\eta) f\, dm + \\ &\quad + 2\int_{\partial M} \nabla_N(\theta,\eta) f\, d\sigma. \end{aligned} \tag{4.9}$$

In fact

$$\begin{aligned} &-\mathcal{E}((\theta,\eta),f) + \mathcal{E}^B_{(p)}(f\theta,\eta) + \mathcal{E}^B_{(p)}(\theta,f\eta) \\ &= \int_M \Delta(\theta,\eta)\cdot f\, dm + 2\int_{\partial M} \nabla_N(\theta,\eta) f\, d\sigma - \\ &\quad - \int_M \{(\Box^B_{(p)}\theta,\eta) f - (\theta,\Box^B_{(p)}\eta) f\}\, dm \\ &= 2\int_M (\nabla\theta,\nabla\eta) f\, dm + 2\int_M R_{(p)}(\theta,\eta) f\, dm + 2\int_{\partial M} \nabla_N(\theta,\eta) f\, d\sigma. \end{aligned}$$

Therefore $\vec{\Gamma}$ in ($\vec{\Gamma}'_\lambda$-2) is given as

$$\vec{\Gamma}(\theta,\eta) = (\nabla\theta,\nabla\eta) + R_{(p)}(\theta,\eta). \tag{4.10}$$

We can easily check the condition ($\vec{D}$) in this case. The rest is to show that the third part of the left-hand side in (4.9) corresponds to $\vec{\gamma}\, d\sigma$ in (2.8).

We take a local frame $\{\omega^1,\ldots,\omega^{d-1}, dr\}$ in T^*M so that it forms a orthonormal basis and ω^j is parallel along the geodesic $t \to (x^1,\ldots,x^{d-1},t)$. We also denote the dual basis of $\{\omega^1,\ldots,\omega^{d-1},dr\}$ by $\{X_1,\ldots,X_{d-1},N\}$. Note that on the other hand,

$$\begin{aligned} d\omega^k[X_j,N] &= \langle\nabla_N\omega^k, X_j\rangle - \langle\nabla_{X_j}\omega^k, N\rangle \\ &= -\langle\nabla_{X_j}\omega^k, N\rangle. \end{aligned}$$

Here we used that $\nabla_N \omega^k = 0$ since ω^k is parallel along the path $t \mapsto (x^1, \ldots, x^{d-1}, t)$. Moreover we note that $\langle \nabla_{X_j} \omega^k, N \rangle = (\alpha(X_j, X_k), N)$ where α is the second fundamental form on ∂M. Thus we have

$$\mathrm{d}\omega^k[X_j, N] = -(\alpha(X_j, X_k), N). \tag{4.11}$$

Using the second fundamental form, we introduce an operator A as follows:

$$A\omega^i = \sum_{j=1}^{d-1} (\alpha(X_i, X_j), N)\omega^j. \tag{4.12}$$

A can be extended to a linear operator from $A^1(\partial M)$ to $A^1(\partial M)$ and it is independent of the choice of $\{\omega^j\}$. Moreover, we define $\mathrm{d}\Gamma(A)$: $A^p(\partial M) \to A^p(\partial M)$ as

$$\mathrm{d}\Gamma(A)(\theta_1 \wedge \cdots \wedge \theta_p) = \sum_{j=1}^{p} \theta_1 \wedge \cdots \wedge A\theta_j \wedge \cdots \wedge \theta_p. \tag{4.13}$$

Then we can have the follolwing.

LEMMA 4.1. *On the boundary ∂M, for*

$$I = \{i_1 < \cdots < i_p\} \subseteq \{1, \ldots, d-1\},$$

it holds that

$$(\mathrm{d}\omega^I)_n = -(-1)^{|I|} \mathrm{d}\Gamma(A)\omega^I. \tag{4.14}$$

Proof. From (4.11) and (4.12), it holds that

$$\sum_{k=1}^{d-1} \mathrm{d}\omega^j[X_k, N]\omega^k = -A\omega^j.$$

Hence

$$\begin{aligned}
\mathrm{d}\omega^I &= \mathrm{d}(\omega^{i_1} \wedge \cdots \wedge \omega^{i_p}) \\
&= \sum_{\mu=1}^{p} (-1)^\mu \omega^{i_1} \wedge \cdots \wedge \mathrm{d}\omega^{i_\mu} \wedge \cdots \wedge \omega^{i_p} \\
&= \sum_{\mu=1}^{p} \sum_{1 \leqslant k \leqslant l \leqslant d-1} (-1)^\mu \omega^{i_1} \wedge \cdots \wedge \mathrm{d}\omega^{i_\mu}[X_k, X_l](\omega^k \wedge \omega^l) \wedge \cdots \wedge \omega^{i_p} + \\
&\quad + \sum_{\mu=1}^{p} \sum_{k=1}^{d-1} (-1)^\mu \omega^{i_1} \wedge \cdots \wedge \mathrm{d}\omega^{i_\mu}[X_k, N](\omega^k \wedge \mathrm{d}r) \wedge \cdots \wedge \omega^{i_p}
\end{aligned}$$

$$= \sum_{\mu=1}^{p} \sum_{1 \leqslant k \leqslant l \leqslant d-1} (-1)^{\mu} \omega^{i_1} \wedge \cdots \wedge d\omega^{i_\mu}[X_k, X_l](\omega^k \wedge \omega^l) \wedge \cdots \wedge \omega^{i_p} -$$

$$- (-1)^p \sum_{\mu=1}^{p} \sum_{k=1}^{d-1} \omega^{i_1} \wedge \cdots \wedge A\omega^{i_\mu} \wedge \cdots \wedge \omega^{i_p} \wedge dr.$$

Now, recalling the definition of $d\Gamma(A)$, we have

$$(d\omega^I)_n = -(-1)^p \, d\Gamma(A)\omega^I$$

which is the desired result. □

Now we can deal with the absolute boundary condition.

THEOREM 4.2. *For p-forms $\theta, \eta \in \mathrm{Dom}(\mathcal{E}^{B_a}_{(p)})$ and $f \in C^\infty(M)$, it holds that*

$$\begin{aligned} -\mathcal{E}((\theta, \eta), f) &+ \mathcal{E}^{B_a}_{(p)}(f\theta, \eta) + \mathcal{E}^{B_a}_{(p)}(\theta, f\eta) \\ &= 2\int_M (\nabla\theta, \nabla\eta) f \, dm + 2\int_M R_{(p)}(\theta, \eta) f \, dm + \\ &\quad + 2\int_{\partial M} (d\Gamma(A)\theta_n, \eta_n) f \, d\sigma. \end{aligned} \tag{4.15}$$

Proof. First we write the boundary condition in terms of local coordinate. We decompose θ as follows:

$$\theta = \sum_{I:|I|=p} f_I \omega^I + \sum_{J:|J|=p-1} g_J \omega^J \wedge dr.$$

Noting that $g_J = 0$ on ∂M, we have,

$$\begin{aligned} d\theta &= \sum_{I:|I|=p} df_I \wedge \omega^I + \sum_{I:|I|=p} f_I \, d\omega^I + \\ &\quad + \sum_{J:|J|=p-1} dg_J \wedge \omega^J \wedge dr + \sum_{J:|J|=p-1} g_J \, d\omega^J \wedge dr \\ &= \sum_{I:|I|=p} df_I \wedge \omega^I + \sum_{I:|I|=p} f_I \, d\omega^I. \end{aligned}$$

Since $(d\theta)_n = 0$, we have

$$\begin{aligned} 0 &= \sum_{I:|I|=p} \langle df_I, N \rangle \, dr \wedge \omega^I + \sum_{I:|I|=p} f_I (d\omega^I)_n \\ &= \sum_{I:|I|=p} (-1)^{|I|} \langle df_I, N \rangle \omega^I \wedge dr - \sum_{I:|I|=p} (-1)^{|I|} f_I \, d\Gamma(A)\omega^I \wedge dr \\ &= \sum_{I:|I|=p} (-1)^{|I|} \langle df_I, N \rangle \omega^I \wedge dr - (-1)^{|I|} d\Gamma(A)\theta_t \wedge dr. \end{aligned}$$

Now we have

$$\sum_{I:|I|=p} \langle \mathrm{d} f_I, N \rangle \omega^I = \mathrm{d}\Gamma(A)\theta_t.$$

Now we calculate $(\nabla_N \theta, \eta)$.

$$\begin{aligned}
(\nabla_N \theta, \eta) &= \left(\nabla_N \left\{ \sum_{I:|I|=p} f_I \omega^I + \sum_{J:|J|=p-1} g_J \omega^J \wedge \mathrm{d}r \right\}, \eta_t \right) \\
&= \left(\sum_{I:|I|=p} (\nabla_N f_I) \omega^I, \eta_t \right) \quad (\because \nabla_N \omega^I = 0) \\
&= (\mathrm{d}\Gamma(A)\theta_t, \eta_t).
\end{aligned}$$

This completes the proof. □

Next we consider the relative boundary condition. This can be easily done by noting that the Hodge operation gives an isometry which interchanges the absolute boundary condition and the relative boundary condition. We denote the Hodge operation on ∂M by $*$. Then we have

THEOREM 4.3. *For p-forms $\theta, \eta \in \mathrm{Dom}(\mathcal{E}_{(p)}^{B_r})$ and $f \in C^\infty(M)$, it holds that*

$$\begin{aligned}
&-\mathcal{E}((\theta, \eta), f) + \mathcal{E}_{(p)}^{B_r}(f\theta, \eta) + \mathcal{E}_{(p)}^{B_r}(\theta, f\eta) \\
&\quad = 2 \int_M (\nabla\theta, \nabla\eta) f \,\mathrm{d}m + 2 \int_M R_{(p)}(\theta, \eta) f \,\mathrm{d}m + \\
&\qquad + 2 \int_{\partial M} (*^{-1} \mathrm{d}\Gamma(A) * \theta_n, \eta_n) f \,\mathrm{d}\sigma.
\end{aligned} \tag{4.16}$$

Now we can apply Theorem 2.1. For example, if $\mathrm{Ric} \geqslant \lambda I$ and A is nonnegative definite, then we have

$$|\vec{T}_t \theta| \leqslant \mathrm{e}^{-\lambda t} |\theta|, \quad \text{for } \theta \in A^1(M),$$

where $\vec{T}_t$ is a semigroup generated by $\Box_{(1)}^{B_a}$. Moreover, we can give an estimate of the constant for the logarithmic Sobolev inequality.

Taking $f = 1$ in (4.15), we have

$$\begin{aligned}
\mathcal{E}_{(p)}^{B_a}(\theta, \eta) &= \int_M (\nabla\theta, \nabla\eta) \,\mathrm{d}m + \int_M R_{(p)}(\theta, \eta) \,\mathrm{d}m + \\
&\quad + \int_{\partial M} (\mathrm{d}\Gamma(A)\theta_t, \eta_t) \,\mathrm{d}\sigma.
\end{aligned} \tag{4.17}$$

This identity is already known (see, e.g., Schwarz [17, Thm 2.1.5]).

In particular, if $p = 1$, it holds that

$$\mathcal{E}_{(1)}^{B_a}(\theta, \eta) = \int_M (\nabla\theta, \nabla\eta)\,\mathrm{d}m + \int_M \mathrm{Ric}(\theta, \eta)\,\mathrm{d}m + \int_{\partial M} (A\theta_t, \eta_t)\,\mathrm{d}\sigma. \quad (4.18)$$

Here Ric denotes the Ricci curvature. Using this identity, we can show the Lichnèrowicz theorem. Let Δ be the Laplacian with Neumann boundary condition on M. We assume that $\mathrm{Ric} \geqslant (d-1)\rho I$ with $\rho > 0$ and A is nonnegative definite. Then we have

$$\lambda_1 \geqslant \rho d, \quad (4.19)$$

where λ_1 is the first nonzero eigenvalue of Δ.

To see this, note that

$$\begin{aligned}(\Delta f, \Delta f) &= -(\mathrm{dd}f, \mathrm{dd}f) - (\delta\,\mathrm{d}f, \delta\,\mathrm{d}f)\\ &= \mathcal{E}_{(1)}^{B_a}(\nabla f, \nabla f)\\ &= \int_M (\nabla^2 f, \nabla^2 f)\,\mathrm{d}m + \int_M \mathrm{Ric}(\nabla f, \nabla f)\,\mathrm{d}m +\\ &\quad + \int_{\partial M} (A(\nabla f)_t, (\nabla f)_t)\,\mathrm{d}\sigma.\end{aligned}$$

Now, by the standard argument, we easily have

$$\frac{d-1}{d}(\Delta f, \Delta f) \geqslant \int_M \mathrm{Ric}(\nabla f, \nabla f)\,\mathrm{d}m \geqslant (d-1)\rho\mathcal{E}(f, f).$$

Now we can get the desired result.

Moreover, it is known that the identity holds if and only if M is isomorphic to the hemisphere (Xia [20]). Similar result holds for the Laplacian with Dirichlet boundary condition, see Reilly [14].

References

1. Aida, S., Masuda, T. and Shigekawa, I.: Logarithmic Sobolev inequalities and exponential integrability, *J. Funct. Anal.* **126** (1994), 83–101.
2. Bakry, D. and Emery, M.: Diffusions hypercontractives, In: *Séminaire de Prob.* XIX, Lecture Notes in Math. 1123, Springer, New York, 1985, pp. 177–206.
3. Bouleau, N. and Hirsch, F.: *Dirichlet Forms and Analysis on Wiener Space*, de Gruyter, Berlin, 1991.
4. Deuschel, J.-D. and Stroock, D. W.: *Large Deviations*, Academic Press, San Diego, 1989.
5. Deuschel, J.-D. and Stroock, D. W.: Hypercontractivity and spectral gap of symmetric diffusions with applications to stochastic Ising models, *J. Funct. Anal.* **92** (1990), 30–48.
6. Donnelly, H. and Li, P.: Lower bounds for the eigenvalues of Rimannian manifolds, *Michigan Math. J.* **29** (1982), 149–161.
7. Fukushima, M.: *Dirichlet Forms and Markov Processes*, North-Holland/Kodansha, Amsterdam/Tokyo, 1980.
8. Gross, L.: Logarithmic Sobolev inequalities, *Amer. J. Math.* **97** (1975), 1061–1083.

9. Hess, H., Schrader, R. and Uhlenbrock, D. A.: Domination of semigroups and generalization of Kato's inequality, *Duke Math. J.* **44** (1977), 893–904.
10. Hess, H., Schrader, R. and Uhlenbrock, D. A.: Kato's inequality and the spectral distribution of Laplacians on compact Riemannian manifolds, *J. Differential Geom.* **15** (1980), 27–37.
11. Ma, Z.-M. and Röckner, M.: *Introduction to the Theory of (Non-symmetric) Dirichlet Forms*, Springer, New York, 1992.
12. Méritet, A.: Théorème d'anulation pour la cohomologie absolue d'une variété riemannienne a bord, *Bull. Sci. Math. (2)* **103** (1979), 379–400.
13. Ouhabaz, E.: Invariance of closed convex sets and domination criteria for semigroups, *Potential Anal.* **5** (1996), 611–625.
14. Reilly, R. C.: Applications of the Hessian operator in a Riemannian manifold, *Indiana Univ. Math. J.* **26** (1977), 459–472.
15. Ray, D. B. and Singer, I. M.: R-torsion and the Laplacian on Riemannian manifolds, *Adv. in Math.* **7** (1971), 145–210.
16. Shigekawa, I.: L^p contraction semigroups for vector valued functions, *J. Funct. Anal.* **147** (1997), 69–108.
17. Schwarz, G.: *Hodge-Decomposition – a Method for Solving Boundary Value Problems*, Lecture Notes in Math. 1607, Springer, New York, 1995.
18. Simon, B.: An abstract Kato's inequality for generators of positivity preserving semigroups, *Indiana Univ. Math. J.* **26** (1997), 1069–1073.
19. Simon, B.: Kato's inequality and the comparison of semigroups, *J. Funct. Anal.* **32** (1979), 97–101.
20. Xia, C.-Y.: The first nonzero eigenvalue for manifolds with Ricci curvature having positive lower bound, In: W.-T. Wu and M.-D. Cheng (eds), *Chinese Mathematics into the 21st Century*, Peking Univ. Press, Beijing, 1991, pp. 243–249.

Acta Applicandae Mathematicae **63:** 411–432, 2000.

The Product of Independent Random Variables with Regularly Varying Tails

Dedicated to Professor Takeyuki Hida on the occasion of his 70th birthday

TAKAAKI SHIMURA
The Institute of Statistical Mathematics, 4-6-7 Minami-Azabu Minato-ku, Tokyo, 106-8569, Japan.
e-mail: shimura@ism.ac.jp

(Received: 30 April 1999)

Abstract. A distribution μ is said to have regularly varying tail with index $-\alpha$ ($\alpha \geqslant 0$) if $\lim_{x\to\infty} \mu(kx,\infty)/\mu(x,\infty) = k^{-\alpha}$ for each $k > 0$. Let X and Y be independent positive random variables with distributions μ and ν, respecitvely. The distribution of product XY is called Mellin–Stieltjes convolution (MS convolution) of μ and ν. It is known that $\mathbf{D}(\alpha)$ (the class of distributions on $(0,\infty)$ that have regularly varying tails with index $-\alpha$) is closed under MS convolution. This paper deals with decomposition problem of distributions in $\mathbf{D}(\alpha)$ related to MS convolution. A representation of a regularly varying function F of the following form is investigated: $F(x) = \sum_{k=0}^{n-1} b_k f(a^k x)$, where f is a measurable function and a and b_k ($k = 1,\ldots,n-1$) are real constants. A criterion is given for these constants in order that f be regularly varying. This criterion is applicable to show that there exist two distributions μ and ν such that neither μ nor ν belongs to $\mathbf{D}(\alpha)$ ($\alpha > 0$) and their MS convolution belongs to $\mathbf{D}(\alpha)$.

Mathematics Subject Classifications (2000): primary 60E05, secondary 60E07, 60F05.

Key words: tail of distribution, regularly varying function, slowly varying function, product of independent random variables.

1. Introduction

Let X and Y be independent positive random variables with distributions μ and ν, respectively. We denote the distribution of the product XY by $\mu\circ\nu$ and call it the Mellin–Stieltjes convolution (MS convolution) of μ and ν. A distribution μ_1 is said to be a factor of a distribution μ, if $\mu = \mu_1\circ\nu$ with some ν. A distribution is said to have regularly varying tail with index $-\alpha$ ($\alpha \geqslant 0$) if the tail $\mu(x,\infty)$ satisfies that $\lim_{x\to\infty} \mu(kx,\infty)/\mu(x,\infty) = k^{-\alpha}$ for each $k > 0$. Such distributions are important in various areas of probability and statistics. Let $\mathbf{D}(\alpha)$ ($\alpha \geqslant 0$) be the class of distributions on $(0,\infty)$ with regularly varying tails with index $-\alpha$. The purpose of this paper is to study properties of distributions in $\mathbf{D}(\alpha)$ related to MS convolution.

Such problems have been studied in [EG80, C86] and so on. In those papers, the main interest is in the properties of MS convolution for given distributions. On the other hand, our purpose is to investigate the properties of factors in case a distribution in $\mathbf{D}(\alpha)$ is written as MS convolution of two factors. We call it the decomposition problem of $\mathbf{D}(\alpha)$. Our main concern is whether some factor has regularly varying tail or not. We introduce some results for another class $\mathbf{M}(\alpha)$ ($\alpha > 0$) (the class of distributions μ on $[0, \infty)$ whose αth truncated moments $\int_0^x t^\alpha \mu(dt)$ are slowly varying). The decomposition problem of $\mathbf{M}(\alpha)$ is studied in [S97]: $\mathbf{M}(\alpha)$ is closed under MS convolution. If $\mu \circ \nu \in \mathbf{M}(\alpha)$ and the αth moment of ν is finite, then $\mu \in \mathbf{M}(\alpha)$. Further, there exist distributions $\mu \in \mathbf{M}(\alpha)$ and $\nu \notin \mathbf{M}(\alpha)$ such that $\mu \circ \nu \in \mathbf{M}(\alpha)$.

The class $\mathbf{D}(\alpha)$ is also closed under MS convolution and, if $\mu \in \mathbf{D}(\alpha)$ and $\int_0^\infty t^{\alpha+\varepsilon} \nu(dt) < \infty$ for some $\varepsilon > 0$, then $\mu \circ \nu \in \mathbf{D}(\alpha)$ ([EG80]).

We shall show the following results on decomposition problem of $\mathbf{D}(\alpha)$, which are different between $\alpha = 0$ (slowly varying tails) and $\alpha > 0$ (regularly varying tails with negative indices). In the case of $\alpha = 0$, we get the converse statement mentioned above: if $\mu \circ \nu \in \mathbf{D}(0)$ and $\int_0^\infty t^\varepsilon \nu(dt) < \infty$ for some $\varepsilon > 0$, then $\mu \in \mathbf{D}(0)$. The general result on the decomposition of nonincreasing slowly varying functions is useful to prove this result. But, in the case of $\alpha > 0$, the similar fact is not true: there exist two distributions μ and ν such that $\mu, \nu \notin \mathbf{D}(\alpha)$ and $\mu \circ \nu \in \mathbf{D}(\alpha)$. In addition, we can make one factor a distribution with finite support. The proof of the case $\alpha > 0$ is also different from that of $\alpha = 0$. We consider a representation of a regularly varying F as $F(x) = \sum_{k=0}^{n-1} b_k f(a^k x)$, where f is a measurable function and a and b_k ($k = 1, \dots, n-1$) are real constants. Though F is regularly varying if f is regularly varying and a and b_k ($k = 1, \dots, n-1$) are positive, f is not necessarily regularly varying even if F is regularly varying. We will give a criterion for f to be regularly varying. This criterion is applied to construct two distributions μ and ν such that neither μ nor ν belongs to $\mathbf{D}(\alpha)$ ($\alpha > 0$) and $\mu \circ \nu$ belongs to $\mathbf{D}(\alpha)$.

2. Preliminaries

In this section, we prepare some definitions, notations and fundamental facts.

$\mathbb{R}$, $\mathbb{C}$ and $\mathbb{N}$ denote the set of real numbers, complex numbers and positve integers, respectively. A positive measurable function f is said to be regularly varying (r.v.) with index $\rho (\in \mathbb{R})$ if $\lim_{x\to\infty} f(kx)/f(x) = k^\rho$ for each $k > 0$. In particular, f is called slowly varying (s.v.) if $\rho = 0$.

The following is a basic fact on $\mathbf{D}(\alpha)$ and MS convolution.

PROPOSITION 2.1. (1) *If μ is in $\mathbf{D}(\alpha)$ ($\alpha \geqslant 0$) and the $(\alpha + \varepsilon)$th moment of ν is finite for some $\varepsilon > 0$, then $\mu \circ \nu$ belongs to $\mathbf{D}(\alpha)$. Then*

$$\lim_{x\to\infty} \mu \circ \nu(x, \infty)/\mu(x, \infty) = \int_0^\infty t^\alpha \nu(dt). \tag{2.1}$$

(2) *If* $\mu \in \mathbf{D}(\alpha)$ *and* $\nu \in \mathbf{D}(\beta)$ $(0 \leqslant \alpha \leqslant \beta)$, *then* $\mu\circ\nu$ *belongs to* $\mathbf{D}(\alpha)$.
(3) *If* μ *is in* $\mathbf{D}(\alpha)$,

$$\liminf_{x\to\infty} \mu\circ\nu(x,\infty)/\mu(x,\infty) \geqslant \int_0^\infty t^\alpha \nu(\mathrm{d}t)$$

for any distribution ν *on* $\mathbb{R}$. *In particular, in case* $\nu = \mu$,

$$\liminf_{x\to\infty} \mu\circ\mu(x,\infty)/\mu(x,\infty) \geqslant 2\int_0^\infty t^\alpha \mu(\mathrm{d}t).$$

Proof (sketch). A distribution is said to have exponential tail $\mu(x,\infty)$ if $\mathrm{e}^{k\alpha x}\mu(x+k,\infty)$ is asymptotically equivalent to $\mu(x,\infty)$ for all k. Since MS convolution of distributions of r.v. tails means convolution of distributions with exponential tails, Lemma 1 in [C86] and Theorem 3 in [EG80] implies (1) and (2), respectively. ([C86] obtains more profound results.) Use Fatou's lemma for (3). The case $\mu = \nu$ implies that the condition $\varepsilon > 0$ in (1) is significant. □

3. Decomposition of Distributions with Slowly Varying Tails

In this section, we deal with distributions with s.v. tails. The following implies that the converse statement of Proposition 2.1(1) is also true for $\alpha = 0$:

THEOREM 3.1. *If* $\mu\circ\nu$ *is in* $\mathbf{D}(0)$ *and* $\int_0^\infty t^\varepsilon \nu(\mathrm{d}t) < \infty$ *for some* $\varepsilon > 0$, *then* μ *belongs to* $\mathbf{D}(0)$. *In this case,* (2.1) *holds.*

We prepare some lemmas.

LEMMA 3.1 ([S94], Proposition 3.6 (2)). *Let* f *be nonincreasing s.v. Assume that it is expressed as* $f = f_1 + f_2$ *by two positive nonincreasing functions* f_1 *and* f_2. *If* $\liminf_{x\to\infty} f_1(x)/f(x) > 0$, *then* f_1 *is s.v.*

LEMMA 3.2. *Let* $\mu\circ\nu \in \mathbf{D}(\alpha)$ $(\alpha \geqslant 0)$. *For any* $\varepsilon > 0$, *there exists* $\delta > 0$ *such that*

$$\limsup_{x\to\infty} \int_0^\delta \mu(x/t,\infty)\nu(\mathrm{d}t) \Big/ \int_0^\infty \mu(x/t,\infty)\nu(\mathrm{d}t) \leqslant \varepsilon.$$

Proof. First we prove

$$\limsup_{x\to\infty} \mu(x,\infty)/\mu\circ\nu(x,\infty) < \infty.$$

Choose $\varepsilon_0 > 0$ such that $\nu(\varepsilon_0,\infty) > 0$. Since

$$\mu\circ\nu(x,\infty)/\mu(x/\varepsilon_0,\infty)$$

$$
\begin{aligned}
&= \int_0^\infty \mu(x/t,\infty)/\mu(x/\varepsilon_0,\infty)\nu(\mathrm{d}t)\\
&= \int_0^{\varepsilon_0} \mu(x/t,\infty)/\mu(x/\varepsilon_0,\infty)\nu(\mathrm{d}t)+\\
&\quad+\int_{\varepsilon_0}^\infty \mu(x/t,\infty)/\mu(x/\varepsilon_0,\infty)\nu(\mathrm{d}t)\\
&\geqslant \nu(\varepsilon_0,\infty)>0,
\end{aligned}
$$

we have

$$
\begin{aligned}
&\limsup_{x\to\infty}\mu(x,\infty)/\mu\circ\nu(x,\infty)\\
&\quad=\limsup_{x\to\infty}\mu(x/\varepsilon_0,\infty)/\mu\circ\nu(x/\varepsilon_0,\infty)\\
&\quad=\limsup_{x\to\infty}\varepsilon_0^{-\alpha}\mu(x/\varepsilon_0,\infty)/\mu\circ\nu(x,\infty)\\
&\quad\leqslant(\varepsilon_0^\alpha\nu(\varepsilon_0,\infty))^{-1}<\infty.
\end{aligned}
$$

Therefore, if we choose positive constants C and δ such that $\mu(x,\infty)/\mu\circ\nu(x,\infty) \leqslant C$ and $C\delta^\alpha\nu(0,\delta]<\varepsilon$, then

$$
\int_0^\delta \mu(x/t,\infty)\nu(\mathrm{d}t)\leqslant C\int_0^\delta \mu\circ\nu(x/t,\infty)\nu(\mathrm{d}t)\leqslant C\nu(0,\delta]\mu\circ\nu(x/\delta,\infty).
$$

The regular variation of $\mu\circ\nu(x,\infty)$ implies that

$$
\begin{aligned}
&\limsup_{x\to\infty}\int_0^\delta \mu(x/t,\infty)\nu(\mathrm{d}t)/\mu\circ\nu(x,\infty)\\
&\quad\leqslant C\nu(0,\delta]\limsup_{x\to\infty}\mu\circ\nu(x/\delta,\infty)/\mu\circ\nu(x,\infty)\\
&\quad= C\delta^\alpha\nu(0,\delta]<\varepsilon.
\end{aligned}
$$

The proof is complete. □

LEMMA 3.3. *Assume that* $\mu\circ\nu\in\mathbf{D}(\alpha)$ $(\alpha\geqslant 0)$ *and* $\int_0^\infty t^{\alpha+\varepsilon_0}\nu(\mathrm{d}t)<\infty$ *for some* $\varepsilon_0>0$. *Then, for any* $\varepsilon>0$, *there exists* $\gamma>0$ *such that*

$$
\limsup_{x\to\infty}\int_\gamma^\infty \mu(x/t,\infty)\nu(\mathrm{d}t)/\mu\circ\nu(x,\infty)<\varepsilon.
$$

Proof. For C and $\varepsilon>0$ in the proof of Lemma 3.2, choose $\gamma>0$ such that $C\int_\gamma^\infty t^\alpha\nu(\mathrm{d}t)<\varepsilon$. Then

$$
\begin{aligned}
&\int_\gamma^\infty \mu(x/t,\infty)\nu(\mathrm{d}t)/\mu\circ\nu(x,\infty)\\
&\quad\leqslant C\int_\gamma^\infty \mu\circ\nu(x/t,\infty)/\mu\circ\nu(x,\infty)\nu(\mathrm{d}t)\\
&\quad= C\int_\gamma^\infty ((x/t)^{\alpha+\varepsilon_0}\mu\circ\nu(x/t,\infty))/(x^{\alpha+\varepsilon_0}\mu\circ\nu(x,\infty))t^{\alpha+\varepsilon_0}\nu(\mathrm{d}t).
\end{aligned}
$$

Noticing that $((x/t)^{\alpha+\varepsilon_0}\mu\circ\nu(x/t,\infty))/(x^{\alpha+\varepsilon_0}\mu\circ\nu(x,\infty))$ is bounded ([S97], Lemma 2.1) and converges to $t^{-\varepsilon_0}$ as $x \to \infty$ and $t^{\alpha+\varepsilon_0}\nu(\mathrm{d}t)$ is finite measure, we see that the last term converges to $C\int_\gamma^\infty t^\alpha \nu(\mathrm{d}t)$ as $x \to \infty$. Thus we get

$$\limsup_{x\to\infty} \int_\gamma^\infty \mu(x/t,\infty)\nu(\mathrm{d}t)/\mu\circ\nu(x,\infty) = C\int_\gamma^\infty t^\alpha \nu(\mathrm{d}t) < \varepsilon. \qquad \square$$

Proof of Theorem 3.1. From Lemmas 3.2 and 3.3, we can choose δ and γ ($0 < \delta < \gamma < \infty$) such that

$$\liminf_{x\to\infty} \int_\delta^\gamma \mu(x/t,\infty)\nu(\mathrm{d}t)/\mu\circ\nu(x,\infty) > 0.$$

It follows from Lemma 3.1 that $\int_\delta^\gamma \mu(x/t,\infty)\nu(\mathrm{d}t)$ is s.v. Since

$$\mu(x/\delta,\infty)\int_\delta^\gamma \nu(\mathrm{d}t) \leqslant \int_\delta^\gamma \mu(x/t,\infty)\nu(\mathrm{d}t) \leqslant \mu(x/\gamma,\infty)\int_\delta^\gamma \nu(\mathrm{d}t)$$

for $k \leqslant 1$, we see that

$$1 \leqslant \mu(kx,\infty)/\mu(x,\infty) \leqslant \int_\delta^\gamma \mu(k\delta x/t,\infty)\nu(\mathrm{d}t)\Big/\int_\delta^\gamma \mu(\gamma x/t,\infty)\nu(\mathrm{d}t).$$

The right-hand side goes to 1 as $x \to \infty$. The case $k > 1$ is proved in the same way. Thus we get $\mu \in \mathbf{D}(0)$. Proposition 2.1 yields (2.1). This completes the proof. $\square$

4. Regularly Varying Tails with Negative Index

The following is a result on the decomposition of distributions in $\mathbf{D}(\alpha)$ ($\alpha > 0$).

THEOREM 4.1. *Let μ and ν be distributions on $(0,\infty)$. Assume that the support of ν is included in a set that consists of a geometric progression*

$$\{t_k = t_0 r^k \ : \ k = 0, 1, \ldots, n-1, t_0 > 0, r > 1\}.$$

Set

$$\mathbf{C} = \left\{|z| \ : \ \sum_{k=0}^{n-1} r^{k\alpha}\nu(\{t_k\})z^k = 0\right\}.$$

(i) *If $1 \notin \mathbf{C}$ and $\mu\circ\nu \in \mathbf{D}(\alpha)$ ($\alpha > 0$), then μ belongs to $\mathbf{D}(\alpha)$.*

(ii) *If $1 \in \mathbf{C}$, then, for any r.v. function F with negative index $-\alpha$, then there exist two distributions $\mu_1 \in \mathbf{D}(\alpha)$ and $\mu_2 \notin \mathbf{D}(\alpha)$ such that*

$$\lim_{x\to\infty} \mu_1\circ\nu(x,\infty)/F(x) = \lim_{x\to\infty} \mu_2\circ\nu(x,\infty)/F(x) = 1.$$

This is a consequence of the following some general result.

THEOREM 4.2. *Let ρ, a and b_k $(k = 0, \dots, n-1)$ be real constants such that $0 < a \neq 1$ and $\sum_{k=0}^{n-1} a^{k\rho} b_k > 0$. Set*

$$\mathbf{E} = \left\{ |z| \ : z \in \mathbb{C} \text{ and } \sum_{k=0}^{n-1} a^{k\rho} b_k z^k = 0 \right\}.$$

(i) *If $1 \notin \mathbf{E}$ and if a r.v. function F with index ρ is represented as*

$$F(x) = \sum_{k=0}^{n-1} b_k f(a^k x) \tag{4.1}$$

by some positive function f satisfying $\sup_{x \geqslant 0} |f(x)/F(x)| < \infty$, then f is r.v.

(ii) *If $1 \in \mathbf{E}$ and $\rho \neq 0$, then, for any r.v. function F with index ρ, there exist a positive monotone r.v. function f_1 and a positive monotone function f_2 that is not r.v. such that*

$$\lim_{x\to\infty} \sum_{k=0}^{n-1} b_k f_1(a^k x)/F(x) = \lim_{x\to\infty} \sum_{k=0}^{n-1} b_k f_2(a^k x)/F(x) = 1. \tag{4.2}$$

Remark. If F is monotone s.v. $(\rho = 0)$ and $b_k \geqslant 0$ for every k, then f in (4.1) is s.v. (see Lemma 3.1). This is a big difference between $\rho = 0$ and $\rho \neq 0$ ([S94], Theorem 3.2, Proposition 3.13).

First we will prove Theorem 4.2.

Proof of Theorem 4.2. (i) Let

$$B = \begin{pmatrix} -b_1 & -b_2 & \dots & -b_{n-2} & -b_{n-1} \\ 1 & 0 & \dots & 0 & 0 \\ 0 & \ddots & & \vdots & \vdots \\ \vdots & & \ddots & 0 & \vdots \\ 0 & \dots & 0 & 1 & 0 \end{pmatrix}.$$

We can assume that $0 < a < 1$, $b_0 = 1$ and $b_{n-1} \neq 0$ without loss of generality: The assertion for $0 < a < 1$ yields that for $1 < a < \infty$, easily. Since the proof goes well for $-F$ instead of F, $b_0 = 1$ could be assumed. The last one $b_{n-1} \neq 0$ is trivial, which implies $|B| \neq 0$. Hence, 0 is not an eigenvalue of B. We shall use induction on n.

From (4.1),

$$\begin{pmatrix} f(x) \\ f(ax) \\ \vdots \\ \vdots \\ f(a^{n-2}x) \end{pmatrix}$$

$$= \begin{pmatrix} -b_1 & -b_2 & \dots & -b_{n-2} & -b_{n-1} \\ 1 & 0 & \dots & 0 & 0 \\ 0 & \ddots & & \vdots & \vdots \\ \vdots & & \ddots & 0 & \vdots \\ 0 & \dots & 0 & 1 & 0 \end{pmatrix} \begin{pmatrix} f(ax) \\ f(a^2x) \\ \vdots \\ \vdots \\ f(a^{n-1}x) \end{pmatrix} + \begin{pmatrix} F(x) \\ 0 \\ \vdots \\ \vdots \\ 0 \end{pmatrix}.$$

It follows from induction that, for any positive integer m,

$$\begin{pmatrix} f(x) \\ f(ax) \\ \vdots \\ f(a^{n-2}x) \end{pmatrix} = \sum_{j=0}^{m-1} B^j \begin{pmatrix} F(a^j x) \\ 0 \\ \vdots \\ 0 \end{pmatrix} + B^m \begin{pmatrix} f(a^m x) \\ f(a^{m+1}x) \\ \vdots \\ f(a^{m+n-2}x) \end{pmatrix}.$$

Dividing the both sides by F, we get

$$\begin{pmatrix} f(x)/F(x) \\ f(ax)/F(x) \\ \vdots \\ f(a^{n-2}x)/F(x) \end{pmatrix}$$

$$= \sum_{j=0}^{m-1} \frac{a^{-j\rho} F(a^j x)}{F(x)} (a^\rho B)^j \begin{pmatrix} 1 \\ 0 \\ \vdots \\ 0 \end{pmatrix} +$$

$$+ (a^\rho B)^m \begin{pmatrix} a^{-m\rho} f(a^m x)/F(x) \\ a^{-m\rho} f(a^{m+1}x)/F(x) \\ \vdots \\ a^{-m\rho} f(a^{m+n-2}x)/F(x) \end{pmatrix}.$$

Setting $g(x) = f(x)/F(x)$, $F(x) = x^\rho L(x)$ (L is s.v.) and $PJP^{-1} = a^\rho B$ (J is Jordan canonical form and P is nonsingular matrix), we have

$$\begin{pmatrix} g(x) \\ g(ax)\frac{F(ax)}{F(x)} \\ \vdots \\ g(a^{n-2}x)\frac{F(a^{n-2}x)}{F(x)} \end{pmatrix} - \sum_{j=0}^{m-1} PJ^jP^{-1} \begin{pmatrix} 1 \\ 0 \\ \vdots \\ 0 \end{pmatrix} -$$

$$- PJ^mP^{-1} \begin{pmatrix} g(a^m x) \\ a^\rho g(a^{m+1}x) \\ \vdots \\ a^{(n-2)\rho} g(a^{m+n-2}x) \end{pmatrix}$$

$$= \sum_{j=0}^{m-1} \left(\frac{L(a^j x)}{L(x)} - 1 \right) PJ^jP^{-1} \begin{pmatrix} 1 \\ 0 \\ \vdots \\ 0 \end{pmatrix} +$$

$$+PJ^mP^{-1}\begin{pmatrix} g(a^m x)\left(\frac{L(a^m x)}{L(x)}-1\right) \\ a^{\rho}g(a^{m+1}x)\left(\frac{L(a^{m+1}x)}{L(x)}-1\right) \\ \vdots \\ a^{(n-2)\rho}g(a^{m+n-2}x)\left(\frac{L(a^{m+n-2}x)}{L(x)}-1\right)\end{pmatrix}. \tag{4.3}$$

Let Λ be the set of eigenvalues of $a^{\rho}B$. Λ is identical with the set of reciprocal numbers of the elements of $\mathbf{E}$ because $|a^{\rho}B - tI| = (-1)^{n-1}\sum_{k=0}^{n-1} a^{k\rho}b_k t^{n-1-k} = 0$. Therefore, the assumption of (i) means that Λ does not contain any numbers with absolute value 1.

For each $\lambda \in \Lambda$, it is easy to see that the dimension of eigenspace for λ is 1. Hence, the unique Jordan block J_λ corresponds to each eigenvalue λ. We can extend each Jordan block J_λ to $(n-1)\times(n-1)$ matrix, replacing other Jordan blocks in J by null matrix and denote it by the same notation J_λ. Then $J = \sum_{\lambda\in\Lambda} J_\lambda$ and $J^j = \sum_{\lambda\in\Lambda} J_\lambda^j$.

Since we assume that g is bounded, the right-hand side in (4.3) converges to zero vector as $x \to \infty$ by the slow variation of L. For any positive integer m,

$$\lim_{x\to\infty}\begin{pmatrix} g(x) \\ g(ax)\frac{F(ax)}{F(x)} \\ \vdots \\ g(a^{n-2}x)\frac{F(a^{n-2}x)}{F(x)}\end{pmatrix} - \sum_{\lambda\in\Lambda} I_{\lambda,m}(x) = 0,$$

where

$$I_{\lambda,m}(x) = \sum_{j=0}^{m-1} PJ_\lambda^j P^{-1}\begin{pmatrix}1\\0\\\vdots\\0\end{pmatrix} + PJ_\lambda^m P^{-1}\begin{pmatrix} g(a^m x) \\ a^{\rho}g(a^{m+1}x) \\ \vdots \\ a^{(n-2)\rho}g(a^{m+n-2}x)\end{pmatrix}.$$

Letting the first row of $I_{\lambda,m}$ be $I_{\lambda,m}^{(1)}$, we have

$$\lim_{x\to\infty} g(x) - \sum_{\lambda\in\Lambda} I_{\lambda,m}^{(1)}(x) = 0. \tag{4.4}$$

In the case of $n = 2$, for any positive integer m,

$$\lim_{x\to\infty} g(x) - (1+a^{\rho}b_1)^{-1} - \{(-a^{\rho}b_1)^m(g(a^m x) - (1+a^{\rho}b_1)^{-1})\} = 0.$$

Since $|a^{\rho}b_1| \neq 1$ is the assumption for $n = 2$, we have $\lim_{x\to\infty} g(x) = (1 + a^{\rho}b_1)^{-1} > 0$. This implies that f is r.v. with the same index as F.

Let us calculate $I_{\lambda,m}$. We denote $k \times k$ null matrix and unit matrix by O_k and I_k, respectively. $J(\lambda, k)$ denotes $k \times k$ Jordan block for eigenvalue λ:

$$J(\lambda, k) = \begin{pmatrix} \lambda & 1 & 0 & \dots & 0 \\ 0 & \ddots & \ddots & & \vdots \\ \vdots & \ddots & \ddots & \ddots & 0 \\ \vdots & & \ddots & \ddots & 1 \\ 0 & \dots & \dots & 0 & \lambda \end{pmatrix}.$$

We assume that J has both real and complex eigenvalues and it is represented as

$$J = J_\beta + J_{re^{i\theta}} + J_{re^{-i\theta}} + \sum_{\lambda \in \Lambda \setminus \{\beta, re^{\pm i\theta}\}} J_\lambda,$$

where $J_\beta = J(\beta, s) \oplus O_{n-s-1}$, $J_{re^{i\theta}} = O_s \oplus J(re^{i\theta}, t) \oplus O_{n-s-t-1}$ and $J_{re^{-i\theta}} = O_{s+t} \oplus J(re^{-i\theta}, t) \oplus O_{n-s-2t-1}$.

First, we start to calculate $I_{\beta,m}$.

Since

$$J_\beta^j = \sum_{l=0}^{(s-1) \wedge j} {}_j\mathrm{C}_l \beta^{j-l} N_\beta^l,$$

where

$$N_\beta = (J(\beta, s) - \beta I_s) \oplus O_{n-s-1} \quad \text{and} \quad N_\beta^0 = I_{n-1},$$

we have

$$P J_\beta^j P^{-1} = \sum_{l=0}^{(s-1) \wedge j} {}_j\mathrm{C}_l \beta^{j-l} P N_\beta^l P^{-1}.$$

Putting $P = (p_{i,j})_{1 \leqslant i,j \leqslant n-1}$ and $P^{-1} = (q_{i,j})_{1 \leqslant i,j \leqslant n-1}$, we get

$$P J_\beta^j P^{-1} = \sum_{l=0}^{(s-1) \wedge j} {}_j\mathrm{C}_l \beta^{j-l} \begin{pmatrix} d_{l,0} & d_{l,1} & \dots & d_{l,n-2} \\ * & * & \dots & * \\ \vdots & \vdots & \vdots & \vdots \\ * & * & \dots & * \end{pmatrix},$$

where $d_{l,k} = \sum_{u=1}^{s-l} p_{1,u} q_{l+u,k+1}$ $(k = 0, \dots, n-2)$. Therefore, the first term of $I_{\beta,m}$ is

$$\sum_{j=0}^{m-1} P J_\beta^j P^{-1} \begin{pmatrix} 1 \\ 0 \\ \vdots \\ 0 \end{pmatrix} = \sum_{j=0}^{m-1} \sum_{l=0}^{(s-1) \wedge j} {}_j\mathrm{C}_l \beta^{j-l} \begin{pmatrix} d_{l,0} \\ * \\ \vdots \\ * \end{pmatrix}.$$

The second term is, for $m \geqslant s-1$,

$$P J_\beta^m P^{-1}\begin{pmatrix} g(a^m x) \\ a^\rho g(a^{m+1}x) \\ \vdots \\ a^{(n-2)\rho} g(a^{m+n-2}x) \end{pmatrix} = \sum_{l=0}^{s-1} {}_m\mathrm{C}_l \beta^{m-l} \begin{pmatrix} \sum_{k=0}^{n-2} d_{l,k} a^{k\rho} g(a^{m+k}x) \\ * \\ \vdots \\ * \end{pmatrix}.$$

Thus, we have

$$I_{\beta,m}^{(1)}(x) = \sum_{j=0}^{m-1} \sum_{l=0}^{(s-1)\wedge j} {}_j\mathrm{C}_l \beta^{j-l} d_{l,0} + + \sum_{l=0}^{s-1} {}_m\mathrm{C}_l \beta^{m-l} \sum_{k=0}^{n-2} d_{l,k} a^{k\rho} g(a^{m+k}x). \tag{4.5}$$

The behaviours of $I_{\beta,m}^{(1)}$ differ between $|\beta| < 1$ and $|\beta| > 1$. In the case of $|\beta| < 1$, $\sum_{j=0}^{\infty} \sum_{l=0}^{(s-1)\wedge j} {}_j\mathrm{C}_l |\beta|^j < \infty$. Hence the first term of the right-hand side of (4.5) converges absolutely as $m \to \infty$. Since g is bounded and $\lim_{m\to\infty} \sum_{l=0}^{s-1} {}_m\mathrm{C}_l \beta^{m-l} = 0$,

$$\lim_{m\to\infty} \limsup_{x\to\infty} \left| \sum_{l=0}^{s-1} {}_m\mathrm{C}_l \beta^{m-l} \sum_{k=0}^{n-2} d_{l,k} a^{k\rho} g(a^{m+k}x) \right| = 0.$$

Therefore we have

$$\lim_{m\to\infty} \limsup_{x\to\infty} \left| I_{\beta,m}^{(1)}(x) - \sum_{j=0}^{\infty} \sum_{l=0}^{(s-1)\wedge j} {}_j\mathrm{C}_l \beta^{j-l} d_{l,0} \right| = 0 \tag{4.6}$$

for $|\beta| < 1$, where $\sum_{j=0}^{\infty} \sum_{l=0}^{(s-1)\wedge j} {}_j\mathrm{C}_l \beta^{j-l} d_{l,0}$ is a finite constant. In the case of $|\beta| > 1$, we split $I_{\beta,m}^{(1)}$ into four terms and consider them by turns.

$$I_{\beta,m}^{(1)}(x) = {}_m\mathrm{C}_{s-1} \beta^{m-s+1} \left\{ \frac{d_{s-1,0}}{\beta - 1} + \sum_{k=0}^{n-2} d_{s-1,k} a^{k\rho} g(a^{m+k}x) \right\} + + \left\{ \sum_{j=0}^{m-1} \sum_{l=0}^{(s-1)\wedge j} {}_j\mathrm{C}_l \beta^{j-l} d_{l,0} + \frac{d_{s-1,0}}{1-\beta} \sum_{l=0}^{s-1} {}_m\mathrm{C}_l \beta^{m-l} \right\} +$$

$$+\sum_{l=0}^{s-2} {}_m\mathrm{C}_l \beta^{m-l}\left\{\frac{d_{s-1,0}}{\beta-1}+\sum_{k=0}^{n-2} d_{s-1,k}a^{k\rho}g(a^{m+k}x)\right\}+$$
$$+\sum_{l=0}^{s-1} {}_m\mathrm{C}_l \beta^{m-l}\left\{\sum_{k=0}^{n-2} d_{l,k}a^{k\rho}g(a^{m+k}x)-\sum_{k=0}^{n-2} d_{s-1,k}a^{k\rho}g(a^{m+k}x)\right\}. \tag{4.7}$$

From the first column of $a^\rho BP = PJ$:

$$a^\rho p_{k,1} = \beta p_{k+1,1} \quad (k=1,\ldots,n-2), \qquad a^\rho \sum_{k=1}^{n-1} b_k p_{k,1} + \beta p_{1,1} = 0,$$

we see $p_{1,1} \neq 0$. Similarly, by the sth row of $a^\rho P^{-1}B = JP^{-1}$:

$$\begin{aligned} &a^\rho b_k q_{s,1} + \beta q_{s,k} = a^\rho q_{s,k+1} \quad (k=1,\ldots,n-2),\\ &a^\rho b_{n-1} q_{s,1} + \beta q_{s,n-1} = 0, \end{aligned} \tag{4.8}$$

$q_{s,1} \neq 0$. Hence, we get $d_{s-1,0} = p_{1,1}q_{s,1} \neq 0$. By (4.8),

$$\left(z-\frac{1}{\beta}\right)\sum_{k=0}^{n-2} d_{s-1,k}a^{k\rho}z^k = -\frac{d_{s-1,0}}{\beta}\sum_{k=0}^{n-1} a^{k\rho}b_k z^k.$$

Noticing this equality, we define the function $H_\beta(x)$ as

$$H_\beta(x) = \frac{1}{\beta-1} + \frac{1}{d_{s-1,0}}\sum_{k=0}^{n-2} d_{s-1,k}a^{k\rho}g(a^k x).$$

The first term of $I^{(1)}_{\beta,m}$ is written as $d_{s-1,0\,m}\mathrm{C}_{s-1}\beta^{m-s+1}H_\beta(a^m x)$. We show that other terms in (4.7) are $o({}_m\mathrm{C}_{s-1}\beta^{m-s+1})$ uniformly in x as $m \to \infty$. It is sufficient for the second term to show the following because

$$\lim_{m\to\infty}\sum_{l=0}^{s-1} {}_m\mathrm{C}_l\beta^{m-l}/{}_m\mathrm{C}_{s-1}\beta^{m-s+1} = 1.$$

$$\lim_{m\to\infty}\frac{\sum_{j=0}^{m-1}\sum_{l=0}^{(s-1)\wedge j} {}_j\mathrm{C}_l\beta^{j-l}d_{l,0}}{{}_m\mathrm{C}_{s-1}\beta^{m-s+1}} + \frac{d_{s-1,0}}{1-\beta} = 0 \quad \text{for } |\beta| > 1.$$

Setting $j = m-k$, we have

$$\sum_{j=0}^{m-1}\frac{\sum_{l=0}^{(s-1)\wedge j} {}_j\mathrm{C}_l\beta^{j-l}d_{l,0}}{{}_m\mathrm{C}_{s-1}\beta^{m-s+1}}$$
$$=\sum_{k=1}^{m}\left(\sum_{l=0}^{(s-1)\wedge(m-k)}\frac{{}_{m-k}\mathrm{C}_l}{{}_m\mathrm{C}_{s-1}}\beta^{s-l-1}d_{l,0}\right)\beta^{-k}.$$

The dominated convergence theorem guarantees that the right-hand side converges to $d_{s-1,0}(\beta - 1)^{-1}$ as $m \to \infty$. It follows from the boundeness of g and $\lim_{m\to\infty} \sum_{l=0}^{s-1} {}_m\mathrm{C}_l\beta^{m-l}/{}_m\mathrm{C}_{s-1}\beta^{m-s+1} = 1$ that both the third and last terms are $o({}_m\mathrm{C}_{s-1}\beta^{m-s+1})$ uniformly in x as $m \to \infty$.

We conclude that

$$I^{(1)}_{\beta,m}(x) = d_{s-1,0\,m}\mathrm{C}_{s-1}\beta^{m-s+1}(H_\beta(a^m x) + o_1(m,x)) \tag{4.9}$$

for $|\beta| > 1$, where $\lim_{m\to\infty} \limsup_{x\to\infty} |o_1(m,x)| = 0$.

Now we consider the complex eigenvalues. The details of the calculation are similar to the case of real one and are omitted. But, in this case, we have to deal with the sum of $I^{(1)}_{re^{i\theta},m}$ and $I^{(1)}_{re^{-i\theta},m}$.

$$\begin{aligned}
&I^{(1)}_{re^{i\theta},m}(x)\\
&= \sum_{j=0}^{m-1}\sum_{l=0}^{(t-1)\wedge j} {}_j\mathrm{C}_l(re^{i\theta})^{j-l}d^+_{l,0} + \sum_{l=0}^{t-1} {}_m\mathrm{C}_l(re^{i\theta})^{m-l}\sum_{k=0}^{n-2} d^+_{l,k}a^{k\rho}g(a^{m+k}x)\\
&= {}_m\mathrm{C}_{t-1}(re^{i\theta})^{m-t+1}\left\{\frac{d^+_{t-1,0}}{re^{i\theta}-1} + \sum_{k=0}^{n-2} d^+_{t-1,k}a^{k\rho}g(a^{m+k}x)\right\} +\\
&\quad + \left\{\sum_{j=0}^{m-1}\sum_{l=0}^{(t-1)\wedge j} {}_j\mathrm{C}_l(re^{i\theta})^{j-l}d^+_{l,0} + \frac{d^+_{t-1,0}}{1-re^{i\theta}}\sum_{l=0}^{t-1} {}_m\mathrm{C}_l(re^{i\theta})^{m-l}\right\} +\\
&\quad + \sum_{l=0}^{t-2} {}_m\mathrm{C}_l(re^{i\theta})^{m-l}\left\{\frac{d^+_{t-1,0}}{re^{i\theta}-1} + \sum_{k=0}^{n-1} d^+_{t-1,k}a^{k\rho}g(a^{m+k}x)\right\} +\\
&\quad + \sum_{l=0}^{t-1} {}_m\mathrm{C}_l(re^{i\theta})^{m-l}\left\{\sum_{k=0}^{n-2} d^+_{l,k}a^{k\rho}g(a^{m+k}x) - \sum_{k=0}^{n-2} d^+_{t-1,k}a^{k\rho}g(a^{m+k}x)\right\},
\end{aligned} \tag{4.10}$$

where

$$d^+_{l,k} = \sum_{u=1}^{t-l} p_{1,s+u}q_{s+l+u,k+1} \quad (k = 0,\dots,n-2).$$

The calculation of $I^{(1)}_{re^{-i\theta},m}$ is similar to that of $I^{(1)}_{re^{i\theta},m}$.

In the case of $r < 1$,

$$\lim_{m\to\infty}\limsup_{x\to\infty} |I^{(1)}_{re^{i\theta},m}(x) - \sum_{j=0}^{\infty}\sum_{l=0}^{(t-1)\wedge j} {}_j\mathrm{C}_l(re^{i\theta})^{j-l}d^+_{l,0}| = 0, \tag{4.11}$$

$$\lim_{m\to\infty}\limsup_{x\to\infty} |I^{(1)}_{re^{-i\theta},m}(x) - \sum_{j=0}^{\infty}\sum_{l=0}^{(t-1)\wedge j} {}_j\mathrm{C}_l(re^{i\theta})^{j-l}d^-_{l,0}| = 0,$$

where

$$d_{l,k}^{-} = \sum_{u=1}^{t-l} p_{1,s+t+u} q_{s+t+l+u,k+1} \quad (k = 0, \ldots, n-2).$$

In the case of $r > 1$,

$$I_{re^{i\theta},m}^{(1)}(x)$$
$$= {}_m\mathrm{C}_{t-1}(re^{i\theta})^{m-t+1} \left\{ \frac{d_{t-1,0}^{+}}{re^{i\theta}-1} + \sum_{k=0}^{n-2} d_{t-1,k}^{+} a^{k\rho} g(a^{m+k}x) + o_2^{+}(m,x) \right\},$$

$$I_{re^{-i\theta},m}^{(1)}(x)$$
$$= {}_m\mathrm{C}_{t-1}(re^{-i\theta})^{m-t+1} \left\{ \frac{d_{t-1,0}^{-}}{re^{-i\theta}-1} + \sum_{k=0}^{n-2} d_{t-1,k}^{-} a^{k\rho} g(a^{m+k}x) + o_2^{-}(m,x) \right\}, \quad (4.12)$$

where

$$\lim_{m\to\infty} \limsup_{x\to\infty} |o_2^{+}(m,x)| = \lim_{m\to\infty} \limsup_{x\to\infty} |o_2^{-}(m,x)| = 0.$$

Let us combine the conjugate ones. To begin with, we prove the following.

$$d_{t-1,0}^{+} G(x) = \frac{d_{t-1,0}^{+}}{re^{i\theta}-1} + \sum_{k=0}^{n-2} d_{t-1,k}^{+} a^{k\rho} g(a^k x), \quad (4.13)$$

where

$$G(x) = H_{re^{\pm i\theta}}(x) - re^{-i\theta} H_{re^{\pm i\theta}}(ax).$$

$H_{re^{\pm i\theta}}(x)$ is defined as

$$H_{re^{\pm i\theta}}(x) = \sum_{k=0}^{n-3} c_k g(a^k x) - \{(1-re^{i\theta})(1-re^{-i\theta})\}^{-1}$$

and c_k $(k = 0, \ldots, n-3)$ are the constants determined by

$$\left(z - \frac{1}{re^{i\theta}}\right)\left(z - \frac{1}{re^{-i\theta}}\right) \sum_{k=0}^{n-3} c_k z^k = \frac{1}{r^2} \sum_{k=0}^{n-1} a^{k\rho} b_k z^k. \quad (4.14)$$

In a similar way to the real case, we have $d_{t-1,0}^{+} \neq 0$.

By the $(s+t)$th row, in the kth columns $(k = 1, \ldots, n-2)$ of $a^\rho P^{-1} B = J P^{-1}$, we get

$$a^\rho q_{s+t,k+1} = re^{i\theta} q_{s+t,k} + a^\rho b_k q_{s+t,1}. \quad (4.15)$$

Seeing the $(n-1)$th column, we have

$$a^{\rho} q_{s+t,1} b_{n-1} + re^{i\theta} q_{s+t,n-1} = 0. \tag{4.16}$$

Since (4.16) yields

$$d^{+}_{t-1,0} G(x) = d^{+}_{t-1,0} H_{re^{\pm i\theta}}(x) + d^{+}_{t-1,n-2} a^{-\rho} r^2 b^{-1}_{n-1} H_{re^{\pm i\theta}}(ax),$$

it suffices to show that

$$\begin{aligned} &\frac{d^{+}_{t-1,0}}{re^{i\theta}-1} + \sum_{k=0}^{n-2} d^{+}_{t-1,k} a^{k\rho} g(a^k x) \\ &\quad = d^{+}_{t-1,0} H_{re^{\pm i\theta}}(x) + d^{+}_{t-1,n-2} a^{-\rho} r^2 b^{-1}_{n-1} H_{re^{\pm i\theta}}(ax). \end{aligned}$$

We compare the coefficients of $g(a^k x)$ $(k = 0, \ldots, n-2)$ and constants of both sides.

From (4.16), the constants of both sides are equal. The coefficients of $g(x)$ $(k = 0)$ are equal from $c_0 = 1$. For other coefficients, we show that

$$\begin{aligned} &a^{k\rho} q_{s+t,k+1} \\ &\quad = q_{s+t,1} c_k + q_{s+t,n-1} a^{-\rho} r^2 b^{-1}_{n-1} c_{k-1} \quad (k = 1, \ldots, n-3), \\ &a^{(n-1)\rho} b_{n-1} = r^2 c_{n-3} \quad (k = n-2). \end{aligned} \tag{4.17}$$

By (4.16), the upper one is equivalent to the following:

$$a^{k\rho} q_{s+t,k+1} = (c_k - re^{-i\theta} c_{k-1}) q_{s+t,1} \quad (k = 1, \ldots, n-3). \tag{4.18}$$

From (4.14),

$$a^{\rho} b_1 = c_1 - 2r\cos\theta, \tag{4.19}$$

$$\begin{aligned} &a^{k\rho} b_k = c_k - 2r\cos\theta c_{k-1} + r^2 c_{k-2} \quad (k = 2, \ldots, n-3), \\ &a^{(n-2)\rho} b_{n-2} = -2r\cos\theta c_{n-3} + r^2 c_{n-4}, \\ &a^{(n-1)\rho} b_{n-1} = r^2 c_{n-3}. \end{aligned} \tag{4.20}$$

The last one of (4.20) is the lower equality of (4.17). We use induction on k $(k = 1, \ldots, n-3)$ to show (4.18). For $k = 1$, it follows from the case of $k = 1$ in (4.15) and (4.19) that

$$a^{\rho} q_{s+t,2} = (re^{i\theta} + a^{\rho} b_1) q_{s+t,1} = (c_1 - re^{-i\theta}) q_{s+t,1}.$$

Assume that the assertion holds for k. Using (4.15), the assumption (the upper equality of (4.18)) and (4.20) in turns, we get

$$\begin{aligned} a^{(k+1)\rho} q_{s+t,k+2} &= a^{k\rho} re^{i\theta} q_{s+t,k+1} + a^{(k+1)\rho} b_{k+1} q_{s+t,1} \\ &= (re^{i\theta}(c_k - re^{-i\theta} c_{k-1}) + a^{(k+1)\rho} b_{k+1}) q_{s+t,1} \\ &= (c_{k+1} - re^{-i\theta} c_k) q_{s+t,1}. \end{aligned}$$

Thus, we get (4.13). Simiarly, the conjugate is obtained. Further, $\overline{d^+_{t-1,0}} = d^-_{t-1,0}$ ($\neq 0$) (proof is omitted, since the detail is complicated and the proof of the theorem does not depend on this).

It follows from (4.12) and (4.13) that

$$\begin{aligned} & I^{(1)}_{re^{i\theta},m}(x) + I^{(1)}_{re^{-i\theta},m}(x) \\ & = 2{}_m\mathrm{C}_{t-1}r^{m-t+1}\{\Re e^{i(m-t+1)\theta}d^+_{t-1,0}G(a^m x) + o_2(m,x)\}, \end{aligned} \tag{4.21}$$

where $\lim_{m\to\infty}\limsup_{x\to\infty}|o_2(m,x)| = 0$.

Thus, we have finished the calculation of $I_{\lambda,m}$. Let r be the absolute value of the eigenvalues of $a^\rho B$ with maximum absolute values and t be the maximum of the rank of Jordan blocks correspoding to such eigenvalues.

If $r < 1$, then, by (4.6) and (4.11), there exists a constant c_λ for each λ such that

$$\lim_{m\to\infty}\limsup_{x\to\infty}|I^{(1)}_{\lambda,m}(x) - c_\lambda| = 0. \tag{4.22}$$

From (4.22) and (4.4),

$$\lim_{x\to\infty} g(x) = \sum_{\lambda\in\Lambda} c_\lambda.$$

By (4.1),

$$1 = \lim_{x\to\infty}\sum_{k=0}^{n-1} b_k\frac{f(a^k x)}{F(x)} = \lim_{x\to\infty}\sum_{k=0}^{n-1} b_k\frac{f(a^k x)}{F(a^k x)}\frac{F(a^k x)}{F(x)} = \sum_{\lambda\in\Lambda} c_\lambda\sum_{k=0}^{n-1} a^{k\rho}b_k.$$

Hence

$$\lim_{x\to\infty} g(x) = \left(\sum_{k=0}^{n-1} a^{k\rho}b_k\right)^{-1} > 0.$$

This implies that f is r.v. with index ρ.

The proof is finished for $r < 1$. If $r > 1$, then divide Λ into the three classes.

$$\begin{aligned} \Lambda_1 &= \{\lambda\in\Lambda : |\lambda| = r \text{ and the rank of } J_\lambda = t\}, \\ \Lambda_2 &= \{\lambda\in\Lambda : 1 < |\lambda| < r \text{ or } |\lambda| = r, \text{ the rank of } J_\lambda \text{ is less than } t\}, \\ \Lambda_3 &= \{\lambda\in\Lambda : |\lambda| < 1\}. \end{aligned}$$

We assume that Λ_1 contains both real and complex eigenvalues and denote them by r, $re^{\pm i\theta_l}$ ($l = 1,\ldots,q$). By (4.9) and (4.21),

$$\begin{aligned} & \lim_{m\to\infty}\limsup_{x\to\infty}\left|\sum_{\lambda\in\Lambda_1} I^{(1)}_{\lambda,m}(x)/{}_m\mathrm{C}_{t-1}r^{m-t+1}\right| \qquad (4.23) \\ & = \lim_{m\to\infty}\limsup_{x\to\infty}\left|d_{t-1,0}H_r(a^m x) + 2\sum_{l=1}^{q}\Re e^{i(m-t+1)\theta_l}d^{l+}_{t-1,0}G_l(a^m x)\right|, \end{aligned}$$

where

$$G_l(x) = H_{re^{\pm i\theta_l}}(x) - re^{-i\theta_l} H_{re^{\pm i\theta_l}}(ax), \qquad d_{t-1,0}^{l+} \quad (l = 1, \ldots, q)$$

are nonzero constants.

Noticing

$$\lim_{m\to\infty} {}_m\mathrm{C}_{k-1} \frac{\lambda^{m-k+1}}{{}_m\mathrm{C}_{t-1} r^{m-t+1}} = 0$$

for $|\lambda| < r$ or $|\lambda| = r,\ k \leqslant t-1$, we have, by (4.9) and (4.21),

$$\lim_{m\to\infty} \limsup_{x\to\infty} \left| \sum_{\lambda\in\Lambda_2} I_{\lambda,m}^{(1)}(x) / {}_m\mathrm{C}_{t-1} r^{m-t+1} \right| = 0. \tag{4.24}$$

By (4.22)

$$\lim_{m\to\infty} \limsup_{x\to\infty} \left| \sum_{\lambda\in\Lambda_3} I_{\lambda,m}^{(1)}(x) / {}_m\mathrm{C}_{t-1} r^{m-t+1} \right| = 0. \tag{4.25}$$

Using

$$\begin{aligned} &\left| \sum_{\lambda\in\Lambda_1} I_{\lambda,m}^{(1)}(x) \right| \\ &\quad \leqslant \left| g(x) - \sum_{\lambda\in\Lambda} I_{\lambda,m}^{(1)}(x) \right| + |g(x)| + \left| \sum_{\lambda\in\Lambda_2} I_{\lambda,m}^{(1)}(x) \right| + \left| \sum_{\lambda\in\Lambda_3} I_{\lambda,m}^{(1)}(x) \right|, \end{aligned}$$

we have

$$\begin{aligned} &\lim_{m\to\infty} \limsup_{x\to\infty} \left| \sum_{\lambda\in\Lambda_1} I_{\lambda,m}^{(1)}(x) / {}_m\mathrm{C}_{t-1} r^{m-t+1} \right| \\ &\quad \leqslant \limsup_{x\to\infty} \left| g(x) - \sum_{\lambda\in\Lambda} I_{\lambda,m}^{(1)}(x) / {}_m\mathrm{C}_{t-1} r^{m-t+1} \right| + \\ &\qquad + \lim_{m\to\infty} \limsup_{x\to\infty} |g(x) / {}_m\mathrm{C}_{t-1} r^{m-t+1}| + \\ &\qquad + \lim_{m\to\infty} \limsup_{x\to\infty} \left| \sum_{\lambda\in\Lambda_2} I_{\lambda,m}^{(1)}(x) / {}_m\mathrm{C}_{t-1} r^{m-t+1} \right| + \\ &\qquad + \lim_{m\to\infty} \limsup_{x\to\infty} \left| \sum_{\lambda\in\Lambda_3} I_{\lambda,m}^{(1)}(x) / {}_m\mathrm{C}_{t-1} r^{m-t+1} \right|. \end{aligned}$$

By (4.4), the boundedness of g, (4.24) and (4.25), the right-hand side is equal to 0. Thus we get

$$\lim_{m\to\infty} \limsup_{x\to\infty} \left| \sum_{\lambda\in\Lambda_1} I_{\lambda,m}^{(1)}(x) / {}_m\mathrm{C}_{t-1} r^{m-t+1} \right| = 0.$$

Combining it and (4.23), we have

$$\lim_{m\to\infty}\limsup_{x\to\infty}\left|d_{t-1,0}H_r(x)+2\sum_{l=1}^{q}\Re e^{i(m-t+1)\theta_l}d_{t-1,0}^{l+}G_l(x)\right|=0.$$

Further, using Lemma 4.3 (see after the proof of the theorem), we have

$$\lim_{x\to\infty}H_r(x)=0,\qquad \lim_{x\to\infty}G_l(x)=0\quad(l=1,\dots,q). \tag{4.26}$$

Moreover, since

$$\begin{aligned}|G(x)|^2 &= (H_{re^{\pm i\theta}}(x)-r\cos\theta H_{re^{\pm i\theta}}(ax))^2+(r\sin\theta H_{re^{\pm i\theta}}(ax)^2)\\ &\geqslant (r\sin\theta H_{re^{\pm i\theta}}(ax))^2,\end{aligned}$$

we have

$$\lim_{x\to\infty}H_r(x)=0,\qquad \lim_{x\to\infty}H_{re^{\pm i\theta_l}}(x)=0\quad(l=1,\dots,q).$$

In case Λ_1 has only real eigenvalue or only complex eigenvalues, we can get the equality similar to the above ones. In case $\lim_{x\to\infty}H_{re^{\pm\theta_l}}(x)=0$ for some l, since

$$\lim_{x\to\infty}\frac{\sum_{k=0}^{n-3}c_k a^{-k\rho}f(a^kx)}{F(x)}-H_{re^{\pm\theta_l}}(x)=\frac{1}{(1-re^{i\theta_l})(1-re^{-i\theta_l})}$$

by the definition of $H_{re^{\pm\theta_l}}$ and the regular variation of F,

$$\lim_{x\to\infty}\frac{\sum_{k=0}^{n-3}c_k a^{-k\rho}f(a^kx)}{F(x)}=\frac{1}{(1-re^{i\theta_l})(1-re^{-i\theta_l})}.$$

This means that $\sum_{k=0}^{n-3}c_k a^{-k\rho}f(a^kx)$ is r.v. By (4.14), we can apply the assumption of induction to show that f is r.v. We can get the same conclusion from $\lim_{x\to\infty}H_r(x)=0$ in the same way. In cases J has only real eigenvalues or only complex eigenvalues, the same argument is available. Thus we have completed the proof of Theorem 4.2(i).

Proof of (ii). By the monotone equivalence of r.v. functions with nonzero indices ([BGT87], Theorem 1.5.3), we see that monotone function f_1 asymptotic to $(\sum_{k=0}^{n-1}a^{k\rho}b_k)^{-1}F$ is a desired one. Let us construct f_2. From the assumption, there exists θ ($\neq 0$) such that

$$\sum_{k=0}^{n-1}a^{k\rho}b_k e^{-ki\theta}=0.$$

Now we define a sequence $\{c_j : j\in\mathbb{N}\}$ as $c_j=\cos j\theta$. This sequence oscillates in the bounded range and satisfies that

$$\sum_{k=0}^{n-1}a^{k\rho}b_k c_{j-k}=0. \tag{4.27}$$

Using this sequence, we define the function h as

$$h(x) = \frac{c_{j+1} - c_j}{a^{-j-1} - a^{-j}}(x - a^{-j}) + c_j \quad \text{for } a^{-j} < x \leqslant a^{-j-1}.$$

Then h satisfies that

$$\sum_{k=0}^{n-1} a^{k\rho} b_k h(a^k x) = 0. \tag{4.28}$$

(4.28) is obtained as follows. For each x, there exists $j = j(x) \in \mathbb{N}$ such that $x \in [a^{-j}, a^{-j-1})$. Then $a^k x \in [a^{k-j}, a^{k-j-1})$ $(k = 0, 1, \ldots, n-1)$ and

$$\begin{aligned} h(a^k x) &= \frac{c_{j+1-k} - c_{j-k}}{a^{k-j-1} - a^{k-j}}(a^k x - a^{k-j}) + c_{j-k} \\ &= \frac{c_{j+1-k} - c_{j-k}}{a^{-j-1} - a^{-j}}(x - a^{-j}) + c_{j-k}. \end{aligned}$$

Using (4.27), we have

$$\begin{aligned} &\sum_{k=0}^{n-1} a^{k\rho} b_k h(a^k x) \\ &\quad = \sum_{k=0}^{n-1} a^{k\rho} b_k \left(\frac{c_{j+1-k} - c_{j-k}}{a^{-j-1} - a^{-j}}(x - a^{-j}) + c_{j-k} \right) = 0. \end{aligned}$$

It is not difficult to show that there exists a positive constant K such that

$$f_K(x) = x^\rho (h(x) + K)$$

is monotone for sufficiently large x (note that $\rho \neq 0$). Then $f_2(x) = C^{-1} f_K(x) L(x)$, where $C = K \sum_{k=0}^{n-1} a^{k\rho} b_k$, is the desired one. We show that f_2 satisfies (4.2) and is not r.v. By (4.28), it is easy to see that $\sum_{k=0}^{n-1} b_k f_K(a^k x) = C x^\rho$. Therefore we have

$$\begin{aligned} &\left| \frac{\sum_{k=0}^{n-1} b_k f_2(a^k x)}{F(x)} - 1 \right| \\ &\quad = \left| \sum_{k=0}^{n-1} \left(\frac{b_k f_K(a^k x)}{C x^\rho} \frac{L(a^k x) - L(x)}{L(x)} \right) \right| \\ &\quad \leqslant \max_{0 \leqslant k \leqslant n-1} \left| \frac{L(a^k x) - L(x)}{L(x)} \right|. \end{aligned}$$

The last term tends to 0 as $x \to \infty$ by the slow variation of L.

$$a^\rho \frac{f_2(a^{-j-1})}{f_2(a^{-j})} = \frac{L(a^{-j-1})}{L(a^{-j})} \frac{c_{j+1} + K}{c_j + K}.$$

The right-hand side does not go to 1 as $j \to \infty$, which means that f_2 is not r.v. We have completed the proof of Theorem 4.2(ii). □

LEMMA 4.3. *Let* $c_j \in \mathbb{C}$, $\theta_i \neq \theta_j$ $(i \neq j)$. *If* $\lim_{m\to\infty} \sum_{j=1}^{n} c_j e^{im\theta_j} = 0$, *then* $c_j = 0$ *for every* j.

Proof. From the assumption,

$$\lim_{m\to\infty} \begin{pmatrix} e^{im\theta_1} & e^{im\theta_2} & \dots & e^{im\theta_n} \\ e^{i(m+1)\theta_1} & e^{i(m+1)\theta_2} & \dots & e^{i(m+1)\theta_n} \\ \vdots & \vdots & \vdots & \vdots \\ e^{i(m+n-1)\theta_1} & e^{i(m+n-1)\theta_2} & \dots & e^{i(m+n-1)\theta_n} \end{pmatrix} \begin{pmatrix} c_1 \\ c_2 \\ \vdots \\ c_n \end{pmatrix} = \begin{pmatrix} 0 \\ 0 \\ \vdots \\ 0 \end{pmatrix}.$$

Hence

$$\lim_{m\to\infty} e^{im\theta_1} e^{im\theta_2} \dots e^{im\theta_n} \begin{pmatrix} 1 & 1 & \dots & 1 \\ e^{i\theta_1} & e^{i\theta_2} & \dots & e^{i\theta_n} \\ \vdots & \vdots & \vdots & \vdots \\ e^{i(n-1)\theta_1} & e^{i(n-1)\theta_2} & \dots & e^{i(n-1)\theta_n} \end{pmatrix} \begin{pmatrix} c_1 \\ c_2 \\ \vdots \\ c_n \end{pmatrix} = \begin{pmatrix} 0 \\ 0 \\ \vdots \\ 0 \end{pmatrix}.$$

Thus we have $c_1 = c_2 = \cdots = c_n = 0$.

Remark. Noticing that the Jordan block is unique for each eigenvalue and applying Lemma 4.3 to 0, $\pm\theta_l$ $(l = 1, \dots, q)$, we get (4.26).

Remark. In Theorem 4.2, it is impossible to constract f_2 such that f_2 is not r.v. and $F(x) = \sum_{k=0}^{n-1} b_k f_2(a^k x)$ for sufficient large x. Let $F(x) = x^\rho L(x)$. In case L is a positive constant, we can make such a function as in the proof of Theorem 4.2. But, for example, if $L(x) = \log x$ and $F(x) = f(x) + b_1 f(ax)$ $(a^\rho b_1 = 1)$, then an easy calculation shows that $\lim_{x\to\infty} f(x)/F(x) = 1/2$.

Remark. Setting $F(x) = x^\rho L(x)$, $f(x) = x^\rho l(x)$ and $x = a^{-j}$ in (4.1), we have

$$L(a^{-j}) = \sum_{k=0}^{n-1} a^{k\rho} b_k l(a^{k-j}).$$

Then the sequence $\{l(a^{-j}) : j \in \mathbb{N}\}$ is defined by a recurrence formula including s.v. function. Therefore, the essence of Theorem 4.2 is the asymptotic behavior of this sequence.

Proof of Theorem 4.1. (i) Apply Theorem 4.2 to

$$\mu \circ \nu(x, \infty) = \sum_{k=0}^{n-1} \mu(x/t_k, \infty)\nu(\{t_k\}).$$

(ii) Proof is essentially the same as that of Theorem 4.2 and omitted. □

Remark. μ_2 in Theorem 4.2(ii) belongs to $\mathbf{M}(\alpha) \setminus \mathbf{D}(\alpha)$ by Proposition 4.4 ([S97]). (Notice that $\mu_2 \circ \nu \in \mathbf{D}(\alpha) \subset \mathbf{M}(\alpha)$.)

The following shows the importance of the assumption $\sum_{k=0}^{n-1} a^{k\rho} b_k = 0$ in Theorem 4.2.

THEOREM 4.4. *If a r.v. function F with index ρ is expressed as $F(x) = \sum_{k=0}^{n-1} b_k f(a^k x)$ by a measurable function f and real constants such that $0 < a \neq 1$, b_k and $\sum_{k=0}^{n-1} a^{k\rho} b_k = 0$, then $\lim_{x\to\infty} |f(x)/F(x)| = \infty$.*

Proof. Assume that $g = f/F$ is bounded. Then we get the (4.4) in a similar way to the proof of Theorem 4.2. Further, we see that it suffices to prove under the assumption that Λ does not contain the numbers whose absolute values are larger than 1. Let s and t be the multiplicity of eigenvalues 1 and $e^{i\theta}$, respectively. Letting $\beta = 1$ in (4.5),

$$I_{1,m}^{(1)}(x) = \sum_{j=0}^{m-1} \sum_{l=0}^{(s-1)\wedge j} {}_j\mathrm{C}_l d_{l,0} + \sum_{l=0}^{s-1} {}_m\mathrm{C}_l \sum_{k=0}^{n-2} d_{l,k} a^{k\rho} g(a^{m+k} x). \tag{4.29}$$

Noticing $\lim_{m\to\infty} \sum_{j=1}^{m-1} j^{s-1}/m^s = s^{-1}$, we see that the first term is asymptotically equal to $d_{s-1,0} m^s/s!$ and the second term is $\mathrm{O}(m^{s-1})$ uniformly in x as $m \to \infty$. Hence

$$I_{1,m}^{(1)}(x) = d_{s-1,0} \frac{m^s}{s!} (1 + \mathrm{o}_3(m,x)), \tag{4.30}$$

where $\lim_{m\to\infty} \limsup_{x\to\infty} |\mathrm{o}_3(m,x)| = 0$. Similarly, letting $r = 1$ in (4.10),

$$I_{e^{i\theta},m}^{(1)}(x) = \sum_{j=0}^{m-1} \sum_{l=0}^{(t-1)\wedge j} {}_j\mathrm{C}_l e^{i(j-l)\theta} d_{l,0}^{+} + \sum_{l=0}^{t-1} {}_m\mathrm{C}_l e^{i(m-l)\theta} \sum_{k=0}^{n-2} d_{l,k}^{+} a^{k\rho} g(a^{m+k} x).$$

Using

$$\lim_{m\to\infty} \sum_{j=1}^{m-1} j^{t-1} e^{ij\theta}/m^{t-1} = e^{im\theta}/(e^{i\theta} - 1)$$

for $\theta \neq 0$, we see that the first and second term are asymptotically equal to

$$\frac{d_{t-1,0}^{+} e^{i(m-t+1)\theta} m^{t-1}}{(e^{i\theta} - 1)(t-1)!}$$

and

$$\frac{e^{i(m-t+1)\theta} m^{t-1}}{(t-1)!} \sum_{k=0}^{n-2} d_{t-1,k}^{+} a^{k\rho} g(a^{m+k} x)$$

as $m \to \infty$, respectively. Hence we get

$$I_{e^{i\theta},m}^{(1)}(x)$$
$$= \frac{m^{t-1}}{(t-1)!} e^{(m-t+1)\theta} \left\{ \frac{d_{t-1,0}^{+}}{e^{i\theta} - 1} + \sum_{k=0}^{n-2} d_{t-1,k}^{+} a^{k\rho} g(a^{m+k} x) \right\} (1 + \mathrm{o}_4^{+}(m,x)),$$

where $\lim_{m\to\infty} \limsup_{x\to\infty} |o_4^+(m,x)| = 0$. Combining it and its conjugate, we have

$$\begin{aligned} &I^{(1)}_{e^{i\theta},m}(x) + I^{(1)}_{e^{-i\theta},m}(x) \\ &\quad = 2\frac{m^{t-1}}{(t-1)!}\{\Re e^{i(m-t+1)\theta} d^+_{t-1,0} G(a^m x) + o_4(m,x)\}, \end{aligned} \tag{4.31}$$

where $\lim_{m\to\infty} \limsup_{x\to\infty} |o_4(m,x)| = 0$. By the same argument as the proof of Theorem 4.2, the case $s < t-1$ is reduced to the case $s = t-1$. By $|\sum_{\lambda\in\Lambda} I^{(1)}_{\lambda,m}(x)| \leqslant |g(x) - \sum_{\lambda\in\Lambda} I^{(1)}_{\lambda,m}(x)| + |g(x)|$ and (4.4), we get

$$\limsup_{x\to\infty} \left|\sum_{\lambda\in\Lambda} I^{(1)}_{\lambda,m}(x)\right| \leqslant \limsup_{x\to\infty} |g(x)| < \infty.$$

Therefore,

$$\limsup_{m\to\infty} \limsup_{x\to\infty} \left|\sum_{\lambda\in\Lambda} I^{(1)}_{\lambda,m}(x)\right| < \infty. \tag{4.32}$$

On the other hand, if $s > t-1$, then it follows from (4.22), (4.30) and (4.31) that

$$\lim_{m\to\infty} \liminf_{x\to\infty} |I^{(1)}_{1,m}(x)| = \infty \quad \text{and} \quad \lim_{m\to\infty} \liminf_{x\to\infty} \left|\frac{\sum_{\lambda\in\Lambda} I^{(1)}_{\lambda,m}(x)}{I^{(1)}_{1,m}(x)}\right| = 1.$$

Therefore,

$$\lim_{m\to\infty} \liminf_{x\to\infty} \left|\sum_{\lambda\in\Lambda} I^{(1)}_{\lambda,m}(x)\right| = \infty,$$

which is contradictory to (4.32). In the case of $s = t-1$,

$$\sum_{\lambda\in\Lambda} I^{(1)}_{\lambda,m}(x) = \frac{m^s}{s!}\left\{d_{s-1,0} + 2\sum_{l=1}^{q} \Re e^{i(m-s)\theta_l} G_l(a^m x) + o_5(m,x)\right\},$$

where $\lim_{m\to\infty} \limsup_{x\to\infty} |o_5(m,x)| = 0$. This is also contrary to (4.32) by Lemma 4.3. Thus we conclude that $\limsup_{x\to\infty} |g(x)| = \infty$.

EXAMPLE. For the function $f(x) = (\log x)^s x^\rho$, where s is the multiplicity of 1 in $a^\rho B$,

$$F(x) = (\log a)^s \left(\sum_{k=0}^{n-1} b_k a^{k\rho} k^s\right) x^\rho \quad \text{and} \quad \sum_{k=0}^{n-1} b_k a^{k\rho} k^s \neq 0.$$

Acknowledgement

The author is grateful to Professor N. Yoshida for the proof of Lemma 4.3.

References

[BGT87] Bingham, N. H., Goldie, C. M. and Teugels, J. L.: *Regular Variation*, Encyclopedia Math. Appl., Cambridge University Press, Cambridge, 1987.

[C86] Cline, D. B. H.: Convolution tails, product tails and domains of attraction, *Probab. Theory Related Fields* **72** (1986), 529–557.

[EG80] Embrechts, P. and Goldie, C. M.: On closure and factorization properties of subexponential distributions, *J. Austral. Math. Soc. (A)* **29** (1980), 243–256.

[S94] Shimura, T.: Decomposition problem of probability measures related to monotone regularly varying functions, *Nagoya Math. J.* **135** (1994), 87–111.

[S97] Shimura, T.: The product of independent random variables with slowly varying truncated moments, *J. Austral. Math. Soc. (A)* **62** (1997), 186–197.

Acta Applicandae Mathematicae **63:** 433–439, 2000.

Entropy in Subordination and Filtering

Dedicated to Professor Takeyuki Hida on the occasion of his 70th birthday

SI SI[1] and WIN WIN HTAY[2]
[1] *Faculty of Science and Technology, Aichi Prefectural University, Aichi Prefecture, Japan*
[2] *Department of Computational Mathematics, University of Computer Studies, Yangon, Myanmar (Burma)*

(Received: 7 July 1999)

Abstract. We propose a stochastic model of transmitting random information at random time. In this model, the signal is observed as a random sampling according to an increasing stable stochastic process. Thus we are given a subordinate stochastic process which is a typical irreversible process. As the characteristic of this phenomea we observe the loss of entropy.

Mathematics Subject Classification (2000): 60H40.

Key words: white noise, subordination, filtering.

1. Introduction

We are interested in various methods of modulation of stochastic processes which are used as models in communication theory. Among them, frequency modulation, amplitude modulation are well known.

Having been inspired by astrophysical data, we propose a stochastic model of transmitting random information. Thus, we are going to introduce another modulation using Bochner's subordination in this note.

Observing typical X-ray light curves which are flickering from a black hole, we see that the curves have self-similarity (cf. [7]). We may therefore guess, in fact conclude, that those curves are sample functions of a stable stochastic process. Now, one may ask, what should be the input signal under the obtained information?

Note that if the information source is random with a limited power, it is taken to be Gaussian and to be optimal in the sense of information theory. Thus, in our model we may admit to white noise $\dot{B}(t)$ being an information source.

Since the accumulation sum of white noise according to time propagation is a Brownian motion, formally writing,

$$\int_0^t \dot{B}(s)\,\mathrm{d}s = B(t), \tag{1.1}$$

the Brownian motion can be taken as an input signal.

Emission of the signal comes out at random times and is subject to the law of the exponential holding time with various intensities. It leads us to think of the emitting time series as a compound Poisson process; in particular an increasing stable process that enjoys self-similarity.

Applying the technique of subordination, the observed data can be expressed as $B(Y_\alpha(t))$, which is to be self-similar. The information loss shows the 'irreversibility' from the view point of information theory.

2. Prerequisite

In this section, we recall some well-known facts which we are going to apply in what follows.

2.1. COMPOUND POISSON PROCESSES

We start with a system of elementary stochastic processes for Lévy processes. Let $\{P_u(t,\omega), 0 < u < \infty\}$ be a system of independent Poisson processes with unit intensity, each of which is elementary. Then let us form a compound Poisson process of the following particular type:

$$\begin{aligned} Y_\alpha(t,\omega) &= \int_0^\infty u P_{\mathrm{d}u}(t) \frac{1}{u^{1+\alpha}}\,\mathrm{d}u, \quad 0 < \alpha < 1 \\ &= \int_{-\infty}^\infty u\{P_{\mathrm{d}u}(t) - E(P_{\mathrm{d}u}(t))\} \frac{1}{|u|^{1+\alpha}}\,\mathrm{d}u, \quad 1 \leqslant \alpha < 2. \end{aligned}$$

We see that the process $Y_\alpha(t)$ is a Lévy process and it enjoys a self-similar property, namely $\{Y_\alpha(at)\}$ and $\{a^{1/\alpha} Y_\alpha(t)\}$ have the same probability distribution.

In fact, $Y_\alpha(t)$ is a stable process of exponent α, initiated by P. Lévy. This property can be seen by its probability distribution. The characteristic function of which is

$$\varphi_t(z) = \exp(t\Psi_\alpha(z)), \tag{2.1}$$

where

$$\Psi_\alpha(z) = -|\Gamma(-\alpha)| \exp\left(\mathrm{sign}(z)\frac{i\alpha\pi}{2}|z|^\alpha\right). \tag{2.2}$$

In Section 3, we shall deal with a stable process $Y_{\frac{1}{2}}(t)$, with exponent $1/2$, for the choice of emission random time, the density function of which is given by

$$f(y(t)) = \frac{t}{\sqrt{2\pi}} \mathrm{e}^{-\frac{t^2}{2y}} y^{-\frac{3}{2}}. \tag{2.3}$$

For $\alpha = 1$, Y_1 is a symmetric stable process. We know precisely that it is a Cauchy process.

2.2. SUBORDINATION

We briefly revise the notion of subordination, due to S. Bochner.

Take an additive process $\{X(t, \omega)\}$ with independent stationary increments and form a new process $\{Z(t, \omega)\}$, such that

$$Z(t, \omega) = X(Y(t), \omega) \tag{2.4}$$

by changing the time variable t to $Y(t)$, where $Y(t)$ is an increasing random function with $Y(0) = 0$.

This process $\{X(Y(t), \omega)\}$ is said to be subordinate to $\{X(t)\}$ using the time $\{Y(t)\}$. It means that the original process is observed only at random times.

In our case, $X(t)$ and the time variable $Y(t)$ are taken to be Brownian motion $B(t)$ and $Y_{\frac{1}{2}}(t)$, respectively.

3. Stochastic Model

As was discussed in Section 1, observation of the emission will be made at a random time which is taken to be a stable process. We choose the stable process with index $\frac{1}{2}$, that is $Y_{\frac{1}{2}}(t)$, by noting that it is the inverse function of the maximum of a Brownian motion which can be considered as a driving force, i.e.

$$P(M(t) \geqslant y) = P(Y_{\frac{1}{2}}(y) \leqslant t), \tag{3.1}$$

where $M(t) = \max_{s \leqslant t} B(s)$.

Remark. This Brownian motion is a different one and is independent of the Brownian motion which is taken as an information source that we mentioned in Section 1.

The reason why we may consider $Y_{\frac{1}{2}}(t)$ is that there is a Brownian motion that attains its maximum value. Such a setup enables us to introduce $M(t)$ and $Y_{\frac{1}{2}}(t)$.

In addition, $Y_{\frac{1}{2}}(t)$ is a compound Poisson process so that each Poisson process is formed by a random waiting time with a probability distribution of the exponential type (note the lack of memory).

For computation of the entropy, we consider our model on a time interval $(0, T)$.

The above discussion yields a stochastic model, with the information source $\{B(t),\ t \in [0, T], Y_{\frac{1}{2}}(t),\ t \in [0, T]\}$, and the observation data $\{B(Y_{\frac{1}{2}}(t)),\ t \in [0, T]\}$.

For the computation, we take only a finite number of points t_i, $i = 1, \ldots, n$ in the interval $[0, T]$. Then the information source is taken to be

$$\left\{B(t_i), i = 1, \ldots, n,\ Y_{\frac{1}{2}}(t_i),\ i = 1, \ldots, n\right\}$$

and the observed data will be obtained in the form

$$\left\{B(Y_{\frac{1}{2}}(t_i)),\ i = 1, \ldots, n\right\}.$$

Since we wish to have independent systems, we transform them as follows:

$$\{B(t_i),\ i=1,\ldots,n\} \to \{B(t_k)-B(t_{k-1}),\ k=1,\ldots,n\},$$
$$\left\{B(Y_{\frac{1}{2}}(t_i)),\ i=1,\ldots,n\right\} \to \left\{B(Y_{\frac{1}{2}}(t_k)) - B(Y_{\frac{1}{2}})(t_{k-1}),\ k=1,\ldots,n\right\}.$$

Then the observed data will be

$$\left\{B(Y_{\frac{1}{2}}(t_k)) - B(Y_{\frac{1}{2}}(t_{k-1})),\ k=1,\ldots,n\right\}.$$

Note that, for each case, according to the transformation, the Jacobian is 1.

Fact 1. First we evaluate the entropy for $\{B(t_1),\ldots,B(t_n)\}$. Since the entropy for a Gaussian random variable with variance σ^2 is

$$\log\sqrt{2\pi e} + \log\sigma, \tag{3.2}$$

for $B(t)$ we have

$$\log\sqrt{2\pi e} + \log t. \tag{3.3}$$

Consequently, the entropy for $\{B(t_k)-B(t_{k-1}),\ k=1,\ldots,n\}$ is

$$H(B) = n\log\sqrt{2\pi e} + \sum_{k=1}^{n}\log(t_k - t_{k-1}). \tag{3.4}$$

Fact 2. The entropy for $Y_{\frac{1}{2}}(t)$ is

$$\log\sqrt{2\pi} + \tfrac{1}{2} - \tfrac{3}{2}\{\log 2 + \psi(\tfrac{1}{2})\} + 2\log t, \tag{3.5}$$

where ψ is Euler's ψ-function.

So the entropy for $\{B(Y_{\frac{1}{2}}(t_i)),\ i=1,\ldots,n\}$ is

$$H(Y) = n\log\sqrt{2\pi} + \frac{n}{2} - \tfrac{3}{2}n\{\log 2 + \psi(\tfrac{1}{2})\} + 2\sum_{k=1}^{n}\log(t_k - t_{k-1}). \tag{3.6}$$

Fact 3. $B(Y_{\frac{1}{2}}(t))$ has Cauchy distribution and its entropy is

$$\log 4\pi + \log t \tag{3.7}$$

and thus the entropy for $\{B(Y_{\frac{1}{2}}(t_k)) - B(Y_{\frac{1}{2}}(t_{k-1})),\ k=1,\ldots,n\}$ is

$$H(B(Y)) = n\log 4\pi + \sum_{k=1}^{n}\log(t_k - t_{k-1}). \tag{3.8}$$

Then the entropy loss is obtained as

$$n\left\{1 + 2\sum_{k=1}^{n}\log(t_k - t_{k-1}) - \tfrac{3}{2}\Psi(\tfrac{1}{2}) - \tfrac{5}{2}\log 2\right\}. \tag{3.9}$$

If we take the time intervals with equal length 1, then the entropy loss is

$$n\{1 - \tfrac{3}{2}\Psi(\tfrac{1}{2}) - \tfrac{5}{2}\log 2\}. \tag{3.10}$$

THEOREM 3.1. *The entropy loss is propotional to n if we take the observations at times with unit intervals.*

4. Filtering

Following the idea of C. E. Shannon, we may introduce the quantity of the information contained in the stationary Gaussian process $X(t)$. However, the absolute entropy is infinite, so we are only interested in the difference of entropy between the white noise that is taken to be the input of the $X(t)$ and the output $X(t)$ itself. To make an actual comparison, we limit the band, say, to $[-W, W]$.

Let a stationary Gaussian process $X(t)$ have the canonical representation of the form

$$X(t) = \int_{-\infty}^{t} F(t-u)\dot{B}(u)\,du. \tag{4.1}$$

Then the spectral representation is

$$X(t) = \int e^{it\lambda}\hat{F}(\lambda)\dot{Z}(\lambda)\,d\lambda, \tag{4.2}$$

where

$$\dot{Z}(\lambda) = \frac{1}{\sqrt{2\pi}}\int e^{it\lambda}\dot{B}(t)\,dt \quad \text{with } E|\dot{Z}(\lambda)|^2 = \frac{1}{d\lambda} \text{ formally.} \tag{4.3}$$

The band limited representations are

$$X_W(t) = \int_{-W}^{W} e^{it\lambda}\hat{F}(\lambda)\dot{Z}(\lambda)\,d\lambda \tag{4.4}$$

and

$$\dot{B}_W(t) = \int_{-W}^{W} e^{-it\lambda}\dot{Z}(\lambda)\,d\lambda, \tag{4.5}$$

respectively. It is known that $\dot{Z}(\lambda)$ can be represented as

$$\dot{Z}(\lambda) = \frac{\operatorname{sgn}(\lambda)}{\sqrt{2}}\{\dot{Z}_1(\lambda) + i\dot{Z}_2(\lambda)\}; \quad \dot{Z}_1(\lambda), \dot{Z}_2(\lambda)\text{: real.}$$

Here we note that $\dot{Z}(\lambda) = -\overline{\dot{Z}(-\lambda)}$.

Based on white noise, we compute its relative entropy or entropy loss by approximation. Since white noise can be expressed as (4.5), we approximate it by $\sum_{i=1}^{N} e^{it\lambda_k}\Delta_k Z$. Similarly, we can approximate $X(t)$ by

$$\sum_{i=1}^{N} e^{it\lambda_k} G(\lambda_k)\Delta_k Z.$$

To compute the entropy of them we see the joint distribution of $\{\Delta_k Z\}$ and that of $\{G(\lambda_k)\Delta_k Z\}$.

Then we are given the information loss

$$\sum_K \log |G(\lambda_k)|^2 \Delta_k$$

by noting $(\Delta_k B_1, \Delta_k B_2)$ is a two-dimensional Gaussian random variable.

THEOREM 4.1. *Based on a band-limited white noise, the entropy loss of the output is given by*

$$\int_{-W}^{W} \log |G(\lambda)|^2 \, d\lambda. \tag{4.6}$$

We now consider a particular case of averaging. The integral operator G_a, $a > 0$ is defined by

$$(G_a f(t)) = a \int_{-\infty}^{t} e^{-a(t-u)} f(u) \, du. \tag{4.7}$$

It is

(i) causal (i.e. operating only on the past values),
(ii) averaging $(G_a 1)(t) = 1$,
(iii) stationarity holds, as is shown below.

Apply G_a to a stationary process, then we see some changing of the characteristic function

$$F(t) \to \hat{F}(\lambda)\hat{G}_a(\lambda), \quad \hat{G}_a(\lambda) = \frac{a}{a + i\lambda}, \tag{4.8}$$

where F is the cannonical kernel.

PROPOSITION 4.1. *The information loss is given by*

$$\int_{-W}^{W} \log \frac{a^2 + \lambda^2}{\lambda^2} \, d\lambda. \tag{4.9}$$

We recall that the transmission function of a stationary N-ple Markov Gaussian process $X(t)$ is expressed in the form

$$\hat{F}(\lambda) = \frac{Q(i\lambda)}{P(i\lambda)}, \tag{4.10}$$

where $\hat{F}$ is the Fourier transform of the canonical kernel F and where P and Q are polynomials with degree $Q <$ degree $P = N$. It is to be noted that all the roots of P and Q are in the lower half of the complex plane.

If the $X(t)$ passes through a filter with characteristic $\hat{G}(\lambda)$, then that of the output is given by $\hat{F} \cdot \hat{G}$.

PROPOSITION 4.2. *The information loss is given by*

$$\int_{-W}^{W} \log \frac{|Q|^2}{|P|^2}\, d\lambda. \tag{4.11}$$

References

1. Bochner, S.: *Harmonic Analysis and the Theory of Probability*, Univ. of California Press, 1955.
2. Hida, T.: *Stationary Stochastic Processes*, Math. Notes, Princeton Univ. Press, 1970.
3. Hida, T.: *Brownian Motion*, Springer-Verlag, New York, 1980.
4. Hida, T. and Hitsuda, M.: *Gaussian Processes*, Transl. Math. Monogr. 12, Amer. Math. Soc., Providence, 1993.
5. Lévy, P.: *Théorie de l'addition des variables aléatoires*, Gauthier-Villars, Paris, 1937.
6. Lévy, P.: Sur certains processus stochastiques homogenes, *Compositio Math.* **7** (1939), 283–339.
7. Oda, M.: Fluctuation in astrophysical phenomena, In: T. Hida (ed.), *Proc. IIAS Workshop, Mathematical Approach to Fluctuations*, Vol. I, 1992, pp. 115–137.
8. Htay, Win Win: Optimalities for random functions: Lee–Wiener's network and non-canonical representations of stationary Gaussian processes, *Nagoya Math. J.* **149** (1998), 9–17.

Acta Applicandae Mathematicae **63**: 441–464, 2000.

Asymptotic Windings of Brownian Motion Paths on Riemann Surfaces

Dedicated to Professor Takeyuki Hida on his 70th birthday

SHINZO WATANABE
Department of Mathematics, Kyoto University, Kyoto, Japan. e-mail: watanabe@kusm.kyoto-u.ac.jp

(Received: 25 December 1998)

Abstract. We study asymptotic winding properties of Brownian motion paths on Riemann surfaces by obtaining limit laws for stochastic line integrals along Brownian paths of meromorphic differential 1-forms (Abelian differentials).

Mathematics Subject Classifications (2000): 60J65, 60H05, 60F99, 32A19.

Key words: Brownian motion on a Riemann surface, stochastic line integral along a Brownian motion path of a 1-differential form, Walsh's Brownian motions, elliptic integrals.

1. Introduction

Asymptotic laws of windings around several points of Brownian sample paths on the complex plane $\mathbf{C}$ (i.e. paths of a planar Brownian motion) have been studied extensively by Pitman and Yor ([PY1, PY2, PY3] cf. also [RY, Yo1, Ya]). In particular, they obtained an 'asymptotic residue theorem' (cf. Theorem 1.1 of [PY3]) which may be regarded as an asymptotic law for stochastic line integrals of meromorphic differential 1-forms on the complex plane $\mathbf{C}$ or on the Riemann sphere $\mathbf{S} \cong \mathbf{C} \cup \{\infty\}$ along the Brownian paths. Since the Riemann sphere is simply connected, there are no homological windings. On a Riemann surface of positive genus g, however, there are homological windings (equivalently, intersections with cycles in the homology basis) of Brownian paths and then the problem is to study the asymptotic law of stochastic line integrals of holomorphic differential 1-forms on the Riemann surface along the Brownian paths (cf. [M] for a general treatment of homological windings). This problem has been studied in the case of a Riemann surface of genus 1 as an asymptotic problem of a stochastic line integral corresponding to the Legendre first order elliptic integral by Kozlov, Pitman and Yor ([KPY1, KPY2]); in the latter, these authors interpreted the stochastic intersection number of Brownian paths with a 1-cycle (*goal-post*) on the surface as the score in what they call *Wiener football* or *Wiener soccer*, cf. [B] for related topics.

In this way, an asymptotic winding property of Brownian paths on a Riemann surface is reflected on the stochastic line integrals of meromorphic differential 1-forms (Abelian differentials) on the surface along the Brownian paths; windings around points and homological windings are related to residues and periods of the forms, respectively. A main purpose of this note is to study possible asymptotic laws for these stochastic line integrals.

It should be remarked that Brownian motions on a Riemann surface R are not unique and the asymptotic laws are different for different Brownian motions accordingly (cf. Th. 2 and Th. 3 in [PY1]). Here we mean, by a Brownian motion on a Riemann surface, a non-singular holomorphic diffusion on the surface (a diffusion on a complex manifold is called a holomorphic diffusion if the coordinate process in each local holomorphic coordinate is a conformal martingale). Since the complex dimension of R is one, Brownian motions on R is uniquely determined up to a time change; any time change of a Brownian motion is a Brownian motion and any two Brownian motions are obtained from each other by a time change. We may assume that R is a finitely sheeted covering surface of the Riemann sphere $\mathbf{S}$ and then $R' = R \setminus \{\text{points at infinity}\}$ is a finitely sheeted covering of $\mathbf{C}$. Although the way of such a covering is not unique, we fix one of possible coverings. Then, fixing a point $o \in R'$ which is over the origin 0 in the complex plane $\mathbf{C}$, we can lift the standard complex Brownian motion z_t on $\mathbf{C}$ starting at the origin uniquely to its covering motion $\tilde{z}_t$ on R' starting at o. $\tilde{z}_t$ is a Brownian motion on $R' \subset R$ and it is mainly for this Brownian motion that we study its asymptotic windings. Our results, Theorem 2.1 and Theorem 2.2 together with Example 1 of Section 4, may be regarded as a natural extension of those by Pitman and Yor ([PY2, PY3]) and Kozlov, Pitman and Yor ([KPY1, KPY2]). A remarkable feature in our results is that we need a Brownian spider or Walsh's Brownian motion on l-rays in order to describe limiting random variables in the asymptotic law when R is an l-sheeted covering over $\mathbf{S}$.

It should be remarked that, in these asymptotic laws, the order of scaling is different for windings around points and homological windings; the law for windings around points corresponds to a limit theorem with a stable limit random variable with index 1 while the law for homological windings corresponds to a cental limit theorem in which a limit random variable is a stable random variable with index 2, i.e., a Gaussian random variable.

2. Main Results

Let R be a compact Riemann surface of genus g. We may consider R as a finitely sheeted, say, l-sheeted covering surface of the Riemann sphere $\mathbf{S}$ and we denote by $\pi\colon R \to \mathbf{S}$ the projection. Then $R' := R \setminus \{\infty_1, \dots, \infty_l\}$ is an l-sheeted covering surface of the complex plane $\mathbf{C}$, where $\infty_1, \dots, \infty_l$ are points at infinity on R, i.e., $\{\infty_1, \dots, \infty_l\} = \pi^{-1}(\text{the north pole of } \mathbf{S})$. We assume that $\infty_1, \dots, \infty_l$ are all regular points (i.e., nonbranch points) with respect to this covering, for simplicity (cf. Remark 2 below for a necessary modification in the general case).

As explained in the Introduction, we consider the case of Brownian motion $\tilde{z}_t$ on $R' := R\setminus\{\infty_1, \ldots, \infty_l\}$ which is the covering motion of the standard Brownian motion z_t on the complex plane $\mathbf{C}$ such that $z_0 = 0$, i.e., by fixing a regular point $o \in R$ such that $\pi(o) = 0$, $\tilde{z}_t$ is the unique continuous curve on R' such that $\pi(\tilde{z}_t) \equiv z_t$ and $\tilde{z}_0 = o$, which is well-defined because the Brownian motion z_t never hits any image under π of branch points in R, a.s. (cf. [IM], 7.18 for the notion of covering motion). We will study the asymptotic property of windings around several points in R and also homological windings of the Brownian motion $\tilde{z}_t$. This study will be achieved by obtaining joint asymptotic laws for stochastic line integrals of meromorphic 1-forms on R along the Brownian path $\tilde{z}_t$ on $R' \subset R$.

In order to describe the limiting random variables, we set up the following random elements $\{C_a = (C_a(t)); a \in R'\}$, $\mathcal{W} = (W(t))$ and $B = (B(t))$ on a suitable probability space such that they are mutually independent, where

(1) $\{C_a = (C_a(t)); a \in R'\}$ is a family of one-dimensional standard Cauchy processes such that, if $a_1, \ldots, a_n \in R'$ are distinct, then processes $C_{a_1}, \ldots, C_{a_n}$ are mutually independent. Here, we mean, by a one-dimensional standard Cauchy process, a right-continuous process $C = (C(t))$ on the line with stationary independent increments such that $E(\mathrm{e}^{i\xi C(t)}) = \mathrm{e}^{-t|\xi|}$, $\xi \in \mathbf{R}$, so in particular, $C(0) = 0$, a.s.

(2) $\mathcal{W} = (W(t))$ is a Brownian spider or Walsh's Brownian motion (cf. [Wal, BPY] or [Yo2], 17.2) on l-rays $L_1 \cup \cdots \cup L_l$ such that $W(0) = 0$ (the origin); $L_1, \ldots, L_l$ are half lines ($\cong [0, \infty)$) in different directions from the origin 0 in a plane. This is a diffusion process moving as one-dimensional Brownian motion on each ray and reflecting at the origin with the same rate of excursions to each ray. This process can be constructed in the same way as [IW], Chap. III, Sect. 4.3, from l-independent Poisson point processes of positive Brownian excursions. An equivalent construction from a given system of l-independent reflecting Brownian motions will be given in Section 3.

(3) $B = (B(t))$ is a one-dimensional Brownian motion with $B(0) = 0$ (denote it by $BM^0(1)$).

If we define a process $|W(t)|$ on $[0, \infty)$ by

$$|W(t)| = \text{the distance from the origin 0 of } W(t), \tag{1}$$

then $|W(t)|$ is a reflecting Brownian motion on $[0, \infty)$. Let

$$|W(t)| = \Xi(t) + \Phi(t) \tag{2}$$

be the Skorokhod equation for $|W(t)|$ (cf. [IW], p. 120) so that $\Xi(t)$ is a $BM^0(1)$ and $\Phi(t)$ is the local time at 0 of $|W(t)|$:

$$\Phi(t) = \lim_{\epsilon\downarrow 0} \frac{1}{2\epsilon} \int_0^t 1_{[0,\epsilon)}(|W(s)|)\, \mathrm{d}s. \tag{3}$$

Let, for $t \geqslant 0$,

$$\sigma(t) = \min\{u \mid |W(u)| = t\} \tag{4}$$

and

$$e(t) = \Phi(\sigma(t)). \tag{5}$$

The process $e(t)$ is the inverse of Dwass's extremal process and $e(t)$ for fixed t is an exponential random variable with mean t (cf. [Wat]).

For each $a \in R'$, define a right-continuous complex process $Z_a = (Z_a(t))$ by

$$Z_a(t) = \frac{1}{l}e(t) + iC_a\left(\frac{e(t)}{l}\right) \left(\stackrel{\mathrm{d}}{=} \frac{1}{l}[e(t) + iC_a(e(t))]\right) \tag{6}$$

and define right-continuous complex processes $Z_\infty^1 = (Z_\infty^1(t)), \ldots, Z_\infty^l = (Z_\infty^l(t))$ by

$$\begin{aligned} Z_\infty^k(t) &= \int_0^{\sigma(t)} 1_{L_k}(W(s))\,\mathrm{d}\Xi(s) + \\ &\quad + i\int_0^{\sigma(t)} 1_{L_k}(W(s))\,\mathrm{d}B(s), \quad k = 1, \ldots, l. \end{aligned} \tag{7}$$

Remark 1. By (2), we have

$$\sum_{k=1}^{l} \int_0^{\sigma(t)} 1_{L_k}(W(s))\,\mathrm{d}\Xi(s) = \Xi(\sigma(t)) = t - e(t). \tag{8}$$

Remark 2. If in the case some points at infinity are branch points so that

$$\pi^{-1}(\text{the north pole of } \mathbf{S}) = \{\infty_{j_1}, \ldots, \infty_{j_p}\},$$

∞_{j_k} being a branch point of order $n_k - 1$ with $\sum_{k=1}^{p} n_k = l$, then we define

$$\widetilde{Z}_\infty^{j_k}(t) = \frac{1}{n_k} \sum_{\nu=n_1+\cdots+n_{k-1}+1}^{n_1+\cdots+n_k} Z_\infty^\nu(t), \quad k = 1, \ldots, p$$

and replace the role of $Z_\infty^k(t)$ by $\widetilde{Z}_\infty^{j_k}(t)$.

Let ω be a meromorphic 1-form on the compact Riemann surface R. Since the Brownian motion $\tilde{z}_t$ never hits the poles of ω, the stochastic line integral of ω along the Brownian path $\tilde{z}(t)$, denoted by $\langle\omega\rangle(t) = \int_{\tilde{z}[0,t]} \omega$, is well-defined as a continuous local conformal martingale (cf. [IW], Chap. VI, Sect. 6 for the definition and Chap. III, Sect. 6 for the reason why a Stranovich stochastic integral in this definition is a conformal martingale). We associate to ω the following right-continuous complex process $Z[\omega] = (Z[\omega](t))$:

$$Z[\omega](t) = \sum_{a \in R'} \text{res}(\omega)_a Z_a(t) - \sum_{k=1}^{l} \text{res}(\omega)_{\infty_k} Z_\infty^k(t), \tag{9}$$

where $\text{res}(\omega)_a$ denotes the residue of ω at point $a \in R$ so that $\text{res}(\omega)_a = 0$ if a is not a simple pole of ω. Since the number of poles of ω is finite, the right-hand side of (9) is actually a finite sum.

THEOREM 2.1. *Let ω be a meromorphic 1-form on the Riemann surface R which does not possess a pole of order $k \geqslant 2$ at points at infinity $\infty_1, \dots, \infty_l$ and let $\langle\omega\rangle$ be the corresponding stochastic line integral along the Brownian path $\tilde{z}(t)$ on $R' \subset R$. Then,*

$$\frac{1}{\lambda}\{\langle\omega\rangle(\mathrm{e}^{2\lambda t} - 1)\} \xrightarrow{f.d.} \{Z[\omega](t)\} \tag{10}$$

as $\lambda \to \infty$. Here 'f.d.' means the convergence in law of every finite-dimensional joint distribution.

COROLLARY 1.

$$\frac{2}{\log t}\langle\omega\rangle(t) \xrightarrow{d} Z[\omega](1) \tag{11}$$

as $t \to \infty$.

Remark 3. Since the both sides of (10) are linear in ω, we can conclude from (10) the following joint convergence: For meromorphic 1-forms $\omega_1, \dots, \omega_m$ on R satisfying the condition of Theorem 2.1,

$$\frac{1}{\lambda}\left\{\langle\omega_1\rangle(\mathrm{e}^{2\lambda t} - 1), \dots, \langle\omega_m\rangle(\mathrm{e}^{2\lambda t} - 1)\right\} \xrightarrow{f.d.} \left\{Z[\omega_1](t), \dots, Z[\omega_m](t)\right\} \tag{12}$$

as $\lambda \to \infty$.

Remark 4. In the case of $l = 1$, i.e., $R = \mathbf{S}$ or $R' = \mathbf{C}$, Theorem 2.1 reduces to the asymptotic residue theorem of Pitman and Yor ([PY3], Th. 1.1, cf. also [RY], Chap. XIII, Th. 3.11, [Yo1], Th. 7.1). In this case, writing $C_j(t) = C_{a_j}(t)$ and $Z_j(t) = Z_{a_j}(t)$,

$$Z_j(t) = e(t) + iC_j(e(t)) \quad \text{and}$$
$$Z_\infty(t) = \Xi(\sigma(t)) + iB(\sigma(t)) = t - e(t) + iB(\sigma(t)),$$

so that, if ω is a meromorphic 1-form on $\mathbf{S}$ with possible poles at $a_1, \ldots, a_n$ and ∞,

$$Z[\omega](t) = \sum_j \operatorname{res}(\omega)_{a_j} Z_j(t) - \operatorname{res}(\omega)_\infty Z_\infty(t).$$

In the above literatures, the description of limit processes $Z_j(t)$ and $Z_\infty(t)$ are seemingly different: In our notation of Section 3 below, their expressions correspond exactly to

$$\begin{aligned} Z_j(t) &= \int_0^{\sigma_j(t)} 1_{[\xi^j(s)<0]}\, \mathrm{d}\xi^j(s) + i \int_0^{\sigma_j(t)} 1_{[\xi^j(s)<0]}\, \mathrm{d}\eta^j(s) \\ &= l_j(\sigma_j(t)) + i \int_0^{\sigma_j(t)} 1_{[\xi^j(s)<0]}\, \mathrm{d}\eta^j(s) \end{aligned}$$

and

$$\begin{aligned} Z_\infty(t) &= \int_0^{\sigma_j(t)} 1_{[\xi^j(s)>0]}\, \mathrm{d}\xi^j(s) + i \int_0^{\sigma_j(t)} 1_{[\xi^j(s)>0]}\, \mathrm{d}\eta^j(s) \\ &= t - l_j(\sigma_j(t)) + i \int_0^{\sigma_j(t)} 1_{[\xi^j(s)>0]}\, \mathrm{d}\eta^j(s), \end{aligned}$$

where

$$\sigma_j(t) = \min\{u \mid \xi^j(u) = t\} \quad \text{and} \quad l_j(t) = \lim_{\epsilon \to 0} \frac{1}{2\epsilon} \int_0^t 1_{[0,\epsilon)}(\xi^j(s))\, \mathrm{d}s.$$

We can see in the proof of Section 3 below that these seemingly different expressions are in fact the same.

Let ω and θ be meromorphic 1-forms on the Riemann surface R satisfying the same condition as in Theorem 2.1. Suppose further that ω has at least one simple pole (an Abelian differential of the third kind) but θ does not have any simple pole (an Abelian differential of the first or the second kind). Then by Theorem 2.1,

$$\frac{1}{\lambda}\{\langle\theta\rangle(\mathrm{e}^{2\lambda t} - 1)\} \xrightarrow{f.d.} \{0\}$$

so that Theorem 2.1 for θ degenerates. However, we can refine Theorem 2.1 as Theorem 2.2 below. Before proceeding, we recall some of fundamental notions concerning Riemann surfaces ([K, S]).

Let $A_1, \ldots, A_g$ and $B_1, \ldots, B_g$ be a canonical homology basis of 1-cycles on R:

$$A_\alpha \times B_\beta = \delta_{\alpha\beta}, \qquad A_\alpha \times A_\beta = 0, \qquad B_\alpha \times B_\beta = 0 \quad \text{for } \alpha, \beta = 1, \ldots, g,$$

where $A \times B$ is thc intersection number of the cycle A with the cycle B. Then there exists a unique system

$$\varphi_1, \ldots, \varphi_g \tag{13}$$

of holomorphic differential 1-forms on R such that

$$\int_{A_\alpha} \varphi_\beta = \delta_{\alpha\beta}, \quad \alpha, \beta = 1, \dots, g.$$

Every holomorphic 1-form φ can be expressed as

$$\varphi = \sum_{\alpha=1}^{g} c_\alpha[\varphi]\varphi_\alpha,$$

where $c_\alpha[\varphi]$ is the A_α-period of φ:

$$c_\alpha[\varphi] = \int_{A_\alpha} \varphi, \quad \alpha = 1, \dots, g. \tag{14}$$

If we set

$$\tau_{\alpha\beta} = \int_{B_\beta} \varphi_\alpha, \quad \alpha, \beta = 1, \dots, g, \tag{15}$$

then the matrix $\tau = (\tau_{\alpha\beta})$ is symmetric and $\Im(\tau)$ is positive definite.

Assume that on our probability space is also given the following system of complex martingales $\zeta^1(t), \dots, \zeta^g(t)$ such that $\zeta^1(0) = \dots = \zeta^g(0) = 0$ and

$$\begin{aligned} &\mathrm{d}\zeta^\alpha(t) \cdot \mathrm{d}\zeta^\beta(t) = 0, \\ &\mathrm{d}\zeta^\alpha(t) \cdot \overline{\mathrm{d}\zeta^\beta(t)} = 2\Im(\tau_{\alpha\beta})\,\mathrm{d}t \quad \text{for } \alpha, \beta = 1, \dots, g. \end{aligned} \tag{16}$$

Furthermore, we assume that this system is so given that it is independent of random elements $\{C_a\}_{a\in R'}$, $\mathcal{W}$ and B given above. Note that the law of such a system is uniquely determined and (16) implies that this system is a g-dimensional conformal and Gaussian martingale (cf. [IW], Chap. III, Sect. 6 for necessary notions and notations). We associate to the differential 1-form θ given above the following right-continuous complex process:

$$\mathcal{Z}[\theta](t) = \sum_{\alpha=1}^{g} c_\alpha[\theta]\zeta^\alpha(e(t)), \tag{17}$$

where $c_\alpha[\theta]$ is the A_α-period of θ (cf. (14)) and the process $e(t)$ is defined by (5).

THEOREM 2.2. *Let ω and θ be meromorphic 1-forms on R which do not possess poles of order $k \geqslant 2$ at points at infinity $\infty_1, \dots, \infty_l$. Assume further that θ does not possess simple poles so that, in particular, it is holomorphic at $\infty_1, \dots, \infty_l$. Then,*

$$\left\{\frac{1}{\lambda}\langle\omega\rangle(\mathrm{e}^{2\lambda t} - 1),\ \frac{1}{\sqrt{\lambda}}\langle\theta\rangle(\mathrm{e}^{2\lambda t} - 1)\right\} \xrightarrow{f.d.} \{Z[\omega](t),\ \mathcal{Z}[\theta](t)\} \tag{18}$$

as $\lambda \to \infty$.

COROLLARY 2.

$$\left(\frac{2}{\log t}\langle\omega\rangle(t),\ \frac{2}{\sqrt{\log t}}\langle\theta\rangle(t)\right) \xrightarrow{d} (Z[\omega](1),\ \mathcal{Z}[\theta](1)) \tag{19}$$

as $t \to \infty$.

Remark 5. The same remark as Remark 3 applies so that we can deduce the joint convergence for several meromorphic differentials $\omega_1, \ldots, \omega_m$ and $\theta_1, \ldots, \theta_n$.

Finally, we study the joint law of limit random variables in the right-hand side of (19). Both the real and imaginary parts of these random variables are linear combinations of the following evaluated at $t = 1$:

$$\begin{aligned}
&e(t),\ C_a\left(\frac{e(t)}{l}\right), a \in R',\ \int_0^{\sigma(t)} 1_{L_k}(W(s))\,\mathrm{d}\Xi(s),\\
&\int_0^{\sigma(t)} 1_{L_k}(W(s))\,\mathrm{d}B(s), \quad k = 1, \ldots, l,\\
&\xi^\alpha(e(t)) := \Re(\zeta^\alpha(e(t))),\ \eta^\alpha(e(t)) := \Im(\zeta^\alpha(e(t))),\ \alpha = 1, \ldots, g.
\end{aligned}$$

In the following, we fix $t > 0$. We have, for $\lambda \geqslant 0$ and real $\gamma_a,\ \delta_k,\ \epsilon_k,\ \theta_\alpha,\ \kappa_\alpha$,

$$\begin{aligned}
&E\left\{\exp\left[-\lambda e(t) - i\left(\sum_a \gamma_a C_a\left(\frac{e(t)}{l}\right) + \sum_{k=1}^{l}\delta_k \int_0^{\sigma(t)} 1_{L_k}(W(s))\,\mathrm{d}\Xi(s) + \right.\right.\right.\\
&\qquad \left.\left.\left. + \sum_{k=1}^{l}\epsilon_k \int_0^{\sigma(t)} 1_{L_k}(W(s))\,\mathrm{d}B(s) + \sum_{\alpha=1}^{g}\theta_\alpha \xi^\alpha(e(t)) + \sum_{\alpha=1}^{g}\kappa_\alpha \eta^\alpha(e(t))\right)\right]\right\}\\
&\quad = E\left\{\exp\left[-\left(\lambda + \frac{1}{l}\sum_a |\gamma_a| + \sum_{\alpha,\beta=1}^{g}\frac{1}{2}\Im(\tau_{\alpha\beta})[\theta_\alpha\theta_\beta + \kappa_\alpha\kappa_\beta]\right)e(t) - \right.\right.\\
&\qquad \left.\left. - \sum_{k=1}^{l}\frac{\epsilon_k^2}{2}\int_0^{\sigma(t)} 1_{L_k}(W(s))\,\mathrm{d}s + i\sum_{k=1}^{l}\delta_k \int_0^{\sigma(t)} 1_{L_k}(W(s))\,\mathrm{d}\Xi(s)\right]\right\}.
\end{aligned} \tag{20}$$

Hence, it is sufficient to compute generally, for given $\alpha \geqslant 0$, $\beta_k \geqslant 0$ and $\gamma_k \in \mathbf{R}$, the following expectation concerning Walsh's Brownian motion $\mathcal{W}$ on $L_1 \cup \cdots \cup L_l$:

$$\begin{aligned}
&I(\alpha, \{\beta_k\}, \{\gamma_k\})\\
&\quad = E\left\{\exp\left[-\alpha e(t) - \sum_{k=1}^{l}\beta_k \int_0^{\sigma(t)} 1_{L_k}(W(s))\,\mathrm{d}s + \right.\right.\\
&\qquad \left.\left. + i\sum_{k=1}^{l}\gamma_k \int_0^{\sigma(t)} 1_{L_k}(W(s))\,\mathrm{d}\Xi(s)\right]\right\}.
\end{aligned} \tag{21}$$

By applying the Feynman–Kac and Girsanov formula, we can easily deduce that

$$I(\alpha, \{\beta_k\}, \{\gamma_k\}) = u_1(0), \tag{22}$$

where $u_1(x)$ is a function on $[0, t]$ determined, together with functions $u_2(x), \ldots, u_l(x)$ on $[0, t]$, as a unique solution of the following system of differential equations with boundary conditions:

$$\frac{1}{2}u_k''(x) + i\gamma_k u_k'(x) - \left[\beta_k + \frac{\gamma_k^2}{2}\right]u_k(x) = 0 \quad \text{on } [0, t], \quad k = 1, \ldots, l, \tag{23}$$

$$\frac{1}{l}[u_1'(0) + \cdots u_l'(0)] = \alpha u_1(0), \tag{24}$$

$$u_1(0) = \cdots = u_l(0) \quad \text{and} \quad u_1(t) = \cdots = u_l(t) = 1. \tag{25}$$

We can easily solve this to obtain

$$\begin{aligned} I(\alpha, \{\beta_k\}, \{\gamma_k\}) &= \left[\alpha + \frac{i}{l}\sum_{k=1}^{l}\gamma_k + \frac{1}{l}\sum_{k=1}^{l}\sqrt{2\beta_k}\frac{\cosh\sqrt{2\beta_k}t}{\sinh\sqrt{2\beta_k}t}\right]^{-1} \times \\ &\quad \times \frac{1}{l}\sum_{k=1}^{l} e^{i\gamma_k t}\frac{\sqrt{2\beta_k}}{\sinh\sqrt{2\beta_k}t}. \end{aligned} \tag{26}$$

So far, we have chosen as our Brownian motion on the Riemann surface R the covering motion $\tilde{z}_t$ of the standard complex Brownian motion on the complex plane. If we instead consider the covering motion $\hat{z}_t$ on the surface R of the standard Brownian motion on the Riemann sphere $\mathbf{S}$ which is, by definition, a Kähler diffusion on $\mathbf{S}$ (cf. [IW], Chapt. V, Sect. 7) corresponding to the Kähler metric on $\mathbf{S}$ given by $(1+|z|^2)^{-2}\, dz\, \overline{dz}$ in the local coordinate $z \in \mathbf{C}$ of the point in $\mathbf{S}$ mapped to z under the stereographic projection, then $\hat{z}_t$ is another Brownian motion on R. Contrary to the Brownian motion $\tilde{z}_t$, the process $\hat{z}_t$ is an ergodic (positively recurrent) diffusion on R with a finite invariant measure. If $\langle\langle\omega\rangle\rangle(t) = \int_{\hat{z}[0,t]}\omega$ is the stochastic line integral of a meromorphic differential 1-form on R, then it is again a local conformal martingale. Now we have the following asymptotic law for a system of such integrals due essentially to Franchi ([F]) in which, contrary to Theorem 2.2, every point in R takes an equal part in the limit law. This fact is natural because we do not distinguish particular points in R as points at infinity when we consider the Brownian motion $\hat{z}_t$.

We extend the above family $\{C_a\}_{a\in R'}$ of standard Cauchy processes to the family $\{C_a\}_{a\in R}$ so that $C_{a_1}, \ldots, C_{a_n}$ are independent if $a_1, \ldots, a_n \in R$ are distinct.

THEOREM 2.3. *Let ω and θ be meromorphic 1-forms on the Riemann surface R and suppose that θ does not possess simple poles. Then*

$$\left\{\frac{1}{\lambda}\langle\langle\omega\rangle\rangle(\lambda t), \frac{1}{\sqrt{\lambda}}\langle\langle\theta\rangle\rangle(\lambda t)\right\} \xrightarrow{f.d.} \{\widehat{Z}[\omega(t)], \hat{\mathcal{Z}}[\theta](t)\} \tag{27}$$

as $\lambda \to \infty$. *Here*

$$\widehat{Z}[\omega](t) = \frac{1}{l} \sum_{a \in R} \operatorname{res}(\omega)_a [t + iC_a(t)] \tag{28}$$

and

$$\hat{\mathcal{Z}}[\theta](t) = \frac{1}{\sqrt{l}} \sum_{\alpha=1}^{g} c_\alpha[\theta] \zeta^\alpha(t), \tag{29}$$

where $\{\zeta^\alpha(t)\}$ *is a system of conformal Gaussian martingales defined as above* (*cf.* (16)).

COROLLARY 3.

$$\left(\frac{1}{t} \langle\langle \omega_1 \rangle\rangle(t), \frac{1}{\sqrt{t}} \langle\langle \theta_1 \rangle\rangle(t) \right) \xrightarrow{d} (\widehat{Z}[\omega_1](1), \ \widehat{\mathcal{Z}}[\theta](1)) \tag{30}$$

as $t \to \infty$.

3. Proofs

In the following, we denote by $CBM^0(1)$ a complex Brownian motion z_t with $z_0 = 0$. First, we would recall some of basic facts due to Pitman and Yor in the study of asymptotic windings around several points of a $CBM^0(1)$ z_t.

Let $a_1, \ldots, a_n$ be n distinct points in the complex plane $\mathbf{C}$ which are all different from the origin 0. Set, for each $i = 1, \ldots, n$,

$$\begin{aligned} Z^i(t) &= \int_0^t \frac{1}{z_s - a_i} \mathrm{d}z_s, \qquad \langle Z^i \rangle(t) = \int_0^t \frac{1}{|z_s - a_i|^2} \mathrm{d}s, \\ \widehat{Z}^i(t) &= Z^i(\langle Z^i \rangle^{-1}(t)), \end{aligned} \tag{31}$$

where $\langle Z^i \rangle^{-1}(t)$ is the inverse function of $t \mapsto \langle Z^i \rangle(t)$.

PROPOSITION 1.

$$\left\{ \frac{1}{\lambda} \widehat{Z}^1(\lambda^2 t), \ldots, \frac{1}{\lambda} \widehat{Z}^n(\lambda^2 t) \right\} \xrightarrow{d} \{\zeta^1(t), \ldots, \zeta^n(t)\} \quad as\ \lambda \to \infty, \tag{32}$$

where d denotes the convergence in law in the Prokhorov topology on $\mathcal{C}([0,\infty) \to \mathbf{C}^n)$. *Here,* $\zeta(t) = \{\zeta^1(t), \ldots, \zeta^n(t)\}$ *is a continuous* $\mathbf{C}^n$*-valued process uniquely characterized by the following property:*

(1) $\zeta^i(t) = \xi^i(t) + \sqrt{-1}\eta^i(t)$ *is a* $CBM^0(1)$ *for each* $i = 1, \ldots, n$.

(2) *If we define, for each* $i = 1, \dots, n$,

$$\zeta_+^i(t) = \int_0^t 1_{[\xi^i(s)>0]}\, d\zeta^i(s), \quad \langle \zeta_+^i \rangle(t) = \int_0^t 1_{[\xi^i(s)>0]}\, ds, \quad \widehat{\zeta}_+^i(t) = \zeta_+^i(\langle \zeta_+^i \rangle^{-1}(t)),$$

and

$$\zeta_-^i(t) = \int_0^t 1_{[\xi^i(s)<0]}\, d\zeta^i(s), \quad \langle \zeta_-^i \rangle(t) = \int_0^t 1_{[\xi^i(s)<0]}\, ds, \quad \widehat{\zeta}_-^i(t) = \zeta_-^i(\langle \zeta_-^i \rangle^{-1}(t)),$$

then,

$$\widehat{\zeta}_+^1(t) \equiv \cdots \equiv \widehat{\zeta}_+^n(t) := \widehat{\zeta}_+^\infty(t)$$

and $(n+1)CBM^0(1)$*'s,* $\{\widehat{\zeta}_-^1(t)\}, \dots, \{\widehat{\zeta}_-^n(t)\}, \{\widehat{\zeta}_+^\infty(t)\}$*, are mutually independent.*

PROPOSITION 2. *Let, for each* $i = 1, \dots, n$, U_i *be a bounded neighborhood of* a_i *and* V_i *be the complement of a bounded neighborhood of* a_i*. For each* λ*, define a continuous increasing process* $T_\lambda^i(t)$ *by the relation*

$$\exp\{2\lambda T_\lambda^i(t)\} - 1 = \langle Z^i \rangle^{-1}(\lambda^2 t).$$

Then the following convergence in law holds in the Prokhorov topology on $\mathcal{C}([0,\infty) \to \mathbf{C}^{2n} \times [0,\infty)^n)$*:*

$$\left\{ \frac{1}{\lambda} \int_0^{\exp\{2\lambda T_\lambda^i(t)\}-1} 1_{U_i}(z_s) \frac{1}{z_s - a_i}\, dz_s, \right.$$
$$\left. \frac{1}{\lambda} \int_0^{\exp\{2\lambda T_\lambda^i(t)\}-1} 1_{V_i}(z_s) \frac{1}{z_s - a_i}\, dz_s, \ T_\lambda^i(t) \right\}_{i=1}^n$$
$$\xrightarrow{d} \left\{ \int_0^t 1_{[\xi^i(s)<0]}\, d\zeta^i(s), \int_0^t 1_{[\xi^i(s)>0]}\, d\zeta^i(s), \max_{0 \leqslant s \leqslant t} \xi^i(s) \right\}_{i=1}^n \tag{33}$$

as $\lambda \to \infty$.

The $\mathbf{C}^n$-valued continuous process in Proposition 1 can be constructed from a given $(n+1)$ system of Poisson point processes of Brownian positive excursions or, equivalently, from a Walsh's Brownian motion on $(n+1)$ rays. Here, we would use Walsh's Brownian motions systematically: It should be remaked that a relevance of Walsh's Brownian motions to the winding problem of planar Brownian motions has been already noticed by Barlow, Pitman and Yor ([BPY]). Before proceeding, therefore, we would recall basic notions concerning Walsh's Brownian motions.

Let $S = L_1 \cup \cdots \cup L_l$ be a collection of l different rays $L_i \cong [0,\infty)$, $i = 1, \dots, l$, each starting at the common end point 0 (origin). It is immaterial whether these rays lie on a same plane or not. If $x_i \geqslant 0$ is the distance of a point $x \in L_i$ from the origin, then each point $x \in S$ can be denoted by $[x_1, \ldots, x_l]$ in which at most one of x_i is different from 0. Obviously, the origin 0 is denoted by $[0, \ldots, 0]$. For $x = [x_1, \ldots, x_l] \in S$, we set $|x| = \sum_{i=1}^{l} x_i$. Let $L_i^o = L_i \setminus \{0\}$ so that $x \in L_i^o$ if and only if $|x| = x_i > 0$.

Let $p_i > 0, i = 1, \ldots, l$, be given such that $\sum_{i=1}^{l} p_i = 1$. Then, for $x \in S$, there exists a continuous process $\mathcal{W}_x = \{W(t) = [W_1(t), \ldots, W_l(t)]\}$ on S such that, with probability one, $W(0) = x$, $\int_0^t 1_{[W(s)=0]}\, ds = 0$ and

$$W_i(t) = x_i + M_i(t) + p_i L(t), \quad i = 1, \ldots, l, \tag{34}$$

where, denoting by $\mathcal{F}$ the natural filtration of $\mathcal{W}_x$,

(1) $L(t)$ is an $\mathcal{F}$-adapted continuous increasing process such that

$$L(0) = 0 \quad \text{and} \quad \int_0^t 1_{[W(s)=0]}\, dL(s) = L(t),$$

(2) $\{M_1(t), \ldots, M_l(t)\}$ is a system of continuous $\mathcal{F}$-martingales such that

$$M_i(0) = 0 \quad \text{and} \quad \langle M_i, M_j\rangle(t) = \delta_{ij} \int_0^t 1_{[W(s)\in L_i^o]}\, ds, \quad i, j = 1, \ldots, l.$$

It is well-known (cf., e.g., [BPY]) that such a process $\mathcal{W}_x$ exists uniquely in law and defines a diffusion process on S. It is called a Walsh Brownian motion on l rays with the initial point at x and with the rate of reflection at the origin given by $p_1, \ldots, p_l$. We denote it by $WBM^x(l; p_1, \ldots, p_l)$. In the following, we consider $WBM^0(l; p_1, \ldots, p_l)$, i.e. the case $x = 0$, exclusively.

We list some of the fundamental properties of $WBM^0(l; p_1, \ldots, p_l)$: Let $\{W(t) = [W_1(t), \ldots, W_l(t)]\}$ be a $WBM^0(l; p_1, \ldots, p_l)$ on $S = L_1 \cup \cdots \cup L_l$.

(1) Let $1 \leqslant i_1 < \cdots < i_k \leqslant l$ and $S' = \bigcup_{j=1}^{k} L_{i_j}$. Set $A(t) = \int_0^t 1_{S'}(W(s))\, ds$ and define a continuous process $\{W'(t) = [W'_{i_1}(t), \ldots, W'_{i_k}(t)]\}$ on S' by

$$W'_{i_j}(t) = W_{i_j}(A^{-1}(t)), \quad j = 1, \ldots, k,$$

where $A^{-1}(t)$ is the right-continuous inverse of $t \to A(t)$. Then $W'(t)$ is a $WBM^0(k; p'_1, \ldots, p'_k)$ on S' where $p'_j = p_{i_j} / \sum_{j=1}^{k} p_{i_j}$.

In particular, if $A_i(t) = \int_0^t 1_{L_i}(W(s))\, ds$, $i = 1, \ldots, n$, then $\widehat{W}_i(t) = W_i(A_i^{-1}(t))$ is a reflecting Brownian motion on $[0, \infty)$ starting at 0 (denote it by $RBM^0(1)$). It is easy to see by Knight's theorem that $\widehat{W}_1, \ldots, \widehat{W}_n$ are mutually independent. Conversely, given n independent $RBM^0(1)$'s, $r_i = \{r_i(t)\}$, $i = 1, \ldots, n$, there exists a unique $WBM^0(n; p_1, \ldots, p_n)$, denoted by $\mathcal{W} = (W(t))$, such that $\widehat{W}_i$ coincide with r_i, $i = 1, \ldots, n$. A construction of $\mathcal{W}$ from $\{r_i\}_{i=1}^{n}$ is as follows: Let $r_i(t) = b_i(t) + l_i(t)$ be the Skorokhod decomposition of $r_i(t)$ and let $a_i(t) = l_i^{-1}(t)$ be the right-continuous inverse of $t \mapsto l_i(t)$. Set $a(t) = \sum_{i=1}^n a_i(p_i t)$. Then $l(t) = a^{-1}(t)$ is a continuous increasing process and $a(l(t)-) \leqslant t \leqslant a(l(t))$. Set

$$\varphi_i(t) = a_i(p_i l(t)-) + (t - a(l(t)-)) \wedge (a_i(p_i l(t)) - a_i(p_i l(t)-))$$

and

$$W_i(t) = r_i(\varphi_i(t)), \quad i = 1, \ldots, n.$$

Then $W(t) = [W_1(t), \ldots, W_n(t)]$ on $S = \bigcup_{i=1}^n L_i$ is a $WBM^0(n; p_1, \ldots, p_n)$ and, by noting that $\varphi_i(t) = \int_0^t 1_{L_i}(W(s))\,\mathrm{d}s$, we can confirm that $\widehat{W}_i(t) = W_i(\varphi_i^{-1}(t)) = r_i(t)$, $i = 1, \ldots, n$.

(2) Let $\{1, 2, \ldots, l\} = I_1 \cup \cdots \cup I_m$ be a disjoint partition of $\{1, 2, \ldots, l\}$ and we associate a ray $\widetilde{L}_j \cong [0, \infty)$ to I_j, $j = 1, \ldots, m$. Define a continuous process $\{\widetilde{W}(t) = [\widetilde{W}_1(t), \ldots, \widetilde{W}_m(t)]\}$ on $\widetilde{S} = \bigcup_{j=1}^m \widetilde{L}_j$ by

$$\widetilde{W}_j(t) = \sum_{i \in I_j} W_i(t) \quad \text{on} \quad \widetilde{L}_j, \quad j = 1, \ldots, m.$$

Then, it is a $WBM^0(m; \tilde{p}_1, \ldots, \tilde{p}_m)$ on $\widetilde{S}$ where $\tilde{p}_j = \sum_{i \in I_j} p_j$, $j = 1, \ldots, m$.

In particular, $|W|(t) = \sum_{i=1}^l W_i(t)$ is a reflecting Brownian motion on $[0, \infty)$ starting at 0 (denote it $RBM^0(1)$). If

$$|W|(t) = M(t) + L(t)$$

is the Skorokhod decomposition, then $L(t)$ coincides with that in (34) (when $x = 0$) and $M_i(t)$ in (34) is given by

$$M_i(t) = \int_0^t 1_{[W(s) \in L_i^o]}\,\mathrm{d}M(s).$$

For the proof of Theorem 2.1, we set up a $WBM^0((n+1)l; 1/(n+1)l, \ldots, 1/(n+1)l)$ on $(n+1)l$ rays which we denote by $W(t) = [W_{ik}(t), W_{\infty k}(t)]_{i=1,k=1}^{n,\ l}$. Remember that our compact Riemann surface R is an l-sheeted covering surface of the Riemann sphere $\mathbf{S} = \mathbf{C} \cup \{\infty\}$ with the projection $\pi\colon R \to \mathbf{S}$ and we denote $\{\infty_1, \ldots, \infty_l\} = \pi_{-1}\{\infty\}$ and $\{a_{i1}, \ldots, a_{il}\} = \pi_{-1}\{a_i\}$, $i = 1, \ldots, n$, where $a_1, \ldots, a_n$ are n distinct points in $\mathbf{C}$ all different from 0. We assume that a_{ik}, ∞_k are all regular points (i.e., nonbranch points, cf. [S], p. 76), for simplicity. First, we discuss a construction of the process $\zeta(t) = (\zeta^1(t), \ldots, \zeta^n(t))$ in Proposition 1.

Set

$$\widetilde{W}_i(t) = \sum_{k=1}^l W_{ik}(t), \quad i = 1, \ldots, n, \qquad \widetilde{W}_\infty(t) = \sum_{k=1}^l W_{\infty k}(t). \tag{35}$$

Then, $\widetilde{W}(t) = [\widetilde{W}_1(t), \dots, \widetilde{W}_n(t), \widetilde{W}_\infty(t)]$ is $WBM^0((n+1); 1/(n+1), \dots, 1/(n+1))$ on $(n+1)$ rays $\widetilde{S} = \bigcup_{i=1}^n \widetilde{L}_i \cup \widetilde{L}_\infty$, as we saw above. Set, for each $i = 1, \dots, n$,

$$A_i(t) = \int_0^t 1_{\widetilde{L}_i \cup \widetilde{L}_\infty}(\widetilde{W}(s))\, ds \tag{36}$$

and define a continuous process $\hat{W}^{(i)}(t) = [\hat{W}_i(t), \hat{W}^i_\infty(t)]$ on $\widehat{S} = \widetilde{L}_i \cup \widetilde{L}_\infty$, by

$$\widehat{W}_i(t) = \widetilde{W}_i(A_i^{-1}(t)) \quad \text{and} \quad \widehat{W}^i_\infty(t) = \widetilde{W}_\infty(A_i^{-1}(t)). \tag{37}$$

Then $\widehat{W}^{(i)}(t)$ is a $WBM^0(2; 1/2, 1/2)$ on $\hat{S}$. Set

$$\xi^i(t) = \widehat{W}^i_\infty(t) - \widehat{W}_i(t). \tag{38}$$

That is, we define a continuous process $\xi^i(t)$ on $\mathbf{R}$ by

$$\xi^i(t) \vee 0 = \widehat{W}^i_\infty(t) \quad \text{and} \quad \xi^i(t) \wedge 0 = -\widehat{W}_i(t).$$

Then, clearly, $\xi^i(t)$ is a $BM^0(1)$ on $\mathbf{R}$ for each $i = 1, \dots, n$. We set up $n + 1$ independent $BM^0(1)$'s: $\beta_1(t), \dots, \beta_n(t), \beta_\infty(t)$, which are also independent of $W(t)$ and set

$$\eta^i(t) = \beta_\infty\left(\int_0^t 1_{[\xi^i(s)>0]}\, ds\right) + \beta_i\left(\int_0^t 1_{[\xi^i(s)<0]}\, ds\right). \tag{39}$$

Now we can deduce that $\zeta(t) = (\zeta^1(t), \dots, \zeta^n(t))$ defined by

$$\zeta^i(t) = \xi^i(t) + \sqrt{-1}\eta^i(t), \quad i = 1, \dots, n, \tag{40}$$

coincides with the one given in Proposition 1.

Set also,

$$\begin{aligned} &\widehat{W}_{ik}(t) = W_{ik}(A_i^{-1}(t)), \\ &\widehat{W}^i_{\infty k}(t) = W_{\infty k}(A_i^{-1}(t)), \quad k = 1, \dots, l,\ i = 1, \dots, n \end{aligned} \tag{41}$$

so that

$$\begin{aligned} &\sum_{k=1}^l \widehat{W}_{ik}(t) = \widehat{W}_i(t) = -\xi^i(t) \wedge 0 \quad \text{and} \\ &\sum_{k=1}^l \widehat{W}^i_{\infty k}(t) = \widehat{W}^i_\infty(t) = \xi^i(t) \vee 0. \end{aligned} \tag{42}$$

Choose sufficient small neighborhoods U_i of a_i, $i = 1, \dots, n$, in $\mathbf{C}$ so that they are mutually disjoint and $\pi^{-1}(U_i) = \bigcup_{k=1}^l U_{ik}$ where U_{ik}, $k = 1, \dots, l$ are mutually disjoint neighborhoods in R', respectively. Choose sufficient small neighborhoods V' in $\mathbf{S} = \mathbf{C} \cup \{\infty\}$ of ∞ so that $\pi^{-1}(V') = \bigcup_{k=1}^l V'_k$ where V'_k,

$k = 1, \ldots, l$, are mutually disjoint neighborhoods of ∞_k in R, respectively. We choose U_i and V' so small that each U_{ik} and V'_k consist of all regular points of R. Set $V = V' \cap \mathbf{C}$ and $V_k = V'_k \cap R'$. Let z_t be $CBM^0(1)$ and $\tilde{z}_t$ be the covering motion on R' as explained above. The follwing proposition is fundamental in the proof of Theorem 2.1:

PROPOSITION 3.

$$\left\{\frac{1}{\lambda}\int_0^{\exp\{2\lambda T^i_\lambda(t)\}-1} 1_{U_{ik}}(\tilde{z}_s)\frac{1}{z_s - a_i}\,\mathrm{d}z_s,\right.$$

$$\left.\frac{1}{\lambda}\int_0^{\exp\{2\lambda T^i_\lambda(t)\}-1} 1_{V_k}(\tilde{z}_s)\frac{1}{z_s - a_i}\,\mathrm{d}z_s,\ T^i_\lambda(t)\right\}_{i=1,k=1}^{n,\ l}$$

$$\xrightarrow{d} \left\{\int_0^t 1_{[\widehat{W}_{ik}(s)>0]}\,\mathrm{d}\zeta^i(s),\ \int_0^t 1_{[\widehat{W}^i_{\infty k}(s)>0]}\,\mathrm{d}\zeta^i(s),\ \max_{0\leqslant s\leqslant t}\xi^i(s)\right\}_{i=1,k=1}^{n,l} \tag{43}$$

as $\lambda \to \infty$. Here, d means the convergence in law in the Prokhorov topology on $\mathcal{C}([0,\infty) \to \mathbf{C}^{2nl} \times [0,\infty)^n)$.

Proof. As usual, we may assume in Proposition 1 that

$$P\left(\frac{1}{\lambda}\{\widehat{Z}^i(\lambda^2 t)\}_{i=1}^n \Rightarrow \{\zeta^i(t)\}_{i=1}^n \text{ as } \lambda \to \infty\right) = 1, \tag{44}$$

where $\Rightarrow$ denotes the uniform convergence in t on each bounded interval. Remember that $\zeta^i(t) = \xi^i(t) + \sqrt{-1}\eta^i(t)$. Now

$$\frac{1}{\lambda}\int_0^{\exp\{2\lambda T^i_\lambda(t)\}-1} 1_{V_k}(\tilde{z}_s)\frac{1}{z_s - a_i}\,\mathrm{d}z_s = \frac{1}{\lambda}\int_0^t 1_{V_k}\big[\tilde{z}\big(\langle Z^i\rangle^{-1}(\lambda^2 s)\big)\big]\,\mathrm{d}\widehat{Z}^i(\lambda^2 s).$$

Note that the local coordinate $z \in \mathbf{C}$ of a point $\tilde{z}$ in each neighborhood V_k or U_{ik} can be taken as $z = \pi(\tilde{z})$ so that the process $\tilde{z}_t$ in each local coodinate coincides with z_t. Also, $z_s = a_i - a_i \cdot \mathrm{e}^{Z^i(s)}$ so that

$$z\big(\langle Z^i\rangle^{-1}(\lambda^2 s)\big) = a_i - a_i \mathrm{e}^{\widehat{Z}^i(\lambda^2 s)}. \tag{45}$$

It is easy to deduce from (45) that, for each $s > 0$,

$$1_V\big[z\big(\langle Z^i\rangle^{-1}(\lambda^2 s)\big)\big] = \sum_{k=1}^{l} 1_{V_k}\big[\tilde{z}\big(\langle Z^i\rangle^{-1}(\lambda^2 s)\big)\big] \longrightarrow 1_{[\xi^i(s)>0]} \tag{46}$$

and

$$1_{U_i}\big[z\big(\langle Z^i\rangle^{-1}(\lambda^2 s)\big)\big] = \sum_{k=1}^{l} 1_{U_{ik}}\big[\tilde{z}\big(\langle Z^i\rangle^{-1}(\lambda^2 s)\big)\big] \longrightarrow 1_{[\xi^i(s)<0]} \tag{47}$$

as $\lambda \to \infty$. From this, we can find a family of nonnegative continuous processes $\omega_{ik}(t)$ and $\theta_{ik}(t)$, $i = 1, \ldots, n$, $k = 1, \ldots, l$, such that $\omega_{ik}(0) = \theta_{ik}(0) = 0$ and the following properties hold:

(i) For each fixed i and for each $t > 0$, at most one among $\{\omega_{ik}(t)\}$ and at most one among $\{\theta_{ik}(t)\}$ are strictly positive.

$$\text{(ii)} \sum_{k=l}^{l} \omega_{ik}(t) \equiv \xi^i(t) \vee 0 \quad \text{and} \quad \sum_{k=l}^{l} \theta_{ik}(t) \equiv -(\xi^i(t) \wedge 0).$$

(iii) For each fixed $s > 0$,

$$1_{V_k}\left[\tilde{z}\left(\langle Z^i\rangle^{-1}(\lambda^2 s)\right)\right] \to 1_{[\omega_{ik}(s)>0]} \quad \text{and}$$

$$1_{U_{ik}}\left[\tilde{z}\left(\langle Z^i\rangle^{-1}(\lambda^2 s)\right)\right] \to 1_{[\theta_{ik}(s)>0]}$$

as $\lambda \to \infty$.

Indeed, if $e_\alpha = (l_\alpha, r_\alpha)$ is an interval of the positive (negative) excursions of $\xi^i(t)$ from the origin (cf. [IW], Chap. III, 4.3 for notions involved), then (44) and (45) imply that, for all $s \in (l_\alpha, r_\alpha)$, $z\left(\langle Z^i\rangle^{-1}(\lambda^2 s)\right) \in V$ (resp. U_i) if λ is large enough and, by the continuity of $\tilde{z}_t$, there exists a unique k such that $\tilde{z}\left(\langle Z^i\rangle^{-1}(\lambda^2 s)\right) \in V_k$ (resp. U_{ik}) for all $s \in (l_\alpha, r_\alpha)$ if λ is large enough. Then set, for $s \in (l_\alpha, r_\alpha)$,

$$\omega_{ik}(s) = \xi^i(s) > 0, \qquad \omega_{ik'}(s) = 0, \quad k' \neq k \quad \text{and} \quad \theta_{ij} = 0, \ \forall j,$$

resp.

$$\theta_{ik}(s) = -\xi^i(s) > 0, \qquad \theta_{ik'}(s) = 0, \quad k' \neq k \quad \text{and} \quad \omega_{ij} = 0, \ \forall j.$$

On the zero-points of ξ^i, ω_{ik} and θ_{ik} are set to be 0. In this way, the processes ω_{ik} and θ_{ik} are uniquely defined and it is easy to verify the above properties (i) $\sim$ (iii). Also, by an obvious reason of symmetry, we must have that

$$\omega_{ik}(t) = \int_0^t 1_{[\omega_{ik}(s)>0]}\, d\xi^i(s) + \frac{1}{l} l_i(t) \quad \text{and}$$

$$\theta_{ik}(t) = -\int_0^t 1_{[\theta_{ik}(s)>0]}\, d\xi^i(s) + \frac{1}{l} l_i(t),$$

where $l_i(t)$ is the local time at 0 of ξ^i:

$$\xi^i(t) \vee 0 = \int_0^t 1_{[\xi^i(s)>0]}\, d\xi^i(s) + l_i(t).$$

Now, it is not difficult to deduce that $\omega_{ik}(t)$ and $\theta_{ik}(t)$ must be given as

$$\omega_{ik}(t) \equiv \widehat{W}^i_{\infty k}(t) \quad \text{and} \quad \theta_{ik}(t) \equiv \widehat{W}_{ik}(t),$$

where $\widehat{W}^i_{\infty k}(t)$ and $\widehat{W}_{ik}(t)$ are defined by (41) from a WBM^0 $W(t)$ on $(n+1)l$ rays as above. The proof of Proposition 3 can be completed by following [PY2, RY] or [Ya] in every technical point.

By Proposition 3, we can deduce (as in, e.g., [Ya]) that

$$\left\{\frac{1}{\lambda}\int_0^{\exp(2\lambda t)-1} 1_{U_{ik}}(\tilde{z}_s)\frac{1}{z_s-a_i}\,\mathrm{d}z_s,\ \frac{1}{\lambda}\int_0^{\exp(2\lambda t)-1} 1_{V_k}(\tilde{z}_s)\frac{1}{z_s-a_i}\,\mathrm{d}z_s\right\}_{i=1,k=1}^{n,l}$$

$$\xrightarrow{f.d.}\left\{\int_0^{\sigma_i(t)} 1_{[\widehat{W}_{ik}(s)>0]}\,\mathrm{d}\zeta^i(s),\ \int_0^{\sigma_i(t)} 1_{[\widehat{W}^i_{\infty k}(s)>0]}\,\mathrm{d}\zeta^i(s)\right\}_{i=1,k=1}^{n,l} \tag{48}$$

as $\lambda \to \infty$, where

$$\sigma_i(t) = \min\{u \mid \xi^i(u) = t\}, \quad i = 1, \dots, n. \tag{49}$$

We now identify the stochastic processes appearing in the right-hand side of (48):

PROPOSITION 4.

$$\left\{\int_0^{\sigma_i(t)} 1_{[\widehat{W}_{ik}(s)>0]}\,\mathrm{d}\zeta^i(s),\ \int_0^{\sigma_i(t)} 1_{[\widehat{W}^i_{\infty k}(s)>0]}\,\mathrm{d}\zeta^i(s)\right\}_{i=1,k=1}^{n,l}$$

$$\stackrel{d}{=}\left\{Z^k_\infty(t),\ Z_{a_{ik}}(t)\right\}_{i=1,k=1}^{n,l}, \tag{50}$$

where $\{Z^k_\infty(t)\}_{k=1}^l$ *and* $\{Z_a(t),\ a \in R'\}$ *are defined by* (7) *and* (6) *in Section* 2.

Proof. Proof can be provided by more or less easy applications of fundamental properties of Walsh's Brownian motions explained above. We only sketch the outline of it. We first note that if we set

$$A^i_+(t) = \int_0^t 1_{[\widehat{W}^i_\infty(s)>0]}\,\mathrm{d}s = \int_0^t 1_{[\xi^i(s)>0]}\,\mathrm{d}s \quad \text{and}$$

$$A^i_-(t) = \int_0^t 1_{[\widehat{W}_i(s)>0]}\,\mathrm{d}s = \int_0^t 1_{[\xi^i(s)<0]}\,\mathrm{d}s,$$

and define

$$R^i_\infty(t) = \widehat{W}^i_\infty((A^i_+)^{-1}(t)) = \xi^i((A^i_+)^{-1}(t)) \quad \text{and}$$

$$R_i(t) = \widehat{W}_i((A^i_-)^{-1}(t)) = \xi^i((A^i_-)^{-1}(t)),$$

then $R^i_\infty(t) = \widetilde{W}_\infty(a_\infty^{-1}(t)) := R_\infty(t)$ is independent of i, where $a_\infty(t) = \int_0^t 1_{[\widetilde{W}_\infty(s)>0]}\,\mathrm{d}s$. Note also that $R_i(t) = \widetilde{W}_i(a_i^{-1}(t))$ where $a_i(t) = \int_0^t 1_{[\widetilde{W}_i(s)>0]}\,\mathrm{d}s$. Then, $R_\infty = \{R_\infty(t)\}$, $R_1 = \{R_1(t)\}, \dots, R_n = \{R_n(t)\}$ are mutually independent $RMB^0(1)$'s.

Similarly, $\omega_\infty(t) = [\widehat{W}^i_{\infty k}((A^i_+)^{-1}(t)) = \widetilde{W}_{\infty k}(a_\infty^{-1}(t))]_{k=1}^l$ is independent of i and, if we set $\omega_i(t) = [\widehat{W}_{ik}((A^i_-)^{-1}(t)) = \widetilde{W}_{ik}(a_i^{-1}(t))]_{k=1}^l$, then $\omega_\infty, \omega_1, \dots, \omega_n$ are mutually independent $WBM^0(l; 1/l, \dots, 1/l)$'s. Note that $|\omega_\infty|(t) = R_\infty(t)$ and $|\omega_i|(t) = R_i(t)$. Let

$$R_\infty(t) = B_\infty(t) + L_\infty(t)$$

be the Skorokhod decomposition of R_∞. Writing $\omega_\infty(t) = [\omega_{\infty k}(t)]_{k=1}^l$, we can deduce easily that

$$\int_0^{\sigma_i(t)} 1_{[\widehat{W}^i_{\infty k}(s)>0]}\, \mathrm{d}\xi^i(s) = \int_0^{\sigma(t)} 1_{[\omega_{\infty k}(s)>0]}\, \mathrm{d}B_\infty(s),$$

where

$$\sigma(t) = \min\{u \mid R_\infty(u) = t\}.$$

Also,

$$\int_0^{\sigma_i(t)} 1_{[\widehat{W}^i_{\infty k}(s)>0]}\, \mathrm{d}\eta^i(s) = \int_0^{\sigma(t)} 1_{[\omega_{\infty k}(s)>0]}\, \mathrm{d}\beta_\infty(s).$$

In particular,

$$\int_0^{\sigma_i(t)} 1_{[\xi^i(s)>0]}\, \mathrm{d}\xi^i(s) = B_\infty(\sigma(t)) = t - e(t)$$

and

$$\int_0^{\sigma_i(t)} 1_{[\xi^i(s)>0]}\, \mathrm{d}\eta^i(s) = \beta_\infty(\sigma(t)),$$

where we set

$$e(t) = L_\infty(\sigma(t)). \tag{51}$$

Since, for each $i = 1, \ldots, n$, $[\widehat{W}^i_{\infty k}(t), \widehat{W}_{ik}(t)]_{k=1}^l$ is $WBM^0(2l; 1/2l, \ldots, 1/2l)$ and

$$\sum_{k=1}^l \widehat{W}^i_{\infty k}(t) = \xi^i(t) \vee 0, \qquad \sum_{k=1}^l \widehat{W}_{ik}(t) = -\xi^i(t) \wedge 0,$$

writing

$$\xi^i(t) \vee 0 = \int_0^t 1_{[\xi^i(s)>0]}\, \mathrm{d}\xi^i(s) + l_i(t) \quad \text{and}$$

$$\xi^i(t) \wedge 0 = \int_0^t 1_{[\xi^i(s)<0]}\, \mathrm{d}\xi^i(s) - l_i(t),$$

we have

$$-\widehat{W}_{ik}(t) = \int_0^t 1_{[\widehat{W}_{ik}(s)>0]}\, \mathrm{d}\xi^i(s) - \frac{1}{l} l_i(t)$$

and hence,

$$\int_0^{\sigma_i(t)} 1_{[\widehat{W}_{ik}(s)>0]}\, \mathrm{d}\xi^i(s) = \frac{1}{l} l_i(\sigma_i(t)) = \frac{1}{l} L_\infty(\sigma(t)) = \frac{1}{l} e(t).$$

If we set $A_-^{ik}(t) = \int_0^t 1_{[\widehat{W}_{ik}(s)>0]}\,\mathrm{d}s$, then we see by Knight's theorem that there exists an independent family $\{B_{ik}, \widetilde{B}_{ik}\}$ of $BM^0(1)$'s which are also independent of ω_∞, such that

$$\int_0^t 1_{[\widehat{W}_{ik}(s)>0]}\,\mathrm{d}\xi^i(s) = B_{ik}(A_-^{ik}(t)) \quad \text{and}$$

$$\int_0^t 1_{[\widehat{W}_{ik}(s)>0]}\,\mathrm{d}\eta^i(s) = \widetilde{B}_{ik}(A_-^{ik}(t)).$$

Then,

$$\int_0^{\sigma_i(t)} 1_{[\widehat{W}_{ik}(s)>0]}\,\mathrm{d}\eta^i(s) = \widetilde{B}_{ik}(A_-^{ik}[\sigma_i(t)]).$$

If we set, from $\hat{\omega}_{ik}(t)$,

$$R_{ik}(t) = \hat{\omega}_{ik}(a_{ik}^{-1}(t)),$$

where $a_{ik}(t) = \int_0^t 1_{[\hat{\omega}_{ik}(s)>0]}\,\mathrm{d}s$, then $\{R_{ik}\}$ are mutually independent $RBM^0(1)$'s. The Skorohod decomposition of R_{ik} is given by

$$R_{ik}(t) = B_{ik}(t) + L_{ik}(t)$$

so that the martingale part coincides with B_{ik}. In particular, we note that $\{R_{ik}, L_{ik}\}$ and $\{\widetilde{B}_{ik}\}$ are mutually independent. It is easy to deduce that

$$A_-^{ik}[\sigma_i(t)] = L_{ik}^{-1}\left(\frac{1}{l}L_\infty(\sigma(t))\right) = L_{ik}^{-1}\left(\frac{1}{l}e(t)\right).$$

Hence,

$$\int_0^{\sigma_i(t)} 1_{[\widehat{W}_{ik}(s)>0]}\,\mathrm{d}\eta^i(s) = \widetilde{B}_{ik}\left(L_{ik}^{-1}\left(\frac{1}{l}e(t)\right)\right) = C_{ik}\left(\frac{1}{l}e(t)\right),$$

where $\{C_{ik}(t) = \widetilde{B}(L_{ik}^{-1}(t))\}$ are mutually independent Cauchy processes which are also independent of $\hat{\omega}_\infty$. Now the proof of Proposition 4 is completed if we understand that the random elements $\{C_{a_{ik}}\}$, $\mathcal{W}$ and $\{B\}$ of Section 2 are taken to be

$$C_{a_{ik}}(t) := C_{ik}(t), \quad \mathcal{W} := \hat{\omega}_\infty = \{\hat{\omega}_\infty(t) = [\hat{\omega}_{\infty k}(t)]_{k=1}^l\}, \quad B := \{\beta_\infty(t)\}.$$

By (48) and (50), the proof of Theorem 2.1 can be given as in [PY2] or [RY]. Theorems 2.2 and 2.3 can be deduced in a similar way as in [PY2, KK, KPY1] by extending Proposition 3 to a joint limit law in which a class of additive functionals (occupation times) of z_t are included and also by an arguments involving the time change: We omit the details.

4. Examples in the Case of Classical Elliptic Integrals

We consider the case of R which is a Riemann surface of an irrational function

$$\sqrt{(1-z^2)(1-k^2z^2)}, \quad 0<k<1.$$

It is well-known (cf. [S]) that R can be realized as a two-sheeted covering surface of the Riemann sphere $\mathbf{S}$ obtained by gluing together two spheres along cuts $\overline{1, 1/k}$ and $\overline{-1/k, -1}$ in each sphere. Then, topologically, it is homeomorphic to a torus and the genus g is 1. There are two points at infinity ∞^+ and ∞^- which are over the north pole of $\mathbf{S}$ and the open Riemann surface $R' = R \setminus \{\infty^+, \infty^-\}$ is a two sheeted covering surface over the complex plane $\mathbf{C}$. As in the previous sections, let $\tilde{z}_t$ be the covering motion on $R' \subset R$ starting at a fixed point o over the origin $0 \in \mathbf{C}$ of $CBM^0(1)$ z_t on $\mathbf{C}$. Then stochastic line integrals along the Brownian path $\tilde{z}_t$ of meromorphic differential 1-forms on R can be realized as stochastic line integrals along the complex Brownian motion z_t of corresponding irrational 1-forms on $\mathbf{C}$. Note that line integrals of these irrational 1-forms are nothing but classical elliptic integrals of Legendre–Jacobi.

Consider the following four types of differential 1-forms on the complex plane $\mathbf{C}$:

(I)
$$\theta = \frac{1}{2iK'}\frac{dz}{\sqrt{(1-z^2)(1-k^2z^2)}}, \tag{52}$$

(II) for $a \in \mathbf{C} \setminus \{1, -1, \frac{1}{k}, -\frac{1}{k}\}$,

$$\vartheta_a = \frac{dz}{(z-a)}, \tag{53}$$

(III) for $a \in \mathbf{C} \setminus \{1, -1, \frac{1}{k}, -\frac{1}{k}\}$,

$$\omega_a = \frac{1}{(z-a)}\sqrt{\frac{(1-a^2)(1-k^2a^2)}{(1-z^2)(1-k^2z^2)}}\, dz, \tag{54}$$

(IV)
$$\omega_\infty = \frac{kz\, dz}{\sqrt{(1-z^2)(1-k^2z^2)}}. \tag{55}$$

Here, we define positive constants K and K', as usual, by

$$K = \int_0^1 \frac{dz}{\sqrt{(1-z^2)(1-k^2z^2)}},$$

$$K' = \int_0^1 \frac{dz}{\sqrt{(1-z^2)(1-k'^2z^2)}} \quad (k'^2 = 1-k^2). \tag{56}$$

Let A be a 1-cycle in R which is a lift on R of the cut joining points 1 and $1/k$ in the complex plane and choose another 1-cycle B such that $A \times B = 1$. Then the pull-back $\tilde{\theta}$ on R of θ is a holomorphic 1-form (Abelian differential of the first kind) on R such that

$$\int_A \tilde{\theta} = 1 \quad \text{and} \quad \int_B \tilde{\theta} = 2i\frac{K}{K'}.$$

For $a \in \mathbf{C} \setminus \{1, -1, 1/k, -1/k\}$, let a^+ and a^- be points in R sitting over a, i.e., $\{a^+, a^-\} = \pi^{-1}(\{a\})$. If $\tilde{\vartheta}_a$, $\tilde{\omega}_a$ and $\tilde{\omega}_\infty$ are the pull-back on R of ϑ_a, ω_a and ω_∞, respectively, then these are meromorphic 1-forms (Abelian differentials of the third kind) on R having simple poles only at $a^+, a^-, \infty^+, \infty^-$. By choosing the branch of square roots in the above suitably, we may assume that residues are given as follows:

$$\begin{aligned}
&\operatorname{res}(\tilde{\vartheta}_a)_{a^+} = \operatorname{res}(\tilde{\vartheta}_a)_{a^-} = 1, && \operatorname{res}(\tilde{\vartheta}_a)_{\infty^+} = \operatorname{res}(\tilde{\vartheta}_a)_{\infty^-} = -1,\\
&\operatorname{res}(\tilde{\omega}_a)_{a^+} = 1,\ \operatorname{res}(\tilde{\omega}_a)_{a^-} = -1, && \operatorname{res}(\tilde{\omega}_a)_{\infty^+} = \operatorname{res}(\tilde{\omega}_a)_{\infty^-} = 0,\\
&\operatorname{res}(\tilde{\omega}_\infty)_{a^+} = \operatorname{res}(\tilde{\omega}_\infty)_{a^-} = 0, && \operatorname{res}(\tilde{\omega}_\infty)_{\infty^+} = 1, \quad \operatorname{res}(\tilde{\omega}_\infty)_{\infty^-} = -1.
\end{aligned} \tag{57}$$

Let z_t be the standard complex Brownian motion (planar Brownian motion) with $z_0 = 0$. Since z_t never hits any of branch points $1, -1, 1/k, -1/k$ and poles of meromorphic forms, the following stochastic line integrals (actually, Itô integrals by dz_s) are well defined and define a family of conformal martingales:

$$\begin{aligned}
\langle\theta\rangle(t) &= \int_{z[0,t]} \theta = \frac{1}{2iK'}\int_0^t \frac{\mathrm{d}z_s}{\sqrt{(1-z_s^2)(1-k^2z_s^2)}},\\
\langle\vartheta_a\rangle(t) &= \int_{z[0,t]} \vartheta_a = \int_0^t \frac{1}{z_s - a}\,\mathrm{d}z_s,\\
\langle\omega_a\rangle(t) &= \int_{z[0,t]} \omega_a = \int_0^t \frac{1}{z_s - a}\sqrt{\frac{(1-a^2)(1-k^2a^2)}{(1-z_s^2)(1-k^2z_s^2)}}\,\mathrm{d}z_s,\\
\langle\omega_\infty\rangle(t) &= \int_{z[0,t]} \omega_\infty = k\int_0^t \frac{z_s\,\mathrm{d}z_s}{\sqrt{(1-z_s^2)(1-k^2z_s^2)}}.
\end{aligned}$$

Remark 6. We choose a branch of $\sqrt{(1-z_0^2)(1-k^2z_0^2)}$ so that it takes the value 1. Then $\sqrt{(1-z_s^2)(1-k^2z_s^2)}$ is uniquely determined by continuity in s so that the above line integrals are uniquely well-defined by Itô integrals (which are also Stratonovich integrals since z_t is a conformal martingale).

It is obvious that

$$\langle\theta\rangle(t) = \int_{\tilde{z}[0,t]} \tilde{\theta}, \qquad \langle\vartheta_a\rangle(t) = \int_{\tilde{z}[0,t]} \tilde{\vartheta}_a, \qquad \langle\omega_a\rangle(t) = \int_{\tilde{z}[0,t]} \tilde{\omega}_a \quad \text{and}$$

$$\langle\omega_\infty\rangle(t) = \int_{\tilde{z}[0,t]} \tilde{\omega}_\infty.$$

Hence by Theorem 2.2, we can obtain the joint asymptotic laws for these line integrals. Before stating it, we recall and modify the notions and notations introduced in Section 2 so that they are more suitable in this particular example: Let

$$\{C_a^+(t), C_a^-(t)\}_{a\in \mathbf{C}\setminus\{1,-1,1/k,-1/k\}}, \quad \{B(t)\}, \quad \{W(t)\}, \quad \{\zeta(t)\}$$

be independent family of processes given as follows:

(1) $C_a^{\pm}(t)$ are standard Cauchy processes on the line starting at 0.
(2) $B(t)$ is a $BM^0(1)$.
(3) $W(t)$ is a $BM^0(1)$. (Note that a $WBM^0(2; 1/2, 1/2)$ coincides with a $BM^0(1)$.)
(4) $\zeta(t)$ is a $CBM^0(1)$.

Set

$$Z_a^+(t) = \left[\frac{1}{2}e(t) + iC_a^+\left(\frac{e(t)}{2}\right)\right], \quad Z_a^-(t) = \left[\frac{1}{2}e(t) + iC_a^-\left(\frac{e(t)}{2}\right)\right], \tag{58}$$

$$Z_\infty^+(t) = \int_0^{\sigma(t)} 1_{[W(s)>0]}\,\mathrm{d}\Xi(s) + i\int_0^{\sigma(t)} 1_{[W(s)>0]}\,\mathrm{d}B(s), \tag{59}$$

$$Z_\infty^-(t) = \int_0^{\sigma(t)} 1_{[W(s)<0]}\,\mathrm{d}\Xi(s) + i\int_0^{\sigma(t)} 1_{[W(s)<0]}\,\mathrm{d}B(s). \tag{60}$$

Here

$$\sigma(t) = \min\{u \mid |W(u)| = t\} \tag{61}$$

and

$$e(t) = \Phi(e(t)), \tag{62}$$

where

$$|W(t)| = \Xi(t) + \Phi(t) \tag{63}$$

is the Skorokhod equation of the reflecting Brownian motion $|W(t)|$ on $[0,\infty)$ so that

$$\Phi(t) = \lim_{\epsilon\downarrow 0}\frac{1}{2\epsilon}\int_0^t 1_{[0,\epsilon)}(|W(s)|)\,\mathrm{d}s \tag{64}$$

and

$$\Xi(\sigma(t)) = t - e(t). \tag{65}$$

Finally, define

$$\begin{aligned} Z[\vartheta_a](t) &= Z_a^+(t) + Z_a^-(t) + Z_\infty^+(t) + Z_\infty^-(t) \\ &= t + i\left[C_a^+\left(\frac{e(t)}{2}\right) + C_a^-\left(\frac{e(t)}{2}\right) + B(\sigma(t))\right], \end{aligned} \tag{66}$$

$$Z[\omega_a](t) = Z_a^+(t) - Z_a^-(t) = i\left[C_a^+\left(\frac{e(t)}{2}\right) - C_a^-\left(\frac{e(t)}{2}\right)\right], \tag{67}$$

$$\begin{aligned} Z[\omega_\infty](t) &= Z_\infty^-(t) - Z_\infty^+(t) = t - e(t) - 2\int_0^{\sigma(t)} 1_{[W(s)>0]}\,\mathrm{d}\Xi(s) + \\ &\quad + i\left[B(\sigma(t)) - 2\int_0^{\sigma(t)} 1_{[W(s)>0]}\,\mathrm{d}B(s)\right], \end{aligned} \tag{68}$$

$$\mathcal{Z}[\theta](t) = 2\sqrt{\frac{K}{K'}}\zeta(e(t)). \tag{69}$$

EXAMPLE 1. For $a_1, \ldots, a_n \in \mathbf{C} \setminus \{0, 1, -1, 1/k, -1/k\}$ which are distinct,

$$\begin{aligned} &\left\{\frac{1}{\lambda}\Big[\langle\vartheta_{a_1}\rangle(\mathrm{e}^{2\lambda t} - 1), \ldots, \langle\vartheta_{a_n}\rangle(\mathrm{e}^{2\lambda t} - 1),\ \langle\omega_{a_1}\rangle(\mathrm{e}^{2\lambda t} - 1), \ldots\right. \\ &\left.\quad \ldots, \langle\omega_{a_n}\rangle(\mathrm{e}^{2\lambda t} - 1),\quad \langle\omega_\infty\rangle(\mathrm{e}^{2\lambda t} - 1)\Big],\quad \frac{1}{\sqrt{\lambda}}\langle\theta\rangle(\mathrm{e}^{2\lambda t} - 1)\right\} \\ &\quad \xrightarrow{f.d.} \big\{Z[\vartheta_{a_1}](t), \ldots, Z[\vartheta_{a_n}](t),\ Z[\omega_{a_1}](t), \ldots \\ &\quad \ldots, Z[\omega_{a_n}](t),\ Z[\omega_\infty](t),\ \mathcal{Z}[\theta](t)\big\} \end{aligned} \tag{70}$$

as $\lambda \to \infty$.

References

[B] Baryshnikov, Y.: Wiener soccer and its generalization, *Elect. Comm. Probab.* **3** (1997), 1–11.

[BPY] Barlow, M., Pitman, J. and Yor, M.: On Walsh's Brownian motions, In: *Séminaire de probabilités XXIII*, Lecture Notes in Math. 1372, Springer, Berlin, 1989, pp. 275–293.

[F] Franchi, J.: Théorème des résidus asymptotique pour le mouvement brownien sur une surface riemannienne compacte, *Ann. Inst. H. Poincaré* **27** (1991), 445–462.

[IW] Ikeda, N. and Watanabe, S.: *Stochastic Diffential Equations and Diffusion Processes*, 2nd edn, North-Holland/Kodansha, Amsterdam/Tokyo, 1988.

[IM] Itô, K. and McKean, H. P., Jr.: *Diffusion Processes and Their Sample Paths*, Springer, Berlin, 1965.

[KK] Kasahara, Y. and Kotani, S.: On limit processes for a class of additive functionals of recurrent diffusion processes, *Z. Wahrsch. verw. Gebiete* **49** (1979), 133–153.

[KPY1] Kozlov, S. M., Pitman, J. W. and Yor, M.: Brownian interpretations of an elliptic integral, In: *Seminar on Stochastic Processes 1991*, Birkhäuser, Boston, 1992, pp. 83–95.

[KPY2] Kozlov, S. M., Pitman, J. B. and Yor, M.: Wiener football, *Theory Probab. Appl.* **37** (1992), 550–552.

[K] Kusunoki, Y.: *Kansûron (Function Theory)*, in Japanese, Asakura-Shoten, Tokyo, 1973.

[M] Manabe, S.: Stochastic intersection number and homological behaviors of diffusion processes on Riemannian manifolds, *Osaka J. Math.* **19** (1982), 429–457.

[PY1] Pitman, J. W. and Yor, M.: The asymptotic joint distribution of windings of planar Brownian motion, *Bull. Amer. Math. Soc.* **10** (1984), 109–111.

[PY2] Pitman, J. W. and Yor, M.: Asymptotic laws of planar Brownian motion, *Ann. Probab.* **14** (1986), 733–779.

[PY3] Pitman, J. W. and Yor, M.: Further asymptotic laws of planar Brownian motion, *Ann. Probab.* **17** (1989), 965–1011.

[RY] Revuz, D. and Yor, M.: *Continuous Martingales and Brownian Motion*, Springer, Berlin, 1991.

[S] Springer, G.: *Introduction to Riemann Surfaces*, Addison-Wesley, Reading, Mass., 1957.

[Wal] Walsh, J. B.: A diffusion with discontinuous local time, *Astérisque* **52–53**, Soc. Math. France (1978), 37–45.

[Wat] Watanabe, S.: A limit theorem for sums of i.i.d. random variables with slowly varying tail probability, In: *Multivariate Analysis* V, North-Holland, Amsterdam, 1980, pp. 249–261.

[Ya] Yamazaki, Y.: On limit theorems related to a class of 'winding-type' additive functionals of complex Brownian motion, *J. Math. Kyoto Univ.* **34** (1992), 801–841.

[Yo1] Yor, M.: *Some Aspects of Brownian Motion*, I, Lectures in Math. ETH Zürich, Birkhäuser, Basel, 1992.

[Yo2] Yor, M.: *Some Aspects of Brownian Motion*, II, Lectures in Math. ETH Zürich, Birkhäuser, Basel, 1997.

Zeitfracht Medien GmbH
Ferdinand-Jühlke-Straße 7
99095 Erfurt, Deutschland
produktsicherheit@kolibri360.de